Volume II
Solvents

Handbook of Envir...

# FATE
# and
# EXPOSURE
# DATA
## For Organic Chemicals

Editor

# Philip H. Howard

Associate Editors

Gloria W. Sage
William F. Jarvis
D. Anthony Gray

CRC Press
Taylor & Francis Group
Boca Raton London New York

CRC Press is an imprint of the
Taylor & Francis Group, an informa business

Published 1990 by CRC Press, LLC
Taylor & Francis Group
6000 Broken Sound Parkway NW, Suite 300
Boca Raton, FL 33487-2742

©1990 by Taylor & Francis Group, LLC
CRC Press is an imprint of Taylor & Francis Group, an Informa business

First issued in paperback 2019

No claim to original U.S. Government works

ISBN-13: 978-0-367-45086-1 (pbk)
ISBN-13: 978-0-87371-204-0 (hbk)

**Visit the Taylor & Francis Web site at**
**http://www.taylorandfrancis.com**

**and the CRC Press Web site at**
**http://www.crcpress.com**

## Library of Congress Cataloging-in-Publication Data

Howard, Philip H.
Handbook of environmental fate and exposure data
for organic chemicals.
Includes bibliographical references.
ISBN 0-87371-151-3 (v. 1)
ISBN 0-87371-204-8 (v. 2)
Contents: 1. Large production and priority
pollutants—v. 2. 2. Solvents.
1. Pollutants—Handbooks, manuals, etc. I. Title.
TD176.4.H69    1990
363.7'38                                      89-2436

Library of Congress Card Number 89-2436

## Associate Editors for Volume II

The following individuals from the Syracuse Research Corporation's Chemical Hazard Assessment Division either were authors of the individual chemical records prepared for the Hazardous Substances Data Bank or edited the expanded and updated chemical chapters in this volume. The order of names, which will vary in each volume, is by the number of chemicals for which the individual was responsible.

**Gloria W. Sage, Ph.D.**

**William F. Jarvis, Ph.D.**

**D. Anthony Gray, Ph.D.**

**Philip H. Howard** joined Syracuse Research Corporation in 1970 and has served as project director for numerous environmental fate and effects projects for federal agencies and industry. Dr. Howard's current research projects include development of structure/biodegradability correlations, development of estimation techniques for environmental fate physical properties and rate constants, and databases of information to support these efforts. He received a B.S. degree in chemistry from Norwich University in 1965 and a Ph.D. in organic chemistry from Syracuse University in 1970.

**Gloria W. Sage** joined Syracuse Research Corporation in 1980 at which time she was the codeveloper of the Environmental Fates Data Bases, Datalog and Chemfate. Later on, she developed the environmental fate section of the Hazardous Substances Data Bank and contributed profiles to this section. Dr. Sage has been involved in many projects at Syracuse Research Corporation concerned with assessing the environmental fate of chemicals and making exposure assessments. She is currently interested in using mosses to measure atmospheric deposition of metals and organics. Dr. Sage holds an A.B. with distinction and honors in Chemistry from Cornell University, an A.M. in Chemistry from Radcliffe College, and a Ph.D. in Physical Chemistry from Harvard University.

**William F. Jarvis** joined Syracuse Research Corporation in 1985 and has participated in the preparation of critical and comprehensive literature reviews on the environmental fate and effects associated with environmental pollutants. Dr. Jarvis' current research projects include risk evaluation projects funded by the U.S. EPA and ATSDR. He received a B.S. degree in chemistry from SUNY College at Cortland in 1974 and a Ph.D. in organic chemistry from Syracuse University in 1988.

**D. Anthony Gray** joined Syracuse Research Corporation in 1979 where he currently is the Director of the Exposure Modeling Center. Since joining, Dr. Gray has served in a number of capacities including technical writer, project manager, and program manager. Dr. Gray's research interests include the assessment of exposure to multiple chemicals at hazardous waste sites by modeling their fate in soil, water, and air, and determining the most commonly found mixtures of chemicals occurring at hazardous waste sites. He received his B.S. degree in chemistry from Drexel University in 1973 and his Ph.D. in organic chemistry from Syracuse University in 1981.

# Preface

Many articles and books have been written on how to review the environmental fate and exposure of organic chemicals (e.g., [11] and [19]). Although these articles and books often give examples of the fate and exposure of several chemicals, rarely do they attempt to review large numbers of chemicals. These "how to" guides provide considerable insight into ways of estimating and using physical/chemical properties as well as mechanisms of environmental transport and transformation. However, when it comes to reviewing the fate and exposure of individual chemicals, there are discretionary factors that significantly affect the overall fate assessment. For example, is it reasonable to use regression equations for estimating soil or sediment adsorption for aromatic amine compounds? Is chemical oxidation likely to be important for phenols in surface waters? These discretionary factors are dependent upon the available data on the individual chemical or, when data are lacking, on chemicals of related structures.

This series of books outlines in detail how individual chemicals are released, transported, and degraded in the environment and how they are exposed to humans and environmental organisms. It is devoted to the review and evaluation of the available data on physical/chemical properties, commercial use and possible sources of environmental contamination, environmental fate, and monitoring data of individual chemicals. Each review of a chemical provides most of the data necessary for either a qualitative or quantitative exposure assessment.

Chemicals were selected from a large number of chemicals prepared by Syracuse Research Corporation (SRC) for inclusion in the National Library of Medicine's (NLM) Hazardous Substances Data Bank (HSDB). Chemicals selected for the first two volumes were picked from lists of high volume commercial chemicals, priority pollutants, and solvents. The chemicals in the first two volumes include most of the non-pesticidal priority pollutants and many of the chemicals on priority lists for a variety of environmental regulations (e.g., RCRA and CERCLA Reportable Quantities, Superfund, SARA). Pesticides, polycyclic aromatic hydrocarbons, and other groups of chemicals will be included in later volumes.

The chemicals are listed in strict alphabetical order by the name considered to be the most easily recognized. Prefixes commonly used in organic chemistry which are not normally considered part of the name, such as ortho-, meta-, para-, alpha-, beta-, gamma-, n-, sec-, tert-, cis-, trans-, N-, as well as all numbers, have not been considered for alphabetical order. Other prefixes which normally are considered part of the name, such as iso-, di-, tri-, tetra-, and cyclo-, are used for alphabetical positioning. For example, 2,4-Dinitrotoluene is under D and tert-Butyl alcohol is under B. In addition, cumulative indices are provided at the end of each volume to allow the reader to find a given chemical by chemical name synonym, Chemical Abstracts Services (CAS) number, and chemical formula.

# Acknowledgments

The following authors of the initial chemical records for the Hazardous Substance Data Bank were or are staff scientists with the Chemical Hazard Assessment Division of the Syracuse Research Corporation: Gloria W. Sage, William Meylan, William F. Jarvis, Erin K. Crosbie, Jeffery Jackson, Amy E. Hueber, Jeffrey Robinson, Edward Michalenko, and Jay Tunkel. The assistance of Heather Taub in editing and incorporating the chemical structures is greatly appreciated. We wish to thank several individuals at the National Library of Medicine (NLM) for their encouragement and support during the project. Special thanks go to our project officer, Vera Hudson, and to Bruno Vasta and Dalton Tidwell.

# Explanation of Data

In the following outline, each field covered for the individual chemicals is reviewed with such information as the importance of the data, the type of data included in each field, how data are usually handled, and data sources. For each chemical, the physical properties as well as the environmental fate and monitoring data were identified by conducting searches of the Environmental Fate Data Bases of Syracuse Research Corporation (SRC) [12].

## SUBSTANCE IDENTIFICATION

**Synonyms:** Only synonym names used fairly frequently were included.

**Structure:** Chemical structure.

**CAS Registry Number:** This number is assigned by the American Chemical Society's Chemical Abstracts Services as a unique identifier.

**Molecular Formula:** The formula is in Hill notation, which is given as the number of carbons followed by the number of hydrogens followed by any other elements in alphabetic order.

**Wiswesser Line Notation:** This is a chemical structure representation that can be used for substructure searching. It was designed back when computer notations had to fit into 80 characters and, therefore, is very abbreviated (e.g., Q is used for a benzene ring).

## CHEMICAL AND PHYSICAL PROPERTIES

The Hazardous Substances Data Bank (HSDB) of the National Library of Medicine was used as a source of boiling points, melting points, and molecular weights. The dissociation constant, octanol/water partition coefficient, water solubility, vapor pressure, and Henry's Law constant were judiciously selected from the many values that were identified in SRC's DATALOG file. All values selected were referenced to the primary literature source when possible.

**Boiling Point:** The boiling point or boiling point range is given along with the pressure. When the pressure is not given it should be assumed that the value is at 760 mm Hg.

**Melting Point:** The melting point or melting point range is given.

**Molecular Weight:** The molecular weight to two decimal points is given.

**Dissociation Constants:** The acid dissociation constant as the negative log (pKa) is given for chemicals that are likely to dissociate at environmental pH's (between 5 and 9). Chemical classes where dissociation is important include, for example, phenols, carboxylic acids, and aliphatic and aromatic amines. Once the pKa is known, the percent in the dissociated and undissociated form can be determined. For example, for an acid with a pKa of 4.75, the following is true at different pH's:

> 1% dissociated at pH 2.75
> 10% dissociated at pH 3.75
> 50% dissociated at pH 4.75
> 90% dissociated at pH 5.75
> 99% dissociated at pH 6.75

The degree of dissociation affects such processes as photolysis (absorption spectra of chemicals that dissociate can be considerably affected by the pH), evaporation from water (ions do not evaporate), soil or sediment adsorption, and bioconcentration. Values from evaluated sources such as Perrin [21] and Serjeant and Dempsey [23] were used when available.

**Log Octanol/Water Partition Coefficient:** The octanol/water partition coefficient is the ratio of the chemical concentration in octanol divided by the concentration in water. The most reliable source of values is from the Medchem project at Pomona College [8]. When experimental values are unavailable, estimated values have been provided using a fragment constant estimation method, CLOGP3, from Medchem. Occasionally chemical octanol/water partition coefficients were not calculated because a necessary fragment constant for the chemical was not available. The octanol/water partition coefficient has been shown to correlate well with bioconcentration factors in aquatic organisms [26] and adsorption to soil or sediment [13], and recommended regression equations have been reviewed [15].

**Water Solubility:** The water solubility of a chemical provides considerable insight into the fate and transport of a chemical in the environment. High water soluble chemicals, which have a tendency to remain dissolved in the water column and not partition to soil or sediment or bioconcentrate in aquatic organisms, are less likely to volatilize from water (depending upon the vapor pressure - see Henry's Law constant) and are generally more likely to biodegrade. Low water soluble chemicals are just the opposite; they partition to soil or sediments and bioconcentrate in aquatic organisms, volatilize more readily from water, and are less likely to be biodegradable. Other fate processes that are, or can be, affected by water solubility include photolysis, hydrolysis, oxidation, and washout from the atmosphere by rain or fog. Water solubility values were taken from either the Arizona Data Base [27] or from SRC's DATALOG or CHEMFATE files. The values were reported in ppm at a temperature at or as close as possible to 25 °C. Occasionally when no values were available, the value was estimated from the octanol/water partition coefficient using recommended regression equations [15].

**Vapor Pressure:** The vapor pressure of a chemical provides considerable insight into the transport of a chemical in the environment. The volatility of the pure chemical is dependent upon the vapor pressure, and volatilization from water is dependent upon the vapor pressure and water solubility (see Henry's Law constant). The form in which a chemical will be found in the atmosphere is dependent upon the vapor pressure; chemicals with a vapor pressure less than $10^{-6}$ mm Hg will be mostly found associated with particulate matter [7]. When available, sources such as Boublik et al [3], Riddick et al [22], and Daubert and Danner [5] were used, since the data in these sources were evaluated and some of them provided recommended values. Vapor pressure was reported in mm Hg at or as close as possible to 25 °C. In many cases, the vapor pressure was calculated from a vapor pressure/temperature equation.

**Henry's Law Constant:** The Henry's Law constant, H, is really the air/water partition coefficient, and therefore a nondimensional H relates the chemical concentration in the gas phase to its concentration in the water phase. The dimensional H can be determined by dividing the vapor pressure in atm by the water solubility in mole/m$^3$ to give H in atm-m$^3$/mole. H provides an indication of the partition between air and water at equilibrium and also is used to calculate the rate of evaporation from water (see discussion under Evaporation from Water/Soil). Henry's Law constants can be directly measured, calculated from the water solubility and vapor pressure, or estimated from structure using the method of Hine and Mookerjee [9], and this same order was used in selecting values. Some critical review data on Henry's Law constants are available (e.g., [16]).

# ENVIRONMENTAL FATE/EXPOSURE POTENTIAL

Data for the following sections were identified with SRC's Environmental Fate Data Bases. Biodegradation data were selected from the DATALOG, BIOLOG, and BIODEG files. Abiotic degradation data were identified in the Hydrolysis, Photolysis, and Oxidation fields in DATALOG and CHEMFATE. Transport processes such as Bioconcentration, Soil Adsorption/Mobility, and Volatilization as well as the monitoring data were also identified in the DATALOG and CHEMFATE files.

**Summary:** This section is an abbreviated summary of all the data presented in the following sections and is not referenced; to find the citations the reader should refer to appropriate sections that follow. In general, this summary discusses how a chemical is used and released to the environment, how the chemical will behave in soil, water, and air, and how exposure to humans and environmental organisms is likely to occur.

**Natural Sources:** This section reviews any evidence that the chemical may have any natural sources of pollution, such as forest fires and volcanos, or may be a natural product that would lead to its detection in various media (e.g., methyl iodide is found in marine algae and is the major source of contamination in the ocean).

**Artificial Sources:** This section is a general review of any evidence that the chemical has anthropogenic sources of pollution. Quantitative data are reviewed in detail in Effluent Concentrations; this section provides a qualitative review of various sources based upon how the chemical is manufactured and used as well as the physical/chemical properties. For example, it is reasonable to assume that a highly volatile chemical which is used mostly as a solvent will be released to the atmosphere as well as the air of occupational settings even if no monitoring data are available. Information on production volume and uses was obtained from a variety of chemical marketing sources including the Kirk-Othmer Encyclopedia of Chemical Technology, SRI International's Chemical Economics Handbook, and the Chemical Profiles of the Chemical Marketing Reporter.

**Terrestrial Fate:** This section reviews how a chemical will behave if released to soil or groundwater. Field studies or terrestrial model ecosystems studies are used here when they provide insight into the overall behavior in soil. Studies which determine an individual process (e.g., biodegradation, hydrolysis, soil adsorption) in soil are reviewed in the appropriate sections that follow. Quite often, except with pesticides, field or terrestrial ecosystem studies either are not available or do not give enough data to make conclusions on the terrestrial fate of a chemical. In these cases, data from the sections on Biodegradation, Abiotic Degradation, Soil Adsorption/Mobility, Volatilization from Water/Soil, and any appropriate monitoring data will be used to synthesize how a chemical is likely to behave if released to soil.

**Aquatic Fate:** This section reviews how a chemical will behave if released to fresh, marine, or estuarine surface waters. Field studies or aquatic model ecosystems are used here when they provide insight into the overall behavior in water. Studies which determine an individual process (e.g., biodegradation, hydrolysis, photolysis, sediment adsorption, and bioconcentration in aquatic organisms) in water are reviewed in the appropriate sections that follow. When field or aquatic ecosystems studies are not available or do not give enough data to make conclusions on the aquatic fate of the chemical, data from the appropriate degradation, transport, or monitoring sections will be used to synthesize how a chemical is likely to behave if released to water.

**Atmospheric Fate:** This section reviews how a chemical will behave if released to the atmosphere. The vapor pressure will be used to determine if the chemical is likely to be in the vapor phase or adsorbed to particulate matter [7]. The water solubility will be used to assess the likelihood of washout with rain. Smog chamber studies or other studies where the mechanism of degradation is not determined will be reviewed in this section; studies of the rate of reaction with hydroxyl radical or ozone or direct photolysis will be reviewed in Abiotic Degradation and integrated into this section.

**Biodegradation:** The principles outlined by Howard and Banerjee [10] are used in this section to review the relevant biodegradation data pertinent to biodegradation in soil, water, or wastewater treatment. In general, the studies have been separated into screening studies (inoculum in defined nutrient media), biological treatment simulations, and grab samples (soil or water sample with chemical added and loss of concentration followed). Pure culture studies are only used to indicate potential metabolites, since the artificial nutrient conditions under which the pure cultures are isolated provide little assurance that these same organisms will be present in any quantity or that their enzymes will be functioning in various soil or water environments. Anaerobic biodegradation studies, which are pertinent to whether a chemical will biodegrade in biological treatment digestors, sediment, and some groundwaters, are discussed separately.

**Abiotic Degradation:** Non-biological degradation processes in air, water, or soil are reviewed in this section. For most chemicals in the vapor phase in the atmosphere, reaction with photochemically generated hydroxyl radicals is the most important degradation process. Occasionally reaction in the atmosphere with ozone (for olefins), nitrate radicals at night, and direct photolysis (direct sunlight absorption resulting in photochemical alteration) are significant for some chemicals [2]. For many chemicals, experimental reaction rate constants for hydroxyl radical are available (e.g., [1]) and are used to calculate an estimated half-life by assuming an average hydroxyl radical concentration of $5 \times 10^{+5}$ molecules/cm$^3$ in non-smog conditions (e.g., [2]). If experimental rate constants are not available, they have been estimated using the fragment constant method of Atkinson [1] and then a half-life estimated using the assumed radical concentration. The reaction rate for ozone reaction with olefins may be experimentally available or can be estimated using the Fate of Atmospheric Pollutants (FAP) from the Graphic Exposure and Modelling System (GEMS) (available from the Exposure Evaluation Division, Office of Toxic Substances, U.S. Environmental Protection Agency). Using either the experimental or estimated rate constant and an assumed concentration of $6.0 \times 10^{+11}$ molecules/cm$^3$ (FAP) or $7.2 \times 10^{+11}$ molecules/m$^3$ [2], an estimated half-life for reaction with ozone can be calculated. Nitrate radicals are significant only with certain classes of chemicals such as higher alkenes, dimethyl sulfide and lower thiols, furan and pyrrole, and hydroxy-substituted aromatics [2].

The possibility of direct photolysis in air or water can be partially assessed by examining the ultraviolet spectrum of the chemical. If the chemical does not absorb light at wavelengths provided by sunlight (>290 nm), the chemical cannot directly photolyze. If it does absorb sunlight, it may or may not photodegrade depending upon the efficiency (quantum yield) of the photochemical process, and unfortunately such data are rarely available. Indirect photolysis processes may be important for some chemicals in water [17]. For example, some chemicals can undergo sensitized photolysis by absorbing triplet state energy from the excited triplet state of chemicals commonly found in water, such as humic acids. Transient oxidants found in water, such as peroxy radicals, singlet oxygen, and hydroxyl radicals, may also contribute to abiotic degradation in water for some chemicals. For example, phenols and aromatic amines have half-lives of less than a day for reaction with peroxy radicals; substituted and unsubstituted olefins have half-lives of 7 to 8 days with singlet oxygen; and dialkyl sulfides have half-lives of 27 hours with singlet oxygen [17].

Chemical hydrolysis at pH's that are normally found in the environment (pH's 5 to 9) can be important for a variety of chemicals that have functional groups that are potentially hydrolyzable, such as alkyl halides, amides, carbamates, carboxylic acid esters, epoxides and lactones, phosphate esters, and sulfonic acid esters [18]. Half-lives at various pH's are usually reported in order to provide an indication of the influence of pH.

**Bioconcentration:** Certain chemicals, due to their hydrophobic nature, have a tendency to partition from the water column and bioconcentrate in aquatic organisms. This concentration of chemicals in aquatic organisms is of concern because it can lead to toxic concentrations being reached when the organism is consumed by higher organisms such as wildlife and humans. Such bioconcentrations are usually reported as the bioconcentration factor (BCF), which is the concentration of the chemical in the organism at equilibrium divided by the concentration of the chemical in water. This unitless BCF value can be determined experimentally by dosing water containing the organism and dividing the concentration in the organism by the concentration in the water once equilibrium is reached, or if equilibration is slow, the rate of uptake can be used to calculate the BCF at equilibrium. The BCF value can also be estimated by using recommended regression equations that have been shown to correlate well with physical properties such as the octanol/water partition coefficient and water solubility [15]; however, these estimation equations assume that little metabolism of the chemical occurs in the aquatic organism, which is not always correct. Therefore, when available, experimental values are preferred.

**Soil Adsorption/Mobility:** For many chemicals (especially pesticides), experimental soil or sediment partition coefficients are available. These values are measured by determining the concentration in both the solution (water) and solid (soil or sediment) phases after shaking for about 24 to 48 hours and using different initial concentrations. The data are then fit to a Freundlich equation to determine the adsorption coefficient, Kd. These Kd values for individual soils or sediments are normalized to the organic carbon content of the soil or sediment by dividing by the organic content (Koc), since of the numerous soil properties that affect sorption (organic carbon content, particle size, clay mineral composition, pH, cation-exchange capacity) [14], organic carbon is the most important for undissociated organic chemicals. Occasionally the experimental adsorption coefficients are reported on a soil-organic matter basis (Kom) and these are converted to Koc by multiplying by 1.724 [15]. When experimental values are unavailable, estimated Koc values are calculated using either the water solubility or octanol/water partition coefficient and some recommended regression equations [15]. The measured or estimated adsorption values are used to determine the likelihood of leaching through soil or adsorbing to sediments using the criteria of Swann et al [24]. Occasionally experimental soil thin-layer chromatography studies are also available and can be used to assess the potential for leaching.

The above discussion applies generally to undissociated chemicals, but there are some exceptions. For example, aromatic amines have been shown to covalently bond to humic material [20] and this slow but non-reversible process can lead to aromatic amines being tightly bound to the humic material in soils. Methods to estimate the soil or sediment adsorption coefficient for dissociated chemicals which form anions are not yet available, so it is particularly important to know the pKa value for chemicals that can dissociate so that a determination of the relative amounts of the dissociated and undissociated forms can be determined at various pH conditions. Chemicals that form cations at ambient pH conditions are generally thought to sorb strongly to clay material, similar to what occurs with paraquat and diquat (pyridine cations).

**Volatilization from Water/Soil:** For many chemicals, volatilization can be an extremely important removal process, with half-lives as low as several hours. The Henry's Law constant can give qualitative indications of the importance of volatilization; for chemicals with values less than $10^{-7}$ atm-m$^3$/mole, the chemical is less volatile than water and as water evaporates the concentration will increase; for chemicals around $10^{-3}$ atm-m$^3$/mole, volatilization will be rapid. The volatilization process is dependent upon physical properties of the chemical (Henry's Law constant, diffusivity coefficient), the presence of modifying materials (adsorbents, organic films, electrolytes, emulsions), and the physical and chemical properties of the environment (water depth, flow rate, the presence of waves, sediment content, soil moisture, and organic content) [15]. Since the overall volatilization rate cannot be estimated for all the various environments to which a chemical may be released, common models have been used in order to give an indication of the relative importance of volatilization. For most chemicals that have a Henry's Law constant greater than $10^{-7}$ atm-m$^3$/mole, the simple volatilization model outlined in Lyman et al [15] was used; this model assumes a 20 °C river 1 meter deep flowing at 1 m/sec with a wind velocity of 3 m/sec and requires only the Henry's Law constant and the molecular weight of the chemical for input. This model gives relatively rapid volatilization rates for this model river and values for ponds, lakes, or deeper rivers will be considerably slower. Occasionally a chemical's measured reaeration coefficient ratio relative to oxygen is available, and this can be used with typical oxygen reaeration rates in ponds, rivers, and streams to give volatilization rates for these types of bodies of water. For chemicals that have extremely high Koc values, the EXAMS-II model has been used to estimate volatilization both with and without sediment adsorption (extreme differences are noted for these high Koc chemicals). Soil volatilization models are less validated and only qualitative statements are given of the importance of volatilization from moist (about 2% or greater water content) or dry soil, based upon the Henry's Law constant or vapor pressure, respectively. This assumes that once the soil is saturated with a molecular layer of water, the volatilization rate will be mostly determined by the value of the Henry's Law constant, except for chemicals with high Koc values.

**Water Concentrations:** Ambient water concentrations of the chemical are reviewed in this section, with subcategories for surface water, drinking water, and groundwater when data are available. In general, the number of samples, the percent positive, the range of concentrations, and the average concentration are reported when the data are available.

**Effluents Concentration:** Air emissions and wastewater effluents are reviewed in this section. In general, the number of samples, the percent positive, the range of concentrations, and the average concentration are reported when the data are available.

**Sediment/Soil Concentrations:** Sediment and soil concentrations are reviewed in this section. In general, the number of samples, the percent positive, the range of concentrations, and the average concentration are reported when the data are available.

**Atmospheric Concentrations:** Ambient atmospheric concentrations are reviewed in this section, with subcategories for rural/remote and urban/suburban when data are available in such sources as Brodzinsky and Singh [4]. In general, the number of samples, the percent positive, the range of concentrations, and the average concentration are reported when the data are available.

**Food Survey Values:** Market basket survey data such as found in Duggan et al [6] and individual studies of analysis of the chemical in processed food are reported in this section. In general, the number of samples, the percent positive, the range of concentrations, and the average concentration are reported when the data are available.

**Plant Concentrations:** Concentrations of the chemical in plants are reviewed in this section. If the plant has been processed for food, it is reported in Food Survey Values.

**Fish/Seafood Concentrations:** Concentrations in fish, seafood, shellfish, etc. are reviewed in this section. If the fish or seafood have been processed for food, the data are reported in Food Survey Values.

**Animal Concentrations:** Concentrations in animals are reviewed in this section. If the animals have been processed for food, the data are reported in Food Survey Values.

**Milk Concentrations:** Since dairy milk constitutes a high percentage of the human diet, concentrations of the chemical found in dairy milk are reviewed in this section and not in Food Survey Values.

**Other Environmental Concentrations:** Concentrations of the chemical found in other environmental media that may contribute to an understanding of how a chemical may be released to the environment or exposed to humans (e.g., detection in gasoline or cigarette smoke are reviewed in this section).

**Probable Routes of Human Exposure:** The monitoring data and physical properties are used to provide conclusions on the routes (oral, dermal, inhalation) of exposure.

**Average Daily Intake:** The average daily intake is a calculated value of the amount of the chemical that is typically taken in daily by human adults. The value is determined by multiplying typical concentrations in drinking water, air, and food by average intake factors such as 2 liters of water, 20 $m^3$ of air, and 1600 grams of food [25].

**Occupational Exposures:** Monitoring data, usually air samples, from occupational sites are reviewed in this section. In addition, estimates of the number of workers exposed to the chemical from the two National Institute for Occupational Safety and Health (NIOSH) surveys are reviewed in this section. The National Occupational Hazard Survey (NOHS) conducted from 1972 to 1974 and the National Occupational Exposure Survey (NOES) conducted from 1981 to 1983 provided statistical estimates of worker exposures based upon limited walk-through industrial hygiene surveys.

**Body Burdens:** Any concentrations of the chemical found in human tissues or fluids is reviewed in this section. Included are blood, adipose tissue, urine, and human milk.

## REFERENCES

1. Atkinson RA; Internat J Chem Kinet 19: 799-828 (1987)
2. Atkinson RA; Chem Rev 85: 60-201 (1985)
3. Boublik T et al; The Vapor Pressures of Pure Substances. Amsterdam: Elsevier (1984)
4. Brodzinsky R, Singh HB; Volatile organic chemicals in the atmosphere: an assessment of available data. SRI Inter EPA contract 68-02-3452 Menlo Park, CA (1982)
5. Daubert TE, Danner RP; Data Compilation Tables of Properties of Pure Compounds. Amer Inst Chem Engr pp 450 (1985)
6. Duggan RE et al; Pesticide Residue Levels in Foods in the U.S. from July 1, 1969 to June 30, 1976. Washington, DC: Food Drug Administ. 240 pp (1983)
7. Eisenreich SJ et al; Environ Sci Technol 15: 30-8 (1981)
8. Hansch C, Leo AJ; Medchem Project Issue No 26. Claremont CA: Pomona College (1985)
9. Hine J, Mookerjee PK; J Org Chem 40: 292-8 (1975)
10. Howard PH, Banerjee S; Environ Toxicol Chem 3: 551-562 (1984)
11. Howard PH et al; Environ Sci Technol 12: 398-407 (1978)
12. Howard PH et al; Environ Toxicol Chem 5: 977-88 (1986)
13. Karickhoff SW; Chemosphere 10: 833-46 (1981)
14. Karickhoff SW; Environ Expos from Chemicals Vol I. ed Neely WB, Blau GE, Boca Raton, FL: CRC Press p 49-64 (1985)
15. Lyman WJ et al; Handbook of Chemical Property Estimation Methods. McGraw-Hill, NY (1982)
16. Mackay D, Shiu WY; J Phys Chem Ref Data 10: 1175-99 (1981)
17. Mill T, Mabey W; In Environ Expos from Chemicals Vol I. ed Neely WB, Blau GE, Boca Raton, FL: CRC Press p 175-216 (1985)
18. Neely WB; In Environ Expos from Chemicals Vol I. ed Neely WB, Blau GE, Boca Raton, FL: CRC Press p 157-73 (1985)
19. Neely WR, Blau GE; Environ Expos from Chemicals Vol I. Boca Raton, FL: CRC Press (1985)
20. Parris GE; Environ Sci Technol 14: 1099-1105 (1980)
21. Perrin DD; Dissociation Constants of Organic Bases in Aqueous Solution. IUPAC Chemical Data Series, London: Buttersworth (1965)
22. Riddick JA et al; Organic Solvents: Physical Properties and Methods of Purification, 4th Edit. New York: J Wiley & Sons (1986)
23. Serjeant EP, Dempsey B; Ionisation Constants of Organic Acids in Aqueous Solution. IUPAC Chemical Data Series No 23, New York: Pergamon Press (1979)
24. Swann RL et al; Residue Reviews 85: 17-28 (1983)
25. U.S. EPA; Reference Values for Risk Assessment. Environ Criteria Assess Office, Off Health Environ Assess, Off Research Devel, ECAO-CIN-477, Cincinnati, OH: U.S. Environ Prot Agency (1986)
26. Veith GD et al; J Fish Res Board Can 36: 1-40-8 (1979)
27. Yalkowsky SH et al; ARIZONA dATABASE of Aqueous Solubility, U. Arizona, Tucson, AZ (1987)

# Contents

**Volume II**
Solvents

# Handbook of Environmental

# FATE
# and
# EXPOSURE
# DATA

## For Organic Chemicals

# Acetic Acid

## SUBSTANCE IDENTIFICATION

**Synonyms:** Methanecarboxylic acid

**Structure:**

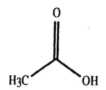

**CAS Registry Number:** 64-19-7

**Molecular Formula:** $C_2H_4O_2$

**Wiswesser Line Notation:** QV1

## CHEMICAL AND PHYSICAL PROPERTIES

**Boiling Point:** 118.5 °C

**Melting Point:** 16.604 °C

**Molecular Weight:** 60.05

**Dissociation Constants:** pKa = 4.75 at 25 °C [59]

**Log Octanol/Water Partition Coefficient:** -0.17 [26]

**Water Solubility:** 6,029,000 mg/L at 25 °C [61]

**Vapor Pressure:** 12 mm Hg at 20 °C [60]

**Henry's Law Constant:** 1 x $10^{-7}$ atm-m³/mole at pH 4 to 1 x $10^{-9}$ atm-m³/mole at pH 7 [18]

## Acetic Acid

## ENVIRONMENTAL FATE/EXPOSURE POTENTIAL

**Summary:** Acetic acid occurs throughout nature as a normal metabolite of both plants and animals. Humans release acetic acid to the environment in a variety of waste effluents, in emissions from combustion processes, and in exhaust from gasoline and diesel engines. If released to the atmosphere, it is degraded in the vapor phase by reaction with photochemically produced hydroxyl radicals (estimated typical half-life of 26.7 days). It occurs in atmospheric particulate matter in acetate form, and physical removal from air can occur via wet and dry deposition. If released to water, acetic acid will biodegrade readily. If released to soil, biodegradation will also occur. Evaporation from dry surfaces is likely to occur. Since acetic acid exists ubiquitously in the environment, the general population is continuously exposed to the compound. Primary routes of exposure to the general population are through oral consumption of foods and inhalation of air. Occupational exposure occurs through inhalation and dermal contact.

**Natural Sources:** Decomposition of solid biological wastes produces acetic acid which is readily metabolized by living organisms [1]; acetic acid occurs as a normal metabolite in both plants and animals [1]. It occurs naturally in various vegetation [23].

**Artificial Sources:** Acetic acid is released to the environment in waste emissions from the manufacture of various chemicals, explosives, lacquers, starch, sugars, wines, and vinegar and from wood distillation plants and textile mills [1,23]. Atmospheric emissions occur from combustion of biomass, plastics, and refuse and in exhaust from gasoline and diesel engines [1,23,33].

**Terrestrial Fate:** The major environmental fate process for acetic acid in soil is biodegradation. A large number of biological screening studies have determined that acetic acid biodegrades readily under both aerobic and anaerobic conditions. Acetic acid has been noted to leach from biological disposal areas; however, it is expected to be efficiently biodegraded during its migration through soil [1]. Based on the vapor pressure, acetic acid is likely to evaporate from dry surfaces.

**Aquatic Fate:** The dominant environmental fate process for acetic acid in water is biodegradation. A large number of biological screening studies have determined that acetic acid biodegrades readily under both aerobic and anaerobic conditions. Aquatic hydrolysis and

2

bioconcentration are not important. Two aqueous adsorption studies found that acetic acid exists primarily in the water column.

**Atmospheric Fate:** Acetic acid is expected to exist almost entirely in the vapor phase in the ambient atmosphere, based on the vapor pressure [15]. In acetate form, however, it has been detected in atmospheric particulate material [24]. It is degraded in the vapor phase by reaction with photochemically produced hydroxyl radicals (estimated half-life of 26.7 days in an average ambient atmosphere). Due to its miscibility in water, acetic acid is likely to be removed physically from the atmosphere by wet deposition. Particulate acetate material may be physically removed by both wet and dry deposition. Formic and acetic acid have been identified as the major sources of free acidity in precipitation from remote regions of the world [34].

**Biodegradation:** Closed bottle test, 5-30 day 51-99% BODT [17]. Warburg respirometer, 30-day 60% BODT, acclimated sewage seed [28]. Zahn-Wellens test, >90% degradation in 3 days using an activated sludge inocula [62]. AFNOR T 90/103 test, 5-day 36% BODT, microbes from 3 polluted surface waters [12]. Standard dilution BOD water, 5-day 57.7% BODT avg [30]. Water die-away tests, 12.3%/hr in estuarine water, 1.0%/hr in Belgian coastal water, 0.06%/hr in open seawater [5]. Standard dilution BOD water, 76-96% BODT in 5-20 days; seawater dilution, 66-100% BODT in 5-20 days, sewage inocula [49]. Batch aeration in sewage, 99.5% degradation in 24 hr [48]. Warburg respirometer, 24-hr 40% BODT, activated sludge inocula [40]. Warburg respirometer, 5-day 77% BODT, sewage inocula [10]. Batch aeration, virtual loss of BOD in 6 hr, settled domestic sewage inocula [27]. Standard dilution BOD water, 5-day 81.3% BODT; seawater dilution, 5-day 77.6% BODT [56]. Standard dilution BOD water, 5-day 63.2% BODT, sewage inocula [50]. Electrolytic respirometer test, 10-day 87% BODT [57]. Laboratory-scale anaerobic digester, microbial decay coefficient of 0.283/day [37]. Modified OECD protocol, 75% degradation in 14 days using garden soil as inocula, >90% degradation in 14 days using sediment from the Rhine R as inocula [36]. Biofilm column study, 95% removal under aerobic conditions, 99% removal under methanogenic conditions [6]. Acetate solutions were readily degraded in anaerobic lake sediments [55].

**Abiotic Degradation:** The experimentally determined rate constant for the vapor phase reaction of acetic acid with photochemically produced hydroxyl radicals has been reported to be $0.6 \times 10^{-12}$ cm$^3$/molecule-sec at 25 °C [4]; the atmospheric half-life for this reaction can be estimated to be 26.7 days, assuming an average atmospheric hydroxyl radical concn of 5 x $10^{+5}$ molecules/cm$^3$ [4]. The rate constant for the reaction of acetic acid with hydroxyl radicals in aqueous solution is approximately 0.48-0.85 x $10^{+8}$ 1/mol-sec [3,13]; if the hydroxyl radical concn of sunlit natural water is assumed to be 1 x $10^{-17}$ moles/L [41], the half-life would be approximately 26-46 years. Carboxylic acids are generally resistant to aqueous environmental hydrolysis [38].

**Bioconcentration:** Based on the log Kow, the BCF for acetic acid can be estimated to be less than 1 [38]. This indicates that bioconcentration is not significant.

**Soil Adsorption/Mobility:** In 24 hr aqueous adsorption studies using montmorillite and kaolinite clay adsorbents, 2.4-30.4% of added acetic acid was observed to be in the adsorbed phase [29]. In adsorption studies using the adsorbent hydroxyapatite (a mineral which occurs in the environment as a result of the diagenesis of skeletal apatite), only 5% of added acetic acid (in aqueous solution) became adsorbed to the hydroxyapatite [22]. Acetic acid has been noted to leach from biological disposal areas [1]. Acetic acid's pKa value indicates that it will exist predominantly in the anionic form in the environment. The adsorption characteristics of an anionic species may be different from the neutral species, and cannot be predicted adequately without experimental data.

**Volatilization from Water/Soil:** The values of Henry's Law constant at various pH's indicates that acetic acid will not volatilize significantly from water [38].

**Water Concentrations:** DRINKING WATER: Acetic acid was qualitatively detected in the District of Columbia drinking water supply [51]. GROUND WATER: Acetic acid was qualitatively detected in ground water from a landfill well in Norman, OK in 1972 [14]. Levels of 0.66-4.60 ppm were identified in ground water below a closed wood treatment facility in Pensacola, FL in 1984 [21]. Acetic acid was qualitatively identified in ground water associated with an Australian quarry where dumping of organic wastes had occurred [54]. SURFACE WATER: Acetic acid was detected at concn of 12-198 ppb in the Scheldt estuary in Belgium during 1977-78 [5]. Levels of 75-300 ppb were found at various depths of Lake Kizaki in Japan [25]. Concn of 13-72, 6-12, and 25 ppb were detected in the Ohio, Little Miami, and

## Acetic Acid

Tannes Rivers, respectively [43]. Concn generally below 0.1 ppb were monitored in the Lee River in Great Britain [58]. SEAWATER: Acetic acid was detected at concn of 2.4-144 ppb near the Belgian coast and 12-240 ppb near Calais on the English Channel during 1977-78 monitoring [5].

**Effluent Concentrations:** Acetic acid was qualitatively detected in wastewater effluents from publicly owned treatment works (POTW) in Decatur and Bensenville IL [16]. Acetic acid concn of 125 ppm were identified in wastewaters from a coal gasification facility in North Dakota [19]. Wastewater from a shale oil process in Australia contained 140 ppm acetic acid [11]. Acetic acid was detected in leachate from a sanitary landfill in Barcelona, Spain [2]. Acetic acid has reportedly been detected in wastewater effluents from chemical, resin, and paper manufacturing plants, from various landfill leachates, and from sewage treatment facilities [52].

**Sediment/Soil Concentrations:** Acetic acid concn of 17.3-48.5 mmol/kg wet mud were detected in bottom sediments of Lake Biwa in Japan; however, no acetic acid was found in the interstitial water [39]. Concn of 0.133-1.836 mg/g (dry wt) were detected in sediments from Loch Eil in Scotland [42]; water removed from sediments contained levels of 0.244-0.251 mg/mL [42].

**Atmospheric Concentrations:** Acetic acid concn of 40-244 ug/m$^3$ were detected in indoor air of homes in Italy [9]. The acetate concn of the atmospheric aerosol collected over a wet tropical forest in Guyana in 1984 ranged from 2-11 ng/m$^3$ [24]. Mean atmospheric concn in Los Angeles, CA between Jul and Sept 1984 were 0.262-3.90 ppb [33]. Levels of 1-6 ppb were reported for ambient air in Tucson, AZ [33].

**Food Survey Values:** Acetic acid was identified as the major volatile constituent of commercial brown sugars [20]; concn ranging from 31-827 ppm were detected in 26 brown sugars collected worldwide [20]. The source of the acetic acid in brown sugar was thought to result from bacterial action on sucrose waters used in its production [20]. Acetic acid was qualitatively detected as a volatile component of fried bacon, smoked pork, baked potatoes, soy sauce, and roasted filbert nuts [8,31,35,53].

## Acetic Acid

**Plant Concentrations:** Acetic acid occurs in various plants, such as in essential oil from juniper [44]. Acetic acid occurs in tobacco [32].

**Fish/Seafood Concentrations:** Acetic acid occurs as a volatile emission product during fish processing [23].

**Animal Concentrations:** Acetic acid was identified as a component of poultry manure which was responsible for a vinegar-like odor [7].

**Milk Concentrations:**

**Other Environmental Concentrations:** Gasoline engine exhaust from a 1982 Toyota Corolla contained 31.81 ppb acetic acid [33]. New motor oil contained 5.3 nmol/mL acetic acid while used motor oil contained 145 nmol/mL [33]. Acetic acid has been identified in tobacco smoke [32].

**Probable Routes of Human Exposure:** Acetic acid occurs ubiquitously and is a normal metabolite in animals; therefore, the general population is continually exposed to the compound. Primary routes of exposure to the general population are through oral consumption of foods and inhalation of air. Occupational exposure occurs through inhalation and dermal contact.

**Average Daily Intake:** AIR INTAKE: assume ambient atmospheric concn of 0.1-1.6 ug/m$^3$ [33]: 2-32 ug/day; WATER INTAKE: insufficient data; FOOD INTAKE: insufficient data.

**Occupational Exposures:** NIOSH (NOHS Survey 1972-74) has statistically estimated that 1,400,824 workers are potentially exposed to acetic acid [46]. NIOSH (NOES Survey 1981-83) has statistically estimated that 595,346 workers are potentially exposed to acetic acid [45].

**Body Burdens:** Acetic acid was qualitatively detected in 2 of 12 human milk samples collected from volunteers in four US cities [47].

### REFERENCES

1.  Abrams EF et al; Identification of Organic Compounds in Effluents from Industrial Sources. USEPA-560/3-75-002 p. 3 (1975)
2.  Albaiges J et al; Water Res 20: 1153-9 (1986)

# Acetic Acid

3. Anbar M, Neta P; Int J Appl Radiation and Isotopes 18: 493-523 (1967)
4. Atkinson RA; Chem Rev 85: 60-201 (1985)
5. Billen G et al; Estuarine Coastal Marine Sci 11: 279-94 (1980)
6. Bouwer EJ, McCarty PL; Ground Water 22: 433-40 (1984)
7. Burnett WE; Environ Sci Technol 3: 744-9 (1969)
8. Coleman EC et al; J Agric Food Chem 29: 42-8 (1981)
9. Debortoli M et al; Environ Int 12: 343-50 (1986)
10. Dias FF, Alexander M; Appl Microbial 22: 1114-8 (1971)
11. Dobson KR et al; Water Res 19: 849-856 (1985)
12. Dore M et al; Trib Cebedeau 28: 3-11 (1975)
13. Dorfman LM, Adams GE; Reactivity of Hydroxyl Radical in Aqueous Solution, NSRD-NBS-46 Washington DC: Natl Bureau of Standards (1973)
14. Dunlap WJ et al; Organic Pollutants Contributed to Ground Water by a Landfill. USEPA-600/0-76-004 p. 106 (1976)
15. Eisenreich SJ et al; Environ Sci Technol 15: 30-8 (1981)
16. Ellis DD et al; Arch Environ Contam Toxicol 11: 373-382 (1982)
17. Fischer WK et al; Wasser-Und Abwasser-Forschung 7: 99-118 (1974)
18. Gaffney JS et al; Environ Sci Technol 21: 519-24 (1987)
19. Giabbai MF et al; Intern J Environ Anal Chem 20: 113-29 (1985)
20. Godshall MA, DeLucca AJ; J Agric Food Chem 32: 390-3 (1984)
21. Goerlitz DF et al; Environ Sci Technol 19: 955-61 (1985)
22. Gordon AS, Millero FJ; Microb Ecol 11: 289-98 (1985)
23. Graedel TE et al; Atmospheric Chemical Compounds. Sources, Occurrence, and Bioassay. Orlando,FL: Academic Press p. 345 (1986)
24. Gregory GL et al; J Geophys Res 91: 8603-12 (1986)
25. Hama T, Handa N; Jap J Limnol 42: 8-19 (1981)
26. Hansch C, Leo AJ; Medchem Project Issue No 26. Claremont CA: Pomona College (1985)
27. Hatfield R; Ind Eng Chem 49: 192-6 (1957)
28. Helfgott TB et al; An Index of Refractory Organics. USEPA-600/2-77-174 (1977)
29. Hemphill L, Swanson WS; Proc of the 18th Industrial Waste Conf, Eng Bull Purdue Univ, Lafayette IN 18: 204-17 (1964)
30. Heukelekian H, Rand MC; J Water Pollut Control Assoc 29: 1040-53 (1955)
31. Ho CT et al; J Agric Food Chem 31: 336-42 (1983)
32. Johnstone RAW, Plimmer JR; Chem Rev 59: 885-936 (1959)
33. Kawamura K et al; Environ Sci Technol 19: 1082-6 (1985)
34. Keene WC, Galloway JN; Atmos Environ 18: 2491-7 (1984)
35. Kinlin TE et al; J Agric Food Chem 20: 1021-8 (1972)
36. Kool HJ; Chemosphere 13: 751-61 (1984)
37. Lin C et al; Water Res 20: 385-94 (1986)
38. Lyman WJ et al; Handbook of Chemical Property Estimation Methods NY:McGraw-Hill (1982)
39. Maeda H, Kawai A; Bull Japan Soc Sci Fisheries 52: 1205-8 (1986)
40. Malaney GW, Gerhold RM; J Water Pollut Control Fed 41: R18-R33 (1963)
41. Mill T et al; Science 207: 886-7 (1980)
42. Miller D et al; Marine Biology 50: 375-83 (1979)
43. Murtaugh JJ, Bunch RL; J Water Pollut Control Fed 37: 410-5 (1965)
44. Nicholas HJ; p. 382-3 in Phytochemistry; Miller LP ed NY: Van Nostrand Reinhold (1973)
45. NIOSH; National Occupational Exposure Survey (NOES) (1983)

46. NIOSH; National Occupational Hazard Survey (NOHS) (1974)
47. Pellizzari ED et al; Bull Environ Contam Toxicol 28: 322-8 (1982)
48. Placak OR, Ruchhoft CC; Sewage Works J 19: 423-40 (1947)
49. Price KS et al; J Water Pollut Control Fed 46: 63-77 (1974)
50. Saito T et al; Fresenius Z Anal Chem 319: 433-4 (1984)
51. Scheiman MA et al; Biomed Mass Spectrom 4: 209-11 (1974)
52. Shackelford WM, Keith LM; Frequency of Organic Compounds Identified in Water, EPA-600/4-76-062 p. 47-8 (1976)
53. Shibamoto T et al; J Agric Food Chem 29: 57-63 (1981)
54. Stepan S et al; Austral Water Resources Council Conf Ser 1: 415-24 (1981)
55. Strayer RF, Tiedje JM; Appl Environ Microbiol 36: 330-40 (1978)
56. Takemoto S et al; Suishitsu Odaku Kenkyu 4: 80-90 (1981)
57. Urano K, Kato Z; J Hazardous Mater 13: 147-59 (1986)
58. Waggot A; Chem Water Reuse 2: 55-9 (1981)
59. Weast RC; CRC Handbook of Chemistry and Physics 66th ed Boca Raton, FL: CR Press p D-161 (1985)
60. Weber RC et al; Vapor Pressure Distribution of Selected Organic Chemicals. USEPA-600/2-81-021 p. 12 (1981)
61. Yalkowsky SH et al; Arizona Database of Aqueous Solubility. U of Arizona (1987)
62. Zahn R, Wellens H; Z Wasser Abwasser Forsch 13: 1-7 (1980)

# Acetone

## SUBSTANCE IDENTIFICATION

**Synonyms:**

**Structure:**

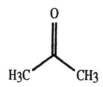

**CAS Registry Number:** 67-64-1

**Molecular Formula:** $C_3H_6O$

**Wiswesser Line Notation:** 1V1

## CHEMICAL AND PHYSICAL PROPERTIES

**Boiling Point:** 56.2 °C at 760 mm Hg

**Melting Point:** -95.35 °C

**Molecular Weight:** 58.09

**Dissociation Constants:**

**Log Octanol/Water Partition Coefficient:** -0.24 [33]

**Water Solubility:** Miscible [60]

**Vapor Pressure:** 231 mm Hg at 25 °C [10]

**Henry's Law Constant:** 3.67 x $10^{-5}$ atm-m³/mole [71]

## ENVIRONMENTAL FATE/EXPOSURE POTENTIAL

**Summary:** Acetone is produced in large quantities and may be released to the environment as stack emissions, fugitive emissions, and in wastewater in its production and use as a chemical intermediate and solvent. Most acetone used in solvents will be ultimately released into the air. Acetone is a product of the photooxidation of some alkanes and alkenes that are found in urban air and it is also released from volcanoes and in forest fires. It is a metabolic product, released by plants and animals. If released on soil, acetone will both volatilize and leach into the ground and probably biodegrade. If released into water, acetone will probably biodegrade. It will also be lost due to volatilization (estimated half-life 20 hr from a model river). Bioconcentration in aquatic organisms and adsorption sediment should not be significant. In the atmosphere, acetone will be lost by photolysis and reaction with photochemically produced hydroxyl radicals. Estimated half-lives from these combines processes averages 22 days and is shorter in summer and longer in winter. It will also be washed out by rain. Occupational exposure to acetone will be via dermal contact with solvents containing the chemical and via inhalation of the vapor. The general population is exposed to acetone in the atmosphere from sources such as auto exhaust, solvents, tobacco smoke, and fireplaces as well as from dermal contact with consumer products such as solvents. In addition, there will be exposure by ingestion of food that may naturally contain acetone or exposure from contaminated drinking water.

**Natural Sources:** Acetone has been identified in vegetation and insects as a naturally occurring, volatile metabolite [28]. Some of the plants it has been identified in are onions [28], apples, grapes, cauliflower, tomato, morning glory, and wild mustard [54]. Acetone is a component of human breath and is of metabolic origin [16]. Volcanos and forest fires are another natural source of acetone emissions [28].

**Artificial Sources:** Acetone is produced in large quantities and may be released to the environment as stack emissions, fugitive emissions, in wastewater in its production and use in the manufacture of methacrylates, as a solvent, and as a chemical intermediate in the manufacture of methyl isobutyl ketone and other chemicals [12]. It is also a by-product of several manufacturing processes including the manufacture of acetaldehyde and acetic acid and wood pulping; releases of acetone to both air and water may be expected from these processes [54]. Most of the 17% of acetone produced that is used as solvents will be ultimately released into the environment. Acetone is a product of the photooxidation of some alkanes and alkenes that are found in urban air and therefore would be found wherever these precursors are present. Acetone is also emitted from wood-burning fireplaces and tobacco smoke [28,50].

**Terrestrial Fate:** If released on soil, acetone will both volatilize and leach into the ground. Acetone readily biodegrades and there is evidence suggesting that it biodegrades fairly rapidly in soils.

**Aquatic Fate:** If released into water, acetone will probably biodegrade. It is readily biodegradable in screening tests, although data from natural water are lacking. It will also be lost due to volatilization (estimated half-life 20 hr from a model river). Adsorption to sediment should not be significant.

**Atmospheric Fate:** In the atmosphere, acetone will be lost by photolysis and reaction with photochemically produced hydroxyl radicals. Half-life estimates from these combined processes are 79 and 13 days in January and June, respectively, for an overall annual average of 22 days. Therefore considerable dispersion should occur. Being miscible in water, wash out by rain should be an important removal process. This process has been confirmed around Lake Shinsei-ko in Japan [41]. There acetone was found in the air and rain as well as the lake, and the amount of acetone in the rain is what would be expected from the air concentration and solubility [41].

**Biodegradation:** Acetone readily biodegrades in screening tests [7,22,26,37,48,57,74,76,78,83]. Typical results using sewage inocula are: 37% and 81% theoretical BOD in 5 and 20 days, respectively [83]; 54% theoretical BOD after 5 days [7]; 71% theoretical BOD in 7 days [37]; 55.4% and 71.8% theoretical BOD after 5 and 10 days, respectively [48]; 38% theoretical BOD in 5 days [78]; 64.5% and 11.9% theoretical BOD in fresh and salt water, respectively [74]; 56% and 38% theoretical BOD after 5 days and 84% and 76% theoretical BOD after 20 days in fresh and salt water, respectively [57]; and 46% theoretical BOD after 5 days [22]. In other tests using activated sludge inocula, acetone degraded at a rate of 0.016 $hr^{-1}$ after a 20 hr lag and had a 42% theoretical BOD after 155 hr when the endogenous respiration period began [76], and half-life was 1 day [26]. A 75% reduction in BOD occurred in 4 hr in a semi-continuous activated sludge test [35]. The chemical was toxic to microorganisms at high concn [27]. Acetone also degrades under anaerobic conditions, with one investigator reporting 100% degradation in 4 days after a 5 day lag [13]. No degradation studies were reported in the environmental waters or soil. However, the fact that acetone was not detected in a well underneath a leaking tank at a paint factory was ascribed to biological removal by aerobic or anaerobic processes [6].

**Abiotic Degradation:** Acetone has a UV absorption band at 270 nm that extends to about 330 nm [53,63]. The photodissociation loss rate in the lower troposphere at 40 Deg N latitude ranges from $1.8 \times 10^{-7}$ $sec^{-1}$ in July to $3.3 \times 10^{-8}$ $sec^{-1}$ in January for an annual average loss rate of $1.0 \times 10^{-7}$ $sec^{-1}$ [53]. This loss rate is roughly comparable to that due to reaction with photochemically produced hydroxyl radicals, ranging from $6.8 \times 10^{-8}$ $sec^{-1}$ in January to $4.4 \times 10^{-7}$ $sec^{-1}$ in July for an annual average loss rate of $3.6 \times 10^{-7}$ $sec^{-1}$. These combined processes result in a half-life of 79 and 13 days in January and July, respectively, or an annual average half-life of 22 days [53]. Smog chamber studies involving acetone conclude that acetone ranks low in photochemical reactivity based on criteria such as ozone production [1,21,24,49,69,82]. In studies where the disappearance of the compounds was investigated, no degradation in 5 hr [82], and 0.9% disappearance per hr [21], was reported. No photodegradation occurred when acetone was exposed to sunlight for 23 hr in distilled water or 15 hr in stream water [59]. When water containing acetone is treated with chlorine for disinfection purposes, the acetone can react with the hypochlorite ion formed by the hydrolysis of chlorine leading to the production of trichloromethane [73]. This reaction is strongly pH dependent and is expected to have a significant effect only at pH's 6-7 [73].

**Bioconcentration:** Using the recommended log octanol/water partition coefficient, the potential for acetone bioconcentration in fish is negligible [52]. One experimental study of bioconcentration in adult haddock at 7-9 °C (static test), resulted in a BCF of 0.69 [62].

**Soil Adsorption/Mobility:** Acetone's miscibility in water would suggest that it does not adsorb appreciably to soil [52]. It displayed no adsorption to montmorillonite or kaolinite clay or stream sediment [59,81]. However when 3 natural clay soils at optimum moisture content were exposed to bulk acetone, 3.5-8% swelling of the soil was observed, indicating that the acetone formed interlayer complexes with clay [30]. Therefore there is adsorption to clay under these conditions that may be applicable to landfills having clay liners. The equilibrium permeability rates of acetone in 3 compacted clay soils ranged from $2.5 \times 10^{-9}$ to $65 \times 10^{-9}$ cm/sec [31].

## Acetone

**Volatilization from Water/Soil:** Using the Henry's Law constant, one can estimate a half-life for volatilization from a model river 1 m deep with 1 m/sec current and a 3 m/sec wind of 20 hr [52]. Based on an experimentally determined mass transfer coefficients relative to oxygen at different mixing rates ranging from 0.074 to 0.335 [59], the half-life of acetone in a model river and lake is estimated to be 2-10 and 16-186 days, respectively [52]. These experiments were performed with mixing only in the liquid phase and since the volatilization of acetone will depend on both the liquid and vapor mass transport coefficients, these coefficients would increase considerably for higher mixing conditions resulting from winds in the air phase [59]. Acetone has a high vapor pressure and low adsorption to soil and should therefore readily evaporate from the soil surface [55]. When a variety of moist soils were treated with acetone and the headspace analyzed over the course of 3 days, acetone was found in the air above all samples [55].

**Water Concentrations:** DRINKING WATER: As part of the US National Organics Reconnaissance Survey, tap water from 10 cities were characterized for organic chemical content [25]. Acetone was found in all 10 drinking waters [25]. It was similarly found, although not quantified, in drinking water in New Orleans [77], Seattle [42], Tuscaloosa [5], and other cities in a 5-city survey [15]. In a listing of organic chemicals identified in water, 30 items referred to detection of acetone in finished drinking water [67]. A sample of finished drinking water from Waterford, NY, a community that obtains its raw water from the Hudson River, did not contain acetone [43]. Six drinking water wells in the vicinity of a landfill contained 0.2 to 0.7 ppb of acetone [19]. A contaminated well in New Jersey contained 3000 ppm of acetone [58]. When an intense taste and odor problem occurred in drinking water in a section of Paris, the problem was identified with the passage of the water through a newly installed section of defective high density polyethylene tubing [2]. Acetone and several other organic chemicals leached from this tubing when it was soaked for 24 hr in mineral water. GROUND WATER: Acetone has been detected but not quantified in ground water in Gastonia, NC [67]. SURFACE WATER: Five of nine sites in Lake Michigan contained 1-4 ppb acetone [46]. In a survey of 14 heavily industrialized river basins in the US (204 samples), 33 contained detectable amounts of acetone including 18 of 31 sites in the Chicago area and the Illinois River basin, 8 of 30 sites in the Delaware River basin, 1 of 45 sites in the Mississippi River basin in Alabama and Texas, 3 of 27 sites in the Ohio River basin, and 3 of 15 West Coast sites [23]. Acetone was identified but not quantified in the Black River in Tuscaloosa, AL [5], and the Cuyahoga River in the Lake Erie basin [29]. SEAWATER: Samples of seawater and surface slicks taken from Biscayne Bay and the Florida Current during June-August 1968 contained 39.6 and 89.7 ppb of acetone, respectively [65]. Grab samples of surface water from the Straits of Florida and the Eastern Mediterranean contained 20 and 28 ppb of acetone, respectively [17]. Samples of ocean water taken at 1200 m depths contained the chemical [17]. RAIN: 50 ppb of acetone was found in one of 6 samples tested at 5 cities in California [32]. An unspecified concn of acetone was found in rain in Japan [41].

**Effluent Concentrations:** In a comprehensive survey of wastewater from 4000 industrial and publicly owned treatment works (POTWs) sponsored by the Effluent Guidelines Division of the USEPA, acetone was identified in discharges of the following industrial category (positive occurrences; median concn in ppb): leather tanning (4; 74.7), petroleum refining (14; 166.9), nonferrous metals (2; 6.6), paint and ink (22; 894.9). printing and publishing (7; 2501.2), coal mining (1; 2260.8), organics and plastics (24; 373.4), inorganic chemicals (8; 13.8), textile mills (4; 11.0), plastics and synthetics (10; 164.1), pulp and paper (6; 59.8), rubber processing (1; 604.4), auto and other laundries (2; 437.5), pesticides manufacture (7; 52.7), photographic industries (1; 94.9), pharmaceuticals (6; 75.4), explosives (23; 388.0), porcelain/enameling (4; 14.7), electronics (12; 441.2), oil and gas extraction (5; 59.2), organic chemicals (1; 113.9), mechanical products (6; 84.4), transportation equipment (6; 616.7), and publicly owned treatment works (40; 96.8) [68]. The highest effluent concn was 37,709 ppb in the paint and ink industry [68]. 50 ppb of acetone was found in the vapor phase effluent of a wood burning stove [45], and wood-burning fireplaces emit 7-145 mg/g wood burned [50]. While the concentration of acetone in the effluent from a community septic tank with a 2 day holding period was 70.3 ppb, its concentration in the influent was 18.2 ppb [20]. Leachate from a landfill in Delaware containing industrial and municipal waste contained 46.6 ppm of acetone [19]. One out of 5 landfills tested in Connecticut had leachate containing detectable amounts of acetone; 3.5 ppm in Southington [64]. An artificial landfill constructed from solid waste separated by soil, flooded with water and maintained for one year contained 0.60 g/L of acetone in the leachate [9]. Acetone has also been identified in effluent from a shale oil wastewater treatment facility [36].

# Acetone

**Sediment/Soil Concentrations:**

**Atmospheric Concentrations:** RURAL: The estimated background concn of acetone in the lower troposphere is 0.111 ppb [70]. Two rural sites outside of Tucson, AZ (18 measurements) 2.6 ppb, mean [72]. Samples from 5 remote sites contained 0.3 to 0.9 ppb of acetone [61]. The acetone concn in air at Pt Barrow, AK (22 measurements) ranged from 0.3 to 2.9 ppb, mean 1.23 ppb [11]. Jones State Forest near Houston, TX - 1.9 ppb [66]; Smokey Mtn, TN - 0.70 to 3.9 ppb [3]; rural site near Tulsa, OK - 1.8 to 6.5 ppb [3]; Rio Blanco County, CO - 5.1 ppb, mean [3]. URBAN/SUBURBAN: Tucson, AZ (17 measurements)- 12 ppb, mean [72]. Two sites in Tulsa, OK - 1.6 to 17.3 ppb [3]. While the mean acetone concn at 5 sites in Stockholm, Sweden ranged from 4.04 to 19.4 ppb with the highest concn being registered on busy streets, there was no statistically significant correlation to exhaust components such as carbon monoxide and benzene [40]. SOURCE AREAS: 22 sites in the US 0.350 ppb median, 53 ppb max [8]. Texaco Refinery in Tulsa. OK - 2.3 to 3.3 ppb [3]. INDOOR AIR: In a study of 36 homes in the Chicago area, the frequency of occurrence of acetone in indoor air, 51, was 1.5 times that in outdoor air [39].

**Food Survey Values:** Acetone has been identified as a volatile component of baked potatoes [14], roasted filberts [44], dried beans and legumes [51], and French cognac [75].

**Plant Concentrations:**

**Fish/Seafood Concentrations:**

**Animal Concentrations:**

**Milk Concentrations:**

**Other Environmental Concentrations:**

**Probable Routes of Human Exposure:** Occupational exposure to acetone will be via dermal contact with solvents containing the chemical and via inhalation of the vapor. The general population is exposed to acetone in the atmosphere from sources such as auto exhaust, solvents, tobacco smoke, and fireplaces as well as from dermal contact with consumer products containing acetone as a solvent. In addition there will be exposure via ingestion of items of food that naturally contain acetone.

**Average Daily Intake:** AIR INTAKE (assume air concn of 0.05-20 ppb): 24-960 mg; WATER INTAKE: insufficient data; FOOD INTAKE: insufficient data.

**Occupational Exposures:** Acetone was found to be one of predominant solvents in the air of leather and shoe factories in Italy [18], and in factories using cutting fluids at elevated temperatures in Russia [4]. In a 1981 study investigating exposure to solvent vapors in spray painting and glue spraying operations, 7 plants belonging to 3 companies were evaluated [80]. The operations studied were believed to be typical of that used for products up to a few feet in size and capable of overhead conveyer transport. The mean concn of acetone for the 89 exposed individuals in the 3 companies were 1.1, and 1.7 and 3.1 ppm and the overall mean and standard deviation were 2.0 and 6.0 ppm, respectively [80]. The average TWA exposure to acetone for higher-aromatic paint spraying, lower-aromatic paint spraying, glue spraying, solvent wiping, and paint mixing was 0.9, 3.2, 2.3, 0.9, and 5.6 ppm, respectively [80].

**Body Burdens:** Acetone was found in all 8 samples of mother's milk analyzed from 4 industrial urban areas in the US [56]. Acetone was ubiquitous in the expired air from a carefully selected urban population of 54 normal, health, non-smoking people (387 samples) with a geometric mean concn of 101.3 ng/L [47]. Elevated levels of acetone on the breath of patients with uncontrolled diabetes is a well known observation. Normal values of acetone (including that derived from acetoacetic acid which is split at the temperature of the test) is 0.5 to 3.0 mg/100 mL in whole blood and about 10% less in serum [38]. Loss in the urine is generally <1 mg/24 hr for a normal adult but is about 50 mg in children [34,79].

## REFERENCES

1. Altshuller AP, Cohen IR; Int J Air Water Pollut 7: 787-97 (1963)
2. Anselme C et al; Sci Total Environ 47: 371-84 (1985)
3. Arnts RR, Meeks SA; Atmos Environ 15: 1643-51 (1981)
4. Berezkin VG, Drugov YS; Zavod Lab 52: 16-9 (1986)
5. Berstch W et al; J Chromatog 112: 701-18 (1975)
6. Botta D et al; Anal Org Micropollut Water p. 261-75 (1984)

# Acetone

7. Bridie Al, et al; Water Res 13: 627-30 (1979)
8. Brodzinsky R, Singh HB; Volatile Organic Chemicals in the Atmos SRI International Contract 68-02-3452 (1982)
9. Burrows WD, Rowe RS; J Water Pollut Control Red 47: 921-3 (1975)
10. Buttery RG et al; J Agric Food Chem 17: 385-9 (1969)
11. Cavanagh LA et al; Environ Sci Technol 3: 251-7 (1969)
12. Chemical Profile: Acetone Chemical Marketing Reporter Aug 27, (1984)
13. Chou WL et al; Bioeng Symp 8: 391-414 (1979)
14. Coleman EC et al; J Agric Food Chem 29: 42-8 (1981)
15. Coleman WE et al; Analysis And Identification of Organic Substances In Water Ann Arbor, MI Chapt 21 pp. 305-27 (1976)
16. Conkle JP et al; Arch Environ Health 30: 290-5 (1975)
17. Corwin JF; Bull Mar Sci 19: 504-9 (1969)
18. Cresci A et al; Nuovi Ann Ing Microbiol 36: 61-76 (1985)
19. Dewalle FB, Chain ESK; J Am Water Works Assoc 73: 206-11 (1981)
20. Dewalle FB et al; Determination of Toxic Chemicals in Effluent from Household Septic Tanks USEPA EPA-600/S2-85-050 p. 4 (1985)
21. Dimitriades B, Joshi SB; USEPA-600/3-77-001b pp. 705-11 (1977)
22. Dore M et al; Trib Cebedeau 28: 3-11 (1975)
23. Ewing BB et al; Monitoring to Detect Previously Unrecognized Pollutants in Surface Waters Appendix USEPA EPA-560/6-77-015 (1976)
24. Farley FF; USEPA-600/3-77-001b pp. 713-27 (1977)
25. Fielding M, Packman RF; J Inst Water Eng Sci 31: 353-75 (1977)
26. Gaudy AF JR et al; J Water Pollut Contr Fed 35: 75-93 (1963)
27. Gerhold RM, Malaney GW; J Water Pollut Contr Fed 38: 562-79 (1966)
28. Graedel TE et al; Atmospheric Chemical Compounds Academic Press NY p. 263 (1986)
29. Great Lakes Water Quality Board Ontario, Canada p. 195 (1983)
30. Green WJ et el; Environ Sci Technol 17: 278-82 (1983)
31. Green WJ et al; J Water Pollut Control Fed 53: 1347-54 (1981)
32. Grosjean D, Wright B; Atmos Environ 17: 2093-6 (1983)
33. Hansch C, Leo AJ; Medchem Project Issue No 26. Claremont CA: Pomona College (1985)
34. Harper HA; Review of Physiological Chemistry 12th ed p. 303 (1969)
35. Hatfield R; Ind Eng Chem 49: 192-6 (1957)
36. Hawthorne SB, Sievers RE; Environ Sci Technol 18: 483-90 (1984)
37. Helfgott TB et al; USEPA-600/2-77-174 Ada, OK (1977)
38. Henry RJ; Clinical Chemistry pp. 693-5 Harper and Row NY (1964)
39. Jarke FH et al; Ashrae Trans 87: 153-77 (1981)
40. Jonsson A et al; Environ Int 11: 383-92 (1985)
41. Kato T, et al; Yokohama Kokuritsu Daigaku Kankyo Kagaku Kenkyu Senta Kiyo 6: 11-20 (1980)
42. Keith LH et al; Identification of Organic Pollut Water Ann Arbor, MI pp. 329-73 (1976)
43. Kim NK, Stone DW; Organic Chemicals And Drinking Water NYS Dept Health pp. 131 (NA)
44. Kinlin TE et al; J Agric Food Chem 20: 1021 (1972)
45. Kleindienst TE et al; Environ Sci Technol 20: 493-501 (1986)
46. Konasewich D et al; Great Lake Water Qual Board (1978)
47. Krotoszynski BK et al; J Anal Toxicol 3: 225-34 (1979)

48. Lamb CB, Jenkins GF; Proc 8th Industrial Waste Conf, Purdue Univ p. 326-9 (1952)
49. Levy A; Adv Chem Ser 124: 70-99 (1977)
50. Lipari F et al; Environ Sci Technol 18: 326-30 (1984)
51. Lovegren NV et al; J Agric Food Chem 27: 851-3 (1979)
52. Lyman WJ et al; Handbook of Chem Property Estimation Methods McGraw-Hill NY (1982)
53. Meyrahn H et al; J Atmos Chem 4: 227-91 (1986)
54. Neal M et al Criteria Document for Acetone SRC TR-81-581 Syracuse NY (1981)
55. Pavlica DA et al; Phytopathology 68: 758-65 (1978)
56. Pellizzari ED et al; Bull Environ Contam Toxicol 28: 322-8 (1982)
57. Price KS, et al; J Water Pollut Contr Fed 46: 63-77 (1974)
58. Rao PSC et al; Soil Crop Sci Soc Fl Proc 44: 1-8 (1985)
59. Rathbun RE et al; Chemosphere 11: 1097-114 (1982)
60. Riddick JA et al; Organic Solvents: Physical Properties and Methods of Purification, 4th Edit. New York: J Wiley & Sons (1986)
61. Robinson E et al; J Geophys Res 78: 5345-51 (1973)
62. Rustung E et al; Biochem Z 242: 366-76 (1931)
63. Sadtler; Sadtler Research Lab Philadelphia, PA
64. Sawhney BL, Kozloski RP; J Environ Qual 13: 349-52 (1984)
65. Seba DB, Corcoran EF; Pestic Monit J 3: 109-3 (1969)
66. Seila RL; Non-urban Hydrocarbon Concentrations in Ambient Air North of Houston, TX USEPA EPA-500/3-79-010 p.38 (1979)
67. Shackelford WM, Keith LH; Frequency of Organic Compounds Identified in Water USEPA EPA-600/4-76-062 (1976)
68. Shackelford WM et al; Analyt Chim Acta 146: 15-27 (1983)
69. Sickles JE II et al; Proc 73rd Annual Meet Air Pollut Control Assoc Paper 80-501 (1980)
70. Singh HB, Hanst PL; Geophys Res Lett 8: 941-4 (1981)
71. Snider JR, Dawson GA; J Geophys Res D Atmos 90: 3797-805 (1985)
72. Speece RE et al; Environ Sci Technol 17: 416a-27a (1983)
73. Stevens AA et al; J Am Water Works Assoc 68: 615-20 (1976)
74. Takemoto S, et al; Suishitsu Odaku Kenkyu 4: 80-90 (1981)
75. TerHeide R et al; Anal Foods Beverages, Chavalambous G(Ed) Academic pp. 249-81 (1978)
76. Urano K, Kato Z; Hazardous Material 13: 147-59 (1986)
77. USEPA; New Orleans Area Water Supply USEPA (1974)
78. Vaishnav DD et al; Chemosphere 16: 695-703 (1987)
79. White WL et al; Chemistry for Medical Technologists 3rd ed Mosby Co St Louis, MO (1970)
80. Whitehead LW et al; Am Ind Hyg Assoc J 45: 767-72 (1984)
81. Wolfe TA et al; J Water Pollut Control Fed 58: 68-76 (1986)
82. Yanagihara S et al; Proc Int Clean Air Congr 4th pp. 472-7 (1977)
83. Young RHF, et al; J Water Pollut Contr Fed 40: 354-68 (1968)

# Adiponitrile

## SUBSTANCE IDENTIFICATION

**Synonyms:** 1,4-Dicyanobutane

**Structure:**

**CAS Registry Number:** 111-69-3

**Molecular Formula:** $C_6H_8N_2$

**Wiswesser Line Notation:** NC4CN

## CHEMICAL AND PHYSICAL PROPERTIES

**Boiling Point:** 295 °C at 760 mm Hg

**Melting Point:** 1 °C

**Molecular Weight:** 108.16

**Dissociation Constants:**

**Log Octanol/Water Partition Coefficient:** -0.32 [16]

**Water Solubility:** 8000 mg/L at 20 °C [14]

**Vapor Pressure:** 0.0023 mm Hg at 25 °C [3]

**Henry's Law Constant:** $7 \times 10^{-9}$ atm-m³/mole at 25 °C [4]

## ENVIRONMENTAL FATE/EXPOSURE POTENTIAL

**Summary:** Adiponitrile could potentially be released to the environment in the effluent or emissions from plants manufacturing adiponitrile, hexamethylenediamine, or nylon-6,6. If released to soil, aerobic biodegradation may be an important removal mechanism. Although adiponitrile has the potential to undergo extensive leaching, biodegradation may limit movement through soil. Volatilization from soil surfaces is not expected to be significant. If released to water, aerobic biodegradation may be an important removal process (half-life in unacclimated river water at 20 °C is about 1 week). Chemical hydrolysis, adsorption to suspended solids and sediments, bioaccumulation in aquatic organisms, and volatilization are not expected to be important fate processes in water. If released to the atmosphere, adiponitrile is expected to exist almost entirely in the vapor phase. Reaction of adiponitrile vapor with photochemically generated hydroxyl radicals is expected to be an important fate process (half-life under typical conditions 11.6 days). Some loss by wet deposition may also occur. Due to the toxicity of adiponitrile, workers take extra precautions to avoid any type of personal contamination when working around this compound.

**Natural Sources:**

**Artificial Sources:** Adiponitrile could potentially be released to the environment in the effluent or emissions from plants manufacturing adiponitrile, hexamethylenediamine, or nylon-6,6 [13,14].

**Terrestrial Fate:** If released to soil, aerobic biodegradation may be an important removal mechanism. Adiponitrile has the potential to undergo extensive leaching; however, biodegradation may limit the movement of this compound through soil. Volatilization from wet and dry soil surfaces is not expected to be significant.

**Aquatic Fate:** If released to water, aerobic biodegradation may be an important removal process. Results of one biodegradation study indicate that in unacclimated river water at 20 °C, adiponitrile had a half-life of about 1 week [7]. Acclimation of microorganisms should increase the rate of degradation, and lower temperatures should decrease the rate of degradation [7]. Chemical hydrolysis, adsorption to suspended solids and sediments, bioaccumulation in aquatic organisms, and volatilization are not expected to be important fate processes in water.

**Atmospheric Fate:** Compounds with a vapor pressure of greater than or equal to $1 \times 10^{-4}$ mm Hg should exist almost entirely in the vapor phase in the atmosphere [2]. Based on the estimated vapor pressure, adiponitrile is expected to exist almost entirely in the vapor phase in the atmosphere. Reaction of adiponitrile vapor with photochemically generated hydroxyl radicals is expected to be an important fate process. Under typical atmospheric conditions, the half-life for this reaction is estimated to be 11.6 days [1]. The water solubility also suggests that some loss by wet deposition may occur.

**Biodegradation:** River die-away, unacclimated Ohio River water with stored sewage as seed, initial substrate concn 0.5-10 mg/L, 2 days negligible degradation, 5 days 40% theoretical BOD, 12 days >100% theoretical BOD [7]. River die-away, unacclimated Ohio River water with stored sewage as seed, initial substrate concn 40 mg/L, 20 °C - 10% theoretical $CO_2$ in 2 days, 60% theoretical $CO_2$ in 9 days; 5 °C - 10% theoretical $CO_2$ in 7.5 days, 60% theoretical $CO_2$ in 33 days [7]. At 20 °C effects of acclimation were examined by redosing, at an initial substrate concn of 40 mg/L; the ratio of time it took to achieve 60% oxidation on 1st and 2nd feeding was 2.1 to 1 [7]. Warburg respirometer; inocula was activated sludge from three different municipal waste treatment plants; initial substrate concn 500 mg/L, 72 hours - toxic to one activated sludge and 2.2-2.8% theoretical BOD in the other two sludges [9]. Bench scale activated sludge unit, influent concn equivalent to 275-350 mg/L BOD, mean aeration detention time 7-13 hours, 93-98% BOD removal achieved [8]. Aeromonas sp. BN 7013 isolated from soil was able to use adiponitrile as its sole source of carbon [5].

**Abiotic Degradation:** Chemical hydrolysis of adiponitrile would not be a significant fate process in the environment [10]. The rate constant for the reaction of adiponitrile vapor with photochemically generated hydroxyl radicals in the atmosphere has been estimated to be 1.4 x $10^{-12}$ cm³/molecule-sec at 25 °C [1]. This value corresponds to a reaction half-life of 11.6 days assuming an ambient hydroxyl radical concentration of 5 x $10^{+5}$ molecules/cm³ under typical atmospheric conditions [1].

**Bioconcentration:** Bioconcentration factors (BCF) of <1 were estimated for adiponitrile using linear regression equations based on the measured log Kow and the measured water solubility [10]. These BCF values suggest that this compound will not bioaccumulate significantly in aquatic organisms.

## Adiponitrile

**Soil Adsorption/Mobility:** Soil adsorption coefficients (Koc) of 9-16 were estimated for adiponitrile using linear regression equations based on the log Kow and the measured water solubility [10]. These Koc values suggest that this compound would be very highly mobile in soil and would not adsorb significantly to suspended solids and sediments in water [15].

**Volatilization from Water/Soil:** The value of Henry's Law constant suggests that volatilization from water would not be an important removal mechanism [10]. The relatively low value for the Henry's Law constant and the low vapor pressure suggest that adiponitrile would not volatilize significantly from wet or dry soil surfaces.

**Water Concentrations:** DRINKING WATER: Adiponitrile was tentatively identified in drinking water collected in New Orleans, LA on Jan 14, 1976 [6].

**Effluent Concentrations:** Identified in effluent from a nylon manufacturing plant during Aug 1973 [13].

**Sediment/Soil Concentrations:**

**Atmospheric Concentrations:**

**Food Survey Values:**

**Plant Concentrations:**

**Fish/Seafood Concentrations:**

**Animal Concentrations:**

**Milk Concentrations:**

**Other Environmental Concentrations:**

**Probable Routes of Human Exposure:**

**Average Daily Intake:**

**Occupational Exposure:** NIOSH has estimated that 477 workers are potentially exposed to adiponitrile based on estimates derived from a survey conducted in 1981-83 in the US [11]. Adiponitrile is extremely toxic, and as a result workers take extra precautions to avoid any type of personal contamination [12].

**Body Burdens:**

# REFERENCES

1. Atkinson R; Intern J Chem Kinet 19: 799-828 (1987)
2. Eisenreich SJ et al; Environ Sci Tech 15: 30-8 (1981)
3. GEMS: Graphical Exposure Modeling System. PCCHEM. USEPA (1987)
4. Hine J, Mookerjee PK; J Org Chem 40: 292-8 (1975)
5. Kuwahara M et al; Hakkokogaku Kaishi 58: 441-8 (1980)
6. Lucas SV; GC/MS Analysis of Organics in Drinking Water Concentrates and Advanced Waste Treatment Concentrates: Vol 2. Computer-Printed Tabulations of Compound Identification Results for Large-Volume Concentrates USEPA-600/1-84-020B NTIS PB-128239 p.10 (1984)
7. Ludzack FJ et al; Sewage and Ind Wastes 31: 33-44 (1959)
8. Ludzack FJ et al; Proc of the 14th Industrial Waste Conf, Eng Ext Ser 104: 547-65 (1959)
9. Lutin PA; J Water Poll Contr Fed 42: 1632-42 (1970)
10. Lyman WJ et al; Handbook of Chemical Property Estimation Methods; NY: McGraw-Hill (1982)
11. NIOSH; National Occupational Exposure Survey (NOES) (1983)
12. Parmeggiani L (ed); Encycl Occup Health and Safety 3rd ed Geneva, Switzerland: Intern Labour Organ 2: 1446-7 (1983)
13. Shackelford WM, Keith LH; Frequency of Organic Compounds Identified in Water USEPA-600/4-76-062 p. 56 (1976)
14. Smiley RA; Kirk-Othmer Encycl Chem Tech 15: 897-901 (1981)
15. Swann RL et al; Res Rev 85: 17-28 (1983)
16. Tanii H, Hashimoto K; Arch Toxicol 57: 88-93 (1985)

# n-Amyl Acetate

## SUBSTANCE IDENTIFICATION

**Synonyms:**

**Structure:**

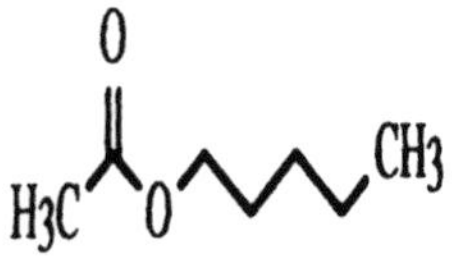

**CAS Registry Number:** 628-63-7

**Molecular Formula:** $C_7H_{14}O_2$

**Wiswesser Line Notation:** 5OV1

## CHEMICAL AND PHYSICAL PROPERTIES

**Boiling Point:** 149.25 °C at 760 mm Hg

**Melting Point:** -70.8 °C

**Molecular Weight:** 130.19

**Dissociation Constants:**

**Log Octanol/Water Partition Coefficient:** 2.258 [5]

**Water Solubility:** 1700 mg/L at 20 °C [17]

**Vapor Pressure:** 9.70 mm Hg at 25 °C [17]

**Henry's Law Constant:** 3.91 x $10^{-4}$ atm-m³/mole at 25 °C [9]

## ENVIRONMENTAL FATE/EXPOSURE POTENTIAL

**Summary:** n-Amyl acetate will be released to the atmosphere and in wastewater during its use as a solvent as well as from foods in which it is used as a flavoring agent. It is also a natural flavor component in

24

some foods. If released on land or in water, volatilization would be important (half-life 5.9 hr in a model river) and biodegradation should be a dominant degradative process. Adsorption to soil or sediment would not occur to any significant extent, so leaching into ground water may occur. Some chemical hydrolysis may occur but only under fairly alkaline conditions. n-Amyl acetate would not be expected to bioconcentrate in aquatic organisms. In air, n-amyl acetate will be scavenged by rain and degrade by reaction with photochemically produced hydroxyl radicals (estimated half-life 2.83 days). Occupational exposure, both dermal and inhalation, would occur primarily from its use as a solvent. The general public will be exposed from ingesting foods in which it occurs naturally or is a flavoring or from contact with products which contain the chemical as a solvent.

**Natural Sources:** n-Amyl acetate is a plant volatile having a banana-like odor [6,7].

**Artificial Sources:** Commercial amyl acetate is a mixture of isomers, the composition depending upon its grade and derivation [7]. n-Amyl acetate is one of the principal isomers contained in it [7]. n-Amyl acetate may be released to the environment as fugitive emissions and in wastewater during the manufacture and use of amyl acetate [7]. These uses include: solvent for lacquers and paints; extraction of penicillin; photographic film; leather polishes; nail polish; warning odor; flavoring agent; printing and finishing fabrics; solvent in dry cleaning; and solvent for phosphors in fluorescent lamps [7,16].

**Terrestrial Fate:** If released on land, n-amyl acetate would volatilize to the atmosphere and leach into the soil. Chemical hydrolysis would not be important except possibly under fairly alkaline conditions (half-life 13.5 days at pH 9). It is probable that biodegradation will be the most important chemical fate process but experimental data for biodegradation in soil are lacking.

**Aquatic Fate:** If released into water n-amyl acetate will be lost by volatilization (half-life 5.9 hr in a model river 1 m deep flowing 1 m/sec with a 3 m/sec wind). Biodegradation should be an important loss process in natural waters (half-life 4.5-10 days). Hydrolysis and adsorption to sediment are not expected to be significant.

**Atmospheric Fate:** Due to its moderately high vapor pressure, n-amyl acetate will exist primarily in the vapor form if released into air.

It will react with photochemically produced hydroxyl radicals (estimated half-life 2.83 days) and also be scavenged by rain.

**Biodegradation:** Two screening tests gave 13% and 38% of theoretical biological oxygen demand values after 5 days [8]. The first test used sewage seed and 1.7-20 ppm of n-amyl acetate and the second test used 440 ppm of the chemical and 10% sewage seed [8]. In a biodegradability screening test using non-acclimated sewage seed, the 5 day biological oxygen demand value for n-amyl acetate was 64% and 35% theoretical in fresh and salt water [15]. After 20 days, the respective values were 72% and 87% [15]. The degradation kinetics of n-amyl acetate in Lake Superior harbor water was first order [19]. When the water was coarsely filtered (suspended solids 5.6 mg/L), the biodegradation half-life was 10.0 days, while when the water was finely-filtered (2.5 mg/L suspended solids), the half-life was only 4.5 days [19].

**Abiotic Degradation:** No information on the hydrolysis of n-amyl acetate could be found in the literature. The base catalyzed process is the dominant hydrolytic process for aliphatic esters. However, since simple esters are resistant to hydrolysis, hydrolysis would not be expected to be a significant degradative process under environmental conditions [12]. The hydrolysis rate for n-amyl acetate at pH > 8 is estimated to be $5.93 \times 10^{-2}$ l/mol-sec at 25 °C, resulting in a half-life at pH 9 of 13.5 days [4]. n-Amyl acetate does not contain any chromaphores which absorb light >290 nm and therefore it will not be subject to direct photolysis. It reacts with photochemically produced hydroxyl radicals by H-atom abstraction, having an estimated half-life of 2.83 days [1].

**Bioconcentration:** Using the estimated value of the log octanol/water partition coefficient, one can estimate a bioconcentration factor of 31 for n-amyl acetate using recommended regression equations [11]. Bioconcentration would therefore not be expected to be a significant transport process in aquatic organisms.

**Soil Adsorption/Mobility:** Using the water solubility, one can estimate a Koc of 73 for n-amyl acetate using recommended regression equations [11]. Adsorption to soil or sediment would therefore not be expected to be an important process.

**Volatilization from Water/Soil:** Using the Henry's Law constant, one can estimate that the volatilization half-life of n-amyl acetate in a model river 1 m deep with a 1 m/s current and a 3 m/s wind speed is 5.9 hr [11]. Given the vapor pressure and soil adsorption of n-amyl acetate, some evaporation from soil and other surfaces might be expected.

**Water Concentrations:** DRINKING WATER: In a survey of drinking water in the United Kingdom, n-amyl acetate was detected in the drinking water in one of 14 treatment plants tested [3]. The source of water for this plant was ground water [3].

**Effluent Concentrations:** n-Amyl acetate has been reported in an effluent of the explosives industry (26 ppm) and of the porcelain/enameling industry (31 ppm) [18].

**Sediment/Soil Concentrations:**

**Atmospheric Concentrations:**

**Food Survey Values:** n-Amyl acetate has been identified in the volatile flavor component of baked potato and fried bacon [2,10].

**Plant Concentrations:**

**Fish/Seafood Concentrations:**

**Animal Concentrations:**

**Milk Concentrations:**

**Other Environmental Concentrations:**

**Probable Routes of Human Exposure:** The general public will be exposed to n-amyl acetate in food in which it occurs naturally or to which it is added as a flavoring agent and possibly in consumer products which contain n-amyl acetate as a solvent.

**Average Daily Intake:**

**Occupational Exposure:** NIOSH (NOES Survey 1981-83) has statistically estimated that 172,440 workers are exposed to n-amyl acetate in the US [14]. NIOSH (NOHS Survey 1972-74) has statistically estimated that 1,259,727 workers are exposed to n-amyl acetate in the US [13]. Workers will be exposed to n-amyl acetate primarily during its use as a solvent.

**Body Burdens:**

## REFERENCES

1. Atkinson R; Internat J Chem Kinetics 19: 799-828 (1987)
2. Coleman EC et al; J Agric Food Chem 29: 42-8 (1981)
3. Fielding M et al; pp 49 in Organic Micropollutants in Drinking Water TR-159 Medmenham, Eng Water Res Cent (1981)
4. GEMS: HYDRO Computer Program (1987)
5. GEMS: Graphical Exposure Modeling System. CLOGP3. (1986)
6. Graedel TE; Chem Compounds in the Atmos, Academic Press New York pp 226 (1978)
7. Hawley GG; Condensed Chemical Dictionary 10th ed Von Nostrand Reinhold NY pp 66 (1981)
8. Heukelekian H, Rand MC; J Water Pollut Control Assoc 29: 1040-53 (1955)
9. Hine J, Mookerjee PK; J Org Chem 40: 292-8 (1975)
10. Ho CT et al; J Agric Food Chem 31: 336-42 (1983)
11. Lyman WJ et al; Handbook of Chemical Property Estimation Methods. Environ Behavior of Organic Compounds. McGraw-Hill NY (1982)
12. Mabey W, Mill T; J Phys Chem Ref Data 7: 383-415 (1978)
13. NIOSH: National Occupational Hazard Survey (1974)
14. NIOSH; National Occupational Exposure Survey (1985)
15. Price KS et al; J Water Pollut Control Fed 46: 63-77 (1974)
16. Reich DA, Cormany CL; Kirk-Othmer Encycl Chem Technol, 3rd ed. 8: 50-68 (1979)
17. Riddick JA et al; Organic Solvents: Physical Properties and Methods of Purification. Techniques of Chemistry. 4th Ed. New York: Wiley-Interscience pp 1325 (1986)
18. Shackelford WM et al; Anal Chem Acta 146: 15-27 (1983)
19. Vaishav DD, Babau L; J Great Lakes Res 12: 184-92 (1986)

# Benzene

SUBSTANCE IDENTIFICATION

**Synonyms:**

**Structure:**

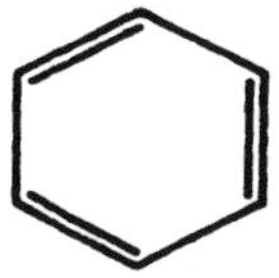

**CAS Registry Number:** 71-43-2

**Molecular Formula:** $C_6H_6$

**Wiswesser Line Notation:** RH

## CHEMICAL AND PHYSICAL PROPERTIES

**Boiling Point:** 80.1 °C

**Melting Point:** 5.5 °C

**Molecular Weight:** 78.11

**Dissociation Constants:**

**Log Octanol/Water Partition Coefficient:** 2.13 [25]

**Water Solubility:** 1791 mg/L [43]

**Vapor Pressure:** 95.19 mm Hg at 25 °C [3]

**Henry's Law Constant:** 5.43 x $10^{-3}$ atm-$m^3$/mole [42]

## ENVIRONMENTAL FATE/EXPOSURE POTENTIAL

**Summary:** Benzene will enter the atmosphere primarily from fugitive emissions and exhaust connected with its use in gasoline. Another important source is emissions associated with its production and use as

29

an industrial intermediate. In addition, there are discharges into water from industrial effluents and losses during spills. If benzene is released to soil, it will be subject to rapid volatilization near the surface and that which does not evaporate will be highly to very highly mobile in the soil and may leach to the ground water. It may be subject to biodegradation based on reported biodegradation of 24% and 47% of the initial 20 ppm benzene in a base-rich para-brownish soil in 1 and 10 weeks, respectively. It may be subject to biodegradation in shallow, aerobic ground waters, but probably not under anaerobic conditions. If benzene is released to water, it will be subject to rapid volatilization; the half-life for evaporation in a wind-wave tank with a moderate wind speed was 5.23 hr; the estimated half-life for volatilization of benzene from a river 1 m deep flowing 1 m/sec with a wind velocity of 3 m/sec is estimated to be 2.7 hr at 20 °C. It will not be expected to significantly adsorb to sediment, bioconcentrate in aquatic organisms, or hydrolyze. It may be subject to biodegradation based on a reported biodegradation half-life of 16 days in an aerobic river die-away test. In a marine ecosystem biodegradation occurred in 2 days after an acclimation period of 2 days and 2 weeks in the summer and spring, respectively, whereas no degradation occurred in winter. Photodegradation, which according to one experiment has a half-life of 17 days could contribute to benzene's removal in situations of cold water, poor nutrients, or other conditions less conducive to microbial degradation. If benzene is released to the atmosphere, it will exist predominantly in the vapor phase. Gas-phase benzene will not be subject to direct photolysis but it will react with photochemically produced hydroxyl radicals with a half-life of 13.4 days calculated using an experimental rate constant for the reaction. The reaction time in polluted atmospheres which contain nitrogen oxides or sulfur dioxide is accelerated with the half-life being reported as 4-6 hr. Products of photooxidation include phenol, nitrophenols, nitrobenzene, formic acid, and peroxyacetyl nitrate. Benzene is fairly soluble in water and is removed from the atmosphere in rain. The primary routes of exposure are inhalation of contaminated air, especially in areas with high traffic and in the vicinity of gasoline service stations, and by consumption of contaminated drinking water.

**Natural Sources:** Volcano, natural constituent of crude oil, forest fires, and plant volatile [21,29].

**Artificial Sources:** Benzene enters the environment from production, storage, transport, venting, and combustion of gasoline, and from production, storage, and transport of benzene itself. Other sources result from its use as an intermediate in the production of other chemicals, and as a solvent; from spills, including oil spills; from its indirect production in coke ovens; from nonferrous metal manufacture, ore mining, wood processing, coal mining and textile manufacture; and from cigarette smoke [21,29].

**Terrestrial Fate:** If benzene is released to soil it will be subject to rapid volatilization near the surface, and that which does not evaporate will be highly to very highly mobile in soil and may leach to the ground water. The effective half-lives for volatilization without water evaporation from soil to benzene uniformly distributed to 1 and 10 cm in soil were 7.2 and 38.4 days, respectively [30]. It may be subject to biodegradation based on reported biodegradation of 24% and 47% of the initial 20 ppm benzene in a base-rich para-brownish silt in 1 and 10 weeks, respectively [24]. It may be subject to biodegradation in shallow, aerobic ground waters, but probably not under anaerobic conditions.

**Aquatic Fate:** If benzene is released to water, it will be subject to rapid volatilization; the half-life for evaporation in a wind-wave tank with a moderate wind speed was 5.23 hr [41]; the estimated half-life for volatilization of benzene from a river 1 m deep flowing 1 m/sec with a wind velocity of 3 m/sec is estimated to be 2.7 hr at 20 °C. It will not be expected to significantly adsorb to sediment, bioconcentrate in aquatic organisms, or hydrolyze. It may be subject to biodegradation based on a reported biodegradation half-life of 16 days in an aerobic river die-away test [75]. In a marine ecosystem biodegradation occurred in 2 days after an acclimation period of 2 days and 2 weeks in the summer and spring, respectively, whereas no degradation occurred in winter [76]. Evaporation was the primary loss mechanism in winter in a mesocosm experiment which simulated a northern bay where the half-life was 13 days [75]. In spring and summer the half-lives were 23 and 3.1 days, respectively [75]. In these cases biodegradation plays a major role and takes about 2 days [75]. However, acclimation is critical and this takes much longer in the colder water in spring [75]. Photodegradation, which according to one experiment has a half-life of 17 days [28], could contribute to benzene's removal in situations of cold water, poor nutrients, or other conditions less conducive to microbial degradation.

# Benzene

**Atmospheric Fate:** If benzene is released to the atmosphere, it will exist predominantly in the vapor phase [14]. Gas-phase benzene will not be subject to direct photolysis but will react with photochemically produced hydroxyl radicals with a half-life of 13.4 days calculated using an experimental rate constant for the reaction. The reaction time in polluted atmospheres that contain nitrogen oxides or sulfur dioxide is accelerated with the half-life being reported as 4-6 hours [36]. Products of photooxidation include phenol, nitrophenols, nitrobenzene, formic acid, and peroxyacetyl nitrate. Benzene is fairly soluble in water and is removed from the atmosphere in rain [31].

**Biodegradation:** No degradation of benzene as measured by BOD was reported in coarse-filtered (through 1 cm cotton layer) Superior harbor water incubated at 21 °C for 12 days [74]. Biodegradation half-lives of 28 and 16 days were reported in die-away for degradation of up to 3.2 ul/L benzene tests using ground water and Lester River water, respectively, under aerobic conditions [74]. The half-life in estuarine water was 6 days [39]. In a marine ecosystem biodegradation occurred in 2 days after an acclimation period of 2 days and 2 weeks in the summer and spring, respectively, whereas no degradation occurred in winter [75]. In a base-rich para-brownish soil, 20 ppm benzene was 24% degraded in 1 week, 44% in 5 weeks, and 47% in 10 weeks [24]. Benzene, in a mixture with toluene and xylenes, is readily biodegraded (total degradation of 7.5 ppm total mixture) in shallow ground water in the presence of oxygen in the unconfined sand aquifer at Canada Forces' Base Borden, Ontario; laboratory batch experiments demonstrated that the degradation could be attributed to biodegradation [2]. Complete biodegradation in 16 days was reported under simulated aerobic ground water conditions at 20 °C [13]. Reported metabolites of benzene using pure cultures of microorganisms include phenol and unidentified phenols [63], catechol and cis-1,2-dihydroxy-1,2-dihydrobenzene [20]. Benzene at 50 ppm was 90% degraded by industrial wastewater seed incubated at 23 °C for 6 hr [12]. Benzene inhibited industrial seed at concn of 100 ppm and above and municipal seed at 50 ppm and above [12]. In a bench scale activated-sludge reactor with an 8 hr retention time, complete degradation occurred with 0.5% of the benzene being lost by air stripping [67]. In laboratory systems, low concentrations of benzene are degraded in 6-14 days [59,69]. 44-100% removal occurred at a sewage treatment plant [17].

**Abiotic Degradation:** Since gas-phase benzene [48] or benzene dissolved in cyclohexane [60] does not absorb light of 290 nm or longer, it will not be expected to directly photolyze in sunlight in these media. However, slight shifts in wavelength of absorption might be expected in more representative environmental media, such as water [27]; e.g., a half-life of 16.9 days was reported for photolysis of benzene dissolved in deionized water saturated with air exposed to sunlight [28]. The rate constant for the vapor phase reaction of benzene with photochemically produced hydroxyl radicals has been reported to be 1.2 x $10^{-12}$ cm$^3$/molecule-sec at 25 °C [53] which corresponds to an atmospheric half-life of 13.4 days at an atmospheric concentration of 5 x $10^{+5}$ hydroxyl radicals per cm$^3$. While benzene is considered to be relatively unreactive in photochemical smog situations (in the presence of nitrogen oxides), its rate of degradation is accelerated with about 16% decrease in concentration in 5 hr [16]. A typical experiment in the presence of active species such as $NO_x$ and $SO_2$ showed that benzene photodegradation was considerably accelerated above that in air alone [79]. Its half-life in the presence of active species was 4-6 hr with 50% mineralization to $CO_2$ in approximately 2 days [36]. Products of degradation include phenol, 2-nitrophenol, 4-nitrophenol, 2,4-dinitrophenol, 2,6-dinitrophenol, nitrobenzene, formic acid, and peroxyacetyl nitrate [26,34,47]. Hydrolysis is not a significant process for benzene [40].

**Bioconcentration:** BCF: eels (Anguilla japonica) 3.5 [51]; pacific herring (Clupea harengus pallasi) 4.4 [35]; goldfish 4.3 [50]. Based on the log Kow, a BCF of 24 was estimated [40]. Based on the reported and estimated BCF, benzene will not be expected to bioconcentrate in aquatic organisms.

**Soil Adsorption/Mobility:** Koc: Woodburn silt loam 31 [9]; 31.7-143 [57]; 83 [32]. Benzene leaches in soil, passing through soil during bank infiltration [22,54]. Based on the log Kow 2.13, a Koc of 98 was estimated [40]. Based on the reported and estimated Koc values, vinyl chloride will be expected to exhibit very high to high mobility in soil [68] and therefore may leach to ground water.

**Volatilization from Water/Soil:** Half-lives for evaporation of benzene from seawater in a mesocosm simulating Narragansett Bay, RI, containing the associated planktonic and microbial communities, varied with the seasons: spring (15 Apr-18 Jun) half-life 23 days, summer (19 Aug-8 Sept) 3.1 days, winter (4 Mar-4 May) 13 days [76]. The effective half-lives for volatilization without water evaporation of benzene uniformly distributed at a rate of 1 kg/ha to 1 and 10 cm in soil with an organic carbon content of 1.25% were 7.2 and 38.4 days, respectively [30]. The half-life for evaporation in a wind-wave tank with a moderate wind speed was 5.23 hr [41]. The estimated half-life for volatilization of benzene from a river 1 m deep flowing 1 m/sec with a wind velocity of 3 m/sec is estimated to be 2.7 hr at 20 °C [40] based on the Henry's Law constant. Based on the vapor pressure, evaporation of benzene from surface soil and other surfaces is expected to be rapid.

**Water Concentrations:** DRINKING WATER: 113 public supplies, 1976, 7 sites pos, avg of positive sites <0.2 ppb [5]. 5 US cities, 1974-5, 0-0.3 ppb [10]. Contaminated drinking water wells in New York, New Jersey, Connecticut, 30-300 ppb; highest concn in drinking water from surface source, 4.4 ppb [7]. 3 surveys of community water supplies: 0 of 111 pos; 7 of 113 pos, mean 4 ppb; 4 of 16 pos (0.95 ppb-max) [44]. US Ground Water Supply Survey (GWS), 1982, finished drinking water, 466 samples selected at random from 1000 in survey, 0.6% pos, 3 ppb median, 15 ppb max [11]. Wisconsin drinking water wells, data through Jun 1984, 1174 community wells, 0.34% pos, 617 private wells, 2.9% pos [38]. GROUND WATER: Chalk Aquifer (UK), 210 m from petrol storage, 1-10 ppb; Chalk Aquifer (UK), 120 m from petrol storage, <250 ppb; Chalk Aquifer (UK), 10 m from petrol storage, 1250 ppb [70]. SURFACE WATER: 14 heavily industrialized with basins, 1975-6, 20% samples > 1 ppb and between 1 and 7 ppb [15]. Lake Erie, 1975-6, 0-1 ppb, 1 of 2 sites positive; Lake Michigan, 1975-6, 0-7 ppb, 5 of 7 sites positive [33]. 700 random sites in US, 1975, 5.4 ppb avg [37]. US EPA STORET data base, 1,271 samples, 15.0% pos, 5.0 ppb median [66]. SEAWATER: 5-15 ppt, Gulf of Mexico, 1977, unpolluted areas; 5-175 ppt, Gulf of Mexico, 1977, anthropogenic influence [58]. RAIN/SNOW: Detected in rainwater in Japan and in the UK (87.2 ppb) [29,31].

**Effluent Concentrations:** Wastewater from coal preparation plants, 0.3-48 ppt [29]; wastewater from plants which manufacture or use benzene <1-179 ppt [29]; stack emissions from coking plants (Czechoslovakia), 15-50 ppm [64]; stack emission estimates from chemical plants using emissions and worst case modeling at 150 m from source, ≤ 5 ppm [18]. Ground water at 178 CERCLA hazardous waste sites, 11.2% pos [56]. US EPA STORET data base, 1,474 samples, 16.4% pos, 2.50 ppb median [66]. Industries in which mean or max levels in raw wastewater exceeded 1 ppm are (number of samples, percent pos, mean, max, in ppm): raw wastewater: auto and other laundries (20 samples, 70% pos, <1.4 ppm mean, 23 ppm max), iron and steel manufacturing (mfg) (9, 77.8%, <8.0, 46), aluminum forming (32, 56.2%, 0.70, 2.1), photographic equipment/supplies (48, 54.2%, 0.16, 2.1), pharmaceutical mfg (9, 100%, 12, 87), organic chemical/plastics mfg (number of samples not reported (NR), 63 detections, 22, NR), paint and ink formulation (36, 63.9%, 1.2, 9.9), petroleum refining (11, NR, <0.10, 2.4), rubber processing (4, 100%, 0.60, 3.4), timber products processing (14, 92.9%, 0.27, 2.8); treated wastewater: auto and other laundries (4 samples, 50% pos, 0.1 ppm mean, 0.2 ppm max), iron and steal manufacturing (mfg) (13, 76.9%, <14, 120), aluminum forming (21, 81.0%, <0.0058, 0.040), photographic equipment/supplies (4, 100%, 0.016, 0.021), pharmaceutical mfg (6, 100%, 1.8, 10), organic chemical/plastics mfg (number of samples not reported (NR), 42 detections, 26, NR), paint and ink formulation (24, 62.5%, 0.39, 3.8), petroleum refining (13, NR, NR, 0.012), rubber processing (5, 100%, <0.0077, 0.010), timber products processing (5, 60%, 0.010, 0.033) [72].

**Sediment/Soil Concentrations:** SEDIMENT: Surface sediments in Walvis Bay (off Capetown, SA) 0-20 ppb [78]. US EPA STORET data base, 355 samples, 9% pos, <5.0 ppb median [66]. SOIL: Soil near factories where benzene was used or produced, 2-191 ug/kg [29].

**Atmospheric Concentrations:** BACKGROUND: Rural background avg concn 0.1-17 ppb [29]. Multi-latitude background concn (ppb/deg North (deg N): Atlantic Ocean 0.07/35 deg N, Pacific Ocean 0.23/45 deg N, Niwot Ridge (Colorado Rockies) 0.16-0.24 ppb [62]; Pacific Ocean: Northern hemisphere 0.05, Southern hemisphere 0.01, Pacific Ocean, 0.581, Pullman, WA, 0.226, Cape Meares, OR, 0.230, Norwegian Arctic, 0.066 [49]. RURAL/REMOTE: US 1977-80, 100 samples, 1.4 ppb avg [6]. Point Barrow, AK, 1967, 24 hr avg is 0.16 ppb in 5 of 25 samples [8]. Remote tropical sites: 5 sites, range not detected to 1.8 ppb, range of avg 0.07-0.65 ppb [23]. SUBURBAN/URBAN: US 1977-80, 2292 samples, 2.8 ppb avg [6]. Toronto, 1971, 98 ppb max, 13 ppb avg, Los Angeles, 1966, 57 ppb max, 15 ppb avg [55]. US cities, 24 hr sampling for 2 weeks in 1979 (ppb): Los Angeles, Apr, 0.72-27.87, 6.04 avg, Phoenix, AZ, Apr-May, 0.39-59.89, 4.74 avg, Oakland, Jun-Jul, 0.06-4.63, 1.55 avg [61]. New Jersey, 1978 (ppb): Rutherford, 149 samples, 107 max, 3.8 avg, Newark, 110 samples, 24 max, 2.6 avg, Piscataway/Middlesex, 18 samples, 1.9 max, 1.0 avg, Somerset County, 30 samples, 33 max, 5.6 avg, Bridgewater Township, 22 samples, 7.9 max, 1.4 avg [4]. The general urban atmosphere, estimated avg, 0.02 ppm [29]. SOURCE DOMINATED: US 1977-80, 487 samples, 3.0 ppb avg [6]. Concn near US chemical factories where benzene is used 0.6-34 ppb; service stations 0.0003-3.2 ppm; cigarette smoke 57-64 ppm [29]. Traffic tunnel in London, poorly ventilated, approx 0.010-0.21 ppb [71].

**Food Survey Values:** Heat treated or canned beef 2 ug/kg; Jamaican rum 120 ug/kg; eggs 500-1900 ug/kg; detected in fruits, nuts, vegetables, dairy products, meat, fish, poultry, eggs, and beverages [73].

**Plant Concentrations:** Plant volatile [21]. 2 species of macroalgae - 20 ppb [77].

**Fish/Seafood Concentrations:** Lake Pontchartrain, LA seafood (ppb wet weight): oysters (Crassostra virginica), from the Inner Harbor Navigational Canal, avg of 5 samples, 220, clams composite samples (Rangia cuneata): from Chef Menteur Pass, 260, from The Rigolets, not detected [19].

**Animal Concentrations:**

**Milk Concentrations:**

**Other Environmental Concentrations:**

**Probable Routes of Human Exposure:** Human populations are primarily exposed to benzene through inhalation of contaminated ambient air particularly in areas with heavy traffic and around filling stations. In addition, air close to manufacturing plants which produce or use benzene may contain high concentrations of benzene [21,29]. Another source of exposure from inhalation is from tobacco smoke [29]. Although most public drinking water supplies are free of benzene or contain <0.3 ppb, exposure can be very high from consumption of contaminated sources drawn from wells contaminated by leaky gasoline storage tanks, landfills, etc. Although benzene has been detected in various food items, data is too scant to estimate exposure from ingestion of contaminated food.

**Average Daily Intake:** Avg exposure for urban/suburban residents: AIR INTAKE (assume typical range of concn 2.8-20 ppb [6,29]) 182-1300 ug; WATER INTAKE: (assume typical concn 0.1 ppb [5])

# Benzene

0.2 ug; FOOD INTAKE: insufficient data.

**Occupational Exposures:** Human exposure to atmospheric benzene from emission sources: Chemical manufacturing numbers in parenthesis are annual average concentration levels in ppb, multiply these concentrations by 10 to get 8 hr worst case levels 7,497 (0.1-1.0), 970,000 (1.1-2.0), 453,000 (2.1-4.0), 644,000 (4.1-10.0), 319,000 (>10.0), 9,883,000 (total). Coke ovens: 15,726,000 (0.1-1.0), 521,000 (1.1-2.0), 50,000 (2.1-4.0), 2,000 (4.1-10.0), 16,299,000 (total). People using gasoline service stations: 37,000,000 (total) - 245 ppb for 1.5 hr/yr/person. People residing near gasoline service stations: 87,000,000 (0.1-1.0), 31,000,000 (1.1-2.0), 118,000,000 (total). Petroleum refineries: 6,529,000 (0.1-1.0), 64,000 (1.1-2.0), 4,000 (2.1-4.0), <500 (4.1-10.0), 6,597,000 (total). Solvent operations (rubber related industries): 208,000 (0.1-10), 5,000 (1.1-2.0), 2,000 (2.1-4.0), 215,000 (total). Storage and Distribution: very few exposures (<0.1 ppb). Urban exposure (auto emissions): 68,337,000 (0.1-1.0), 45,353,000 (1.1-2.0), 113,690,000 (total) [64]. NIOSH (NOES Survey 1981-83) has statistically estimated that 70,983 workers are exposed to benzene in the US [45]. NIOSH (NOHS Survey 1972-74) has statistically estimated that 1,495,706 workers are exposed to benzene in the US [46].

**Body Burdens:** Detected in all 8 samples of mothers' milk from 4 US urban areas [52]. Breath of persons without specific exposure to benzene 8-20 ppb [29]. Whole blood, 250 subjects (121 males, 129 females), not detected-5.9 ppb, 0.8 ppb avg [1]. US FY82 National Human Adipose Tissue Survey specimens, 46 composites, 96% pos, (>4 ppb, wet tissue concn), 97 ppb max [65].

# REFERENCES

1. Antoine SR et al; Bull Environ Contam Toxicol 36: 364-71 (1986)
2. Barker JF et al; Ground Water Monit Rev 7: 64-72 (1987)
3. Boublik TK et al; The Vapor Pressures of Pure Substances, Vol 17 Amsterdam, Netherslands: Elsevier Sci Publ (1984)
4. Bozzelli JW, Kebbekus BB; Analysis of selected volatile organic substances in ambient air. Newark, NJ: NJ Inst Technol 80 pp (1979)
5. Brass HJ et al; Drinking Water Qual Enhancement Source Prot pp 393-416 (1977)
6. Brodzinsky R, Singh HB; Volatile organic chemicals in the atmosphere: an assessment of available data 198 pp SRI Inter 68-02-3452 (1982)
7. Burmaster DE; Environ 24: 6-13, 33-6 (1982)
8. Cavanagh LA et al; Environ Sci Technol 3: 251-7 (1969)

# Benzene

9. Chiou CT et al; Environ Sci Technol 17: 227-31 (1983)
10. Coleman WE et al; Analysis and Identification of Organic Substances in Water. L Keith ed, Ann Arbor MI: Ann Arbor Press Chapt 21, pp 305-27 (1976)
11. Cotruvo JA; Sci Total Environ 47: 7-26 (1985)
12. Davis EM et al; Water Res 15: 1125-7 (1981)
13. Delfino JJ, Miles CJ; Soil Crop Sci Soc FL Proc 44: 9-14 (1985)
14. Eisenreich SJ et al; Environ Sci Technol 15: 30-8 (1981)
15. Ewing BB et al; Monitoring to Detect Previously Unrecognized Pollutants in Surface Waters. 75 pp USEPA-560/6-77-015 (1977)
16. Farley FF; Inter Conf on Photochemical Oxidant Pollution and Its Control. pp 713-27 USEPA-600/3-77-001B (1977)
17. Feiler HD et al; Proc Natl Conf Munic Sludge Manag 8th, pp 72-81 (1979)
18. Fentiman AF et al; Environmental Monitoring Benzene pp 105-10. (PB-295641) (1979)
19. Ferrario JB et al; Bull Environ Contam Toxicol 34: 246-255 (1985)
20. Gibson DT et al; Biochem 7: 2653-62 (1968)
21. Graedel TE; Chem Compounds in the Atmos, New York, NY Academic Press (1978)
22. Green WJ et al; J Water Pollut Control Fed 53: 1347-54 (1981)
23. Greenberg JP, Zimmerman PR; Am Chem Soc Div Environ Chem 192nd Natl Mtg 26: 10-3 (1986)
24. Haider K et al; Arch Microbiol 96: 183-200 (1974)
25. Hansch C, Leo AJ; Medchem Project Issue No.26 Claremont, CA: Pomona College (1985)
26. Hoshino M et al; Kokuritsu Kogai Kekyusho Kenkyu Hokoku 5: 43-59 (1978)
27. Howard PH, Durkin PR; Sources of Contamination, ambient levels, and fate of benzene in the environment. pp 65 USEPA-560/5-75-005 (1974)
28. Hustert K et al; Chemosphere 10: 995-8 (1981)
29. IARC; Monograph. Some Industrial Chemicals and Dyestuffs 29: 99-106 (1982)
30. Jury WA et al; J Environ Qual 13: 573-9 (1984)
31. Kato T et al; Yokohama Kokuritsu Daigaku Kankyo Kagaku Kenkyu Senta Kiyo 6: 11-20 (1980)
32. Kenaga EE; Ecotox Environ Safety 4: 26-38 (1980)
33. Konasewich D et al; Great Lake Water Qual Board (1978)
34. Kopczynski SL; Int J Air Water Pollut 8: 107-20 (1964)
35. Korn S et al; Fish Bull Natl Marine Fish Ser 75: 633-6 (1977)
36. Korte F, Klein W; Ecotox Environ Safety 6: 311-27 (1982)
37. Kraybill HF; NY Acad Sci Annals 298: 80-9 (1977)
38. Krill RM, Sonzogni WC; J Am Water Works Assoc 78: 70-5 (1986)
39. Lee RF, Ryan C; Microbial Degradation of Pollutants in Marine Environments. pp 443-50 USEPA-600/9-72-012 (1979)
40. Lyman WJ et al; Handbook of Chem Property Estimation Methods McGraw-Hill NY pp 15-9 to 15-31 (1982)
41. Mackay D, Yeun ATK; Environ Sci Technol 9: 1178-80 (1973)
42. Mackay D, Shiu WY; J Phys Chem Ref Data 19:1175-99 (1981)
43. May WE; In: Petroleum in the Marine Environment, Adv in Chemistry Series 185: 142-92 Washington DC: American Chemical Society (1980)
44. NAS; Drinking Water and Health, Vol 3 (1980)
45. NIOSH; The National Occupational Exposure Survey (NOES) (1983)
46. NIOSH; The National Occupational Hazard Survey (NOHS) (1974)
47. Nojima K et al; Chemosphere. 4: 77-82 (1975)

# Benzene

48. Noyes WA et al; J Chem Phys 44: 2100-6 (1966)
49. Nutmagul W, Cronn DR; J Atmos Chem 2: 415-33 (1985)
50. Ogata M et al; Bull Environ Contam Toxicol 33: 561-7 (1984)
51. Ogata M, Miyake Y; Water Res 12: 1041-4 (1978)
52. Pellizzari ED et al; Environ Sci Technol 16: 781-5 (1982)
53. Perry RA et al; J Phys Chem 81: 296-304 (1977)
54. Piet GJ, Morra CF; In: Artificial Groundwater Recharge L Huisman, TL Olsthorn eds Marshfield MA: Pitman Pub pp 31-42 (1983)
55. Pilar S, Graydon WF; Environ Sci Technol 7: 628-712 (1973)
56. Plumb HJr; Ground Water Monit Rev 7: 94-100 (1987)
57. Sabljic A; J Agric Food Chem 32: 243-6 (1984)
58. Sauer TC Jr; Org Geochem 3: 91-101 (1981)
59. Setzkorn EA, Huddleston RL; J Amer Oil Chem Soc 42: 1081-4 (1965)
60. Silverstein RM, Bassler GC; p 166 in: Spectrometric Identification of Organic Compounds 2nd ed. (1968)
61. Singh WB et al; Atmos Environ 15: 601-20 (1981)
62. Singh HB et al; Atmos Environ 19: 1911-9 (1985)
63. Smith RV, Rosazza SP; Arch Biochem Biophys 161: 551-8 (1974)
64. Stanford Research Institute; Human Exposure to Atmospheric Benzene. Center for Resource and Environmental Studies Report No. 30, p 3, Menlo Park, CA: SRI (1977)
65. Stanley JS; Broad Scan Analysis of the FY82 National Human Adipose Tissue Survey Specimens Vol. I Executive Summary p 5 USEPA-560/5-86-035 (1986)
66. Staples CA et al; Environ Toxicol Chem 4: 131-42 (1985)
67. Stover EL, Kincannon DF; J Water Pollut Control Fed 55: 97-109 (1983)
68. Swann RL et al; Res Rev 85: 17-28 (1983)
69. Tabak HH et al; J Water Pollut Control Fed 53: 1503-18 (1981)
70. Tester DJ, Harker RJ; Water Pollut Control 80: 614-631 (1981)
71. Tsani-Bazaca E et al; Environ Technol Lett 2: 303-16 (1981)
72. US EPA; Treatability Manual. p.I.9.1-1 to I.9.1-5 USEPA-600/2-82-001A (1981)
73. USEPA; Ambient Water Quality Criteria: Benzene p.C-5 EPA-440/5-80-018 (1980)
74. Vaishnav DD, Babeu L; J Great Lakes Res 12: 184-91 (1986)
75. Wakeham SG et al; Bull Environ Contam Toxicol 31: 582-4 (1983)
76. Wakeham SG et al; Environ Sci Technol 17: 611-7 (1983)
77. Whelan JK et al; Nature 299: 50-2 (1982)
78. Whelan JK et al; Geochim Cosmochim Acta 44: 1767-85 (1980)
79. Yanagihara S et al; Proc Int Clean Air Cong 4th, pp 472-7 (1977)

# Bromodichloromethane

## SUBSTANCE IDENTIFICATION

**Synonyms:**

**Structure:**

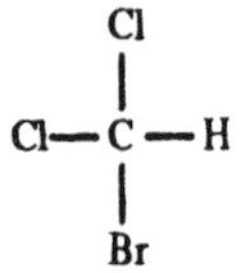

$$Cl—\underset{\underset{Br}{|}}{\overset{\overset{Cl}{|}}{C}}—H$$

**CAS Registry Number:** 75-27-4

**Molecular Formula:** $CHBrCl_2$

**Wiswesser Line Notation:** GYGE

## CHEMICAL AND PHYSICAL PROPERTIES

**Boiling Point:** 90 °C at 760 mm Hg

**Melting Point:** -57.1 °C

**Molecular Weight:** 163.83

**Dissociation Constants:**

**Log Octanol/Water Partition Coefficient:** 2.10 [23]

**Water Solubility:** 4700 mg/L at 22 °C [23]

**Vapor Pressure:** 50 mm Hg at 20 °C [8]

**Henry's Law Constant:** $1.6 \times 10^{-3}$ atm-m³/mole [25]

## ENVIRONMENTAL FATE/EXPOSURE POTENTIAL

**Summary:** The predominant anthropogenic source of bromodichloromethane released to the environment is its inadvertent formation during chlorination treatment processes of water. In addition to human sources, it is biosynthesized and emitted to the environment by various species of marine macroalgae which are abundant in various locations of the world's oceans. If released to surface water, volatilization will be the dominant environmental fate process. The volatilization half-life from rivers and streams has been estimated to range from 33 min to 12 days with a typical half-life being 35 hours. In aquatic regions where volatilization is not viable, anaerobic biodegradation may be the major removal process. Aquatic hydrolysis, oxidation, direct photolysis, adsorption, and bioconcentration are not environmentally important. If released to soil, volatilization is again likely to be the dominant removal process where exposure to air is possible. Bromodichloromethane is moderately to highly mobile in soil and can therefore leach into ground waters. If released to air, the only identifiable transformation process in the troposphere is reaction with hydroxyl radicals which has an estimated half-life of 3.92 months. This relatively persistent half-life indicates that long-range global transport is possible. The general population is exposed to bromodichloromethane through oral consumption of contaminated drinking water, beverages, and food products, through inhalation of contaminated ambient air, and through dermal exposure to chlorinated swimming pool water.

**Natural Sources:** Bromodichloromethane is biosynthesized and emitted to seawater (and eventually to the atmosphere) by various species of marine macroalgae which are abundant in the various locations of the world's oceans [9,18].

**Artificial Sources:** Although environmental releases can result from production and use processes, bromodichloromethane is not produced or used on a large commercial scale indicating that significant releases do not occur from these practices [28]. The predominant environmental release of bromodichloromethane results from its inadvertent formation during chlorination treatment processes of drinking, waste, and cooling waters [28,38]. The amount of bromodichloromethane which may be produced during chlorination processes depends upon a variety of parameters which include temperature, pH, bromide ion concn of the water, fulvic and humic substance concn, and actual chlorination treatment practices [38].

**Terrestrial Fate:** In soils where exposure to the atmosphere can occur, volatilization is likely to be the dominant environmental fate process due to the high vapor pressure of bromodichloromethane [38].

# Bromodichloromethane

Bromodichloromethane is moderately to highly mobile in soil and can therefore leach into ground water and subsurface regions. Laboratory studies have indicated that significant biodegradation can occur under anaerobic conditions; therefore, in soil regions where volatilization is not viable, biodegradation may be the major removal process.

**Aquatic Fate:** Volatilization of bromodichloromethane is the dominant removal mechanism from environmental surface waters. The volatilization half-life from rivers and streams has been estimated to range from 33 min to 12 days with a typical half-life being 35 hours. Laboratory studies have indicated that significant biodegradation can occur under anaerobic conditions; therefore, in aquatic regions where volatilization is not viable, biodegradation may be the major removal process. Aquatic hydrolysis, oxidation, direct photolysis, adsorption, and bioconcentration are not environmentally important.

**Atmospheric Fate:** Due to its high vapor pressure, bromodichloromethane should exist entirely in the vapor phase in the ambient atmosphere. The only identifiable transformation process in the troposphere is reaction with hydroxyl radicals which has an estimated half-life of 3.92 months in typical air. Direct photolysis does not occur below the ozone layer. This relatively persistent tropospheric half-life suggests that a small percentage of the bromodichloromethane present in air will eventually diffuse to the stratosphere where it will be destroyed by photolysis. In addition, long-range global transport is possible. The detection of bromodichloromethane in rainwater indicates that atmospheric removal via washout can occur; however, any bromodichloromethane which is removed by rainfall is likely to revolatilize into the atmosphere.

**Biodegradation:** Loss of bromodichloromethane was observed to be 51-59% utilizing a static flask screening procedure and 28 days of incubation, which was interpreted as significant biodegradation with gradual adaptation [35]. In anaerobic tests using mixed methanogenic bacterial cultures from sewage effluents, bromodichloromethane was totally degraded within 2 weeks while only 43-50% was lost in sterile controls after 6 weeks; no degradation was noted in aerobic tests in either sterile or seeded conditions [5]. Studies conducted under anoxic conditions with denitrifying bacteria found >50% degradation in bacterial cultures after 8 weeks but no degradation in sterile controls [4]. Rapid degradation was observed in a continuous-flow methanogenic fixed-film laboratory-scale column using seeded cultures, but only slow degradation was noted in sterile controls [6]. In a ground water recharge field experiment, bromodichloromethane passed through the biologically active zone (approximate time 0.75 days) with no appreciable degradation [30].

**Abiotic Degradation:** The aqueous hydrolysis half-life of bromodichloromethane at 25 °C and pH 7 has been estimated to be 137 years [24]. Direct photolysis or aquatic oxidation (via peroxy radicals or singlet oxygen) are not environmentally relevant or significant fate processes with respect to bromodichloromethane [23]. The rate constant for the vapor-phase reaction with hydroxyl radicals has been estimated to be $8.522 \times 10^{-14}$ cm$^3$/molecule-sec at 25 °C which corresponds to an atmospheric half-life of 3.92 months in a typical atmosphere [17].

**Bioconcentration:** Using the water solubility and octanol/water partition coefficient, the log BCF of bromodichloromethane can be estimated to be 0.72-1.37 from recommended regression equations [22].

**Soil Adsorption/Mobility:** Bromodichloromethane was observed to have significant mobility in laboratory soil column experiments utilizing a sandy soil [41]. Relatively high soil mobility was noted during a water infiltration study conducted in the Netherlands along the Rhine River [29]. Koc values for bromodichloromethane have been estimated to range from 53-251 based on water solubility and octanol/water partition coefficient; these Koc values are indicative of medium-to-high soil mobility [34].

**Volatilization from Water/Soil:** The Henry's Law constant of bromodichloromethane would suggest significant volatilization from water. Based on experimentally determined gas transfer rates, the volatilization half-life of bromodichloromethane from rivers and streams has been estimated to range from 33 min to 12 days with a typical half-life of about 35 hours [20]. Significant volatilization (half-life of about one hour) was observed from laboratory tanks [14]. Approximately 50% of applied bromodichloromethane volatilized from soil columns during laboratory studies which monitored transport and fate mechanisms [41].

**Water Concentrations:** DRINKING WATER: As part of the USEPA Ground Water Supply Survey, bromodichloromethane was positively detected in 445 of 945 US finished water supplies that use ground water sources at a median level of about 1.8 ppb [40]. Median levels of 9.2-18 ppb were detected in the water supplies of over 80% of the 113 US cities monitored during the three phases (1976-77) of USEPA National Organic Monitoring Survey [37]. A Canadian national survey of 70 drinking water supplies found bromodichloromethane levels of 0-33 ppb with an overall median level of 1.4 ppb [37]. In a survey of drinking waters from 12 areas of the world (China, Taiwan, north and south Philippines, Egypt, Indonesia, Australia, England, Brazil, Nicaragua, Venezuela, and Peru), bromodichloromethane was found in 9 of the 12 waters at levels ranging from 1.7-10 ppb [36]. Positive detections were made in 35 of 40 Michigan drinking water supplies at a median concn of 2.7 ppb [15]. GROUND WATER: Bromodichloromethane was one of 27 organic compounds identified in ground water collected from 315 wells in the area of the Potomac-Raritan-Magothy aquifer system adjacent to the Delaware River [16]. Levels of 0.3 ppb have been detected in ground water from the Netherlands [42]. SURFACE WATER: An analysis of the USEPA STORET data base found that bromodichloromethane had been positively detected in 14.0% of 8885 water observation stations at a median concn of 0.1 ug/L [33]. Bromodichloromethane was positively detected in 20.9% of 4972 samples collected from 11 stations on the Ohio River during 1980-81 with most concn between 0.1-1.0 ppb [26]. Positive identifications were determined at 24 of 204 sites (1-12 ppb) in 14 heavily industrialized river basins in the US [13]. Concentrations ranging from a trace-25 ng/L and not detected-20 ng/L were reported for 16 stations on the Niagara River and 95 stations on Lake Ontario, respectively, for 1981 monitoring [21]. SEAWATER: Bromodichloromethane concentrations of 0.1-1 ng/L have been detected in the North Atlantic while a concn of 0.1 ng/L was detected in the South Atlantic during 1985 monitoring [9]. Qualitative detection has been reported for the Narragansett Bay off Rhode Island in 1979-80 [39]. RAIN/SNOW: A concn of 0.4 ng/L was detected in rain collected in southern Germany in 1985 [9]. OTHER WATER: A bromodichloromethane concn of 2 ppb was detected in stormwater runoff from Eugene, OR as part of the USEPA Nationwide Urban Runoff Program [10]. Bromodichloromethane has been detected in swimming pool water at levels of 6-10 ppb [2].

**Effluent Concentrations:** An analysis of the USEPA STORET data base found that bromodichloromethane had been positively detected in 11.0% of 1375 effluent observation stations at a median concn of 5 ug/L [33]. Bromodichloromethane was detected in 14 of 63 industrial wastewater discharges in the US at levels ranging from <10-100 ppb [27]. Three municipal wastewater treatment facilities in Cincinnati, OH were found to be discharging levels as high as 3.2 ppb in 1982 [11].

**Sediment/Soil Concentrations:** Bromodichloromethane has been qualitatively detected in soil-sediment-water samples collected from the Love Canal near Niagara Falls, NY [19].

**Atmospheric Concentrations:** Mean bromodichloromethane levels of 0.76, 1.4, 120, and 180 ppt have been detected in the ambient air of Magnolia (AR), El Dorado (AR), Chapel Hill (NC), and Beaumont (TX), respectively [7]. Atmospheric bromodichloromethane levels ranging from 0.1-1.0 ppt were found in ambient air samples collected from the north and south Atlantic Ocean, the beaches of Bermuda, and southern Germany between 1982-85 [9]. Monitoring of four California sites between 1982-83 found mean composite concentrations of 20-100 ppt in ambient air [31].

**Food Survey Values:** In accordance with the US FDA's Total Diet Market Basket Study, an analysis of 39 food items purchased at retail markets (from Elizabeth, NJ, Chapel Hill, NC, and Washington, DC) detected bromodichloromethane in one dairy composite (1.2 ppb), in butter (7 ppb), and in two beverages (0.3 and 0.6 ppb); subsequent analysis of cola soft drinks resulted in three positive detections at levels of 2.3, 3.4, and 3.8 ppb [12]. An analysis of various soft drinks found cola beverages to average 0.9-5.9 ppb bromodichloromethane while clear soft drinks had levels of 0.1-0.2 ppb; municipal water supplies from which the soft drinks were manufactured were found to contain up to 20 ppb trihalomethanes [1].

**Plant Concentrations:** Mean bromodichloromethane levels of 7-22 ng/g (dry wt) have been detected in various species of marine algae [18].

**Fish/Seafood Concentrations:**

**Animal Concentrations:**

**Milk Concentrations:**

**Other Environmental Concentrations:**

**Probable Routes of Human Exposure:** The general population is exposed to bromodichloromethane through oral consumption of contaminated drinking water, beverages, and food products. The contamination is a result of inadvertent formation during chlorination treatment of the drinking water and subsequent use of chlorinated tap water to produce food products. Exposure can also occur through inhalation of background levels in ambient air and through dermal exposure in chlorinated swimming pool water.

**Average Daily Intake:** DRINKING WATER INTAKE: (assume concn of 1-10 ug/L) 2-20 ug; AIR INTAKE: (assume 1-100 ppt [6.7-670 ng/m$^3$]) 0.134-13.4 ug; FOOD INTAKE: insufficient data.

**Occupational Exposures:**

**Body Burdens:** A bromodichloromethane concn of 14 ng/mL was found in a blood sample from a resident living near the Love Canal near Niagara Falls, NY [3]. Bromodichloromethane was not detected in any sample from the USEPA National Human Adipose Tissue Survey for fiscal year 1982 [32].

## REFERENCES

1. Abdel-Rahman MS; J Appl Toxicol 2: 165-6 (1982)
2. Aggazzotti G, Predieri G; Wat Res 20: 959-63 (1986)
3. Barkley J et al; Biomed Mass Spectrom 7: 139-47 (1980)
4. Bouwer EJ, McCarty PL; Appl Environ Microbiol 45: 1295-99 (1983)
5. Bouwer EJ et al; Environ Sci Technol 15: 596-9 (1981)
6. Bouwer EJ, McCarty PL; Appl Environ Microbiol 45: 1286-94 (1983)
7. Brodzinsky R,Singh HB; Volatile Organic Chemicals in the Atmosphere: An Assessment of Available Data, Menlo Park,CA:SRI International (1982)
8. Callahan MA et al; Water-Related Environmental Fate of 129 Priority Pollutants-Vol II EPA-440/4-79-029b Chapter 42 (1979)
9. Class T et al; Chemosphere 15: 429-36 (1986)
10. Cole RH et al; J Wat Pollut Control Fed 56: 898-908 (1984)
11. Dunovant VS et al; J Water Pollut Control Fed 58: 886-95 (1986)
12. Entz RC et al; J Agric Food Chem 30: 846-9 (1982)
13. Ewing BB et al; Monitoring to Detect Previously Unrecognized Pollutants in Surface Waters USEPA-560/6-77-015a (1977)
14. Francois C et al; Trav Soc Pharm Montpellier 39: 49-50 (1979)
15. Furlong EAN, Ditri FM; Ecol Modeling 32: 215-25 (1986)
16. Fusillo TV et al; Groundwater 23: 354-60 (1985)
17. GEMS; Graphical Exposure Modeling System. FAP.  Fate of Atmospheric Pollutants (1987)
18. Gschwend PM et al; Science 227: 1033-5 (1985)

# Bromodichloromethane

19. Hauser TR, Bromberg SM; Environ Monitor Assess 2: 249-72 (1982)
20. Kaczmar SW et al; Environ Toxicol and Chem 3: 31-5 (1984)
21. Kaiser KLE et al; J Great Lakes Res 9: 212-23 (1983)
22. Lyman WJ et al; Handbook of Chemical Property Estimation Methods NY:McGraw-Hill p.5-4,5-10 (1982)
23. Mabey WR et al; p.179-82 in Aquatic Fate Process Data for Organic Priority Pollutants USEPA-440/4-81-014 (1981)
24. Mabey W, Mill T; J Phys Chem Ref Data 7: 383-415 (1978)
25. Nicholson BC et al; Environ Sci Technol 18: 518-23 (1984)
26. Ohio River Valley Sanit Comm; Assessment of Water Quality Conditions. Ohio River Mainstream, Cincinnati,OH (1982)
27. Perry DL et al; Identification of Organic Compounds in Industrial Effluent Discharges EPA- 600/4-79-016 p.42-3 (1979)
28. Perwak J et al; Exposure and Risk Assessment for Trihalomethanes (Chloroform, Bromoform, Bromodichloromethane, Dibromochloromethane) EPA-440/8- 81-018 p.13 (1980)
29. Piet GJ et al; Studies Environ Sci 17: 557-64 (1981)
30. Rittman BE et al; Ground Water 18: 236-43 (1980)
31. Shikiya J et al; Proc-APCA 77th Annual Mtg Vol 1,84-1.1 (1984)
32. Stanley JS; Broad Scan Analysis of Human Adipose Tissue:Volume II: Volatile Organic Compounds USEPA-560/5-86-036 p.74 (1986)
33. Staples CA et al; Environ Toxicol Chem 4: 131-42 (1985)
34. Swann RL et al; Res Rev 85: 17-28 (1983)
35. Tabak HH et al; J Water Pollut Control Fed 53: 1503-18 (1981)
36. Trussel AR et al; Water Chlorination Environ. Impact Health Effects 3: 39-53 (1980)
37. USEPA; Ambient Water Quality Criteria for Halomethanes USEPA-440/5-80-051 p.C-7 (1980)
38. USEPA; Health and Environmental Effects Profile for Bromochloromethanes ECAO-CIN-P122 (Final Draft) p.12,20,21 (1985)
39. Wakeham SG et al; Can J Fish Aquatic Sci 40: 304-21 (1983)
40. Westrick JJ et al; J Amer Water Works Assoc 76: 52-9 (1984)
41. Wilson JT et al; J Environ Qual 10: 501-6 (1981)
42. Zoeteman BCJ et al; Sci Total Environ 21: 187-202 (1981)

# n-Butanol

**Synonyms:** n-Butyl alcohol

**Structure:**

H3C ∕\∕\ OH

**CAS Registry Number:** 71-36-3

**Molecular Formula:** $C_4H_{10}O$

**Wiswesser Line Notation:** Q4

## CHEMICAL AND PHYSICAL PROPERTIES

**Boiling Point:** 117.2 °C

**Melting Point:** -89.5 °C

**Molecular Weight:** 74.12

**Dissociation Constants:**

**Log Octanol/Water Partition Coefficient:** 0.88 [14]

**Water Solubility:** 77,000 mg/L [3]

**Vapor Pressure:** 7.024 mm Hg at 25 °C [8]

**Henry's Law Constant:** $5.57 \times 10^{-6}$ atm-m³/mole [21]

## ENVIRONMENTAL FATE/EXPOSURE POTENTIAL

**Summary:** Release of n-butanol to the environment is expected to result from its use as a solvent in a variety of products. It may also be released by the action of anaerobic microorganisms. Release of

48

n-butanol to soil may result in volatilization from the soil surface and biodegradation is expected to be significant. n-Butanol should not bind strongly to soil and so is expected to leach to ground water. Release of n-butanol to water is expected to result in biodegradation and in volatilization from the water surface. Photooxidation by hydroxyl radicals in water is expected to be slow. Bioconcentration is not expected to be significant. Vapor phase n-butanol in the atmosphere is expected to react with photochemically generated hydroxyl radicals with a half-life of 1.2-2.3 days. Human exposure to n-butanol is expected to result primarily from contact with products containing the compound.

**Natural Sources:** Two strains of bacteria, Clostridium butylicum and Clostridium acetobutylicum, can ferment starch with the production of n-butanol [6].

**Artificial Sources:** n-Butanol is used as a solvent for fats, waxes, resins, shellac, varnish, and gums, in the manufacture of lacquers, rayon, detergents, and other butyl compounds [22], used in plasticizers, as a dyeing assistant, in hydraulic fluids, and as a dehydrating agent [15]. An unidentified petrochemical company discharged about 90 lb. n-butanol/day [16].

**Terrestrial Fate:** When released to soil, n-butanol is expected to leach to ground water or to biodegrade. Volatilization from the soil surface may also occur.

**Aquatic Fate:** In water, n-butanol is expected to biodegrade. Volatilization from the water surface is expected to occur with estimated half-lives of 2.4 hr, 3.9 hr, and 125.9 days in streams, rivers, and lakes [20]. The actual tendency of n-butanol to volatilize depends upon the temperature, turbulence, wind speed, current velocity, and depth of the water bodies. n-Butanol is not expected to bind strongly to suspended sediments. Bioconcentration is not expected to be significant. The rate of the reaction between hydroxyl radicals and n-butanol in water is $2.2 \times 10^{+9}$ 1/mol-sec [1]. Assuming an hydroxyl radical concentration of $1 \times 10^{-17}$ M in water, this corresponds to a half-life of about 1 year [1].

**Atmospheric Fate:** The rate constant for the reaction between n-butanol and hydroxyl radicals is $6.8 \times 10^{-10}$ $cm^3$/molecule sec [5]. Using a hydroxyl radical concentration of $1 \times 10^{+6}$ molecule/$cm^3$, the half-life was calculated to be 1.2 days [5]. The half-life of n-butanol in a sunlit urban atmosphere was estimated to be 5 hr [5]. The half-life for the reaction of vapor phase n-butanol in the atmosphere with photochemically generated hydroxyl radicals was estimated to be 2.30 days [11]. In the presence of NO, the half-life of n-butanol is 6.5 hr [9].

**Biodegradation:** A 3 ppm solution of n-butanol was incubated in river water at 18-19 °C exerted a biological oxygen demand (BOD) of about 4.5 ppm after about 4 days [13]. After 5 days, 33% of the theoretical BOD was exerted in a solution of n-butanol containing an inoculum from polluted surface water [10]. In a batch system, n-butanol was dissolved to give a concentration corresponding to a chemical oxygen demand (COD) of 200 mg/L [26]. Sufficient adapted, activated sludge was added to make the dry matter of the inoculum 100 mg/L and the system was incubated at 20 °C. Under these conditions, a total of 98.9% of the initial n-butanol was removed at a rate of 84.0 mg COD/g hr [13]. After 5 days incubation at 20 °C, 66% of the theoretical oxygen demand had been exerted in a BOD test [4].

**Abiotic Degradation:** n-Butanol reacted with hydroxyl radical at 19 °C with a rate constant of $6.8 \times 10^{-10}$ cm$^3$/molecule sec [5]. Using a hydroxyl radical concentration of $1 \times 10^{+6}$ molecule/cm$^3$, the half-life was calculated to be 1.2 days [5]. The half-life of n-butanol in a sunlit urban atmosphere was estimated to be 5 hr [5]. The half-life for the reaction of vapor phase n-butanol in the atmosphere with photochemically generated hydroxyl radicals was estimated to be 2.30 days [11]. Ten ppm of n-butanol irradiated with sunlamps (>290 nm) in the presence of 5 ppm NO at a relative humidity of 35% displayed a half-life of 6.5 hr [9]. The intensity of the radiation was 2.6 times that of natural sunlight [9]. The rate constant of the reaction between hydroxyl radicals and n-butanol in water is $2.2 \times 10^{+9}$ 1/mol-sec [1]. Assuming an hydroxyl radical concentration of $1 \times 10^{-17}$ M in water, this corresponds to a half-life of about 1 year [1].

**Bioconcentration:** No data concerning the bioconcentration of n-butanol were available. With a log octanol/water partition coefficient of 0.88 [14], however, very little bioconcentration is expected.

**Soil Adsorption/Mobility:** Using a measured log octanol/water partition coefficient of 0.88 [14], a soil sorption coefficient (Koc) of 71.6 was estimated [20]. A Koc of this magnitude suggests that n-butanol will be moderately to highly mobile in the soil [17].

**Volatilization from Water/Soil:** Using a 6 m X 0.06 m X 0.61 m deep wind-wave tank, a mass transfer coefficient of 3.58 m/sec for n-butanol was determined [21]. The wind speed was 8.57 m/sec. This translates into a volatilization half-life for a 1 m depth of 53.9 hr [21]. Using the Henry's Law constant value, the rates of volatilization from streams, rivers, and lakes were estimated to be 2.4 hr, 3.9 hr, and 3022.5 hr (125.9 days), respectively [20]. The depths of the lakes were assumed to be 50 m and that of the streams and rivers 1 m. A wind velocity of 3 m/sec was assumed and the current velocities of the lakes, rivers, and streams were assumed to be 0.01, 1, and 2 m/sec, respectively.

**Water Concentrations:** SURFACE WATER: Tatsuno City, Japan - 318 ppb [27]. n-Butanol was detected but not quantified in water samples from Lake Ontario [12].

**Effluent Concentrations:** Effluents from an unidentified petrochemical company contained about 16.0 mg/L n-butanol and discharged approximately 90 lb. n-butanol/day [16].

**Sediment/Soil Concentrations:**

**Atmospheric Concentrations:** Point Barrow, AK - 34-445 ppb, 190 ppb mean [6].

**Food Survey Values:** n-Butanol was detected but not quantified in volatiles from roasted filbert nuts [18].

**Plant Concentrations:**

**Fish/Seafood Concentrations:**

**Animal Concentrations:**

**Milk Concentrations:** n-Butanol was detected but not quantified one time in a total of 12 human milk samples [25].

**Other Environmental Concentrations:**

**Probable Routes of Human Exposure:**

**Average Daily Intake:**

**Occupational Exposures:** Eight individuals were exposed to an average of 1.2 mg/m$^3$ during spraying operations [2]. NIOSH (NOES Survey 1981-1983) has statistically estimated that 794,284 workers are exposed to n-butanol in the United States [24]. NIOSH (NOHS Survey 1972-1974) has statistically estimated that 1,778,571 workers are exposed to n-butanol in the United States [23].

**Body Burdens:** n-Butanol was detected but not quantified one time in a total of 12 human milk samples [25]. The concentration of n-butanol in expired air from 8 individuals ranged from 1.3-35.0 ug/hr [7]. The n-butanol was said to be of metabolic origins [7]. n-Butanol was 0.02-0.08 ng/L in the expired air of 54 individuals and detected 2 times (11.1% pos) [19].

## REFERENCES

1. Anbar M, Neta P; Int J Appl Rad Isot 18: 493-23 (1967)
2. Angerer J, Wulf H; Int Arch Occup Environ Health 56: 307-21 (1985)
3. Barton, AFM; Alcohols with Water. IUPAC Solubility Data Series Vol 15 pp438 (1984)
4. Bridie AL et al; Water Res 13: 627-30 (1979)
5. Campbell IM et al; Chem Phys Lett 38: 362-4 (1976)
6. Cavanagh LA et al; Environ Sci Technol 3: 251-7 (1969)
7. Conkle JP et al; Arch Environ Health 30: 290-5 (1975)
8. Daubert TE, Danner RP; Data Compilation Tables of Properties of Pure Compounds. pp 450 American Institute of Chemical Engineers (1985)
9. Dilling WL et al; Environ Sci Technol 10: 351-6 (1976)
10. Dore M et al; Trib Cebedeau 28: 3-11 (1975)
11. GEMS; Graphical Exposure Modeling System. Fate of atmospheric pollutants. FAP. Fate of Atmos Pollut (1986)
12. Great Lakes Water Quality Board; An Inventory of Chemical Substances Identified in the Great Lakes Ecosystem Vol.1 (1983)
13. Hammerton C; J Appl Chem 5: 517-24 (1955)
14. Hansch C, Leo AJ; Medchem Project Issue No 26 Pomona College Claremont CA (1985)
15. Hawley GG; Condensed Chemical Dictionary 10th ed Van Nostrand Reinhold NY p 160 (1981)
16. Keith LH; Sci Total Environ 3: 87-102 (1974)
17. Kenaga EE; Ecotox Env Safety 4: 26-38 (1980)
18. Kinlin TE et al; J Agr Food Chem 20: 1021-8 (1972)
19. Krotoszynski BK et al; J Anal Toxicol 3: 225-34 (1979)
20. Lyman WJ et al; Handbook of Chemical Property Estimation Methods. Environmental Behavior of Organic Compounds. McGraw-Hill NY (1982)
21. Mackay D, Yeun ATK; Environ Sci Technol 17: 211-7 (1983)

22. Merck Index; An Encyclopedia of Chemicals, Drugs and Biologicals 10th ed p 214 (1983)
23. NIOSH; Natl Occup Hazard Survey (1974)
24. NIOSH; Natl Occupational Exposure Survey (1982)
25. Pellizzari ED et al; Bull Environ Contam Toxicol 28: 322-8 (1982)
26. Pitter P; Water Res 10: 231-5 (1976)
27. Yasuhara A et al; Environ Sci Technol 15: 570-3 (1981)

# t-Butanol

## SUBSTANCE IDENTIFICATION

**Synonyms:** t-Butyl alcohol

**Structure:**

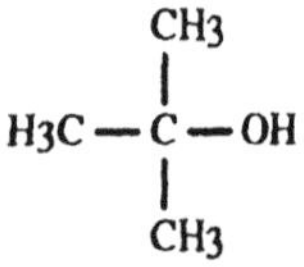

$$\begin{array}{c} CH_3 \\ | \\ H_3C - C - OH \\ | \\ CH_3 \end{array}$$

**CAS Registry Number:** 75-65-0

**Molecular Formula:** $C_4H_{10}O$

**Wiswesser Line Notation:** QX1&1&1

## CHEMICAL AND PHYSICAL PROPERTIES

**Boiling Point:** 82.41 °C

**Melting Point:** 25.7 °C

**Molecular Weight:** 74.12

**Dissociation Constants:**

**Log Octanol/Water Partition Coefficient:** 0.35 [9]

**Water Solubility:** Miscible [2]

**Vapor Pressure:** 41.67 mm Hg at 25 °C [23]

**Henry's Law Constant:** $1.175 \times 10^{-5}$ atm-m$^3$/mole [12]

## ENVIRONMENTAL FATE/EXPOSURE POTENTIAL

**Summary:** t-Butanol release to the environment is likely to result from use of solvents and paint removers containing the compound. Release to the soil is expected to result in volatilization from the soil surface and biodegradation. t-Butanol is not expected to strongly adsorb to soil and so is expected to leach to ground water. When released to water, t-butanol is expected to volatilize and possibly biodegrade. Aqueous photooxidation will be a slow process. Bioconcentration in fish is not expected to be significant. When released to the atmosphere, t-butanol will react with NO with a half-life of above a day. The estimated half-life of the reaction between vapor phase t-butanol and photochemically generated hydroxyl radicals is 1.09 months.

**Natural Sources:**

**Artificial Sources:** t-Butanol is used as a denaturant for ethanol, in the manufacture of flotation agents, flavors and perfumes, as a solvent, in paint removers, and as an octane booster in gasoline [16]. It is also used as a solvent for pharmaceuticals, as a dehydrating agent, and in the manufacture of methyl methacrylate [11]. Vapor is liberated during application of lacquer surface coatings, during use of industrial cleaning compounds, during use as a chemical intermediate in the manufacture of t-butyl chloride and t-butyl phenol, during the mixing of perfumes, lacquers, and denatured alcohol, and in open-surface tanks. It also may be released when used as a solvent for drug extraction, water removal, wax solvent, extraction of hypochlorous acid, lube oil, and laboratory procedures.

**Terrestrial Fate:** t-Butanol may volatilize from the soil surface and is expected to biodegrade. t-Butanol is not expected to bind strongly to soils and may leach to ground water.

**Aquatic Fate:** t-Butanol is expected to volatilize from the water surface and may biodegrade. Photooxidation by hydroxyl radicals is expected to be a slow process (half-life about 8.8 years) [1]. Bioconcentration is not expected to be significant.

**Atmospheric Fate:** t-Butanol present at 10 ppm in air reacted with 5 ppm NO with a half-life of 34.5 hr [5]. The half-life for the reaction of vapor phase t-butanol with photochemically generated hydroxyl radicals was estimated to be 1.09 months [6].

**Biodegradation:** t-Butanol present at 3 ppm in river water at 18-19 °C exerted a biochemical oxygen demand of <0.5 ppm after 12 days [8]. t-Butanol was incubated with subsurface soils from Pennsylvania, New York, and Virginia [19]. When present at 1-10 mg/L, t-butanol degradation was very rapid and non-detectable levels were achieved within about 25 days. At higher concentrations, however, >200 days were required to achieve complete degradation of t-butanol. The degradation rate was lower in saturated zone soil samples than in unsaturated zone soil samples. The New York and Virginia soil samples were both from the saturated zone and degraded t-butanol less rapidly than did the Pennsylvania samples. In the New York and Virginia samples, the degradation rate was directly proportional to the t-butanol concentration [19]. t-Butanol was added at a concentration corresponding to 200 mg/L COD, inoculated with adapted, activated sludge and incubated at 20 °C in a batch test; the rate of biodegradation was 30.0 mg COD/g hr and 98.5% of the t-butanol was removed, based on COD [21]. t-Butanol at 500 mg/L was added to samples of three activated sludges and incubated at 20 °C [7]. After 6, 12, and 24 hr, 0.5%, 0.7%, and 0.8% of the theoretical oxygen demand had been exerted, respectively [7]. Results of other investigations using sewage seeds, however, indicate that t-butanol is resistant to biodegradation [4,10,13,22,24].

**Abiotic Degradation:** The rate constant for the reaction between t-butanol and hydroxyl radicals in water is $2.5 \times 10^{+8}$ 1/mol-sec at pH 7 [1]. Assuming a hydroxyl radical concentration of $1 \times 10^{-17}$ M in water, a half-life of 8.8 years was estimated from this rate constant. t-Butanol present at 10 ppm in air reacted with 5 ppm NO with a half-life of 34.5 hr [5].

**Bioconcentration:** Using a measured log octanol/water partition coefficient of 0.35 [9], a log bioconcentration factor (BCF) of 0.036 was estimated [15]. A log BCF of this magnitude suggests that bioconcentration in fish will not be significant.

**Soil Adsorption/Mobility:** Using a measured log octanol/water partition coefficient of 0.35 [9], a soil-sorption coefficient for t-butanol of 36.9 was estimated [15]. Based on this estimated Koc, t-butanol is not expected to bind strongly to soils [14] and may leach to ground water.

**Volatilization from Water/Soil:** Using the Henry's Law constant, the half-lives for volatilization of t-butanol from streams, rivers, and lakes were estimated. The wind velocity was assumed to be 3 m/sec, the current velocities of the streams, rivers, and lakes 2, 1, and 0.01 m/sec, respectively, and the depths of the lakes 50 m and that of the streams and rivers 1 m. The half-life values were 51.6 hr, 65.37 hr, and 3104.3 hr (129 days) for the streams, rivers, and lakes, respectively [15].

**Water Concentrations:** DRINKING WATER: t-Butanol was detected but not quantified in drinking water samples from at least one of the following cities: Cincinnati, OH, Miami, FL, Ottumwa, IA, Philadelphia, PA, and Seattle, WA [3].

**Effluent Concentrations:**

**Sediment/Soil Concentrations:**

**Atmospheric Concentrations:**

**Food Survey Values:**

**Plant Concentrations:**

**Fish/Seafood Concentrations:**

**Animal Concentrations:**

**Milk Concentrations:**

**Other Environmental Concentrations:**

**Probable Routes of Human Exposure:** Ingestion, inhalation, skin, and eye contact.

**Average Daily Intake:**

**Occupational Exposures:** NIOSH (NOES Survey 1981-1983) has statistically estimated that 149,918 workers are exposed to t-butanol in the United States [18]. NIOSH (NOHS Survey 1972-1974) has statistically estimated that 8,519 workers are exposed to t-butanol in the United States [17].

**Body Burdens:** t-Butanol was detected but not quantified in human milk [20].

# REFERENCES

1. Anbar M, Neta P; Int J Appl Rad Isot 18: 493-523 (1967)
2. Barton AFM; Alcohols with Water IUPAC Solubility Data Series Vol 15: 438 (1984)
3. Coleman WE et al; pp 305-27 in Analysis and Identification of Organic Substances in Water, Keith L ed. Ann Arbor Sci (1976)
4. Dias FF, Alexander M; Appl Microbiol 22: 1114-8 (1971)
5. Dilling WL et al; Environ Sci Technol 10: 351-6 (1976)
6. GEMS; Graphical Exposure Modeling System. Fate of atmospheric pollutants (FAP) Office of Toxic Substances. USEPA (1986)
7. Gerhold RM, Malaney GW; J Water Pollut Cont Fed 38: 562-79 (1966)
8. Hammerton C; J Appl Chem 5: 517-24 (1955)
9. Hansch C, Leo AJ; Medchem Project Issue No.19 Pomona, CA (1985)
10. Hatfield R; Ind Eng Chem 49: 192-6 (1957)
11. Hawley GG; Condensed Chemical Dictionary 10th ed Van Nostrand Reinhold NY p 161 (1981)
12. Hine J, Mookerjee PK; J Org Chem 40: 292-9 (1975)
13. Kawasaki M; Ecotox Environ Safety 4: 444-54 (1980)
14. Kenaga EE; Ecotox Env Safety 4: 24-38 (1980)
15. Lyman WJ et al; Handbook of Chemical Property Estimation Methods. Environmental Behavior of Organic Compounds. McGraw-Hill NY (1982)
16. Merck Index; An Encyclopedia of Chemicals, Drugs and Biologicals 10th ed p 215 (1983)
17. NIOSH; Natl Occupational Hazard Survey (1974)
18. NIOSH; Natl Occupational Exposure Survey (1982)
19. Novak JT et al; Water Sci Technol 17: 71-85 (1985)
20. Pellizzari ED et al; Bull Environ Contam Toxicol 28: 322-8 (1982)
21. Pitter P; Water Res 10: 2331-5 (1976)
22. Wagner R; Vom Wasser 42: 271-305 (1974)
23. Wilhout RC, Zwolinski BJ; Physical and thermodynamic properties of aliphatic alcohols, J Phys Chem Ref Data 2: p1-114 Supplement No 1 (1973)
24. Young RHF et al; J Water Pollut Cont Fed 40: 354-68 (1968)

# n-Butyl Acetate

## SUBSTANCE IDENTIFICATION

**Synonyms:**

**Structure:**

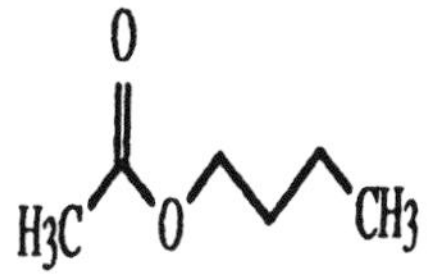

$$\underset{H_3C}{}\overset{O}{\underset{\parallel}{C}}\underset{O}{}\diagup\diagdown\diagup\underset{CH_3}{}$$

**CAS Registry Number:** 123-86-4

**Molecular Formula:** $C_6H_{12}O_2$

**Wiswesser Line Notation:** 4OV1

## CHEMICAL AND PHYSICAL PROPERTIES

**Boiling Point:** 125-126 °C

**Melting Point:** -77 °C

**Molecular Weight:** 116.16

**Dissociation Constants:**

**Log Octanol/Water Partition Coefficient:** 1.82 [27]

**Water Solubility:** 6710 mg/L at 25 °C [27]

**Vapor Pressure:** 12.48 mm Hg at 25 °C [23]

**Henry's Law Constant:** $3.2 \times 10^{-4}$ atm-m³/mole [8]

## ENVIRONMENTAL FATE/EXPOSURE POTENTIAL

**Summary:** Evaporation of n-butyl acetate solvent from lacquers, coatings, inks, and adhesives is the dominant anthropogenic emission source of n-butyl acetate into the environment. n-Butyl acetate has

59

been identified as a naturally occurring component of apples. If released to soil, n-butyl acetate may be susceptible to significant biodegradation based on its demonstrated biodegradability with a natural river-water seed. Chemical hydrolysis in moist alkaline soils (pH approaching 9 or higher) is expected to be important, but not in neutral or acidic soils. n-Butyl acetate may be subject to moderate-to-high leaching based on estimated Koc values of 34 and 233. Volatilization from dry soil surfaces is likely to be rapid. If released to water, biodegradation and volatilization are expected to be the important removal mechanisms. BOD studies using either a sewage inoculum or a natural river-water inoculum have demonstrated that n-butyl acetate is significantly biodegraded. The volatilization half-life from a model river 1 m deep flowing 1 m/sec with a wind velocity of 3 m/sec has been estimated to be 6.1 hr. The hydrolysis half-lives of n-butyl acetate at pHs 7.0, 8.0, and 9.0 are about 3.1 years, 114 days, and 11.4 days, respectively, at 20 °C indicating that hydrolysis will be important only in very alkaline environmental waters. Aquatic adsorption and bioconcentration are not expected to be significant. If released to air, n-butyl acetate will exist almost entirely in the vapor phase in the ambient atmosphere. The dominant removal mechanism in the atmosphere will be the vapor-phase reaction with photochemically produced hydroxyl radicals which has an estimated half-life of about 6 days in an average atmosphere. General population exposure to n-butyl acetate can occur through oral consumption of food (in which it occurs naturally) and by inhalation of contaminated air, especially in the vicinity of usage of lacquers, paints, inks, or adhesives containing n-butyl acetate solvent. Occupational exposure by inhalation and dermal routes may be significant.

**Natural Sources:** n-Butyl acetate has been identified as a naturally occurring component of apples [16].

**Artificial Sources:** n-Butyl acetate is predominantly used as a solvent for nitrocellulose-based lacquers and inks and adhesives, with the top industrial consumers being furniture lacquers and automotive coatings [1]; from these uses, n-butyl acetate evaporates directly into the surrounding air thereby becoming the dominant anthropogenic emission source of n-butyl acetate to the environment.

**Terrestrial Fate:** n-Butyl acetate has been shown to be significantly biodegraded in a screening test using a natural river-water seed suggesting that microbial decomposition in soil may occur. Chemical hydrolysis of n-butyl acetate in moist alkaline soils (pH approaching 9 or higher) is expected to be important, but hydrolysis in neutral or acidic soils is not expected to be important. Based on estimated Koc values of 34 and 233, n-butyl acetate may be subject to moderate-to-high leaching. Volatilization from dry soil surfaces is likely to be rapid.

**Aquatic Fate:** n-Butyl acetate may be susceptible to significant biodegradation in natural water based on the results of BOD studies using either a sewage inoculum or a natural river-water inoculum. The volatilization half-life from a model river 1 m deep flowing 1 m/sec with a wind velocity of 3 m/sec has been estimated to be 6.1 hours; the volatilization half-life from a similar river 10 m deep has been estimated to be 7.4 days. The hydrolysis half-lives of n-butyl acetate at pHs 7.0, 8.0, and 9.0 are about 3.1 years, 114 days, and 11.4 days, respectively, at 20 °C indicating that hydrolysis will be important only in very alkaline environmental waters. Aquatic adsorption and bioconcentration are not expected to be significant.

**Atmospheric Fate:** n-Butyl acetate will exist almost entirely in the vapor phase in the ambient atmosphere due to its relatively high vapor pressure. The half-life for the vapor-phase reaction of n-butyl acetate with photochemically produced hydroxyl radicals has been estimated to be about 6 days in an average atmosphere indicating that this reaction will be the dominant removal mechanism.

**Biodegradation:** Using a settled sewage seed, 5-day and 20-day theoretical BOD's of 23.5% and 57.4% were measured for n-butyl acetate via standard dilution water tests; using river water as seed, 5-day and 20-day TBOD's of 20.9% and 55.4% were measured [12]. Using a filtered sewage seed, 5-day and 20-day TBOD's of 58% and 83% were measured in fresh water dilution tests; 5-day and 20-day TBOD's of 40% and 61% were measured in salt water [22]. A 5-day TBOD of 56.8% and 51.8% were measured for n-butyl acetate in distilled water and sea water, respectively, via the standard dilution method [26].

**Abiotic Degradation:** Based on experimentally derived acid- and base-catalyzed hydrolysis rate constants at 20 °C, the hydrolysis half-lives of n-butyl acetate at pHs 7.0, 8.0, and 9.0 are estimated to be 3.1 years, 114 days, and 11.4 days, respectively [15]. The half-life for the vapor phase reaction of n-butyl acetate with photochemically produced hydroxyl radicals in the atmosphere has been estimated to be about 6 days assuming an average atmospheric hydroxyl radical concn of 8 x $10^{+5}$ molecules/cm$^3$ [7].

**Bioconcentration:** Based on a log Kow of 1.82 [27] and a water solubility of 6710 ppm at 25 °C [27], the BCF value for n-butyl acetate can be estimated to be 14 and 4, respectively, by regression derived equations [14]. These BCF values suggest that bioconcentration is not significant.

**Soil Adsorption/Mobility:** Based on a log Kow of 1.82 and a water solubility of 6710 ppm at 25 °C [27], the Koc value for n-butyl acetate can be estimated to be 233 and 34, respectively, by regression derived equations [14]. These Koc values indicate a high-to-medium soil mobility [25].

**Volatilization from Water/Soil:** Based on experimental vapor-liquid equilibrium data [8], the Henry's Law constant for n-butyl acetate at 25 °C is about 3.2 x $10^{-4}$ atm-m$^3$/mole; this value of Henry's Law constant suggests that volatilization is probably significant from environmental bodies of water [14]. The volatilization half-life from a model river 1 m deep flowing 1 m/sec with a wind velocity of 3 m/sec is estimated to be 6.1 hours [14]; the volatilization half-life from a similar model river 10 meters deep is estimated to be 7.4 days [14]. n-Butyl acetate evaporates relatively rapidly from solid surfaces with a half-life of about 48 min as measured by a standard solvent coating test method at 19 °C [18]; the evaporative half-life of a thin film of n-butyl acetate from a solid surface at 25 °C has been reported to be 4.2 min [4].

**Water Concentrations:** DRINKING WATER: Butyl acetate was detected in less than 5% of finished waters using ground water supplies based on US federal studies [5]. Butyl acetate was qualitatively detected in finished drinking water from England and Dordrecht, The Netherlands [6,24]. GROUND WATER: Butyl acetate was detected in ground waters in the Netherlands at a maximum concn of 1 ppb [29]. SURFACE WATER: Butyl acetate was qualitatively detected in lake water from England [24].

**Effluent Concentrations:** Butyl acetate was detected in 1 of 63 US industrial wastewater effluents at a concn below 10 ppb [21]. It was detected in effluents from advanced waste treatment facilities in Lake Tahoe, CA and Washington, DC [13].

**Sediment/Soil Concentrations:**

**Atmospheric Concentrations:** Butyl acetate was detected in the ambient air near an industrial site in Newark, NJ in 1976 at a concentration of 3000 ng/m$^3$, but was not detected in the air near the Kin-Buc Waste Disposal Site in Edison, NJ during June 29-July 1, 1976 [19,20]. Qualitative detection of butyl acetate was reported for the ambient air of Leningrad, USSR during 1976 monitoring [9].

**Food Survey Values:** Butyl acetate has been detected as a volatile flavor component of baked potatoes [2] and roasted filbert nuts [10]. It has been identified as a naturally occurring flavor and fragrance component of apples [16].

**Plant Concentrations:** Butyl acetate has been identified as a naturally occurring component of apples [16].

**Fish/Seafood Concentrations:**

**Animal Concentrations:**

**Milk Concentrations:**

**Other Environmental Concentrations:**

**Probable Routes of Human Exposure:** General population exposure to n-butyl acetate can occur through consumption of food and inhalation of contaminated air. n-Butyl acetate has been identified as a naturally occurring component of apples and as a flavor component of baked potatoes and roasted filbert nuts. Inhalation of contaminated air in the vicinity of usage of lacquers, paints, inks, or adhesives containing n-butyl acetate solvent may provide significant exposure. Dermal exposure may occur upon contact with lacquers, paints, inks, or adhesives containing n-butyl acetate. Occupational exposure by inhalation and dermal routes related to the use of n-butyl acetate as a solvent may be significant.

**Average Daily Intake:**

## n-Butyl Acetate

**Occupational Exposures:** NIOSH (NOES Survey 1981-1983) has statistically estimated that 720,812 workers are exposed to n-butyl acetate in the United States [17]. NIOSH (NOHS Survey 1972-1974) has statistically estimated that 1,512,312 workers are exposed to n-butyl acetate in the United States [17]. The average TWA of n-butyl acetate detected in worker breathing zones of 3 companies using spray painting and gluing was 0.8 ppm; the highest single concn detected was 6.8 ppm [28].

**Body Burdens:** n-Butyl acetate was detected in the expired air of 3 of 8 human subjects at an expiration rate of 8.7-41.0 ug/hr [3]. n-Butyl acetate was detected in 30.8% of 387 expired air samples collected from 54 human volunteers at a mean concentration of 0.20 ng/L [11].

## REFERENCES

1. Chemical Marketing Reporter; Chem Profile Oct. 8 (1984)
2. Coleman EC et al; J Agric Food Chem 29: 42 (1981)
3. Conkle JP et al; Arch Environ Health 30: 290 (1975)
4. Davis DS; Am Perfumer Cosmet 81: 32 (1966)
5. Dyksen JE, Hess AF III; J Amer Water Works Assoc 74: 394 (1982)
6. Fielding M et al; Organic Micropollut in Drinking Water Medmenham, Eng Water Res Cent TR-159 (1981)
7. GEMS; Graphical Exposure Modeling System. Fate of atmospheric pollutants (FAP) data base. Office of Toxic Substances. USEPA (1986)
8. Hine J, Mookerjee PK; J Org Chem 40: 292-8 (1975)
9. Ioffe BV et al; J Chromatogr 142: 787 (1977)
10. Kinlin TE et al; J Agric Food Chem 20: 1021 (1972)
11. Krotoszynski BK et al; J Anal Toxicol 3: 225 (1979)
12. Lamb CB, Jenkins GF; Proc 8th Industrial Waste Conf, Purdue Univ pp 326-9 (1952)
13. Lucas SV; GC/MS Analysis of Org in Drinking Water Concentrates and Advanced Waste Treatment Concentrates pp 321 USEPA-600/1-84-020A NTIS PB85-128221 (1984)
14. Lyman WJ et al; Handbook of Chemical Property Estimation Methods. Environmental Behavior of Organic Compounds. McGraw-Hill NY (1982)
15. Mabey W, Mill T; J Phys Chem Ref Data 7: 383 (1978)
16. Nicholas HJ; p 381 in Phytochemistry Vol.II New York Van Nostrand Reinhold (1973)
17. NIOSH; National Occupational Hazard Survey (1974): NIOSH; National Occupational Exposure Survey (1983)
18. Park JG, Hofmann HE; Ind Eng Chem 24: 132 (1932)
19. Pellizzari ED; Environ Sci Technol 16: 781 (1982)
20. Pellizzari ED; The Measurement of Carcinogenic Vapors in Ambient Atmosphere USEPA 600/7-77-055 (1977)

# n-Butyl Acetate

21. Perry DL et al; Identification of Org Compounds in Industrial Effluent Discharges USEPA 600/4-79-016 NTIS PB-294794 (1979)
22. Price KS et al; J Water Pollut Contr Fed 46: 63 (1974)
23. Riddick JA et al; Organic Solvents: Physical Properties and Methods of Purification, New York, J Wiley & Sons (1986)
24. Shackelford WM, Keith LH; Frequency of Org Compounds Identified in Water USEPA 600/4-76-062 (1976)
25. Swann RL et al; Res Rev 85: 17 (1983)
26. Takemoto S et al; Suishitsu Odaku Kenkyu 4: 80 (1981)
27. Tewari YB et al; J Chem Eng Data 27: 451 (1982)
28. Whitehead LW et al; Am Ind Hyg Assoc J 45: 767 (1984)
29. Zoeteman BCJ et al; Sci Total Environ 21: 187 (1981)

# n-Butylamine

## SUBSTANCE IDENTIFICATION

**Synonyms:**

**Structure:**

H₃C⌃⌄NH₂

**CAS Registry Number:** 109-73-9

**Molecular Formula:** $C_4H_{11}N$

**Wiswesser Line Notation:** Z4

## CHEMICAL AND PHYSICAL PROPERTIES

**Boiling Point:** 77.8 °C at 760 mm Hg

**Melting Point:** -50 °C

**Molecular Weight:** 73.14

**Dissociation Constants:** pKa = 10.77 [13]

**Log Octanol/Water Partition Coefficient:** 0.97 [4]

**Water Solubility:** Miscible [14]

**Vapor Pressure:** 91.75 mm Hg at 25 °C [14]

**Henry's Law Constant:** $1.69 \times 10^{-5}$ atm-m³/mole at 25 °C [5]

## ENVIRONMENTAL FATE/EXPOSURE POTENTIAL

**Summary:** Release of n-butylamine to the environment may occur as a result of its manufacture and use as an intermediate for pharmaceuticals, dyestuffs, rubber chemicals, emulsifying agents, insecticides, and synthetic tanning agents, and as a result of fertilizer manufacture, fish processing, rendering plants, and sewage treatment. It has been reported to be a component of animal waste. If n-butylamine is released to the soil, it will not adsorb to the soil, and it will be expected to leach rapidly to ground water. Hydrolysis will not be a significant removal process. No information on biodegradation of n-butylamine in soils or ground water was found, but screening studies suggest that biodegradation may be important. Evaporation from soil surfaces and other surfaces may be a significant removal process. If n-butylamine is released to water, it will not be expected to adsorb to sediment, hydrolyze, or bioconcentrate in aquatic organisms. Based on very limited data from laboratory tests of biodegradation of n-butylamine with activated sludges and settled sewage seed, n-butylamine may possibly be subject to biodegradation in natural waters. Evaporation will be expected to be an important removal process with a half-life of 1.95 days predicted for evaporation from a model river 1 m deep, flowing at 1 m/sec with a wind velocity of 3 m/sec. No information was found on photolysis, but n-butylamine should not absorb sunlight. Reaction with hydroxyl radicals will be the fastest chemical removal process for n-butylamine in the atmosphere (estimated half-life of 0.479 days). Dissolution into rain droplets may be the most important physical removal process in the atmosphere. Human exposure will occur mainly through occupational exposure. Minor general exposure may occur through the ingestion of food.

**Natural Sources:** n-Butylamine has been reported to be a component of animal waste [3].

**Artificial Sources:** Reported sources of n-butylamine include: fertilizer manufacture, fish processing, rendering plant, and sewage treatment [3]. Release of n-butylamine to the environment may also occur as a result of its manufacture and use as an intermediate for pharmaceuticals, dyestuffs, rubber chemicals, emulsifying agents, insecticides, and synthetic tanning agents [9].

**Terrestrial Fate:** If n-butylamine is released to the soil, it will not adsorb to the soil; it may leach to the ground water due to its lack of adsorption and high water solubility. Hydrolysis will not be a significant removal process. No information on biodegradation of n-butylamine in soils or ground water was found, but screening studies suggest that biodegradation may be important. Evaporation from soil surfaces and other surfaces may be a significant removal process based on the vapor pressure.

**Aquatic Fate:** If n-butylamine is released to water it will not be expected to adsorb to the sediment, hydrolyze, or bioconcentrate in aquatic organisms. Based on limited data concerning laboratory tests of biodegradation of n-butylamine with activated sludges and settled sewage seed, n-butylamine may be subject to biodegradation in natural waters. Evaporation will be expected to be an important removal process with a half-life of 1.95 days predicted for evaporation from a model river 1 m deep, flowing at 1 m/sec with a wind velocity of 3 m/sec. No information was found on photolysis, but n-butylamine should not absorb sunlight.

**Atmospheric Fate:** Reaction with hydroxyl radicals will be the fastest chemical removal process for n-butylamine in the atmosphere (estimated half-life of 0.479 days). No information was found on photolysis, but n-butylamine should not absorb sunlight. Dissolution into rain droplets may be the most important physical removal process in the atmosphere.

**Biodegradation:** Percent theoretical BOD/duration (inoculum): 67%/12 days (activated sludge) [16]; 50%/approx 6 days (aniline-acclimated activated sludge) [8]; and 26.5%/5 days, 48.8%/10 days, 50%/15 days, and 52.3%/50 days (settled sewage seed) [1].

**Abiotic Degradation:** Based on the stability of methylamine, n-butylamine will not hydrolyze under normal environmental conditions [15]. The estimated vapor phase half-life in the atmosphere is 0.479 days as a result of reaction with photochemically produced hydroxyl radicals at a concentration of $5 \times 10^5$ radicals/cm$^3$ [2].

**Bioconcentration:** Using the octanol water partition coefficient, an estimated BCF of 3.2 was calculated [7]. Based on this estimated BCF, n-butylamine will not bioconcentrate in aquatic organisms.

**Soil Adsorption/Mobility:** Using the octanol/water partition coefficient, an estimated Koc of 80 was calculated [7]. Based on this estimated Koc, n-butylamine will not adsorb to soils or sediments.

**Volatilization from Water/Soil:** Using the Henry's Law constant, a half-life of 1.95 days was predicted for evaporation from a model river

1 m deep, flowing at 1 m/sec with a wind velocity of 3 m/sec [7].

**Water Concentrations:** SURFACE WATER: Detected in the River Elbe in 1 of 2 sites at 1.5 ppb; not detected at 7 other sites (detection limit not given) [10].

**Effluent Concentrations:** Identified, not quantified in advanced water treatment concentrates [6].

**Sediment/Soil Concentrations:**

**Atmospheric Concentrations:**

**Food Survey Values:** Kale, 7 ppm; Pickles: cucumbers in aromatic vinegar, 0.6 ppm, cucumbers pickled with mustard, 5.3 ppm; Cheese: Tilsiter, 3.7 ppm, Camembert and Limburger, not detected; Brown bread, 1.1 ppm [10]. Not detected in: spinach, red cabbage, cauliflower, white beet, carrot, red beet, large radish, red radish, or celery [10].

**Plant Concentrations:**

**Fish/Seafood Concentrations:** Not detected in herring or cod roe (detection limit not given) [10].

**Animal Concentrations:**

**Milk Concentrations:**

**Other Environmental Concentrations:**

**Probable Routes of Human Exposure:**

**Average Daily Intake:**

**Occupational Exposure:** NIOSH estimates that 413 employees are exposed to n-butylamine based upon a 1981-83 survey [11]. A National Occupational Hazard Survey (NOHS) from 1972-74 estimated that 1252 workers were exposed to n-butylamine [12].

**Body Burdens:**

# n-Butylamine

## REFERENCES

1. Ettinger MB; Ind Eng Chem 48: 256-9 (1956)
2. FAP; Fate of Atmospheric Pollut Off of Toxic Substances USEPA (1986)
3. Graedel TE; Chemical Compounds in the Atmosphere Academic Press, NY p 289 (1978)
4. Hansch C, Leo AJ; Medchem Project Issue No.26 Claremont CA Pomona College (1985)
5. Hine J, Mookerjee PK; J Org Chem 40: 292-8 (1975)
6. Lucas SV; GC/MS (Gas Chromatography-Mass Spectrometry) Analysis of Organics in Drinking Water Concentrates and Advanced Waste Treatment Concentrates Vol 2 Battelle Columbus Labs, OH p 397 USEPA 600/1-84-020B (1984)
7. Lyman WJ et al; Handbook of Chem Property Estimation Methods Environ Behavior of Organic Compounds McGraw-Hill (1982)
8. Malaney GW; J Water Pollut Control Fed 32: 1300 (1960)
9. Merck Index; An Encyclopedia of Chemicals, Drugs and Biologicals 10th ed p 215 (1983)
10. Neurath GB et al; Food Cosmet Toxicol 15: 275-82 (1977)
11. NIOSH; National Occupational Exposure Survey (1981-83) (1985)
12. NIOSH; National Occupational Hazard Survey (1972-74) (1974)
13. Perrin DD; Dissociation Constants of Organic Bases in Aqueous Solution. IUPAC Chemical Data Series. Buttersworth: London (1965)
14. Riddick JA et al; Organic Solvents: Physical Properties and Methods of Purification. Techniques of Chemistry. 4th Ed. New York: Wiley-Interscience pp 1325 (1986)
15. Schmidt-Bleek F et al; Chemosphere 11: 383-415 (1982)
16. Yoshimura K et al; J Amer Oil Chem Soc 57: 238-41 (1980)

# t-Butylamine

## SUBSTANCE IDENTIFICATION

**Synonyms:**

**Structure:**

CH3
H3C — C — NH2
CH3

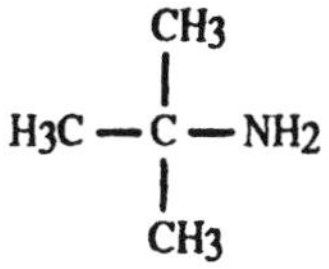

**CAS Registry Number:** 75-64-9

**Molecular Formula:** $C_4H_{11}N$

**Wiswesser Line Notation:** ZX1&1&1

## CHEMICAL AND PHYSICAL PROPERTIES

**Boiling Point:** 44-46 °C

**Melting Point:** -72.65 °C (freezing point)

**Molecular Weight:** 73.14

**Dissociation Constants:** pKa = 10.68 at 25 °C [13]

**Log Octanol/Water Partition Coefficient:** 0.40 [4]

**Water Solubility:** Miscible [14]

**Vapor Pressure:** 362 mm Hg at 25 °C [12]

**Henry's Law Constant:** $1.66 \times 10^{-5}$ atm-m$^3$/mole at 25 °C [6]

## ENVIRONMENTAL FATE/EXPOSURE POTENTIAL

**Summary:** t-Butylamine may be released to the environment in wastewater and air effluents generated at sites of its commercial production or use as an intermediate. If released to soil, t-butylamine

may be susceptible to significant leaching; it has been detected in a leachate from a municipal waste disposal site. t-Butylamine may be expected to evaporate quite rapidly from dry impervious surfaces. Sufficient data are not available to predict the significance of biodegradation in soil or natural water. The results of several screening studies suggest that t-butylamine may be relatively resistant to environmental biodegradation or may undergo significant biodegradation under properly acclimated conditions. If released to water, t-butylamine is not expected to significantly adsorb to sediments or to bioconcentrate. The volatilization half-life from a model river 1 m deep flowing 1 m/sec with a wind velocity of 3 m/sec is estimated to be 1.95 days. The volatilization half-life from a similar river 10 m deep is estimated to be 23.5 days. If released to air, t-butylamine should exist almost entirely in the vapor phase due to its high vapor pressure. The half-life for the vapor-phase reaction with photochemically produced hydroxyl radicals has been estimated to be 0.78 days in a typical atmosphere. Due to the complete water solubility of t-butylamine, physical removal from the atmosphere by washout or by dissolution into clouds with subsequent rainfall may be possible. Occupational exposure by inhalation or dermal routes may occur.

**Natural Sources:** t-Butylamine has been found to occur in Latakia tobacco leaves [7].

**Artificial Sources:** t-Butylamine is used as an intermediate for rubber accelerators, insecticides, fungicides, dyestuffs, and pharmaceuticals [1]. t-Butylamine may be released to the environment in waste effluents (both wastewater and air) generated during the production of these compounds. Similar release may be possible at its commercial production site. It has also been detected in municipal waste dumps at the acidification stage in the Netherlands [5].

**Terrestrial Fate:** When released to soil, t-butylamine may be susceptible to significant leaching due to its complete water solubility and low octanol/water partition coefficient value; it has been detected in a leachate from municipal waste disposal sites. Sufficient data are not available to predict the significance of biodegradation in soil. The results of several screening studies suggest that t-butylamine may be relatively resistant to environmental biodegradation or may undergo significant biodegradation under properly acclimated conditions. t-Butylamine may be expected to evaporate quite rapidly from dry surfaces.

# t-Butylamine

**Aquatic Fate:** t-Butylamine is not expected to significantly adsorb to sediment or to bioconcentrate in aquatic organisms. The volatilization half-life from a model river 1 m deep flowing 1 m/sec with a wind velocity of 3 m/sec is estimated to be 1.95 days. The volatilization half-life from a similar river 10 m deep is estimated to be 23.5 days. Sufficient data are not available to predict the significance of biodegradation in natural water; the results of several screening studies suggest that t-butylamine may be relatively resistant to environmental biodegradation or may undergo significant biodegradation under properly acclimated conditions.

**Atmospheric Fate:** t-Butylamine should exist almost entirely in the vapor phase in the atmosphere due to its relatively high vapor pressure. The half-life for the vapor-phase reaction of t-butylamine with photochemically produced hydroxyl radicals has been estimated to be 0.78 days in a typical atmosphere. Due to the complete water solubility of t-butylamine, physical removal from the atmosphere by dissolution into clouds with subsequent rainfall or by washout may be possible.

**Biodegradation:** No biodegradation of t-butylamine (10-100 ppm) was observed over a 12-day incubation period using Sapromat respiration assays with river mud bacteria inocula, treatment plant sludge inocula, or adapted bacteria inocula [3]. t-Butylamine was bio-oxidized by aniline-acclimated activated sludge using a Warburg respirometer [9]. The bacteria <u>Alcaligenes faecalis</u>, isolated from activated sludge, was not able to biodegrade t-butylamine using the Warburg technique [10].

**Abiotic Degradation:** The half-life for the vapor-phase reaction of t-butylamine with photochemically produced hydroxyl radicals has been estimated to be 0.78 days at 25 °C assuming an average atmospheric hydroxyl radical concn of $5 \times 10^{+5}$ molecules/cm$^3$ [2].

**Bioconcentration:** Based on the log Kow, the BCF value for t-butylamine can be estimated to be 1.2 from a recommended regression-derived equation [8]. The complete water miscibility of t-butylamine also suggests that t-butylamine will not bioconcentrate significantly.

**Soil Adsorption/Mobility:** Based on the log Kow and a complete solubility in water, t-butylamine is not expected to adsorb significantly to soil or sediment; therefore, significant leaching may be possible.

# t-Butylamine

**Volatilization from Water/Soil:** The value of the Henry's Law constant suggests that volatilization may be significant from shallow rivers [8]. The volatilization half-life from a model river 1 m deep flowing 1 m/sec with a wind velocity of 3 m/sec is estimated to be 1.95 days; the volatilization half-life from a similar model river 10 m deep is estimated to be 23.5 days [8]. Based on the vapor pressure, t-butylamine may be expected to evaporate quite rapidly from dry impervious surfaces.

**Water Concentrations:**

**Effluent Concentrations:** t-Butylamine was detected in the leachate from a municipal refuse waste disposal site in the Netherlands at a concn of 41 ppm [5].

**Sediment/Soil Concentrations:**

**Atmospheric Concentrations:**

**Food Survey Values:**

**Plant Concentrations:**

**Fish/Seafood Concentrations:**

**Animal Concentrations:**

**Milk Concentrations:**

**Other Environmental Concentrations:**

**Probable Routes of Human Exposure:** Occupational exposure to t-butylamine may occur through inhalation of contaminated air or by dermal contact.

**Average Daily Intake:**

**Occupational Exposure:** NIOSH (NOES Survey 1981-1983) has statistically estimated that 4354 workers are exposed to t-butylamine in the United States [11].

**Body Burdens:**

# t-Butylamine

## REFERENCES

1. American Chemical Society; Chemcyclopedia 86: 60 (1985)
2. Atkinson R; Internat J Chem Kinetics 19: 799-828 (1987)
3. Calamari D; Chemosphere 9: 753 (1980)
4. Hansch C, Leo AJ; Medchem Project Issue No.26 Pomona College, Claremont CA (1985)
5. Harmsen J; Water Res 17: 699-706 (1983)
6. Hine J, Mookerjee PK; J Org Chem 40: 292 (1975)
7. Irvine WJ, Saxby MJ; Phytochem 8: 473 (1979)
8. Lyman WJ et al; Handbook of Chemical Property Estimation Methods. Environmental Behavior of Organic Compounds. McGraw-Hill NY (1982)
9. Malaney GW; J Water Pollut Control Fed 35: 1300 (1960)
10. Marion CV, Malaney GW; J Water Pollut Control Fed 35: 1269 (1963)
11. NIOSH; National Occupational Exposure Survey (1983)
12. Parrish CF; Kirk-Othmer Encycl Chem Technol 3rd ed. Wiley NY 21: 381 (1983)
13. Perrin DD; Dissociation Constants of Organic Bases in Aqueous Solution. IUPAC Chemical Data Series. Buttersworth: London (1965)
14. Riddick JA et al; Organic Solvents: Physical Properties and Methods of Purification. Techniques of Chemistry. 4th Ed. New York: Wiley-Interscience pp 1325 (1986)

# Carbon Disulfide

## SUBSTANCE IDENTIFICATION

**Synonyms:** Dithiocarbonic anhydride

**Structure:**

$$S = C = S$$

**CAS Registry Number:** 75-15-0

**Molecular Formula:** $CS_2$

**Wiswesser Line Notation:** CS2

## CHEMICAL AND PHYSICAL PROPERTIES

**Boiling Point:** 46.5 °C at 760 mm Hg

**Melting Point:** -111.5 °C

**Molecular Weight:** 76.13

**Dissociation Constants:**

**Log Octanol/Water Partition Coefficient:** 1.70-4.16 [13]

**Water Solubility:** 2,100 mg/L at 20 °C [31]

**Vapor Pressure:** 297 mm Hg at 20 °C [37]

**Henry's Law Constant:** 0.0014 atm-m³/mole (calculated from the water solubility and vapor pressure)

## ENVIRONMENTAL FATE/EXPOSURE POTENTIAL

**Summary:** Carbon disulfide is a natural product of anaerobic biodegradation and is released to the atmosphere from oceans and land masses. Geothermal sources also contribute to carbon disulfide emissions. It also may be released as emissions and in wastewater during its production and use, in the production of viscose rayon, cellophane, and carbon tetrachloride, and as a solvent and fumigant. If released on land, carbon disulfide will be primarily lost by volatilization. It may also readily leach into the ground where it may biodegrade. If released into water, carbon disulfide will be primarily lost due to volatilization (half-life 2.6 hr in a model river). Adsorption to sediment and bioconcentration in fish should not be significant. In the atmosphere, carbon disulfide degrades by reacting with atomic oxygen and photochemically produced hydroxyl radicals (half-life 9 days). The soil may be a natural sink for the chemical by adsorbing and subsequently biodegrading it. Exposure to carbon disulfide is mostly occupational and primarily by inhalation. Only workers in the viscose rayon industry are exposed to high concn. The general population may be exposed to carbon disulfide from ambient air as well as food items containing grain that has been fumigated with the chemical.

**Natural Sources:** The ocean appears to be a major global source of carbon disulfide [6,20,22]. Current data suggests that coastal areas and other areas of high biological productivity have greater fluxes of carbon disulfide than the open ocean [20]. Emissions from the oceans have been estimated to be $6 \times 10^{+11}$ g/yr [20]. The microbial reduction of sulfates in soil produces fluxes of carbon disulfide. The annual global emission from this source has been estimated to be $9 \times 10^{+11}$ g [20]. Other natural sources include volcanic emissions, estimated to be $2 \times 10^{+10}$ g/yr, and marshlands, estimated emissions $1 \times 10^{+11}$ g/yr [20]. Fluxes of carbon disulfide from a salt marsh were measured as 0.2 and 1.13 g sulfur/sq m-yr and inland soil 0.001 g sulfur/sq m-yr [2]. While the emission rate from salt marshes is quite high, the overall contribution of inland soil is much higher [2]. Laboratory experiments also show that tidal marsh soil emits carbon disulfide and that the amount volatilized is greatest when the soil moisture is at field capacity rather than in the completely anaerobic, saturated state [9]. Carbon disulfide was only emitted briefly from a normally aerobic loam soil and only in the saturated state [9]. A less organic forest silty loam soil did not emit any carbon disulfide [9].

**Artificial Sources:** Carbon disulfide may be released to the environment in emissions and wastewater during its production and use in the manufacture of viscose rayon, carbon tetrachloride, cellophane, and other regenerated celluloses and rubber chemicals and as a solvent [7,37]. For every kilogram of viscose used, 20-30 g of carbon disulfide is emitted [35]. The emissions are highest in the production of stable fiber and cellophane where quantities as high as 3000 and 60,000 g may be emitted per hour [35]. It may also be formed and released during sewage treatment [1] and from landfills containing municipal refuse and wastewater sludges [40]. Carbon disulfide is also used for insect control in stored grain and for soil fumigation to control soil fungi and deep-rooted perennial weeds [7]. Volatile emissions can be expected to result from these applications. Small quantities of carbon disulfide are also found in coal tar [24]. Annual inputs of carbon disulfide from 5 US manufacturers of regenerated cellulose rayon and cellophane is estimated to be 14 x $10^{+10}$ g [20]. Total global input of anthropogenic carbon disulfide is estimated to be 36 x $10^{+10}$ g/yr and arises principally from the chemical industry and sulfur recovery processes [20].

**Terrestrial Fate:** If released on land, carbon disulfide will be primarily lost by volatilization. Since it has a low adsorptivity to soil, it should also readily leach into the ground where there is some evidence that it may biodegrade.

**Aquatic Fate:** If released into water, carbon disulfide will be primarily lost due to volatilization (half-life 2.6 hr in a model river). Adsorption to sediment should not be significant.

**Atmospheric Fate:** In the atmosphere, carbon disulfide reacts with atomic oxygen and photochemically produced hydroxyl radicals with a half-life of around 9 days. The action of soil in adsorbing and degrading gaseous carbon disulfide demonstrates that soil may be a natural sink for the chemical [4].

**Biodegradation:** Carbon disulfide has been classified as difficult to degrade, indicating the need for prolonged biological treatment, and has a suggested persistence in unadapted soil of 3 mo to a yr [1]. No data were presented supporting this classification. It has been demonstrated that the adsorption of carbon disulfide by moist unsterilized soil increases sharply after approximately 3 hr and the time for complete sorption of the gas decreases with repeated dosing [4]. This behavior does not occur with air-dried or sterilized soil and has been ascribed to microbial utilization of the chemical [4].

# Carbon Disulfide

**Abiotic Degradation:** Carbon disulfide has a weak UV adsorption band at 317 nm [32]; however, photolysis is not considered to be a significant loss mechanism for the chemical [20]. In the presence of oxygen, carbon monoxide, carbonyl sulfide, sulfur dioxide, and a polymer are formed during irradiation [14]. Observed temporal variability and vertical gradients suggest that the tropospheric lifetime of carbon disulfide is quite short [6]. If the atmospheric concn of carbon disulfide at 6.1 km altitude is typical for marine boundary layer levels, the sharp decrease in concn at higher altitudes supports the concept of a photochemical lifetime of a month or less in the troposphere [6]. The troposphere half-life determined from estimates of sources and global burdens of the chemical is 8.9 days [20]. Carbon disulfide reacts with photochemically produced hydroxyl radicals and atomic oxygen in the atmosphere. The half-lives for these processes are each 9 days [3,12], assuming the atmospheric concentrations of hydroxyl radicals and atomic oxygen as $5 \times 10^{+5}$ and $2.5 \times 10^{+5}$ radicals/cm$^3$, respectively [10]. The product of this oxidation is carbonyl sulfide [20]. Carbon disulfide hydrolysis to carbon dioxide and hydrogen disulfide in alkaline solutions [30]. The half-life for hydrolysis at pH 9 extrapolated from measurements at higher pH is 1.1 yr [30]. It is stable in oxygenated seawater for >10 days [22].

**Bioconcentration:** The BCF for carbon disulfide calculated from its water solubility by regression analysis is 7.9 [19]. Carbon disulfide would therefore not be expected to bioconcentrate significantly in aquatic organisms.

**Soil Adsorption/Mobility:** The Koc for carbon disulfide calculated from its water solubility by regression is 63 [19]. Carbon disulfide in solution would therefore not be expected to adsorb significantly to soil. The average adsorption of carbon disulfide after 10 minutes by 4 air-dried soils was 46% but only 12% by the same soils at 50% water-holding capacity [4]. However after 8 hr, the rate of adsorption was greater by moist soil but only when the soil was unsterilized [4]. Further experiments suggest that this "adsorption" in moist soils is the result of microbial action [4].

**Volatilization from Water/Soil:** Using the Henry's Law constant, one can estimate the volatilization half-life of carbon disulfide from a model river 1 m deep with a 1 m/sec current and a 3 m/sec wind to be 2.6 hr [23]. Due to its high vapor pressure and low adsorption to soil, carbon disulfide would be expected to volatilize readily from soil.

**Water Concentrations:** DRINKING WATER: Drinking water samples from nine cities in the US and one rural well contained no carbon disulfide (no detection limit stated) [16]. It was detected, but no quantified in New Orleans [39] as well as Miami and Cincinnati drinking water [38]. SURFACE WATER: The mean concn of carbon disulfide in the open waters of the Atlantic Ocean and the Atlantic Ocean of Ireland are 0.52 and 0.78 ppt, respectively [22]. The mean concn in stagnant bay water was 5.4 ppt [22]. Water samples from 82 stations in Lake Ontario and 17 in the lower Niagara River were analyzed for volatile organics [18]. Two river samples contained 25 ppt of carbon disulfide while the other station contained <20 ppt, the detection limit [18]. Eleven lake samples contained quantifiable amounts of carbon disulfide whose median and max concn were 400 and 3900 ppt, respectively [18]. Half of the other samples contained trace quantities of the chemical and the other stations contained <80 ppt, the detection limit [18]. Carbon disulfide was prominent in Toronto Harbour, with lower levels in Hamilton Harbor and Oak Orchard Creek [18]. In another study, carbon disulfide was detected but not quantified in the central basin of Lake Erie, the Niagara River, and open waters of Lake Ontario, but was absent from the western basin of Lake Ontario [11].

**Effluent Concentrations:** In a comprehensive survey of wastewater from 4000 industrial and publicly owned treatment works (POTWs) sponsored by the Effluent Guidelines Division of the US EPA, carbon disulfide was identified in discharges of the following industrial category (frequency of occurrence; median concn in ppb): leather tanning (1; 7.5), paint and ink (4; 1078.6), organics and plastics (30; 1654.3), plastics and synthetics (4; 7075.4), pulp and paper (2; 215.6), pesticides manufacture (1; 88.8), and publicly owned treatment works (11; 45.8) [34]. The highest effluent concn was 18,943 ppb in the plastics and synthetics industry [34]. In a survey of 63 industrial wastewater effluents, carbon disulfide was identified in 8 samples, 6 of which were <10 ppb and 2 between 10 and 100 ppb [29]. The concn of carbon disulfide in offgas from two oil shale retorting processes were 24 ppm and 13 ppm [36]. Carbon disulfide was found in both the influent and effluent of a large community septic tank [8]. The combined concentration of carbon disulfide and dichloromethane in the effluent, which was 10 ppb, was much higher than that in the influent and reflected the presence of anaerobic processes in the sewer line or septic tank [8].

**Sediment/Soil Concentrations:** The mean concn of carbon disulfide in 5 samples of mud from the sea bottom was 29.5 ppt [22].

**Atmospheric Concentrations:** RURAL/REMOTE: Levels of carbon disulfide in 61 air samples taken over the course of 2 days at Wallops Island, VA ranged from 29-84 ppt, with a medium of 41 ppt [6]. At Harwell, England concns ranged from 70-370 ppt, with an average of 190 ppt [33]. Concns of carbon disulfide varied with altitude, measuring: 115 ppt mean at 6.1 km altitude, 23 ppt mean at 7.3 km altitude, and 26 ppt at 7.3-7.9 km altitude [6]. The variation with altitude has been ascribed to the updrafting of boundary layer air by cumulus and cumulonimbus activity to the upper troposphere [6]. URBAN/SUBURBAN: The mean, minimum, and maximum carbon disulfide concn in Philadelphia measured over a 40 day period (88 samples) were 65, 25, and 340 ppt [5]. It was detected, but not quantified in air in Leningrad, USSR [17].

**Food Survey Values:** Seven samples of lima beans, five of common beans, two of lentils, and one sample of mung beans, soybeans, and split peas that were analyzed contained a mean carbon disulfide concn of 2.3 ppb [21]. The range was 1.8-3.1 ppb [21]. Some of these samples were obtained from health stores and one was from a home garden where it was grown without the use of herbicides or insecticides [21]. Stored grain is often fumigated with carbon disulfide. The carbon disulfide concn in 9 samples of wheat range from 64 to 7500 ppb although none was found in samples of corn, oats, corn meal, and corn grits [15]. While one sample of bleached flour contained 23 ppb. Other samples tested including corn muffin mix, cake mixes, dried lima beans, noodles, and rice, were free of the fumigant [15]. With the exception of granola, which contained 11 ppb of carbon disulfide, none of the 18 samples of table-ready food items tested contained carbon disulfide [16]. The items tested contained representative samples of most classes of food [16]. In addition, samples of butter and margarine, cheese, peanut butter, and highly processed foods from the US FDA's Market Basket Survey contained no carbon disulfide [16]. However some ready-to-eat food products, specifically two samples of corn chips and oat ring cereal, contained 80, 230, and 95 ppb of carbon disulfide, respectively [16].

**Plant Concentrations:**

Carbon Disulfide

**Fish/Seafood Concentrations:**

**Animal Concentrations:**

**Milk Concentrations:**

**Other Environmental Concentrations:**

**Probable Routes of Human Exposure:** Exposure to carbon disulfide is mostly occupational via inhalation and dermal contact with the vapor or dermal contact with the liquid. Inhalation is the principal route of absorption [41]. While workers engaged in any process using carbon disulfide may be exposed to some degree, in practice only workers in the viscose rayon industry are exposed to high concn [41]. The general population may be exposed to carbon disulfide from ambient air as well as food items that contain grain that has been fumigated with the chemical.

**Average Daily Intake:** AIR INTAKE (assume mean concn of 41-65 ppt): 2.6-4.1 ug; WATER INTAKE: insufficient data; FOOD INTAKE: insufficient data.

**Occupational Exposures:** 728,783 workers are potentially exposed to carbon disulfide based on statistical estimates derived from the NIOSH Survey conducted 1972-74 in the United States [26]. 44,441 workers are potentially exposed to carbon disulfide based on statistical estimates derived from the NIOSH Survey conducted 1981-83 in the United States [25]. Workplace concn of carbon disulfide in the viscose rayon industry ranged from <3 ppm to peaks exceeding 2,000 ppm [27]. 12 of 36 air samples of carbon disulfide in the breathing zone of workers in the spinning and cutting rooms of a viscose rayon plant contained >20 ppm 8-hr TWA, the OSHA standard, and 7 samples exceeded 100 ppm, the acceptable max peak concn [27]. In a further study, 10-20 min breathing zone concn of 8 cutters ranged from <20 to >2,000 ppm and exceeded 100 ppm in more than half of the 196 samples taken [27]. The concns for 6 workers in the spinning area were far lower. The overall TWA concn was 11.2 ppm with a range of 0.9-127 ppm [27]. Shift TWA concns for the cutters ranged from 9.5-129 ppm for the cutters and 4.3-11.1 ppm for the spinners [27]. Seven of the eight cutters were exposed to concn higher than the OSHA standard for an 8-hr day [27]. Half of the general room samples exceed the 30-ppm ceiling limit [27]. Extensive long term monitoring in a Finnish viscose rayon plant showed that concn levels of carbon disulfide that generally exceeded 40 ppm before 1950 have been dropping and had fallen below 5 ppm by 1972 [27]. Environmental concn in a US viscose rayon plant were reported to be between 10-15 ppm [27].

**Body Burdens:** All 8 samples of human milk from 4 urban areas of the US contained carbon disulfide [28].

## REFERENCES

1. Abrams EF et al; Identification of Organic Compounds in Effluents from Industrial Sources USEPA-560-3-75-002 (1975)
2. Aneja VP et al; J Air Pollut Control Assoc 32: 803-7 (1982)
3. Baulch DL et al; J Phys Chem Ref Data 13: 1259-1380 (1984)
4. Bremner JM, Banwart WL; Soil Biol Biochem 8: 79-83 (1976)
5. Brodzinsky R, Singh HB; Volatile Organic Chemicals In The Atmosphere: An Assessment Of Available Data Menlo Park, CA: Atmospheric Science Center SRI International Contract 68-02-3452 198 pp. (1982)
6. Carroll MA; J Geophys Res 90: 10483-6 (1985)
7. Chemical Profiles; Chemical Marketing Reporter Jan 13 (1986)
8. Dewalle FB et al; Determination Of Toxic Chemicals In Effluent From Household Septic Tanks USEPA-600/S2-75-050 p. 4 (1985)
9. Farwell SO et al; Soil Biol Biochem 11: 411-5 (1979)
10. Graedel TE; Chemical Compounds in the Atmosphere Academic Press NY (1978)
11. Great Lakes Water Quality Board; An Inventory Of Chemical Substances Identified In The Great Lakes Ecosystem Volume 1 Windsor Ontario, Canada p. 195 (1983)
12. Hampson RF; Chemical, kinetic and photochemical data sheets for atmospheric reactions FAA-EE-80-17 (1980)
13. Hansch C, Leo AJ; Medchem Project Issue No 26. Claremont CA: Pomona College (1985)
14. Heicken J Et al; pp. 191-222 in Proc Symp 1969 (1971)
15. Heikes DL, Hopper ML; J Assoc Off Anal Chem 69: 990-8 (1986)
16. Heikes DL; J Assoc Off Anal Chem 70: 215-77 (1987)
17. Ioffe BV et al; J Chromatogr 142: 787-95 (1977)
18. Kaiser KLE et al; J Great Lakes Res 9: 212-23 (1983)
19. Kenaga EE; Ecotoxicol Environ Safety 4: 26-38 (1980)
20. Khalil MAK, Rasmusseen RA; Atmos Environ 18: 1805-31 (1984)
21. Lovegren NV et al; J Agric Food Chem 27: 851-3 (1979)
22. Lovelock JE; Nature 248: 625-6 (1974)
23. Lyman WJ et al; Handbook of Chem Property Estimation Methods Environ Behavior of Organic Compounds McGraw-Hill NY pp. 15-1 to 15-34 (1982)
24. Merck Index; 10th ed p. 251 (1983)
25. NIOSH; National Occupational Exposure Survey (1985)
26. NIOSH; National Occupational Health Survey (1975)
27. NIOSH; Criteria for a recommended standard...occupational exposure to carbon disulfide DHEW (NIOSH) Publ No 77-156 (1977)
28. Pellizzari ED et al; Bull Environ Contam Toxicol 28: 322-8 (1982)

29. Perry DL et al; Identification Of Organic Compound In Industrial Effluent Discharges USEPA-600/4-79-016 NTIS PB-294784 p 230 (1979)
30. Peyton TO et al; Carbon Disulfide, Carbonyl Sulfide Literature Review And Environmental Assessment USEPA-600/9-78-009 p. 163 (1976)
31. Riddick JA et al; Organic Solvents: Physical Properties and Methods of Purification, 4th Edit. New York: J Wiley & Sons (1986)
32. Sadlter Index; Philadelphia, PA: Sadlter Research Lab
33. Sandalls FJ, Penkett SA; Atmos Environ 11: 197-9 (1977)
34. Shackelford WM et al; Analyt Chim Acta 146: 15-27 (1983)
35. Sine C; p. C52 in Farm Chemicals Handbook Meister Publishing Co:Willoughby OH (1987)
36. Sklarew DS et al; Environ Sci Technol 18: 592-600 (1984)
37. Timmerman R; Kirk-Othmer Encyclopedia of Chemical Technology 3rd ed 4: 742-59 (1978)
38. USEPA; Preliminary Assessment of Suspected Carcinogens in Drinking Water Interim Report to Congress, June, 1975 Washington, DC (1975)
39. USEPA; New Orleans Area Water Supply Study Draft Analytical Report by the Lower Mississippi River Facility, Slidell, LA, Dallas, Texas: USEPA (1974)
40. Vogt WG, Walsh JJ; in Proc-APCA Annu Meet 78th(vol 6) (1985)
41. World Health Organization; Environmental Health Criteria 10: Carbon Disulfide Geneva (1979)

# Carbon Tetrachloride

## SUBSTANCE IDENTIFICATION

**Synonyms:**

**Structure:**

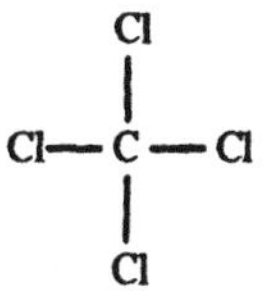

**CAS Registry Number:** 56-23-5

**Molecular Formula:** CCl₄

**Wiswesser Line Notation:** GXGGG

## CHEMICAL AND PHYSICAL PROPERTIES

**Boiling Point:** 76.54 °C

**Melting Point:** -23 °C

**Molecular Weight:** 153.84

**Dissociation Constants:**

**Log Octanol/Water Partition Coefficient:** 2.83 [24]

**Water Solubility:** 805 mg/L at 20 °C [26]

**Vapor Pressure:** 113.8 mm Hg at 25 °C [14]

**Henry's Law Constant:** 3.04 x 10⁻² atm-m³/mole at 24.8 °C [21]

## ENVIRONMENTAL FATE/EXPOSURE POTENTIAL

**Summary:** Large quantities of carbon tetrachloride are produced each year (586 million pounds); most of it is used for chemical synthesis of fluorocarbons. However, 5% (30 million pounds) is used as a degreaser, fire extinguisher, or grain fumigant. Its release to the environment occurs from these minor uses as it has a reasonably high vapor pressure at ambient temperatures. In the troposphere, carbon tetrachloride is extremely stable (residence time of 30-50 years). The primary loss process is by escape to the stratosphere where it photolyzes. As a result of its emission into the atmosphere and slow degradation, the amount of carbon tetrachloride in the atmosphere has been increasing. Some carbon tetrachloride released to the atmosphere is expected to partition into the ocean. In water systems, evaporation appears to be the most important removal process, although biodegradation may occur under aerobic and anaerobic conditions (limited data). Releases or spills on soil should result in rapid evaporation due to high vapor pressure and leaching in soil resulting in ground water contamination due to its low adsorption to soil. Bioconcentration is not significant.

**Natural Sources:** No natural sources are known [61] but ambient levels may not be totally explained by anthropogenic sources [35].

**Artificial Sources:** Wastewater from iron and steel manufacturing, foundries, metal finishing, paint and ink formulations, petroleum refining, and nonferrous metal manufacturing industries contain carbon tetrachloride [70]. Its use as a solvent is a major contributor to atmospheric concentrations [27,61].

**Terrestrial Fate:** Carbon tetrachloride is slightly removed during infiltration of river water into adjacent wells [74]. However, carbon tetrachloride is expected to evaporate rapidly from soil due to its high vapor pressure and migrate into ground water due to its low soil adsorption coefficient. No data are available on biodegradation in soil.

**Aquatic Fate:** Evaporation from water is a significant removal process. Based upon field monitoring data, the estimated half-life in rivers is 3-30 days; in lakes and ground water, 3-300 days [74]. Biodegradation may be important under aerobic or anaerobic conditions, but the data is limited. Adsorption to sediment should not be an important process.

**Atmospheric Fate:** Carbon tetrachloride is very stable in the troposphere with residence times of 30-50 years. Its main loss mechanism is diffusion to the stratosphere where it photolyzes. It is estimated that <1% of the carbon tetrachloride released to the air is partitioned into the oceans [20].

**Biodegradation:** Biodegradation in screening tests has been noted [68], but acclimation may be necessary [25]. Degradation does occur in 16 days under anaerobic conditions [3].

**Abiotic Degradation:** Hydrolysis half-life in water is 7000 years at 25 °C [37]. Direct photolysis is not important in the troposphere, but irradiation at higher energies (195-254 nm) such as found in the stratosphere results in degradation [15,42]. Carbon tetrachloride is stable in the troposphere with residence time of 30-50 years [42]. It does not react significantly with any active species in the atmosphere [20]. The half-life for reaction with hydroxyl radicals is >330 years [13].

**Bioconcentration:** Carbon tetrachloride has a low potential to bioconcentrate [44]. Log of the bioconcentration factor in trout is 1.24 [44,72], in bluegill sunfish - 1.48 [2].

**Soil Adsorption/Mobility:** Carbon tetrachloride is expected to be moderately mobile in soil [29] and only slightly adsorbed to sediment. Calculated Koc of 110 (assumed water solubility of 800 ppm) was reported [29]. Estimated retardation factor in breakthrough sampling in ground water is 1.44 [39].

**Volatilization from Water/Soil:** The high vapor pressure suggests rapid evaporation from soil. The Henry's Law value suggests that carbon tetrachloride would volatilize rapidly from water and moist soil surfaces. Based on this value of Henry's Law constant the volatilization half-life from a model river 1 m deep flowing 1 m/sec with a wind speed of 3 m/sec has been estimated to be 3.7 hours [36]. Measured half-lives of evaporation from water have varied from minutes to hours [10,17,36,38,58,64].

**Water Concentrations:** MARINE WATERS: at the surface - 0.12-0.85 ppt, at 300 m depth - 0.15 ppt [43,61,62]. SURFACE WATER: 0-9 ppb [12,18,19,23,49,50,51]. 14 heavily industrialized rivers 1-3 ppb, 6 of 204 samples pos [48]. Great Lakes - 9-47 ppt [28,30]. US EPA STORET data base - 8,858 water samples, 12% pos., median concn 0.10 ug/L [65]. GROUND WATER: 3-20 ppb - raw water for 27 US cities [12], 5 ppb Netherlands [74]. As of June 1984, analyzed but not found in 1174 community wells and 617 private wells in Wisconsin [31]. DRINKING WATER: 0.1-30 ppb in 181 US cities - surface water source [12], 0.2-13 ppb in 39 US cities - ground water source

[12], 0-190 ppt in 9 homes - Love Canal [1], 135-400 ppb wells in NJ and NY [8], 0-4 ppb - 80 US cities [67]. RAIN: La Jolla, CA 2.8 ppt, Industrial area in England 300 ppt [66]. SNOW: Southern and Central California .33-.36 ppt, Alaska 2.2 ppt [66].

**Effluent Concentrations:** Industries with mean concentrations >90 ppb - non-ferrous metals manufacturing, paint and ink formulation, rubber processing: mean range 90-700, max range 1700-1800 [70]. US EPA STORET data base - 1,343 effluent samples, 5.5% pos, median concn 5.0 ug/L [65].

**Sediment/Soil Concentrations:** US EPA STORET data base - 361 sediment samples, 0.8% pos, median concn <5.0 mg/kg dry wt basis [65].

**Atmospheric Concentrations:** NORTHERN HEMISPHERE: 110.9-142.3 ppt. SOUTHERN HEMISPHERE: 68.9-125.4 ppt [13,57,60,61]. MARINE REMOTE: Pacific - 4.1-70 ppt and Atlantic - 0.2-71.2 ppt [35,43,71]. Atlantic 1984-1985, 140 ppt [11]. URBAN/SUBURBAN: 0.19 ppb mean 1747 samples [6], 0-42.4 ppb [4,5,6,33,34,54,57,59,63,66]. RURAL/REMOTE: 0.13 ppb mean 728 samples [6], 0.082- 0.24 ppb [6,52,53,57,62,66]. INDUSTRIALIZED AND SOURCE DOMINATED AREAS: 0.59 ppb mean 285 samples [6], 0-70 ppb [1,22,56]. The amount of carbon tetrachloride has been increasing in the atmosphere [20]. US EPA TEAM Study: New Jersey, outdoor air: 86 samples Fall 1981, weighted median 0.87 ug/m³; 60 samples Summer 1982, weighted median 0.68 ug/m³ [73]. US EPA TEAM Study - New Jersey, personal air: 344 samples Fall 1981, weighted median 1.5 ug/m³; 148 samples Summer 1982, weighted median 0.85 ug/m³ [73].

**Food Survey Values:** A range of 3-18 ng/g in fats, fruits and vegetables, meat, tea, and bread with oils and fats being at the high end of range (16-18 ng/g) [40]. Possible residues in grain products - 50 ppm (raw cereals), 10 ppm (milled cereals), and 0.05 ppm (cooked cereal products) [27]. 7 samples of grains had residue ranges of 2.9-20.1 ppm [41].

**Plant Concentrations:**

**Fish/Seafood Concentrations:** Mollusks - 2-114 ppb, fish - 3-209 ppb, with medians of 11 and 19 ppb, respectively [16]. US EPA STORET data base - 97 biota samples, 0% pos., detection limit 0.05 mg/kg wet weight basis [65].

**Animal Concentrations:**

**Milk Concentrations:** 3 ppm milk from cows treated with veterinary medication containing carbon tetrachloride [27].

**Other Environmental Concentrations:**

**Probable Routes of Human Exposure:** There are 3 primary routes of exposure - water and other fluids, inhalation, and ingestion of foodstuffs [69]. Significant amounts (30 million pounds) each year are used for degreasing products, fire extinguishers, grain fumigants, etc [9]. Some humans may receive high exposure from these minor uses. Ambient air levels are low (ppt) with urban/suburban and rural values being similar. Drinking water levels are also generally low (ppb) unless contaminated. Carbon tetrachloride has also been found in some veterinary medicines [27].

**Average Daily Intake:** AIR INTAKE: (assume 0.1-4 ppb) 12-511 ug; WATER INTAKE: (assume range of 0.1-30 ppb) - 0.2-60 ug; FOOD: insufficient data.

**Occupational Exposures:** Inhalation of high concentrations is largely restricted to occupational environments. It has been estimated that 160,000 workers have potential exposure to carbon tetrachloride [45]. NIOSH (NOES Survey 1981-83) has statistically estimated that 92,143 workers are potentially exposed to carbon tetrachloride in the United States [46]. NIOSH (NOHS Survey 1972-74) has statistically estimated that 1,380,232 workers are potentially exposed to carbon tetrachloride in the United States [47]. In breathing zone air of workers involved in the production of chlorinated rubbers for road paint (in Italy), mean concn was 3.5-6.9 mg/m$^3$ [7].

**Body Burdens:** Detected but not quantified in 5 of 6 samples of mother's milk in 4 urban sites: Pennsylvania - 1, New Jersey - 2, Louisiana - 1 [55]. In blood from workers exposed to carbon tetrachloride during production of chlorinated rubbers for road paint (in Italy), mean concn was 3.3-6.5 ug/L [7]. Detected in expired air of carefully selected normal, healthy human subjects (non-smokers), 387 samples from 54 subjects, 29.7% samples pos, mean concn 1.4

ng/L [32]. US EPA TEAM Study: New Jersey, 322 breath samples Fall 1981, weighted median 0.69 ug/m³, 148 breath samples Summer 1982 - 0.17 ug/m³ [73].

# REFERENCES

1. Barkley J et al; Biomed Mass Spectrom 7:139-47 (1980)
2. Barrows ME et al; Dyn Exp Hazard Assess Toxic Chem Ann Arbor MI: Ann Arbor Science p 379-92 (1980)
3. Bower EJ, McCarty PL; Appl Environ Microbiol 45:1286-94 (1983)
4. Bozzelli JW, Kebbekus BB; J Environ Sci Health 17:693-713 (1982)
5. Bozzelli JW, Kebbekus BB; Final report New Jersey Inst Technol 80 pp (1979)
6. Brodzinski R, Singh HB; Volatile organics in the atmosphere: an assessment of available data SRI Inter Contract 68-02-3452 p 13 (1982)
7. Brugnone F et al; pp. 575-8 in Developments in the Science and Practice of Toxicology; Hayes AW et al eds: Elsevier Science (1983)
8. Burmaster DE; Environ 24:6-13,33-6 (1982)
9. Chemical Marketing Reporter February 21 Chemical Profile (1983)
10. Chiou CT et al; Environ Inter 3:231-6 (1980)
11. Class T, Ballschmitter K; Chemosphere 15: 413-27 (1986)
12. Coniglio WA et al; Occurrence of volatile organics in drinking water (1980)
13. Cox RA et al; Atmos Environ 10:305-8 (1976)
14. Daubert TE, Danner RP; Data Compilation Tables of Properties of Pure Compounds. American Institute of Chemical Engineers, New York pp 450 (1985)
15. Davis DD et al; J Phys Chem 79:11-7 (1975)
16. Dickson AG, Riley JP; Mar Pollut Bull 7:167-9 (1976)
17. Dilling WL; Environ Sci Technol 11:405-9 (1977)
18. Dreisch FA et al; Survey of Huntington and Philadelphia River water supplies for purgable organic contaminates. p10-11 EPA-903/9- 81-003 (1980)
19. Ewing BB et al; Monitoring to detect previously unrecognized pollutants in surface waters. p 72 EPA-560/6-77-015 (1977)
20. Galbally IE; Science 193:573-6 (1976)
21. Gossett JM, Environ Sci Tech 21: 202-8 (1987)
22. Grimsrud EP, Rasmussen RA; Atmos Environ 9:1014-7 (1975)
23. Haberer K, Normann S; Gas-wasserfach: Wasser/Abwasser 120:302-7 (1979)
24. Hansch C, Leo AJ; Medchem Project, Claremont, CA: Pomona College (1985)
25. Heukelekian H, Rand MC; J Water Pollut Control Assoc 29: 1040-55 (1955)
26. Horvath AL; Halogenated Hydrocarbons: Solubility-Miscibility with Water. NY: Marcel Dekker pp839 (1982)
27. IARC Monographs on the evaluation of carcinogenic risk of chemicals to man 1:53-60 (1972)
28. Kaiser KLE, Valdmanis I; J Great Lakes Res 5:106-9 (1979)
29. Kenaga EE; Ecotox Environ Saf 4:26-38 (1980)
30. Konasewich D et al; Great Lakes Quality Review Board Report (1978)
31. Krill RM, Sonzogni WC; J Am Water Works Assoc 78: 70-5 (1986)
32. Krotoszynski BW et al; J Anal Tox 3: 225-34 (1979)
33. Lillian D et al; Environ Sci Technol 9: 1042-8 (1975)
34. Lillian D et al; Amer Chem Soc Symp Ser 17:152-8 (1975)
35. Lovelock JE et al; Nature 241:194-6 (1973)

36. Lyman WJ et al; Handbook of Chemical Property Estimation Methods. Environmental Behavior of Organic Compounds, New York NY: Mcgraw-Hill (1982)
37. Mabey W, Mill T; J Phys Chem Ref Data 7:383-415 (1978)
38. Mackay D, Yeun ATK Environ Sci Technol 17:211-7 (1983)
39. Mackay DM et al; Amer Chem Soc 186th Natl Mtg Preprint Div Environ Chem 23:368-71 (1983)
40. McConnell G et al; Endeavor 34:13-8 (1975)
41. McMahon BM; J Assoc Off Anal Chem 54:964-5 (1971)
42. Molina MJ, Rowland FS; Geophys Res Lett 1:309-12 (1974)
43. Murray AJ, Riley JP; Nature 242:37-8 (1973)
44. Neely WB et al; Environ Sci Technol 8:1113-5 (1974)
45. NIOSH; Criteria for a recommended standard. Occupational exposure to carbon tetrachloride NTIS PB-250-424 p 16 (1975)
46. NIOSH; National Occupational Exposure Survey (NOES) (1983)
47. NIOSH; National Occupational Hazard Survey (NOHS) (1974)
48. Ohio R Valley Water Sanit Comm; 1980-81 Main stream Assessment (1982)
49. Ohio R Valley Water Sanit Comm; 1978-9 Mainstream Assessment (1980)
50. Ohio R Valley Water Sanit Comm; 1977 Mainstream Assessment (1978)
51. Ohio R Valley Water Sanit Comm; EPA grant R-804615 (1979)
52. Pack DH; Atmos Environ 11:329-44 (1977)
53. Pearson CR, McConnell G; Proc Roy Soc London Ser B 189:305-32 (1975)
54. Pellizzari ED et al; Formulation of preliminary assessment of halogenated organic compounds in man and environmental media EPA-560/13-79-006 (1979)
55. Pellizzari ED et al; Bull Environ Contam Toxicol 28:322-8 (1982)
56. Pellizzari ED; Environ Sci Technol 16:781-5 (1982)
57. Rasmussen RA et al; Science 211:285-7 (1981)
58. Roberts PV, Dandliker PK; Environ Sci Technol 17:484-9 (1983)
59. Singh HB et al; Atmos Environ 15:601-12 (1981)
60. Singh HB; Geophys Res Lett 4: 101-4 (1977)
61. Singh HB et al; Atmospheric distribution, sources and sinks of selected halocarbons, hydrocarbons, $SF_6$ and $N_2O$ p 65-73 EPA-600/3-79-107 (1979)
62. Singh HB et al; J Air Pollut Control Assoc 27:332-6 (1977)
63. Singh HB et al; Atmospheric measurements of selected hazardous organic chemicals EPA-600/S3-81-032 (1981)
64. Smith JH et al; Environ Sci Technol 14:1332-7 (1980)
65. Staples CA et al Environ Tox Chem 4: 131-42 (1985)
66. Su C, Goldberg ED; Mar Pollut Transfer. Windom HL et al eds Lexington, MA: DC Heath Co p 353-74 (1976)
67. Symons JM et al; J Amer Water Works Assoc 67:634-47 (1975)
68. Tabak HH; J Water Pollut Control Fed 53:1503-18 (1981)
69. USEPA; Ambient water quality criteria for carbon tetrachloride EPA-440/5-80-026 p C-1 (1980)
70. USEPA; Treatability Manual p 12.4-1 - 4-5 EPA-600/2-82-001A (1981)
71. Vedder JF et al; Geophys Res Lett 5:33-6 (1978)
72. Veith GD et al; J Fish Res Board Can 36:1040-8 (1979)
73. Wallace LA et al; Environ Res 43: 290-307 (1987)
74. Zoeteman BCJ et al; Chemosphere 9:231-49 (1980)

# Chlorodibromomethane

## SUBSTANCE IDENTIFICATION

**Synonyms:**

**Structure:**

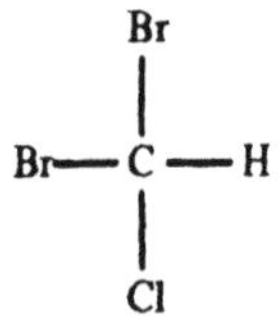

**CAS Registry Number:** 124-48-1

**Molecular Formula:** $CHBr_2Cl$

**Wiswesser Line Notation:** GYEE

## CHEMICAL AND PHYSICAL PROPERTIES

**Boiling Point:** 119-120 °C at 748 mm Hg

**Melting Point:** -20 °C (FP)

**Molecular Weight:** 208.28

**Dissociation Constants:**

**Log Octanol/Water Partition Coefficient:** 2.24 [21]

**Water Solubility:** 4400 mg/L at 22 °C [21]

**Vapor Pressure:** 15 mm Hg at 10.5 °C [7]

**Henry's Law Constant:** $8.5 \times 10^{-4}$ atm-m$^3$/mole at 20 °C [22]

## ENVIRONMENTAL FATE/EXPOSURE POTENTIAL

**Summary:** The predominant anthropogenic source of chlorodibromomethane release to the environment is its inadvertent formation during chlorination treatment processes of water. In addition

to human sources, it is biosynthesized and emitted to the environment by various species of marine macroalgae which are abundant in various locations of the world's oceans. If released to surface water, volatilization will be the dominant environmental fate process. The volatilization half-life from rivers and streams has been estimated to range from 43 min to 16.6 days with a typical half-life being 46 hours. In aquatic media where volatilization is not viable (e.g. ground water), anaerobic biodegradation may be the major removal process. Aquatic hydrolysis, oxidation, direct photolysis, adsorption, and bioconcentration are not environmentally important. If released to soil, volatilization is again likely to be the dominant removal process where exposure to air is possible. Chlorodibromomethane is moderately to highly mobile in soil and can therefore leach into ground waters. If released to air, the only identifiable transformation process in the troposphere is reaction with hydroxyl radicals which has an estimated half-life of 3.92 months. This relatively persistent half-life indicates that long-range global transport is possible. The general population is exposed to chlorodibromomethane through oral consumption of contaminated drinking water, beverages, and food products, through inhalation of contaminated ambient air, and through dermal exposure to chlorinated swimming pool water.

**Natural Sources:** Chlorodibromomethane is biosynthesized and emitted to seawater (and eventually to the atmosphere) by various species of marine macroalgae which are abundant in the various locations of the world's oceans [8,15].

**Artificial Sources:** Although environmental release can result from production and use processes, chlorodibromomethane is not produced or used on a large commercial-scale indicating that significant releases do not occur from these practices [26]. The predominant environmental release of chlorodibromomethane results from its inadvertent formation during chlorination treatment processes of drinking, waste, and cooling waters [26,36]. The amount of chlorodibromomethane which may be produced during chlorination processes depends upon a variety of parameters which include temperature, pH, bromide ion concn of the water, fulvic and humic substance concn, and actual chlorination treatment practices [36].

**Terrestrial Fate:** In soils where exposure to the atmosphere can occur, volatilization is likely to be the dominant environmental fate process due to the high vapor pressure of chlorodibromomethane [36].

# Chlorodibromomethane

Chlorodibromomethane is moderately to highly mobile in soil and can therefore leach into ground water and subsurface regions. Laboratory studies have indicated that significant biodegradation can occur under anaerobic conditions; therefore, in soil regions where volatilization is not viable, biodegradation may be the major removal process.

**Aquatic Fate:** Volatilization of chlorodibromomethane is the dominant removal mechanism from environmental surface waters. The volatilization half-life from rivers and streams has been estimated to range from 43 min to 16.6 days with a typical half-life being 46 hours. Laboratory studies have indicated that significant biodegradation can occur under anaerobic conditions; therefore, in aquatic regions where volatilization is not viable, biodegradation may be the major removal process. Aquatic hydrolysis, oxidation, direct photolysis, adsorption, and bioconcentration are not environmentally important.

**Atmospheric Fate:** Due to its high vapor pressure, chlorodibromomethane should exist entirely in the vapor phase in the ambient atmosphere. The only identifiable transformation process in the troposphere is reaction with hydroxyl radicals which has an estimated half-life of 8.5 months in typical air. Direct photolysis does not occur below the ozone layer. This relatively persistent tropospheric half-life suggests that a small percentage of the chlorodibromomethane present in air may eventually diffuse to the stratosphere where it will be destroyed by photolysis. In addition, long-range global transport is possible. The detection of chlorodibromomethane in rainwater indicates that atmospheric removal via washout can occur; however, any chlorodibromomethane which is removed by rainfall is likely to revolatilize into the atmosphere.

**Biodegradation:** Loss of chlorodibromomethane was observed to be 25-39% utilizing a static flask screening procedure and 28 days of incubation, which was interpreted as relative resistance to biodegradation under aerobic conditions [33]. In anaerobic tests using mixed methanogenic bacterial cultures from sewage effluents, chlorodibromomethane was totally degraded within 2 weeks while only 43-50% was lost in sterile controls after 6 weeks; no degradation was noted in aerobic tests in either sterile or seeded conditions [4]. Studies conducted under anoxic conditions with denitrifying bacteria found >50% degradation in bacterial cultures after 8 weeks but no degradation in sterile controls [3]. Rapid degradation was observed in a continuous-flow methanogenic fixed-film laboratory-scale column using seeded cultures, but only slow degradation was noted in sterile controls [5].

**Abiotic Degradation:** The aqueous hydrolysis half-life of chlorodibromomethane at 25 °C and pH 7 has been estimated to be 274 years [20]. Direct photolysis or aquatic oxidation (via peroxy radicals or singlet oxygen) are not environmentally relevant or significant fate processes with respect to chlorodibromomethane [21]. The rate constant for the vapor-phase reaction with hydroxyl radicals has been estimated to be $3.931 \times 10^{-14}$ m³/molecule-sec at 25 °C which corresponds to an atmospheric half-life of 8.50 months in a typical atmosphere [14].

**Bioconcentration:** Using the water solubility and log Kow values, the log BCF can be estimated to be 0.74-1.47 from recommended regression equations [19].

**Soil Adsorption/Mobility:** Bromodichloromethane, which is similar in structure to chlorodibromomethane, has been observed to have significant mobility in laboratory soil column experiments utilizing a sandy soil [39]. Relatively high soil mobility was noted for chlorodibromomethane during a water infiltration study conducted in the Netherlands along the Rhine River [27]. A soil retardation factor of 6 (indicating significant mobility) was estimated during a ground water recharge project [28]. Koc values for chlorodibromomethane have been estimated to range from 95-468 based on water solubility and log Kow [36]; these Koc values are indicative of medium-to-high soil mobility [32].

**Volatilization from Water/Soil:** The Henry's Law constant would suggest rapid volatilization from water. Based on experimentally determined gas transfer rates, the volatilization half-life of bromodichloromethane from rivers and streams has been estimated to range from 43 min to 16.6 days with a typical half-life of about 46 hours [17]. Significant volatilization (half-life of about 1 hr) was observed from laboratory tanks [11]. Approximately 50% of applied bromodichloromethane, which is similar in structure to chlorodibromomethane, volatilized from soil columns during laboratory studies which monitored transport and fate mechanisms [39].

**Water Concentrations:** DRINKING WATER: As part of the USEPA Ground Water Supply Survey, chlorodibromomethane was positively detected in 405 of 945 US finished water supplies that use ground water sources at a median level of about 3.3 ppb [38]. Median levels of 7.5-17 ppb were detected in the water supplies of over 40% of the 113 US cities monitored during the three phases (1976-77) of USEPA National Organic Monitoring Survey [35]. A Canadian national survey of 70 drinking water supplies found chlorodibromomethane levels of 0-33 ppb with an overall median level of 1.4 ppb [35]. In a survey of drinking waters from 12 areas of the world (China, Taiwan, north and south Philippines, Egypt, Indonesia, Australia, England, Brazil, Nicaragua, Venezuela, and Peru), chlorodibromomethane was found in 7 of the 12 waters at levels ranging from 1.1-13 ppb [34]. Positive detections were made in 30 of 40 Michigan drinking water supplies at a median concn of 2.2 ppb [12]. GROUND WATER: Chlorodibromomethane was one of 27 organic compounds identified in ground water collected from 315 wells in the area of the Potomac-Raritan-Magothy aquifer system adjacent to the Delaware River [13]. Levels of 0.3 ppb have been detected in ground water from the Netherlands [40]. SURFACE WATER: An analysis of the US EPA STORET data base found that chlorodibromomethane had been positively detected in 8.0% of 8515 water observation stations at a median concn below 0.1 ug/L [31]. Chlorodibromomethane was positively detected in 9.8% of 4972 samples collected from 11 stations on the Ohio River during 1980-81 with most concn between 0.1-1.0 ppb [23]. Concentrations ranging from a trace-15 ng/L and not detected-630 ng/L were reported for 16 stations on the Niagara River and 95 stations on Lake Ontario, respectively, for 1981 monitoring [18]. SEAWATER: Chlorodibromomethane concentrations of 0.1-2.2 ng/L have been detected in the North Atlantic while a concn of 0.12 ng/L was detected in the South Atlantic during 1985 monitoring [8]. Qualitative detection has been reported for the Narragansett Bay off Rhode Island in 1979-80 [37]. RAIN/SNOW: A concn of 0.4 ng/L was detected in rain collected in southern Germany in 1985 [8]. OTHER WATER: A chlorodibromomethane concn of 2 ppb was detected in stormwater runoff from Eugene, OR as part of the USEPA Nationwide Urban Runoff Program [9]. Chlorodibromomethane has been detected in swimming pool water at levels of 6-10 ppb [1].

**Effluent Concentrations:** An analysis of the US EPA STORET data base found that chlorodibromomethane had been positively detected in 6.5% of 1298 effluent observation stations at a median concn below 2.4 ug/L [31]. Chlorodibromomethane was detected in 8 of 63 industrial wastewater discharges in the US at levels ranging from

# Chlorodibromomethane

<10-100 ppb [25]. Three municipal wastewater treatment facilities in Cincinnati, OH were found to be discharging levels as high as 25 ppb in 1982 [10].

**Sediment/Soil Concentrations:** Chlorodibromomethane has been qualitatively detected in soil-sediment-water samples collected from the Love Canal near Niagara Falls, NY [16].

**Atmospheric Concentrations:** Mean chlorodibromomethane levels of 0.0, 0.48, 14, 14 and 19 ppt have been detected in the ambient air of Magnolia, AR, El Dorado, AR, Chapel Hill, NC, Beaumont, TX, and Lake Charles, LA, respectively [6]. An analysis of ambient air of several German cities found chlorodibromomethane concn generally ranging from not detectable to 0.1 ug/m$^3$, although one industrial city had a level of 0.9 ug/m$^3$ [2]. Atmospheric chlorodibromomethane levels ranging from 0.06-10 ppt (median of about 0.4) were found in ambient air samples collected from the north and south Atlantic Ocean, the beaches of the Azore Islands and Bermuda, and southern Germany between 1982-85 [8]. Monitoring of four California sites between 1982-83 found mean composite concn of 10-50 ppt in ambient air [29].

**Food Survey Values:**

**Plant Concentrations:** Mean chlorodibromomethane levels of 150-590 ng/g (dry wt) have been detected in various species of marine algae [15].

**Fish/Seafood Concentrations:**

**Animal Concentrations:**

**Milk Concentrations:** An analysis of 12 samples of various German milk products (ice cream, yogurt, curds, and buttermilk) found chlorodibromomethane levels ranging from not detectable to 0.3 ug/kg with an overall mean concn of 0.1 ug/kg [2].

**Other Environmental Concentrations:** Chlorodibromomethane has been identified in various German cosmetic products at maximum levels of 0.2 ug/L [2].

**Probable Routes of Human Exposure:** The general population is exposed to chlorodibromomethane through oral consumption of contaminated drinking water, beverages, and food products. The contamination is a result of inadvertent formation during chlorination treatment of the drinking water and subsequent use of chlorinated tap water to produce food products. Exposure can also occur through inhalation of background levels in ambient air and through dermal exposure in chlorinated swimming pool water.

**Average Daily Intake:** DRINKING WATER INTAKE: (assume a concn of 1-10 ug/L) = 2-20 ug; AIR INTAKE: (assume a concn of 1-50 ppt, 8.5-425 ng/m$^3$) = 0.170-8.5 ug; FOOD INTAKE: insufficient data.

**Occupational Exposures:**

**Body Burdens:** Chlorodibromomethane was identified in one of 12 human milk samples collected from volunteers from four US cities (Bridgeville, PA; Bayonne, NJ; Jersey City, NJ; and Baton Rouge, LA) [24]. Chlorodibromomethane was not detected in any sample from the USEPA National Human Adipose Tissue Survey for fiscal year 1982 [30].

# REFERENCES

1. Aggazzotti G, Predieri G; Wat Res 20: 959-63 (1986)
2. Bauer U; Zentralbl Bakteriol Mikrobiol Hyg ABT 1 Orig B Hyg Umwelthyg Krankenhaushyg Arbeitshyg Praev Med 174: 200-37 (1981)
3. Bouwer EJ, McCarty PL; Appl Environ Microbiol 45: 1295-1299 (1983)
4. Bouwer EJ et al; Environ Sci Technol 15: 596-9 (1981)
5. Bouwer EJ, McCarty PL; Appl Environ Microbiol 45: 1286-94 (1983)
6. Brodzinsky R, Singh HB; Volatile Organic Chemicals in the Atmosphere: An Assessment of Available Data, Menlo Park,CA:SRI International (1982)
7. Callahan MA et al; Water-Related Environmental Fate of 129 Priority Pollutants - Vol II. EPA-440/4-79-029B (1979)
8. Class T et al; Chemosphere 15: 429-36 (1986)
9. Cole RH et al; J Wat Pollut Control Fed 56: 898-908 (1984)
10. Dunovant VS et al; J Water Pollut Control Fed 58: 886-95 (1986)
11. Francois C et al; Trav Soc Pharm Montpellier 39: 49-50 (1979)
12. Furlong EAN, Ditri FM; Ecol Modeling 32: 215-25 (1986)
13. Fusillo TV et al; Groundwater 23: 354-60 (1985)
14. GEMS; Graphical Exposure Modeling System. FAP. Fate of Atmospheric Pollutants (1987)
15. Gschwend PM et al; Science 227: 1033-5 (1985)
16. Hauser TR, Bromberg SM; Environ Monitor Assess 2: 249-72 (1982)
17. Kaczmar SW et al; Environ Toxicol and Chem 3: 31-5 (1984)
18. Kaiser KLE et al; J Great Lakes Res 9: 212-23 (1983)

# Chlorodibromomethane

19. Lyman WJ et al; Handbook of Chemical Property Estimation Methods NY:McGraw-Hill p.5-4,5-10 (1982)
20. Mabey W, Mill T; J Phys Chem Ref Data 7: 383-415 (1978)
21. Mabey WR et al; in Aquatic Fate Process Data for Organic Priority Pollutants USEPA-440/4-81-014 p.179-82 (1981)
22. Nicholson BC et al; Environ Sci Technol 18: 518-23 (1984)
23. Ohio River Valley Sanit Comm; Assessment of Water Quality Conditions. Ohio River Mainstream, Cincinnati,OH (1982)
24. Pellizzari ED et al; Bull Environ Contam Toxicol 28: 322-8 (1982)
25. Perry DL et al; p 42-3 in Identification of Organic Compounds in Industrial Effluent Discharges USEPA-600/4-79-016 (1979)
26. Perwak J et al; Exposure and Risk Assessment for Trihalomethanes (Chloroform, Bromoform, Bromodichloromethane, Dibromochloromethane) USEPA-440/8-81-018 p.13 (1980)
27. Piet GJ et al; Studies Environ Sci 17: 557-64 (1981)
28. Roberts PV, Valocchi AJ; Sci Total Environ 21: 161-72 (1981)
29. Shikiya J et al; Proc-APCA 77th Annual Mtg Vol 1,84-1.1 (1984)
30. Stanley JS; Broad Scan Analysis of Human Adipose Tissue:Volume II: Volatile Organic Compounds USEPA-560/5-86-036 p.74 (1986)
31. Staples CA et al; Environ Toxicol Chem 4: 131-42 (1985)
32. Swann RL et al; Res Rev 85: 17-28 (1983)
33. Tabak HH et al; J Water Pollut Control Fed 53: 1503-18 (1981)
34. Trussel AR et al; Water Chlorination Environ. Impact Health Effects 3: 39-53 (1980)
35. USEPA; Ambient Water Quality Criteria for Halomethanes US EPA-440/5-80-051 p.C-7 (1980)
36. USEPA; Health and Environmental Effects Profile for Bromochloromethanes ECAO-CIN-P122 (Final Draft) p.12,18-21 (1985)
37. Wakeham SG et al; Can J Fish Aquatic Sci 40: 304-21 (1983)
38. Westrick JJ et al; J Amer Water Works Assoc 76: 52-9 (1984)
39. Wilson JT et al; J Environ Qual 10: 501-6 (1981)
40. Zoeteman BCJ et al; Sci Total Environ 21: 187-202 (1981)

# Chloroform

## SUBSTANCE IDENTIFICATION

**Synonyms:** Trichloromethane

**Structure:**

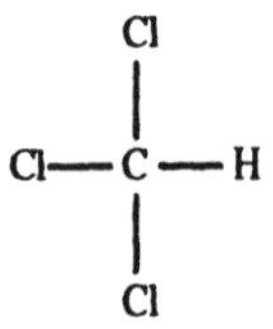

$$Cl-\underset{\underset{Cl}{|}}{\overset{\overset{Cl}{|}}{C}}-H$$

**CAS Registry Number:** 67-66-3

**Molecular Formula:** CHCl₃

**Wiswesser Line Notation:** GYGG

## CHEMICAL AND PHYSICAL PROPERTIES

**Boiling Point:** 61.7 °C at 760 mm Hg

**Melting Point:** -63.5 °C

**Molecular Weight:** 119.39

**Dissociation Constants:**

**Log Octanol/Water Partition Coefficient:** 1.97 [25]

**Water Solubility:** 7950 mg/L [44]

**Vapor Pressure:** 246 mm Hg at 25 °C [22]

**Henry's Law Constant:** $4.35 \times 10^{-3}$ atm-m³/mole at 25 °C [30]

## ENVIRONMENTAL FATE/EXPOSURE POTENTIAL

**Summary:** Chloroform is likely to enter the environment associated with its use as an industrial solvent, extractant, and chemical intermediate as well as from its indirect production in the chlorination of drinking water, municipal sewage, and cooling water. The majority of the environmental releases from industrial uses are to the atmosphere; releases to water and land will be primarily lost by evaporation and will end up in the atmosphere. Release to the atmosphere may be transported long distances and will react in the gas-phase with photochemically produced hydroxyl radicals with a half-life for this reaction of a few months. Spills and other releases on land will also leach into the ground water where they may reside for long periods of time. It will be expected to evaporate rapidly from near-surface soils based on its high vapor pressure. Chloroform released to water will evaporate rapidly with the half-life for evaporation determined in the laboratory to be several hours; modeling studies have estimated the half-life for evaporation to be 36 hours in a river, 40 hours in a pond, and 9-10 days in a lake. Chloroform may be subject to significant biodegradation based on laboratory experiments, although the reported data are somewhat conflicting. It will not significantly adsorb to sediment. Chloroform will not be expected to bioconcentrate in aquatic organisms but contamination of food is likely due to its use as an extractant and its presence in drinking water. It will not significantly hydrolyze in water or soil under normal environmental conditions. Major human exposure is from ingestion of contaminated drinking water and inhalation of contaminated ambient air, the latter particularly in the vicinity of industrial sources. Exposure may also occur via ingestion of contaminated food and by cutaneous contact with contaminated water.

**Natural Sources:** Plant volatile [23].

**Artificial Sources:** Emissions from its production and indirect production (in the manufacture of ethylene dichloride); chlorination of drinking water, municipal sewage, cooling water in electric power generating plants; produced during the atmospheric photodegradation of trichloroethylenes; auto exhaust; from its use as an extractant or solvent, chemical intermediate, dry cleaning agent, or fumigant ingredient; in fluorocarbon 22 production; and in synthetic rubber production [33,79].

**Terrestrial Fate:** If chloroform is released to soil, it will be expected to evaporate rapidly into the atmosphere from near-surface soils due to its high vapor pressure. It is poorly adsorbed to soil, especially soil with low organic carbon content such as subsoils and can leach into the ground water. It may be subject to slow biodegradation in soil and ground water based on aqueous laboratory screening studies, although the data are conflicting. It will not significantly hydrolyze under normal environmental conditions.

**Aquatic Fate:** If chloroform is released into water, it will be primarily lost by evaporation into the atmosphere. Laboratory experiments have measured the half-life for evaporation to be several hours and modeling studies suggest that the volatilization half-life is 36 hours in a river, 40 hours in a pond, and 9-10 days in a lake [79]. Field monitoring data suggest the half-life of chloroform to be 1.2 days in the Rhine River and 31 days in a lake in the Rhine basin [85]. Chloroform from a municipal treatment plant injected into an estuarine arm of Chesapeake Bay entirely disappeared within 4 km in the spring and within 11 km in winter under ice, and the decrease in concentration cannot be entirely due to dilution [28]. Chloroform may be subject to significant biodegradation based on laboratory experiments, although the reported data are somewhat conflicting. Little chloroform will be adsorbed to sediment. It will not significantly hydrolyze under normal environmental conditions.

**Atmospheric Fate:** If chloroform is released to the atmosphere, it will exist predominantly in the vapor phase [17]. Chloroform released to the atmosphere will degrade by reaction with hydroxyl radicals with a half-life of 80 days. It may be transported long distances and will partially return to earth in precipitation [38].

**Biodegradation:** There are conflicting data on the biodegradation of chloroform. Slow but substantial biodegradation apparently can occur when the proper microbial populations exist and are acclimated to the chemical. Under aerobic conditions, some investigators report little or no degradation in up to 25 wk [5,29,39] while others report considerable degradation: 49% in 7 days, 100% in 28 days; however, a large fraction of this loss was due to volatilization [75]: 25% in 14 days [7] and 67% in 24 days [20]. Under anaerobic conditions, slow degradation has been reported after acclimation [6] and degradation was reported in river bank (31% in <1 yr) and dune (100% in <3 mo) infiltration [85]. However, another investigator reported no degradation in 27 weeks in aquifer material in the laboratory [83].

**Abiotic Degradation:** Chloroform has a negligible rate of hydrolysis [43]. Photodegradation does not appear to be a significant loss process in aquatic systems [34]. The key gas-phase photochemical reaction in the atmosphere is the reaction of chloroform with photochemically produced hydroxyl radicals. The half-life for this reaction is 80 days which amounts to a 0.9% loss per sunlit day [24,65]. This value compares reasonably well with the 23-week half-life measured when a flask of chloroform filled with ambient air was exposed to sunlight outdoors [55]. Chloroform is more reactive in photochemical smog situations (presence of $NO_x$) with an avg degradation rate of 0.8%/hr [15].

**Bioconcentration:** Little or no tendency to bioconcentrate: BCF: 6, bluegill sunfish (<u>Lepomis macrochirus</u>) [4]; 3.34-10.35, rainbow trout (<u>Salmo gairdneri</u>), 1.6-2.5, bluegill (<u>Lepomis macrochirus</u>), 2.9-3.1, largemouth bass (<u>Micropterus salmoides</u>), 3.3-3.7, catfish (<u>Ictalurus punctatus</u>) [1].

**Soil Adsorption/Mobility:** Chloroform is adsorbed most strongly to peat moss, less strongly to clay, very slightly to dolomite limestone and not at all to sand [14]. The Koc values measured for 2 soils was 34; however, 3 other soils with the lowest organic carbon content in the same study gave no appreciable adsorption [32]. Field experiments in which chloroform was injected into an aquifer and the concentration in a series of observation wells determined, demonstrated that chloroform is very poorly retained by aquifer material (retardation factor 2-4), less so than other $C_1$- and $C_2$-halogenated compounds studied [32,62]. Laboratory percolation studies with a sandy soil gave similar results (retardation factor <1.5) [84].

**Volatilization from Water/Soil:** Four laboratory studies of the evaporation of chloroform from water gave half lives of 3-5.6 hours with moderate mixing conditions [42,60,61,71]. A modelling study of chloroform predicts a volatilization half-life of 36 hr in a river, 40 hr in a pond, and 9-10 days in a lake [79]. A volatile compound such as chloroform would be expected to volatilize rapidly from near surface soils at spill sites. The half-life for volatilization of chloroform from a river 1 m deep flowing 1 m/sec with a wind velocity of 3 m/sec is estimated to be 4 hr at 20 °C [71] based on the Henry's Law constant.

**Water Concentrations:** DRINKING WATER: US Federal Survey of Finished Waters find a 70.3% occurrence in ground water supplies [16]; 30 Canadian treatment facilities (treated water) 35 ppb avg summer, 21 ppb avg winter (93-97% pos, 110 ppb max - raw water had 2-6 ppb avg concn) [54]; US 5 city survey 1-301 ppb [11]; drinking water wells in New York and New Jersey 67-490 ppb [10]; Other cities report values between 0-190 ppb [36,71,77,82] with the values highest in summer and lowest in winter [36] and increasing on contact with residual chlorine [77].  DRINKING WATER: National Organic Reconnaissance Survey (80 US water supplies, 1975) 0-311 ppb, National Organics Monitoring Survey (113 finished water supplies, 1976-77) 32-68 ppb median of positive supplies, 92-100% pos [74]. Drinking water from 9 US cities, 88.9% pos, not detected to 58 ppb, 22.9 avg of pos [26]. Drinking water, 12 sites from around the world, 91.7% pos, not detected to 57 ppb, 10.4 ppb avg of pos [76]. 40 Michigan treatment plants, chlorinated drinking water, 80% pos, not detected-201.4 ppb, 33.4 ppb avg of pos, 16.7 ppb median [21]. GROUND WATER: Wisconsin ground water survey, as of June 1984, 1174 community wells, 1.1% pos, 617 private wells, 0.32% pos [41]. Contaminated wells in New York and New Jersey, 67-490 [10]. Federal Survey of Finished waters found a 70.3% occurrence in ground water supplies [16]. Ground water in the Netherlands 5 ppb [85].  SURFACE WATER: Ohio River Basin (1980-81, 11 stations, 4972 samples) 72% pos, 832 samples 1-10 ppb, 27 samples >10 ppb [53].  14 heavily industrialized river basins in US (204 sites) 1-120 ppb, 79% pos [19]. US - 5 industrial cities 9-31 ppb avg, 394 ppb max [56].  11 water utilities on Ohio River 0.8 ppb avg, 4.8 ppb max, 68% pos [52]; Delaware River and tributaries - 30 sites 93% of samples >1 ppb [12]. SURFACE WATER: Ohio River and tributaries 232 samples 0.1-22 ppb, 72% pos [51]; Lake Erie, Michigan and Huron 1-30 ppb, 11 of 13 sites pos [40]. US EPA STORET data base, 11,928 data points, 64.0% pos, 0.300 ppb median [73].  SEAWATER: Pacific Ocean <0.05 ppt [70]; Northeast Atlantic Ocean 4-13 ppt, avg 8 ppt [47]; Point Reyes (near shore) 2.8 ppb [69]. Gulf of Mexico 4-200 ppb [64]. RAIN/SNOW: Detected in rain and snow in Japan [37,46] and 250 ppt in rain in West Los Angeles [38].

**Effluent Concentrations:** Rubber and chemical companies - Louisville, KY 22 ppm max [78]. Industries whose wastewater levels of chloroform exceed a mean level of 500 ppb are auto and other laundries, aluminum forming, pharmaceuticals, and pulp and paper mills; the pharmaceutical industry contributes the largest amount of chloroform with mean and max wastewater concn of 49 and 280 ppb, respectively [80]. Auto exhausts typically emit 27 ug/m$^3$ [78]. US EPA STORET data base, 1,513 data points, 37.2% pos, 10.0 ppb median

[73]. Ground water at 178 US CERCLA hazardous waste sites, 28.4% pos [59].

**Sediment/Soil Concentrations:** SEDIMENT: Not detected at an industrial location on US river [35]. Not detected in Back River sediment off Baltimore [28]. US EPA STORET data base, 425 data points, 8.0% pos, <5.0 ppb median [73].

**Atmospheric Concentrations:** RURAL/REMOTE: US, 532 samples 40 ppt avg [9]; Northern Hemisphere, background 17.1 ppt avg [68]. Avg concn (ppt), Norwegian Arctic air, Jul 1982, 7 samples, 16.5, Spring 1983, 10 samples, 26.6 [31]. US URBAN/SUBURBAN - 1739 samples 72 ppt avg [9]. US source dominated areas - 306 samples 820 ppt avg [9]. 11 highly industrialized US locations 0-10.9 ppb [58]; 10 US cities 32-703 ppt avg, 5112 ppt max [65-67]; 3 areas in New Jersey 710 ppt avg, 15% pos avg of pos samples approx 4 ppb [8]. URBAN/SUBURBAN: USEPA Total Exposure Assessment Methodology study, 2 cities in NJ, Fall 1981, concn (ug/m$^3$), personal air, approx 345 samples (night/day), maximum concn, 220/89; weighted means: geometric, 3.3/3.01, arithmetic, 10/7.8; outdoor air, approx 88 samples (night/day), maximum concn, 22/9, weighted means: geometric, 0.55/0.58, arithmetic, 1.2/1.6 [81].

**Food Survey Values:** In pilot market basket survey of 4 food groups at 5 sites, the results for chloroform were: dairy composite 17 ppb (1 of 5 sites); meat composite not detected; oil and fat composite trace (1 of 5 sites); beverage composite 6-32 ppb (4 of 5 sites); high values for individual foods were: soft drinks 9-178 ppb; butter 56 ppb; cheese 15-17 ppb; mayonnaise 34 ppb [18]. England: various samples of food including dairy products, eggs, bread, meat, oils and fats, beverages, fruits and vegetables 0.4-33 ppb; cheese, butter, and tea were high [45]. Residues were found in fumigated sorghum, barley, and corn but generally disappeared within 60 days when aired at 17 °C [33]. 15 table-ready food items, 53.3% pos; butter 670 ppb; cheddar cheese 80 ppb; plain granola 57 ppb; peanut butter 29 ppb; chocolate chip cookies 22 ppb; frozen fried chicken dinner 29 ppb; high meat dinner 17 ppb [27]. Butter, 7 samples, 100% pos, 80-1100 ppb, 422 ppb avg; margarine, 7 samples, 100% pos, 3-740 ppb, 306 ppb avg [27]. 4 types of cheese, 8 samples, 100% pos, 18-810 ppb, 182 ppb avg [27]. Wheat, 10 samples, 105-3000 ppb, 829 ppb avg [26].

**Plant Concentrations:**

# Chloroform

**Fish/Seafood Concentrations:** Great Britain: various species of marine fish 5-851 ppb [13,55]; marine invertebrates 2-1040 ppb [13,55].

**Animal Concentrations:** England: grey seal 7.6-22 ppb (blubber), 0-12 ppb (liver); marine and freshwater birds 0.7-65 ppb [55].

**Milk Concentrations:**

**Other Environmental Concentrations:**

**Probable Routes of Human Exposure:** Humans are exposed to chloroform primarily from ingestion of chlorinated drinking water supplies, although exposure from inhalation of contaminated air which is typically one tenth of that from water would be comparable in source dominated areas. Although data for chloroform in food are fragmentary, they suggest that intake from ingestion of contaminated food may be substantive.

**Average Daily Intake:** AIR INTAKE: (assume typical concentration 72 ppt [1]) 7.1 ug. WATER INTAKE: (assume typical concentration of 32-68 ppb [74]) 64-136 ug; FOOD INTAKE: insufficient data.

**Occupational Exposures:** NIOSH (NOES Survey 1981-83) has statistically estimated that 95,778 workers are exposed to chloroform in the US [49]. NIOSH (NOHS Survey 1972-74) has statistically estimated that 215,000 workers are exposed to chloroform in the US [50]. Plant manufacturing medicinals 23-128 ppm; plant manufacturing film (5 year survey) 7-170 ppm, 47 ppm avg although 2 day sampling by another method gave 30-585 ppm; pharmaceutical plant in Poland 2-205 ppm [48]. Shell Chem Co, Rocky Mountain Arsenal - mean TWA were 2.6, 0.4, and 0.2 ppm for production operators, drummers/bottle fillers, and maintenance/utility personnel (pesticide plant), respectively [63]. Polish pharmaceutical plant 2-205 ppm [63]; police forensic lab - 8 hr TWA - 15.8 ppm (range 2.6-46.4 ppm) [63]; film manufacturing plant using a solvent containing 22% chloroform 1968-72 - 7-170 ppm (mean 47 ppm, 79 samples) [63].

**Body Burdens:** Old Love Canal, Niagara Falls, NY - 9 individuals: breath 3.9-95 ug/m$^3$, 26 ug/m$^3$ median; blood 1.1-3.0 ng/ml, 1.6 ng/mL median; urine 460-1500 ng/L, 860 ng/L median [3]. England - 8 individuals: body fat 5-68 ppb; various organs 1-10 ppb [45]; US - 4 urban sites: human milk 7 of 8 samples pos, detected, not quantified [57]. Blood from 250 US patients, not detected-7.0 ppb, 1.5 ppb avg [2]. USEPA Total Exposure Assessment Methodology study, New Jersey residents from 2 cities, Fall 1981, concn in breath (ug/m³), 295-339 samples, 60% of subjects pos, weighted geometric mean, 1.3; weighted arithmetic mean, 0.3; 29 max concn [81]. US FY82 National Human Adipose Tissue Survey specimens, 46 composites, 76% pos, (>2 ppb, wet tissue concn), 580 ppb max [72].

## REFERENCES

1. Anderson DR, Lusty EB; Acute Toxicity and Bioaccumulation of Chloroform to Four Species of Fresh Water Fish. NUREG/CR-089 Richland, WA: Pacific NW Labs p 8-26 (1980)
2. Antoine SR et al; Bull Environ Contam Toxicol 36: 364-71 (1986)
3. Barkley J et al; Biomed Mass Spectrom 7: 139-47 (1980)
4. Barrows ME et al; Dyn Exposure Hazard Assess Toxic Chem Ann Arbor, MI: Ann Arbor Press p 379-92 (1980)
5. Bouwer EJ et al; Environ Sci Technol 15: 569 (1981)
6. Bouwer EJ, McCarty PL; Appl Environ Microbiol 45: 1286-94 (1983)
7. Bouwer EJ et al; Water Res 15: 151-9 (1981)
8. Bozzelli JW, Kebbekus BB; J Environ Sci Health 17: 693-713 (1982)
9. Brodzinsky R, Singh HB; Volatile organic chemicals in the atmosphere: an assessment of available data. SRI Inter EPA contract 68-02-3452 Menlo Park, CA (1982)
10. Burmaster DE; Environ 24: 6-13, 33-6 (1982)
11. Coleman WE et al; Analysis and Identification of Organic Substances in Water; L Keith Ed, Ann Arbor, MI: Ann Arbor Press p 305-27 (1976)
12. DeWalle FB, Chain ESK; Proc Ind Waste Conf 32: 908-19 (1978)
13. Dickson AG, Riley JP; Marine Pollut Bull 7: 167-9 (1976)
14. Dilling WL et al; Environ Sci Technol 9: 833-8 (1975)
15. Dimitriades B, Joshi SB; Inter Conf on Photochemical oxidant pollution and its control. EPA-600/3-77-001b p 705-11 (1977)
16. Dyksen JE, Hess AF III; J Amer Water Works Assoc p 394-403 (1982)
17. Eisenreich SJ et al; Environ Sci Technol 15: 30-8 (1981)
18. Entz RC et al; J Agric Food Chem 30: 846-9 (1982)
19. Ewing BB et al; Monitoring to detect previously unrecognized pollutants in surface waters. EPA-560/6-77-015 75p (1977)
20. Flathman PE, Dahlgran JR; Environ Sci Technol 16:130 (1982)
21. Furlong EAN, D'Itri FM; Ecological Modelling 32: 215-25 (1986)
22. Gallant RW; Hydrocarbon Process 45: 161-9 (1966)
23. Graedel TE; Chemical Compounds in the Atmosphere. p.324 Academic Press, New York, NY (1978)
24. Hampson RF; Chemical Kinetics and Photochemical Data Sheets for Atmospheric reactions. USDOT report FAA-EE-80-17 (1980)
25. Hansch C et al; J Med Chem 18: 546-48 (1975)
26. Heikes DL, Hopper ML; J Assoc Off Anal Chem 69: 990-8 (1986)
27. Heikes DL; J Assoc Off Anal Chem 70: 215-26 (1987)

28. Helz GR, Hsu RY; Limnol Oceanogr 23: 858-69 (1978)
29. Heukelekian H, Rand MC; J Water Pollut Control Assoc 29: 1040-53 (1955)
30. Hine J, Mookerjee PK; J Org Chem 40: 292-8 (1975)
31. Hov H et al; Geophys Res Lett 11: 425-8 (1984)
32. Hutzler NJ et al; Amer Chem Soc 186th Mtg Div Environ Chem Preprint 23: 499-502 (1983)
33. IARC; Monograph. Some Halogenated Hydrocarbons 20: 402-7 (1979)
34. Jensen S, Rosenberg R; Water Res 9: 659-61 (1975)
35. Jungclaus GA et al; Environ Sci Technol 12: 88-96 (1978)
36. Kasso WB, Wells MR; Bull Environ Contam Toxicol 27: 295-302 (1981)
37. Kato T et al; Yokohama Kokuritsu Daigaku Kankyo Kagaku Kenkya Senta Kiyo 6: 11-20 (1980)
38. Kawamura K, Kaplan IR; Environ Sci Technol 17: 497-501 (1983)
39. Kawasaki M; Ecotox Environ Safety 4: 444-54 (1980)
40. Konasewich D et al; Status report on organic and heavy metal contaminants in the Lakes Erie, Michigan, Huron and Superior basins, Great Lakes Qual Board 373p (1978)
41. Krill RM, Sonzogni WL; J Am Water Works Assoc 78: 70-5 (1986)
42. Lyman WJ et al; Handbook of Chem Property Estimation Methods Environ Behavior of Org Compounds. McGraw-Hill NY (1982)
43. Mabey W, Mill T; J Phys Chem Ref Data 7: 383-415 (1978)
44. MacKay D et al; Chemosphere 9: 701-11 (1980)
45. McConnell G et al; Endeavor 34: 13-8 (1975)
46. Morita M et al; Kokyo Toritsu Eisei Kenkyusho Kenkyu Nempo 25: 399-403 (1974)
47. Murray AJ, Riley JP; Nature 242: 37-8 (1973)
48. NIOSH; Criteria for a recommended standard - occupational exposure to chloroform. NIOSH 75-114 p 57-59 (1974)
49. NIOSH; The National Occupational Exposure Survey (NOES) (1983)
50. NIOSH; The National Occupational Hazard Survey (NOHS) (1974)
51. Ohio R Valley Water Sanit Comm; Assessment of Water Quality Conditions, Ohio River Mainstream 1978-9, Cincinnati, OH p T-53 (1980)
52. Ohio R Valley Water Sanit Comm; Water treatment process modifications for trihalomethane control and organic substances in the Ohio River. EPA grant no. R-804615 Cincinnati, OH (1979)
53. Ohio R Valley Water Sanit Comm; Assessment of water quality conditions 1980-81. Cincinnati, OH table 13 (1982)
54. Otson R et al; J Assoc Off Analyt Chem 65: 1370-4 (1982)
55. Pearson CR, McConnell G; Proc Roy Soc London Ser B 189: 305-32 (1975)
56. Pellizzari ED et al; Formulation of preliminary assessment of halogenated organic compounds in man and environmental media. EPA-560/13-79-006 469p (1979)
57. Pellizzari ED et al; Bull Environ Contam Toxicol 28: 322-8 (1982)
58. Pellizzari ED; Quantification of chlorinated hydrocarbons in previously collected air samples. EPA-450/3-78-112 (1978)
59. Plumb RH; Ground Water Monitoring Rev7: 94-100 (1987)
60. Rathbun RE, Tai DY; Water Res 15: 243-50 (1981)
61. Robert PV, Dandliker PG; Environ Sci Technol 17: 484-9 (1983)
62. Roberts PV et al; Water Res 16: 1025-35 (1982)

# Chloroform

63. Santodonato J et al; Monograph on Human Exposure to Chemicals in the Workplace: Chloroform. NCI contract N01-CP-26002-03, Syracuse Research Corp. July (1985)
64. Sauer TC Jr; Org Geochem 3: 91-101 (1981)
65. Singh HB et al; Atmos Environ 15: 601-12 (1981)
66. Singh HB et al; Environ Sci Technol 16: 372-80 (1982)
67. Singh HB et al; Atmospheric measurements of selected hazardous organic chemicals. EPA-600/53-81-032 (1981)
68. Singh HB; Geophys Res Lett 4: 101-4 (1977)
69. Singh HB et al; J Air Pollut Control Assoc 27: 332-6 (1977)
70. Singh HB et al; Atmospheric distributions, sources and sinks of selected halocarbons, hydrocarbons, $SF_6$ and $NO_2$ EPA-600/3-79-107 134 p (1979)
71. Smith VL et al; Environ Sci Technol 14: 190-6 (1980)
72. Stanley JS; Broad Scan Analysis of the FY82 National Human Adipose Tissue Survey Specimens Vol. I Executive Summary p 5 USEPA-560/5-86-035 (1986)
73. Staples CA et al; Environ Toxicol Chem 4: 131-42 (1985)
74. Symon JM et al; J Amer Water Works Assoc (1982)
75. Tabak HH et al; J Water Pollut Control Fed 53: 1503-18 (1981)
76. Trussell AR et al; Environ Impact Health Effects 3: 39-53 (1980)
77. Uden PC, Miller JW; J Amer Water Works Assoc 75: 524-7 (1983)
78. USEPA; Ambient Water Quality Criteria for Chloroform. EPA-440/5-80-033 pp C-1 to C-5 (1980)
79. USEPA; Health Assessment Document for Chloroform. External Review Draft EPA-600/8-84-004A p 3-4 to 3-28 (1984)
80. USEPA; Treatability Manual. EPA-600/2-82-001a pp I.12.3-1 to I.12.3-5 (1980)
81. Wallace LA et al; Atmos Environ 19: 1651-61 (1985)
82. Williams DT et al; Chemosphere 11: 263-76 (1982)
83. Wilson JT et al; Devel Indust Microbiol 24: 225-33 (1983)
84. Wilson JT et al; J Environ Qual 10: 501-6 (1981)
85. Zoeteman BCJ et al; Chemosphere 9: 231-49 (1980)

# 1-Chloro-1,1-Difluoroethane

## SUBSTANCE IDENTIFICATION

**Synonyms:** Freon 142B

**Structure:**

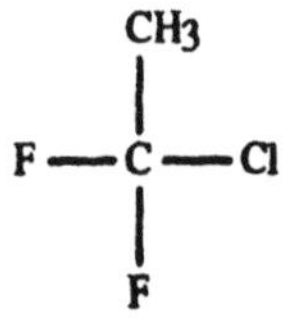

$$CH_3$$
$$|$$
$$F - C - Cl$$
$$|$$
$$F$$

**CAS Registry Number:** 75-68-3

**Molecular Formula:** $C_2H_3ClF_2$

**Wiswesser Line Notation:**

## CHEMICAL AND PHYSICAL PROPERTIES

**Boiling Point:** -9.2 °C

**Melting Point:** -130.8 °C

**Molecular Weight:** 100.50

**Dissociation Constants:**

**Log Octanol/Water Partition Coefficient:** 1.60 [7]

**Water Solubility:** 1400 mg/L at 21 °C [10]

**Vapor Pressure:** 2,523 mm Hg at 25 °C [13]

**Henry's Law Constant:** 0.239 atm-m³/mole at 25 °C (estimated by bond contribution method) [9]

# 1-Chloro-1,1-Difluoroethane

## ENVIRONMENTAL FATE/EXPOSURE POTENTIAL

**Summary:** 1-Chloro-1,1-difluoroethane may be released to the environment as emissions during its production, storage, transport, or use as a refrigerant and as a propellant in aerosol sprays, or it may be released to soil from the disposal of aerosol cans and refrigeration units containing this compound. If released to soil, 1-chloro-1,1-difluoroethane should rapidly volatilize from soil surfaces or leach through soil, possibly into ground water. If released to water, volatilization (half-life of 3 hr from a model river) would be the dominant fate process. If released to the atmosphere, essentially all 1-chloro-1,1-difluoroethane is expected to exist in the vapor phase. In the troposphere, 1-chloro-1,1-difluoroethane would react very slowly with photochemically generated hydroxyl radicals (half-life of 5-12 years) or diffuse into the stratosphere (half-life of 20 years). The overall tropospheric half-life has been estimated to range between 4-7.5 years. In the stratosphere, 1-chloro-1,1-difluoroethane would undergo direct photolysis or react with singlet oxygen. Due to its stability, transport long distances from its sources of emissions will take place. The most probable route of exposure to 1-chloro-1,1-difluoroethane by the general population is inhalation of aerosol sprays in which this compound is used as a propellant.

**Natural Sources:**

**Artificial Sources:** 1-Chloro-1,1-difluoroethane may be released to the environment as emissions or in wastewater during its production, use in the manufacture of fluoropolymers, storage, transport, and use as a propellant in aerosol sprays [3,8]. This compound may be released to soil from the disposal of aerosol cans and refrigeration units, which includes home appliances, mobile air conditioning units, retail food refrigeration systems, and centrifugal and reciprocating chillers [2,3].

**Terrestrial Fate:** If released to soil, 1-chloro-1,1-difluoroethane may rapidly volatilize from soil surfaces or leach through soil, possibly into ground water.

**Aquatic Fate:** If released to water, essentially all 1-chloro-1,1-difluoroethane is expected to be lost by volatilization (half-life of 2.93 hours from a model river). Bioaccumulation and adsorption to sediments are not significant fate processes in water.

111

# 1-Chloro-1,1-Difluoroethane

**Atmospheric Fate:** If released to the atmosphere, essentially all 1-chloro-1,1-difluoroethane is expected to exist in the vapor phase due to its extremely high vapor pressure. In the troposphere, 1-chloro-1,1-difluoroethane reacts slowly with photochemically generated hydroxyl radicals (half-life of 5-12 years) or diffuses into the stratosphere (half-life of 20 years [5]). The overall tropospheric half-life has been estimated to range between 4-7.5 years. In the stratosphere, 1-chloro-1,1-difluoroethane may slowly photolyze, producing chlorine atoms which in turn would participate in the catalytic removal of stratospheric ozone, or it may slowly react with singlet oxygen. The relatively high water solubility of 1-chloro-1,1-difluoroethane suggests that some loss by wet deposition occurs, but any loss by this mechanism is probably returned to the atmosphere by volatilization. As a result of its persistence in the atmosphere, long distance transport occurs.

**Biodegradation:**

**Abiotic Degradation:** Chemical hydrolysis of 1-chloro-1,1-difluoroethane is not an environmentally significant fate process [6]. The half-life for 1-chloro-1,1-difluoroethane reacting with photochemically generated hydroxyl radicals in the troposphere has been estimated to range between 5 and 12 years based on reaction rate constants of $3.7 \times 10^{-15}$ and $8.4 \times 10^{-15}$ cm$^3$/molecule-sec at 20 °C and an ambient hydroxyl radical concentration of $5 \times 10^{+5}$ molecules/cm$^3$ [1]. In the stratosphere, 1-chloro-1,1-difluoroethane may slowly photolyze, producing chlorine atoms which in turn would participate in the catalytic removal of stratospheric ozone, or it may slowly react with singlet oxygen [4]. By analogy to other Freon compounds, the stratospheric lifetime of 1-chloro-1,1-difluoroethane is expected to be on the order of a few decades [4].

**Bioconcentration:** Based on the Kow, a bioconcentration factor (BCF) of 42 was estimated for 1-chloro-1,1-difluoroethane [11]. This BCF value suggests that 1-chloro-1,1-difluoroethane would not bioaccumulate significantly in aquatic organisms.

**Soil Adsorption/Mobility:** A soil adsorption coefficient (Koc) of 35 was estimated using linear regression equations based on the Kow [11]. These Koc values suggest that 1-chloro-1,1-difluoroethane would be moderately to highly mobile in soil and that slight adsorption to suspended solids and sediments in water would take place [14].

# 1-Chloro-1,1-Difluoroethane

**Volatilization from Water/Soil:** The value of Henry's Law constant suggests that 1-chloro-1,1-difluoroethane would volatilize rapidly from all bodies of water and from soil surfaces [11]. Based on this value, the volatilization half-life of 1-chloro-1,1-difluoroethane from a model river 1 m deep flowing 1 m/sec with a wind velocity of 3 m/sec has been estimated to be approximately 2.93 hr [11].

**Water Concentrations:**

**Effluent Concentrations:**

**Sediment/Soil Concentrations:**

**Atmospheric Concentrations:**

**Food Survey Values:**

**Plant Concentrations:**

**Fish/Seafood Concentrations:**

**Animal Concentrations:**

**Milk Concentrations:**

**Other Environmental Concentrations:**

**Probable Routes of Human Exposure:** The most probable route of exposure to 1-chloro-1,1-difluoroethane by the general population is inhalation of aerosol sprays in which this compound is used as a propellant. In occupational settings, it is expected that exposure can occur by inhalation of contaminated air or dermal contact with this compound.

**Average Daily Intake:**

**Occupational Exposure:** NIOSH has statistically estimated that 17,721 workers are exposed to 1-chloro-1,1-difluoroethane in the US based upon a 1972-74 survey [12].

**Body Burdens:**

# REFERENCES

1. Atkinson R; Chem Rev 85: 69-201 (1985)
2. Aviado DM, Micozzi MS; pp. 3104-5 in Patty's Industrial Hygiene and Toxicology 3rd ed Vol IIB; Clayton GD, Clayton FE eds NY: John Wiley & Sons (1981)
3. Chemical Marketing Reporter; Chemical Profile: Fluorocarbons NY: Schnell Publish Co. March 10 (1986)
4. Chou CC et al; J Phys Chem 82: 1-7 (1978)
5. Dilling WL; Environmental Risk Analysis for Chemicals; Conway RA ed NY: Van Nostrand Reinhold Co pp 154-97 (1982)
6. Du Pont de Nemours Co; Freon Products Information B-2; A98825 12/80 (1980)
7. GEMS; Graphical Exposure Modeling System. CLOG3 (1986)
8. Graedel TE; Chemical Compounds in the Atmosphere NY: Academic Press p. 326 (1978)
9. Hine J, Mookerjee PK; J Org Chem 40: 292-8 (1975)
10. Horvath AL; Halogenated Hydrocarbons: Solubility Miscibility with Water. New York: Marcel Dekker pp 889 (1982)
11. Lyman WJ et al; Handbook of Chemical Property Estimation Methods. NY: McGraw-Hill (1982)
12. NIOSH; National Occupational Hazard Survey (1974)
13. Riddick JA et al; Organic Solvents: Physical Properties and Methods of Purification 4th ed NY: Wiley-Interscience pp. 569-70 (1986)
14. Swann RL et al; Res Rev 85: 17-28 (1983)

# Chloropentafluoroethane

## SUBSTANCE IDENTIFICATION

**Synonyms:**

**Structure:**

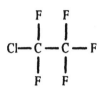

**CAS Registry Number:** 76-15-3

**Molecular Formula:** $C_2ClF_5$

**Wiswesser Line Notation:** GXFFXFFF

## CHEMICAL AND PHYSICAL PROPERTIES

**Boiling Point:** -38 °C at 760 mm Hg

**Melting Point:** -106 °C

**Molecular Weight:** 154.47

**Dissociation Constants:**

**Log Octanol/Water Partition Coefficient:** 2.10 [9]

**Water Solubility:** 58 mg/L at 25 °C [11]

**Vapor Pressure:** 7200 mm Hg at 25 °C [4]

**Henry's Law Constant:** 3.0 atm-m³/mole at 25 °C [10]

## ENVIRONMENTAL FATE/EXPOSURE POTENTIAL

**Summary:** Chloropentafluoroethane is released to the atmosphere in emissions from its use as an aerosol in food dispensers and as a refrigerant gas. If released to the atmosphere, chloropentafluoroethane will diffuse gradually into the stratosphere above the ozone layer where it will degrade slowly due to photolysis. The stratospheric lifetime has been estimated to range from 230 to 550 years. Decomposition will not occur in tropospheric air. If released to soil, chloropentafluoroethane is expected to have low mobility and, therefore, very little leaching into ground water is expected. Rapid volatilization should occur from terrestrial surfaces. If released to water, volatilization is expected to occur rapidly and be the dominant fate process. Bioconcentration in aquatic organisms is not expected to be important. Exposure of chloropentafluoroethane to the general population is possible by inhalation of air contaminated by release from aerosol dispensers or refrigerants in which the compound is or has been used. Significant occupational exposure may be possible.

**Natural Sources:** There are no natural sources of chloropentafluoroethane [5].

**Artificial Sources:** Chloropentafluoroethane is released to the atmosphere in emissions from its use as an aerosol in food dispensers and as a refrigerant gas [5,8].

**Terrestrial Fate:** The vapor pressure of chloropentafluoroethane suggests that volatilization from dry surfaces will occur very rapidly and probably be a dominant fate mechanism. Chloropentafluoroethane is expected to have low mobility in soil which suggests that leaching into ground water should be a slow process.

**Aquatic Fate:** When released into water, chloropentafluoroethane will volatilize rapidly. Based on experiments in a model estuarine ecosystem, degradation is not important in relation to volatilization [2]. Bioconcentration in aquatic organisms is not expected to be important.

**Atmospheric Fate:** Due to its very high vapor pressure, chloropentafluoroethane will exist entirely in the vapor phase in the ambient atmosphere. Chloropentafluoroethane is very stable in the troposphere; the only important decomposition process is UV photolysis which occurs in the stratosphere [6]. The stratospheric lifetime of chloropentafluoroethane has been estimated to range from 230 to 550 years. Therefore, when chloropentafluoroethane is released in the troposphere, it will diffuse gradually into the upper limits of the troposphere, and eventually, into the stratosphere above the ozone layer where it will degrade slowly via photolysis.

## Biodegradation:

**Abiotic Degradation:** No degradation of chloropentafluoroethane occurred in a photochemical smog system during 8 hours of irradiation at wavelengths above 310 nm [13]. Chloropentafluoroethane is not expected to react with hydroxyl radicals in the troposphere [1]. Chloropentafluoroethane does not absorb ultraviolet light above 290 nm [3,12]. The stratospheric lifetime of chloropentafluoroethane has been estimated to range from 230 to 550 years with direct photolysis being the dominant removal mechanism and reaction with singlet oxygen the secondary removal mechanism [3].

**Bioconcentration:** Based on the water solubility, the bioconcentration factor for chloropentafluoroethane can be estimated to be 61 from a recommended regression equation [14]. A bioconcentration factor of 61 does not indicate significant bioconcentration in aquatic organisms.

**Soil Adsorption/Mobility:** The mean Koc value of chloropentafluoroethane has been estimated to be 708 which is indicative of low soil mobility [18].

**Volatilization from Water/Soil:** The exchange of chloropentafluoroethane from water to gas phase was observed to occur at a relatively rapid rate during experiments conducted in a model estuarine ecosystem [2]. Based on the Henry's Law constant, rapid volatilization from water is expected [14].

**Water Concentrations:**

**Effluent Concentrations:**

**Sediment/Soil Concentrations:**

**Atmospheric Concentrations:** The chloropentafluoroethane concentration of a tropospheric air sample collected in 1979 was found to be 4.1 ppt/volume [17]. The concentration of chloropentafluoroethane in the atmosphere at altitudes ranging from about 10 to 34 km was found to be approximately 1-5 ppt/volume during 1979 and 1980 monitoring [5,7].

**Food Survey Values:**

**Plant Concentrations:**

**Fish/Seafood Concentrations:**

**Animal Concentrations:**

**Milk Concentrations:**

**Other Environmental Concentrations:**

**Probable Routes of Human Exposure:** Exposure to chloropentafluoroethane to the general population is possible by inhalation of air contaminated by releases from aerosol dispensers or refrigerants in which the compound is or has been used. Significant occupational exposure may be possible.

**Average Daily Intake:** AIR INTAKE: Assuming the average tropospheric concentration of chloropentafluoroethane is about 4.1 ppt $(25.9 \text{ ng/m}^3)$: 518 ng/day.

**Occupational Exposure:** 124,094 workers are potentially exposed to chloropentafluoroethane based on statistical estimates derived from the NIOSH survey conducted between 1972-74 in the United States [15]. 14,350 US workers are potentially exposed to chloropentafluoroethane based on statistical estimates derived from the NIOSH survey during 1981-83 [16].

**Body Burdens:**

## REFERENCES

1.  Atkinson R; Internat J Chem Kinetics 19: 799-828 (1987)
2.  Bopp RF et al; Org Geochem 3: 9-14 (1981)
3.  Chou CC et al; J Phys Chem 82: 1-7 (1978)
4.  Engineering Sciences Data Unit; Eng Sci Data Item 76004 pp 43 (1976)
5.  Fabian P, Gomer D; Fresenius Z Anal Chem 319: 890-7 (1984)
6.  Fabian P; pp. 23-5 in The Handbook of Environ Chem Hutzinger O, ed Vol 4/Part A Springer-Verlag NY (1986)
7.  Fabian P et al; Nature 294: 733 (1981)
8.  Graedel T; Chemical Compounds in the Atmosphere Academic Press NY p. 326 (1978)

# Chloropentafluoroethane

9.  Hansch C, Leo AJ; Medchem Project Issue No.26 Pomona College, Claremont CA (1985)
10. Hine J, Mookerjee PK; J Org Chem 40: 292-8 (1975)
11. Horvath AL; Halogenated Hydrocarbons: Solubility-Miscibility with Water. New York: Marcel Dekker pp 889 (1982)
12. Hubrich C, Stuhl F; J Photochem 12: 93-107 (1980)
13. Japer S et al; The Photostability of Fluorocarbons. Unpublished Report Univ of Riverside, CA (1974)
14. Lyman WJ et al; Handbook of Chem Property Estimation Methods. Compounds. McGraw-Hill NY (1982)
15. NIOSH; National Occupational Hazard Survey (1984)
16. NIOSH; National Occupational Exposure Survey (1985)
17. Penkett SA et al; J Geophys Res 86: 5172-8 (1981)
18. Roy WR, Griffin RA; Environ Geol Water Sci 7: 241-7 (1985)

# Cyclohexane

## SUBSTANCE IDENTIFICATION

**Synonyms:**

**Structure:**

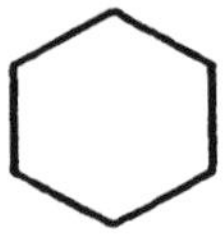

**CAS Registry Number:** 110-82-7

**Molecular Formula:** $C_6H_{12}$

**Wiswesser Line Notation:** L6TJ

## CHEMICAL AND PHYSICAL PROPERTIES

**Boiling Point:** 80.7 °C at 760 mm Hg

**Melting Point:** 6.47 °C

**Molecular Weight:** 84.18

**Dissociation Constants:**

**Log Octanol/Water Partition Coefficient:** 3.44 [15]

**Water Solubility:** 54.8 mg/L at 25 °C [28]

**Vapor Pressure:** 97.6 mm Hg at 25 °C [57]

**Henry's Law Constant:** 0.193 atm-m³/mole [16]

## ENVIRONMENTAL FATE/EXPOSURE POTENTIAL

**Summary:** Cyclohexane occurs naturally in crude oil and may be released wherever petroleum products are refined, stored, and used. Another large source of general release is in exhaust gases from motor vehicles. It is also produced in large quantities primarily as an intermediate in the manufacture of nylon, and releases in wastewater, and as fugitive emissions, can be expected in connection with its manufacture and use. If released on land, cyclohexane will be lost through volatilization and should leach into ground water. While cyclohexane is resistant to biodegradation, degradation occurs slowly in ground water in the presence of other petrochemicals. Volatilization from water (estimated half-life 2 hr in a model river) should be the most important fate process occurring in aquatic systems. While bioconcentration in aquatic organisms and adsorption to sediment is estimated to occur to a moderate extent, vaporization should be so rapid that they will not contribute significantly to cyclohexane's fate in water. In the atmosphere, cyclohexane will degrade by reaction with photochemically produced hydroxyl radicals (half-life 52 hr). The half-life is much faster under photochemical smog conditions with half-lives as low as 6 hr being reported. Human exposure will be primarily via inhalation especially in areas of high traffic and where petroleum products are used.

**Natural Sources:** Cyclohexane occurs naturally in crude oil and as a plant volatile [11]. It is also released into the atmosphere from volcanos [11].

**Artificial Sources:** Cyclohexane is manufactured in large quantities (2.2 billion lbs in 1986) and may be released to the environment as fugitive emissions and in wastewater during its manufacture and use as a chemical intermediate and solvent [5,11]. Ninety percent of this use is in the manufacture of adipic acid and caprolactam for the manufacture of nylon [5]. It is a component of petroleum and its primary release will be fugitive emissions from petroleum refining, vaporization of gasoline, oil spills, and in gasoline exhaust [11]. It was estimated that in 1980, 92.4-92.8 million pounds of cyclohexane were released from land transportation vehicles and over 1 million pounds may have entered the environment as fugitive emissions or evaporative losses from refineries [3]. Another source of release is in tobacco smoke [3].

**Terrestrial Fate:** If released on land, cyclohexane will be lost through volatilization and should leach into the ground. Cyclohexane is resistant to biodegradation but may biodegrade slowly in the presence of other hydrocarbons that are themselves degraded.

# Cyclohexane

**Aquatic Fate:** Volatilization from water (estimated half-life 2 hr in a model river) should be the most important fate process occurring in aquatic systems [44].

**Atmospheric Fate:** In the atmosphere, cyclohexane will degrade by reaction with photochemically produced hydroxyl radicals (half-life 52 hr). The half-life is much faster under photochemical smog conditions with half-lives as low as 6 hr being reported.

**Biodegradation:** Cyclohexanes are highly resistant to biodegradation and are catabolized chiefly by cooxidation [36,44]. Thus they do not support growth of the degrading organism themselves, but rather are metabolized during the course of the microorganisms growth on another, usually similar, substrate. Initial attack involves oxygenation and subsequent ring cleavage to simple, readily degradable acids [44]. 10% degradation in 12 hr was reported by microorganisms isolated from a brackish creek in an area continuously exposed to oil [52]. When incubated at 12 °C with natural flora in ground water in the presence of other components of high octane gasoline, 45% degradation was reported after 8 days [50]. Only slight degradation of cyclohexane was noted in a screening test utilizing a benzene-acclimated activated sludge inoculum [27] and it was listed as degradation resistant according to the MITI test, a screening test of the Japanese Ministry of International Trade and Industry [19]. Only one biodegradability test was reported in soil in which it was listed as nondegradable with only 0.3% mineralization occurring in 10 wk [13].

**Abiotic Degradation:** In the atmosphere, cyclohexane will react with photochemically produced hydroxyl radicals with a half-life of 52 hr based on a recommended rate constant of $7.38 \times 10^{-12}$ cm$^3$/molecule-sec [2] and a hydroxyl radical concentration of $5 \times 10^{+5}$ cm$^3$/sec. Photodegradation is much faster, however, in the presence of nitrogen oxides (photochemical smog conditions). 39% degradation occurred in 6 hr when Los Angeles air was exposed to sunlight [20]. Other investigators reported 50% loss in 6.7 hr [7], 29% loss in 5 hr [56], and 24-46% degradation in 2-4 hr [54]. Its reactivity compared with other solvents measured by ozone forming potential is relatively low (2 on a scale of 5) [9]. The products of reaction are cyclohexanone, cyclohexyl nitrate, and unidentified carbonyl compounds resulting from ring cleavage [53]. Cyclohexane does not have any chromophores that absorb UV radiation >290 nm, so it would not be subject to direct photolysis. When solutions of cyclohexane in distilled water or natural filtered salt or fresh water were exposed to sunlight for 21 days, no significant photodegradation was noted [43].

**Bioconcentration:** No experimental data are available on the bioconcentration of cyclohexane in aquatic organisms. Using the octanol/water partition coefficient, one can estimate a BCF of 242 using a recommended regression equation [25].

**Soil Adsorption/Mobility:** The Koc for cyclohexane estimated from its water solubility by a recommended regression equation is 480, indicating a moderate adsorptivity to soil [25]. The adsorption of components of JP4 jet fuel were tested for their adsorption to three sediments and two clays [26]. The results showed a rather small interaction with these adsorbents and the adsorptivity was only casually related to the organic carbon content of the sediment [26]. The adsorption constant in (mg/g)/(mg/L) for cyclohexane in the three sediments ranged from 13.0 to 61.1 and 20.8 and 0.6 in montmorillonite and illite, respectively [26].

**Volatilization from Water/Soil:** The very high Henry's Law constant for cyclohexane indicates that it will volatilize rapidly from water with the rate being controlled by diffusion through the liquid phase [25]. Using the Henry's Law constant, one can calculate that the volatilization half-life from a model river 1 m deep with a 1 m/sec current and a 3 m/sec wind is 2.8 hr [25]. In the course of performing biodegradability screening tests on cyclohexane, evaporation was so rapid that the chemical was undetected after 4 hr [44]. In view of its high vapor pressure and moderate adsorption to soil, volatilization from soil and surfaces should be considerable.

**Water Concentrations:** DRINKING WATER: A contaminated drinking water well in New York contained 540 ppb of cyclohexane [4]. While a public drinking water supply in East Anglia, England that was 210 m from a leaking petroleum storage tank contained <0.01 ppb of cyclohexane, the ground water 10 and 110 m from the tank were 10 and <1 ppb, respectively [48]. Cyclohexane was found in 5 of 14 waterworks sampled in the United Kingdom. [10]. The sources of the cyclohexane-tainted drinking water were rivers, reservoirs and ground water, and the contaminant was present in raw as well as treated water [10]. GROUND WATER: Trace quantities of cyclohexane were found in 1 of 11 bedrock domestic wells near the municipal landfill in Granby, CT [40]. SURFACE WATER: In a survey of 14 heavily industrialized river basins in the United States (204 sites), 1 site in the Hudson River Basin contained 1 ppb of cyclohexane and 13 sites in the Mississippi River Basin contained 1.0-4.0 ppb of cyclohexane [8]. Grab samples of surface water in the Gulf of Mexico in which there was anthropogenic influence contained 0-0.02 ppb of cyclohexane [39]. Open water and unpolluted coastal water did not contain detectable quantities of the chemical [39]. Surface water at an offshore oil production operation in the Gulf of Mexico contained 0.47 ppb [38].

**Effluent Concentrations:** In a comprehensive survey of wastewater from 4000 industrial and publicly owned treatment works sponsored by the Effluent Guidelines Division of the US EPA, cyclohexane was identified in discharges of the following industrial categories (frequency of occurrence; median concn in ppb): timber products (3; 23.0), petroleum refining (7; 46.0), coal mining (3; 20.0), organics and plastics (33; 14.0), inorganic chemicals (9; 267.0), textile mills (3; 6.0), plastics and synthetics (2; 6.0), rubber processing (5; 125.0), auto and other laundries (1; 3.0), pesticides manufacture (5; 48.0), photographic industry (1; 47.0), pharmaceuticals(1; 3.0), oil and gas extraction (7; 4.0), mechanical products (1; 7.0), and publicly owned treatment works (14; 6.0) [42]. Cyclohexane was one of the volatile organic hydrocarbons emitted from simulated municipal landfills [51]. These "simulated landfills" consisted of steel tanks packed with municipal waste and various loading rates of municipal wastewater sludge. Analysis started 8 months after loading and continued for a year. Cyclohexane sampled in formation water and in an underwater vent plume at an offshore oil production operation in the Gulf of Mexico was 100 ppb and 1 umol/L of gas, respectively [38].

**Sediment/Soil Concentrations:**

**Atmospheric Concentrations:** RURAL/REMOTE: Rio Blanco County, CO July 1978 - 0.10-0.60 ppb-C [1]; Smokey Mountain National Park, Sept 1978 - 0-1.20 ppb-C [1]. Jones State Forest, TX, Jan 1978 - 0.3 to 3.3 ppb-C, 2.5 ppb-C mean [41]. Concentrations of 9 samples of air from the Norwegian Arctic (July 1982) were <20 ppt while 10 samples (Spring 1983) averaged 54 ppt [17]. URBAN/SUBURBAN: The concentration of cyclohexane in Tulsa, OK on July 27, 1978 was 0-3.30 ppb-C whereas at a rural site 37 km downwind it was 2.70 ppb-C [1]. Various locations in Houston, TX, including industrial areas and tunnels - 0-22.4 ppb, 9.37 ppb, mean with 9 of 16 samples pos [24]. Early morning concn of cyclohexane in the central business district in Los Angeles in the summer and fall of 1961 - 0-6 ppb (6th floor level) [29]. A study of Los Angeles air carried out in the fall of 1981 reported cyclohexane concn ranging from 7-31 ppb [12]. The average concn of cyclohexane over a 30 day period (140 samples) in Sydney, Australia and environs was 0.9 ppb [31]. The concn of the hydrocarbons reaches a maximum at 8 AM suggesting the influence of rush hour traffic, and than falls to a minimum at 12 noon [31]. It has been reported in the air in six Russian cities of widely different geography and climate [18]. The mean (standard deviation) of cyclohexane 350-600 m aloft over Tokyo was 0.1 (0.19) ppb [49]. SOURCE AREA: Texaco refinery in Tulsa, OK - 9.0-14.8 ppb-C [1]. Mobil natural gas facility, Rio Blanco County, CO - 31.9 ppb-C [1]. The mean concn (standard deviation) of cyclohexane in the Lincoln Tunnel in New York City was 57.3 (13.0) and 45.6 (13.9) ppb-C in 1970 and 1982, respectively [23]. It was detected in the Allegheny Mountain Tunnel, PA [14] and roadside ambient air samples [47]. Ambient samples in the vicinity of an oil fire contained 0.2 mg/m$^3$ of cyclohexane [35].

**Food Survey Values:** Muscat oil (from Muscat grapes) was found to contain cyclohexane [46]. It was suggested that this may have been due to its presence as a solvent in a pesticide sprayed on the grapes [46].

**Plant Concentrations:**

**Fish/Seafood Concentrations:**

**Animal Concentrations:**

**Milk Concentrations:**

**Other Environmental Concentrations:** Cyclohexane occurs in crude oil at concentrations ranging from 0.5-1.0% (weight) and the concentration in gasoline ranges from 5-15% (volume) [45]. The average concn in the exhaust of 67 vehicles was 0.6% (weight) [30]. One pesticide grade n-hexane sample used as a solvent contained 0.1-1.0% cyclohexane [55]. Trace amounts were also identified in cigarette smoke [37]. JP4 synthetic jet fuel mixture contains 6.5% (weight) of cyclohexane [26].

# Cyclohexane

**Probable Routes of Human Exposure:** The general population will be exposed to cyclohexane from ambient air, especially in areas with heavy traffic, near filling stations, and in other areas.

**Average Daily Intake:** ATMOSPHERIC INTAKE: (assume 0.9 to 9 ppb mean intake) 62.8-628 ng/day; WATER INTAKE: insufficient data; FOOD INTAKE: insufficient data.

**Occupational Exposures:** Based on statistical estimates derived from the NIOSH National Occupational Hazard Survey (NOHS) conducted in 1972-74, 1,139,397 workers are potentially exposed to cyclohexane [33]. This survey sampled 5000 businesses in 67 metropolitan areas throughout the U.S. for the manufacture and use of chemicals, trade name products known to contain the compound, and generic products suspected of containing the compound. According to statistical estimates derived from the NIOSH National Occupational Exposure Survey (NOES) conducted in 1981-83, 51,611 workers are potentially exposed to cyclohexane [32]. Cyclohexane was a predominant pollutant in shoe and leather factories in Italy and was associated with the use of glue [6].

**Body Burdens:** Five of eight samples of human milk collected from 4 urban/industrial areas in the United States were positive for cyclohexane [34]. A population of 54 normal, healthy, nonsmoking subjects from urban environments (387 samples) had a 64.6% occurrence of cyclohexane in expired air with a geometric mean of 5.0 ng/L [21]. In a heterogenous, nonsmoking population of 62 subjects, cyclohexane was a common constituent in the expired air of control, diabetic, and prediabetic subjects [22].

## REFERENCES

1. Arnts RR, Meeks SA; Biogenic Hydrocarbon Contribution to the Ambient Air of Selected Areas EPA-600/3-80-023 (1980)
2. Atkinson R; Chem Rev 85: 69-201 (1985)
3. Beals SM et al; Tech Support Doc Cyclohexane SRC TR-86-030 Syracuse Res Corp, Syracuse NY (1986)
4. Burmaster DE; Environ 24: 6-13 to 33-36 (1982)
5. Chemical and Engineering News; Key Chemicals p. 14 (1986)
6. Cresci A et al; Nuori Ann Ig Microbiol 36: 61-76 (1985)
7. Dilling WL et al; Environ Sci Technol 10: 351-6 (1976)
8. Ewing BB et al; Monitoring to Detect Previously Unrecognized Pollut in Surface Waters EPA-560/6-77-015 p. 75 (1977)

# Cyclohexane

9.  Farley FF; Photochem Reactivity Classification of Hydrocarbons and Other Organic Cmpds EPA-600/3-77-001B pp. 713-27 (1977)
10. Fielding M et al; Organic Micropollutants in Drinking Water TR-159 Eng Water Res Cent p. 49 (1981)
11. Graedel TE; Chem Cmpd in the Atmos, Academic Press NY p. 99 (1978)
12. Grosjean D, Fung K; J Air Pollut Contr Assoc 34: 537-43 (1984)
13. Haider K et al; Arch Microbiol 96: 183-200 (1974)
14. Hampton CV et al; Environ Sci Technol 16: 287-98 (1982)
15. Hansch C, Leo AJ; Medchem Project Issue No 26. Claremont CA: Pomona College (1985)
16. Hine J, Mookerjee PK; J Org Chem 40: 292-8 (1975)
17. Hov O et al; Organic Fases in the Norwegian Arctic Geophys Res Lett 11: 425-8 (1984)
18. Ioffe BV et al; Environ Sci Technol 13: 864-8 (1979)
19. Kawaski M; Ecotoxic Environ Safety 4: 444-54 (1980)
20. Kopczynski SL et al; Environ Sci Technol 6: 342 (1972)
21. Krotoszynski BK et al; J Anal Toxicol 3: 225-34 (1979)
22. Krotoszynski BK, Oneill HJ; J Environ Sci Health 17: 855-83 (1982)
23. Lonneman WA et al; Environ Sci Technol 20: 790-6 (1986)
24. Lonneman WA et al; Hydrocarbons in Houston Air EPA 600/3-79-018 (1979)
25. Lyman WJ et al; Handbook of Chem Property Estimation Methods Environ Behavior of Organic Compounds McGraw-Hill (1982)
26. MacIntyre WG et al; Hydrocarbon Fuel Chem Sediment Water Interaction NTIS AD-A117928 VA Inst Mar Sci p. 53 (1982)
27. Malaney GW, McKinney RE; Water Sewage Works 113: 302-9 (1966)
28. McAuliffe C; J Phys Chem 70: 1267-75 (1966)
29. Neligan RE; Arch Environ Health 5: 581-91 (1962)
30. Nelson PF, Quigley SM; Atmos Environ 18: 79-87 (1984)
31. Nelson PF, Quigley SM; Environ Sci Technol 16: 650-5 (1982)
32. NIOSH; National Occupational Exposure Survey (1985)
33. NIOSH; National Occupational Health Survey (1975)
34. Pellizzari ED et al; Bull Environ Contam Toxicol 28: 322-8 (1982)
35. Perry R; Mass Spectrometry in the Detection and Identification of Air Pollutants. Int. Sympos Ident Meas Environ Pollut pp 130-7 (1971)
36. Perry JJ; Microbial Rev 43: 59-72 (1979)
37. Phillippe RJ et al; J Chromatography 20: 250-9 (1965)
38. Sauer TC, JR; Environ Sci Technol 15: 917-23 (1981)
39. Sauer TC, JR; Org Geochem 3: 91-101 (1981)
40. Sawhney BL, Raabe JA; Ground Water Contamination Movement of Organic Pollut in the Granby Landfill, Bulletin 833, CT Agric Experiment Station p. 9 (1986)
41. Seila RL; Non-Urban Hydrocarbon Concn in Ambient Air North of Houston, Texas EPA-500/3-79-010 (1979)
42. Shackelford WM et al; Analyt Chem Acta 146: 15-27 (1983)
43. Smith JH, Harper JC; in Proceed 12th Conf on Environ Tox 3, 4, and 5, Behavior of Hydrocarbon Fuels in Aquatic Systems, Airforce Aerospace Med Res Lab, Ohio pp. 336-53 (1982)
44. Spain JC et al; Degradation of Jet Fuel Hydrocarbons by Aquatic Microbial Communities NTIS AD-A139791/8 Air Force Eng Serv Ctr p. 226 (1983)
45. SRI International; Chem Economics Handbook, US Menlo Park, CA (1983)
46. Stevens KL et al; J Agric Food Chem 14: 249-52 (1966)

# Cyclohexane

47. Stump FD, Dropkin DL; Anal Chem 57: 2629-34 (1985)
48. Tester DJ, Harker RJ; Water Pollut Control 80: 614-31 (1981)
49. Uno T et al; Atmos Environ 19: 1283-93 (1985)
50. Verschueren K; Handbook of Environ Data on Organic Chemicals 2nd ed Van Nostrand Reinhold NY pp. 417-9 (1983)
51. Vogt WG, Walsh JJ; Volatile Organic Compounds in Gases from Landfill Simulators, Proc APCA Annu Meet Vol 6 pp. 17 (1985)
52. Walker JD, Colwell RR; Appl Environ Microbiol 31: 189-97 (1976)
53. Washida N et al; Kokuritsu Kogai Kenkyusho Kenkyu Hokoku 59: 49-61 (1984)
54. Washida N et al; Bull Chem Soc Japan 51: 2215-21 (1978)
55. William IH; J Chromatographic Sci 11: 593-6 (1973)
56. Yanagihara S et al; Photochemical Reactivities of Hydrocarbons, Proc Int Clean Air Congr 4th ed pp. 472-7 (1977)
57. Zwolinski BJ, Wilhoit RC; Handbook of Vapor Pressures and Heats of Vaporization of Hydrocarbons and Related Compounds AP144-TRC101 (1971)

# Cyclohexanone

## SUBSTANCE IDENTIFICATION

**Synonyms:**

**Structure:**

**CAS Registry Number:** 108-94-1

**Molecular Formula:** $C_6H_{10}O$

**Wiswesser Line Notation:** L6VTJ

## CHEMICAL AND PHYSICAL PROPERTIES

**Boiling Point:** 155.6 °C at 760 mm Hg

**Melting Point:** -32.1 °C

**Molecular Weight:** 98.14

**Dissociation Constants:**

**Log Octanol/Water Partition Coefficient:** 0.81 [8]

**Water Solubility:** 23,000 mg/L at 25 °C [24]

**Vapor Pressure:** 4.8 mm Hg at 25 °C [24]

**Henry's Law Constant:** 1.2 x $10^{-5}$ atm-m³/mole [9]

## ENVIRONMENTAL FATE/EXPOSURE POTENTIAL

**Summary:** Cyclohexanone can be released to the environment through air and wastewater emissions involved with its industrial production and production of derivatives such as nylon. Various solvent uses result in direct evaporation into the atmosphere. If released to the atmosphere, cyclohexanone will degrade relatively rapidly by reaction with sun light-produced hydroxyl radicals (half-life of about 1 day) and by direct photolysis (half-life of about 4.3 days). If released to water, cyclohexanone may degrade significantly through biodegradation and photolysis. Volatilization from environmental waters will not be rapid except from rapidly moving, shallow streams. If released to soil, cyclohexanone will be susceptible to significant leaching. Volatilization and photodegradation will occur on soil surfaces. Humans will be primarily exposed to cyclohexanone by inhalation or dermal contact in occupational settings. The general population may be exposed through consumption of contaminated drinking water or inhalation of contaminated air.

**Natural Sources:**

**Artificial Sources:** Cyclohexanone is emitted to the atmosphere during its industrial production from venting of "spent air" and other vent streams [21]. Use of cyclohexanone as a solvent for lacquers, inks, resins, paint and spot removers, and insecticides results in its evaporation into the air. Cyclohexanone releases in wastewater and air may also occur from its primary use as an intermediate to manufacture nylon.

**Terrestrial Fate:** When released to soil, cyclohexanone can be expected to leach significantly based on a predicted mean Koc value of 10. Its detection in ground water and in leachate from a waste site indicate that leaching does occur in the environment. Various biological screening studies have found that cyclohexanone is biodegradable in various test systems, including natural water. This suggests that biodegradation in soil is possible. Cyclohexanone on soil surfaces will be susceptible to significant volatilization and photodegradation.

**Aquatic Fate:** The important environmental fate processes for cyclohexanone in water appear to be biodegradation, photolysis, and volatilization. Various biological screening studies have found that cyclohexanone is biodegradable in various test systems, including natural water. Cyclohexanone photolyzes in ambient air (half-life of about 4.3 days) which suggests that direct photolysis in water is likely to occur; however, the photolysis rate in water will be slower. Volatilization from shallow, rapidly moving streams should be significant (estimated half-life of 3.1 days from a model stream 1 m deep); however, volatilization from deeper and less rapidly moving bodies of water such as lakes and ponds will be much slower. Aquatic hydrolysis, bioconcentration, and adsorption to sediment are not important.

**Atmospheric Fate:** Based on the vapor pressure, cyclohexanone should exist almost entirely in the vapor phase in the ambient atmosphere [6]. It will degrade relatively rapidly by reaction with sun light-produced hydroxyl radicals (half-life of about 1 day) and by direct photolysis (half-life of about 4.3 days).

**Biodegradation:** Cyclohexanone was found to be significantly biodegradable using the Japanese MITI test protocol [11,28]. A 5-day 32% BODT was determined using the AFNOR T 90/103 test and microbes from 3 polluted surface waters [5]. A 96% removal (based on COD) in a mineral salts solution was observed using an acclimated activated sludge inoculum [22]. A 5-day 68.2% BODT was noted using a standard dilution method and a 5-day 62.4% BODT was noted using a seawater dilution method [31]. In Warburg respirometer studies, a 50% BOD was observed in 20 hr using adapted cultures and a 50% BOD in 50 hr using nonadapted cultures [33]. Cyclohexanone was degradable in an activated sludge system [19].

**Abiotic Degradation:** Cyclohexanone absorbs UV light in the environment up to an approximate cutoff point of 325 nm [16]. The measured direct photolysis rate constant in air is approximately 0.16 days$^{-1}$, which corresponds to a half-life of 4.3 days [16]. The rate constant for the vapor-phase reaction with hydroxyl radicals in an average atmosphere ($5 \times 10^{+5}$ hydroxyl radicals/cm$^3$) has been estimated to be $1.56 \times 10^{-11}$ cm$^3$/molecule-sec at 25 °C which corresponds to a half-life of about 1 day [1]. Ketones are generally resistant to aqueous hydrolysis in the environment [14]; therefore, hydrolysis of cyclohexanone is not expected to be important.

**Bioconcentration:** Based on the octanol/partition coefficient, the log BCF for cyclohexanone has been estimated to be 0.39 from recommended regression equations [14].

**Soil Adsorption/Mobility:** Cyclohexanone has been classified as very highly mobile in soil based on a predicted mean Koc value of 10 (estimated from water solubility, log Kow, and regression equations) [25].

**Volatilization from Water/Soil:** The value of Henry's Law constant indicates that volatilization from environmental waters is possibly a significant loss process, but is not rapid [14]. Using this value of Henry's Law constant, the volatilization half-life of cyclohexanone from a model river 1 m deep flowing 1 m/sec with a wind velocity of 3 m/sec has been estimated to be 3.1 days [14].

**Water Concentrations:** DRINKING WATER: Cyclohexanone was qualitatively detected in drinking water from Miami, FL; Ottumwa, IA; Philadelphia, PA; and Cincinnati, OH [32]. Ottumwa, IA tap water contained 0.1 ppb cyclohexanone [12]. Positive identifications in drinking water from Cincinnati, OH (Jan 1980), Miami, FL (Feb 1976), New Orleans, LA (Jan 1976), Ottumwa, IA (Sept 1976), and Seattle, WA (Nov 1976) [13]. GROUND WATER: Maximum concn of 30 ppb detected in ground waters from the Netherlands [34]. Qualitative detection in a ground water aquifer in Australia [30]. SURFACE WATER: Cyclohexanone was qualitatively detected in surface waters within the blast zone of Mount St. Helens, WA following the volcanic eruption of May 18, 1980 [15]. Qualitative detection reported for the Rhine River [29].

**Effluent Concentrations:** Cyclohexanone detections reported for wastewaters from chemical, latex, and textile manufacturing facilities [29]. Wastewater detections for effluents from advanced treatment facilities in Lake Tahoe, CA; Dallas, TX; and Washington, DC (Blue Plains Treatment Plant) [13]. A final effluent grab sample from the Danville, IL POTW collected in Jun 1980 contained cyclohexanone [7]. Wastewaters from processing shale oil have been found to contain cyclohexanone [4,9]. Cyclohexanone was identified in leachate collected near a municipal landfill in Ontario, Canada [23].

**Sediment/Soil Concentrations:**

**Atmospheric Concentrations:** Cyclohexanone detected in 4 of 12 air samples collected near American Cyanamide in Linden, NJ (1976-78) at levels of 22-629 ng/m$^3$ (270 ng/m$^3$ avg) [20]. Maximum ground level concn of 153 ug/m$^3$ detected on the boundary of the Holston Army Munitions Plant [26]. Qualitative identification made for forest air sampled from the Southern Black Forest in Germany in 1983 [10]. Positive identifications in 2 of 6 indoor air samples (1-6 ug/m$^3$) collected from homes in Italy in 1983-84 [3].

# Cyclohexanone

**Food Survey Values:**

**Plant Concentrations:**

**Fish/Seafood Concentrations:**

**Animal Concentrations:**

**Milk Concentrations:**

**Other Environmental Concentrations:**

**Probable Routes of Human Exposure:** Humans will be primarily exposed to cyclohexanone by inhalation or dermal contact in occupational settings. The general population may be exposed through consumption of contaminated drinking water or inhalation of contaminated air.

**Average Daily Intake:**

**Occupational Exposures:** 839,199 Workers are potentially exposed to cyclohexanone based on statistical estimates derived from the NIOSH Survey conducted between 1972-74 in the United States [18]. 373,986 workers are potentially exposed to cyclohexanone based on preliminary statistical estimates derived from the NIOSH Survey conducted between 1981-83 in the United States [17]. Cyclohexanone levels of 0-10 ug/m³ detected in the vulcanization area of a shoe-sole factory [2]. Mean TWA levels ranging from 6-28 ppm detected in breathing zone of workers at various job sites in a screen printing plant [27].

**Body Burdens:**

## REFERENCES

1.  Atkinson RA; Chem Rev 85: 69-201 (1985)
2.  Cocheo V et al; Am Ind Hyg Assoc J 44: 521-7 (1983)
3.  DeBortoli M et al; Environ Internat 12: 343-50 (1986)
4.  Dobson KR et al; Water Res 19: 849-56 (1985)
5.  Dore M et al; Trib Cebedeau 28: 3-11 (1975)
6.  Eisenreich SJ et al; Environ Sci Technol 15: 30-8 (1981)
7.  Ellis DD et al; Arch Environ Contam Toxicol 11: 373-82 (1982)
8.  Hansch C, Leo AJ; Medchem Project Issue No 26. Claremont CA: Pomona College (1985)
9.  Hawthorne SB et al; Environ Sci Technol 19: 992-7 (1985)

# Cyclohexanone

10. Juttner F; Chemosphere 15: 985-92 (1986)
11. Kawasaki M; Ecotoxic Environ Safety 4: 444-54 (1980)
12. Keith LH et al; pp.329-73 in Ident Anal Organic Pollut Water; Keith LH ed Ann Arbor,MI: Ann Arbor Press (1976)
13. Lucas SV; GC/MS Analysis of Organics in Drinking Water Concentrates and Advanced Waste Treatment Concentrates: Vol 1. USEPA-600/1-84-020a p.45,133 (1984)
14. Lyman WJ et al; Handbook of Chemical Property Estimation Methods NY:McGraw-Hill p.5-4,5-10 (1982)
15. McKnight DM et al; Org Geochem 4: 85-92 (1982)
16. Mill T, Davenport J; ACS Div Environ Chem 192nd Natl Mtg 26: 59-63 (1986)
17. NIOSH; National Occupational Exposure Survey (NOES) (1983)
18. NIOSH; National Occupational Hazard Survey (NOHS) (1974)
19. Patel MD, Patel DR; Indian J Environ Health 19: 310-8 (1977)
20. Pellizzari ED; Quantification of Chlorinated Hydrocarbons in Previously Collected Air Samples. USEPA-450/3-78-112 (1978)
21. Pervier JW et al; Survey Reports on Atmospheric Emissions from the Petrochemical Industry, Vol II. USEPA-450/3-73-005b p.38-61 (1974)
22. Pitter P; Water Res 10: 231-5 (1976)
23. Reinhard M et al; Environ Sci Technol 18: 953-61 (1984)
24. Riddick JA et al; Organic Solvents: Physical Properties and Methods of Purification, 4th Edit. New York: J Wiley & Sons (1986)
25. Roy WR, Griffin RA; Environ Geol Water Sci 7: 241-7 (1985)
26. Ryon MG et al; Database Assessment of the Health and Environmental Effects of Munitions Production Waste Products. Final Report. ORNL-6018. Oak Ridge,TN: Oak Ridge Natl Lab p.164 (1984)
27. Samimi B; Am Ind Hyg Assoc J 43: 43-8 (1982)
28. Sasaki S; p.283-98 in Aquatic Pollutants: Transformation and Biological Effects; Hutzinger O et al eds, Oxford: Pergamon Press (1980)
29. Shackelford WM, Keith LH; Frequency of Organic Compounds Identified in Water. USEPA-600/4-76-062 p.105 (1976)
30. Stepan S et al; Austral Water Resources Council Conf Ser 1: 415-24 (1981)
31. Takemoto S et al; Suishitsu Odaku Kenkyu 4: 80-90 (1981)
32. USEPA; Preliminary Assessment of Suspected Carcinogens in Drinking Water. Interim Report to Congress June, 1975 p.9 (1975)
33. Wotzka J et al; Acta Hydrochim Hydrobiol 13: 583-90 (1985)
34. Zoeteman BCJ et al; Sci Total Environ 21: 187-202 (1981)

# Cyclohexylamine

**Synonyms:**

**Structure:**

**CAS Registry Number:** 108-91-8

**Molecular Formula:** $C_6H_{13}N$

**Wiswesser Line Notation:** L6TJ AZ

## CHEMICAL AND PHYSICAL PROPERTIES

**Boiling Point:** 134.5 °C at 760 mm Hg

**Melting Point:** -17.7 °C

**Molecular Weight:** 99.17

**Dissociation Constants:** pKa = 10.66 at 24 °C [13]

**Log Octanol/Water Partition Coefficient:** 1.49 [6]

**Water Solubility:** Miscible [14]

**Vapor Pressure:** 8.8 mm Hg at 25 °C [14]

**Henry's Law Constant:** 2.13 x $10^{-5}$ at 25 °C (estimated) [7]

## ENVIRONMENTAL FATE/EXPOSURE POTENTIAL

**Summary:** Cyclohexylamine will enter the environment as emissions or in wastewater during its manufacture and use in boiler water treatment, production of rubber chemicals, and as a chemical

# Cyclohexylamine

intermediate. If released to the atmosphere, cyclohexylamine would be expected to photooxidize by reaction with hydroxyl radicals (calculated half-life of 0.279 days). If released on land, it would be subject to evaporation and leaching to ground water where its fate is unknown. It should not adsorb to soil but may be subject to biodegradation. If released to water, it may be subject to evaporation and hydrolysis, but will not adsorb to sediments or bioconcentrate in aquatic organisms. It is biodegraded in river muds and sewage inocula and, therefore, may be susceptible to biodegradation in water. Human exposure is difficult to estimate in the absence of monitoring data, but may be primarily occupational.

**Natural Sources:** Cyclohexylamine is not known to occur as a natural product [8].

**Artificial Sources:** Cyclohexylamine is a metabolite of the artificial sweetener cyclamate [8]. It may also be released as a result of its major use in boiler water treatment, as well as in rubber chemicals and as a chemical intermediate in the production of dyes, insecticides, and pharmaceuticals [15]. Estimated cyclohexylamine demand was 8.2 million lb. in 1983 and 8.5 million lbs. in 1984 [4].

**Terrestrial Fate:**   No information concerning the fate and transport of cyclohexylamine in soil was found. If released to soils, it will probably be subject to evaporation as well as leaching to ground water, with little adsorption to soils expected. Since cyclohexylamine is biodegraded in river muds and sewage inocula, biodegradation in soils may be significant.

**Aquatic Fate:**   Very little information concerning the fate and transport of cyclohexylamine in water was found. If released to water, cyclohexylamine will not be expected to adsorb appreciably to sediments or bioconcentrate in aquatic organisms. Since it is biodegraded in river muds and sewage inocula, it may be biodegraded in water, but is not likely to directly photolyze. Its estimated half-life of evaporation from a model river 1 m deep flowing 1 m/sec with a wind velocity of 3 m/sec is estimated to be 1.8 days.

**Atmospheric Fate:** No information concerning the atmospheric fate and transport of cyclohexylamine was found. If released to the atmosphere, cyclohexylamine is not likely to directly photolyze but should react with photochemically produced hydroxyl ions with an estimated half-life of 0.279 days. Due to its high water solubility, cyclohexylamine may undergo significant washout from the atmosphere.

**Biodegradation:** Biodegradation of cyclohexylamine in 3 inocula over 14 days: acclimated sewage sludge, 100 mg/L - 100% degradation, 79.1% theoretical BOD (TBOD); sewage sludge, 50 mg/L - 100% degradation, 67.8% TBOD; sewage sludge, 100 mg/L - 0% degradation; river mud, 50 mg/L - 0% degradation; river mud, 10 mg/L, 100% degradation, 82.1% TBOD [2].

**Abiotic Degradation:** Aliphatic amines do not absorb radiation above 250 nm [3] so cyclohexylamine would not be expected to directly photolyze. The principal loss mechanism for amines in the atmosphere is by reaction with photochemically produced hydroxyl radicals [5] which results in an estimated half-life of 0.279 days assuming a concentration of $5 \times 10^5$ radicals/cm$^3$ [1].

**Bioconcentration:** Using the recommended value for the log octanol-water partition coefficient, a BCF of 7.99 was estimated [10]. Based on this estimated BCF value, cyclohexylamine is not expected to bioconcentrate significantly in aquatic organisms.

**Soil Adsorption/Mobility:** Using the recommended value for the log octanol/water partition coefficient, a Koc value of 154 was estimated [10]. Based on this Koc value and the high water solubility of cyclohexylamine, extensive leaching and very little adsorption to soil or sediments is expected.

**Volatilization from Water/Soil:** The volatilization half-life from a model river 1 m deep flowing 1 m/sec with a wind velocity of 3 m/sec is estimated to be 1.8 days [10].

**Water Concentrations:**

**Effluent Concentrations:** Effluent from a tire manufacturing plant contained cyclohexylamine at approximately 0.01 ppm [9].

**Sediment/Soil Concentrations:**

**Atmospheric Concentrations:**

**Food Survey Values:**

# Cyclohexylamine

**Plant Concentrations:**

**Fish/Seafood Concentrations:**

**Animal Concentrations:**

**Milk Concentrations:**

**Other Environmental Concentrations:**

**Probable Routes of Human Exposure:**

**Average Daily Intake:**

**Occupational Exposure:** A National Occupational Hazard Survey (1973-74) estimates that 9532 people were exposed to cyclohexylamine [11]. An 1981-83 NIOSH survey estimates that 48,028 total workers are exposed to cyclohexylamine [12].

**Body Burdens:**

## REFERENCES

1. Atkinson R; Internat J Chem Kinetics 19: 799-828 (1987)
2. Calamari D et al; Chemosphere 9: 753-62 (1980)
3. Calvert JG, Pitts JN Jr; Photochemistry. John Wiley & Sons New York pp.899 (1966)
4. Chemical Marketing Reporter Chem Profiles Jan 30 (1984)
5. Graedel TE; p.283-92 in Chemical Compounds in the Atmosphere. Academic Press New York (1978)
6. Hansch C, Leo AJ; Medchem Project Issue No.26, Pomona College, Claremont CA (1985)
7. Hine J, Mookerjee PK; J Org Chem 40: 292-8 (1975)
8. IARC; Some Non-nutritive Sweetening Agents 22: 55-109 (1980)
9. Jungclaus GA et al; Anal Chem 48: 1894-96 (1976)
10. Lyman WJ et al; Handbook of Chemical Property Estimation Methods. McGraw-Hill New York (1982)
11. NIOSH; National Occupational Hazard Survey (NOHS) (1973-74) (1974)
12. NIOSH; National Occupational Exposure Survey (NOES) (1981-83) (1988)
13. Perrin DD; Dissociation Constants of Organic Bases in Aqueous Solution. IUPAC Chemical Data Series. Buttersworth: London (1965)

# Cyclohexylamine

14. Riddick JA et al; Organic Solvents: Physical Properties and Methods of Purification. Techniques of Chemistry. 4th Ed. New York: Wiley-Interscience pp 1325 (1986)
15. Syracuse Research Corporation; Information Profiles on Potential Occupational Hazards. Vol. 1 Single Chemicals TR79-607 (1979)

# Di-n-butylamine

## SUBSTANCE IDENTIFICATION

**Synonyms:**

**Structure:**

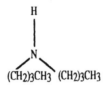

$$H$$
$$N$$
$$(CH_2)_3CH_3 \quad (CH_2)_3CH_3$$

**CAS Registry Number:** 111-92-2

**Molecular Formula:** $C_8H_{19}N$

**Wiswesser Line Notation:**

## CHEMICAL AND PHYSICAL PROPERTIES

**Boiling Point:** 159-160 °C

**Melting Point:** -60 to -59 °C

**Molecular Weight:** 129.24

**Dissociation Constants:** pKa = 11.25 at 25 °C [18]

**Log Octanol/Water Partition Coefficient:** 2.83 [7]

**Water Solubility:** 4700 mg/L at 20 °C [20]

**Vapor Pressure:** 2.28 mm Hg at 20 °C [20]

**Henry's Law Constant:** 1.04 x $10^{-4}$ atm-m³/mole at 25 °C (estimated by group method) [10]

# Di-n-butylamine

## ENVIRONMENTAL FATE/EXPOSURE POTENTIAL

**Summary:** Di-n-butylamine may be released to the environment as emissions and in wastewater during its production and use as a chemical intermediate, corrosion inhibitor, and polymerization inhibitor for butadiene. It also occurs naturally in food and as a metabolic product. If released on land, di-n-butylamine will adsorb strongly to soil. The chemical would be expected to readily biodegrade; however, no estimate of degradative rates in soil are available. If released into water, di-n-butylamine will adsorb to sediment and particulate matter in the water column and probably readily biodegrade. Bioconcentration in aquatic organisms will not be appreciable. In the atmosphere, di-n-butylamine will react with photochemically produced hydroxyl radicals (estimated half-life 4.4 hr). Because of its high water solubility, di-n-butylamine should also be scavenged by rain. Dialkylamines are of particular environmental concern because they are precursors of nitrosamines. The latter are formed in the atmosphere in the presence of nitrogen oxides but are destroyed by sunlight. In aqueous, acidic solutions, di-n-butylnitrosamine may be formed when nitrite ions are present. Humic acids catalyze this reaction. Exposure to di-n-butylamine is primarily occupational. However di-n-butylamine occurs naturally and the general public may be exposed from these sources.

**Natural Sources:** There is some indication that di-n-butylamine may occur naturally in some soils [4] and food items [6]. Amines are produced by microbial processes in decaying organic matter and di-n-butylamine's emission from sewage treatment plants [5] suggests that it may also be a product of microbial metabolism.

**Artificial Sources:** Di-n-butylamine may be released to the environment in air emissions and in wastewater during its production and use as a chemical intermediate for emulsifiers, rubber accelerators, dyes, insecticides, and floatation agents, and corrosion inhibitor and polymerization inhibitor for butadiene [9]. It is also emitted from sewage treatment plants [5].

**Terrestrial Fate:** If released on soil, di-n-butylamine will adsorb strongly to soil. Based on screening studies, biodegradation should occur, but experimental data in soil are lacking.

**Aquatic Fate:** If released into water, di-n-butylamine may sorb to sediment and particulate matter in the water column and slowly volatilize (estimated half-life 12.9 hr in a model river). Based on screening studies, di-n-butylamine should readily biodegrade; however, experimental data are lacking in natural waters.

**Atmospheric Fate:** Di-n-butylamine released into the atmosphere should react rapidly with photochemically produced hydroxyl radicals (estimated half-life 4.4 hr). Due to its high water solubility, washout by rain may also be an important removal process.

**Biodegradation:** In a screening study, di-n-butylamine completely degraded within 14 days at 10 ppm with both an activated sludge and freshwater sediment inoculum [2]. BOD values obtained during this time period indicated that mineralization was essentially complete [2]. River mud bacteria and activated sludge were inhibited by 50 and 100 ppm di-n-butylamine, respectively [2]. In another study that utilized 100 ppm of di-n-butylamine and an activated sludge inoculum, no oxygen consumption was observed until about three days, when the BOD increased sharply to about 30% of theoretical [27]. A third screening test resulted in >90% degradation in 9 days including a 4 day lag period [28]. While low concn of the free diamine were degraded in 10 hr by acclimated mixed cultures, only 25% of the di-n-butylamine adsorbed on bentonite clay was degraded in this time [26]. The sorbed diamine degraded in 2 days [26]. The rate of degradation of the sorbed molecule does not depend on its desorption rate, but rather may be due to restricted access by microorganisms [26]. Under anaerobic conditions with high nitrate loads (denitrification conditions), di-n-butylamine shows little tendency to form nitrosamines [12].

**Abiotic Degradation:** Reaction of aliphatic amines with photochemically produced hydroxyl radicals are rapid and the estimated half-life for the reaction of di-n-butylamine with photochemically produced hydroxyl radicals is 4.4 hr [1]. In the presence of air, water vapor, NO and $NO_2$, dialkylamines reacts to form nitrosamines, products which are destroyed by sunlight [8,19]. In aqueous systems containing nitrite ions, di-n-butylamine reacts to form di-n-butylnitrosamine [23]. A plateau is reached after 48 hr. The reaction is enhanced below pH 5 and increases with temperature. The presence of humic acid has a considerable catalytic affect. While the reaction is not normally affected by light, light greatly enhanced the nitrosation when humic acid was present [23]. Di-n-butylamine does not contain any chromaphores which absorb radiation >290 nm so direct photolysis will not be significant.

# Di-n-butylamine

**Bioconcentration:** Using the octanol/water partition coefficient, an estimated BCF of 83 was calculated for di-n-butylamine using a recommended regression equation [14]. Di-n-butylamine would therefore not be expected to bioconcentrate appreciably in aquatic organisms.

**Soil Adsorption/Mobility:** In dilute aqueous solution, amines interact at surfaces of bentonite clay and humic material by ion exchange forces [24]. Di-n-butylamine strongly adsorbs to environmental particulate matter [24].

**Volatilization from Water/Soil:** Using the estimated Henry's Law constant, a volatilization half-life of 12.9 hr was estimated for di-n-butylamine in a model river 1 m deep, flowing at 1 m/sec with a wind velocity of 3 m/sec [14]. While di-n-butylamine has a moderate vapor pressure, it is strongly bound to soil and therefore would not be expected to volatilize rapidly from soil.

**Water Concentrations:** SURFACE WATER: Di-n-butylamine was not detected in 8 rivers in Germany [15].

**Effluent Concentrations:** In a comprehensive survey of wastewater from 4000 industrial and publicly owned treatment works (POTWs) sponsored by the Effluent Guidelines Division of the US EPA, di-n-butylamine was identified in discharges of the following industrial category (frequency of occurrence, median concn in ppb): iron and steel mfg (4, 170.4); plastics and synthetics (1, 89.1); electronics (1, 6723.3) [21].

**Sediment/Soil Concentrations:** Di-n-butylamine was identified in uncultivated loamy soil from the Moscow region [3]. Since this soil is uncultivated, it is possible that the amines are formed naturally rather than being a contaminant or a metabolite of a fertilizer or pesticide [3]

**Atmospheric Concentrations:**

**Food Survey Values:** An average di-n-butylamine concn of 10 ppb was found in 3 samples of cod roe; however, the amine was not detected in samples of fish sausage, baked ham, spinach, or miso [6]. In a search for the presence of primary and secondary amines in samples of fresh vegetables, preserves, mixed pickles, fish and fish

143

products, bread, cheese, stimulant beverages, and animal feedstuffs, di-n-butylamine was only found in pepperoni (3.4 ppm) [15]. It was similarly not reported in a search for secondary amines in foodstuffs such as fish, ham, frankfurters, and beverages [22]. The interest for the presence of amines in food arises in part because they are regarded as possible precursors of carcinogenic N-nitroso compounds [15].

**Plant Concentrations:**

**Fish/Seafood Concentrations:**

**Animal Concentrations:**

**Milk Concentrations:**

**Other Environmental Concentrations:** Thirteen of fourteen lots of pacifier nipples that were subject to a single artificial-saliva extraction contained extractable di-n-butylamine ranging from 9-3840 ppb, median 267 ppb [25]. Di-n-butylamine has been identified in Latakia tobacco leaf [11], but not in another tobacco or in cigarette smoke [22]. The presence of a large number of amines in tobacco is a function of the curing process, rather than the leaf [11].

**Probable Routes of Human Exposure:** Exposure to di-n-butylamine will primarily be occupational via inhalation and dermal contact. Since di-n-butylamine occurs naturally in some foods and as a metabolic product, the general public may be exposed to the chemical from these sources.

**Average Daily Intake:**

**Occupational Exposure:** NIOSH (NOHS Survey 1972-74) has statistically estimated that 380,525 workers are exposed to di-n-butylamine in the United States [16]. NIOSH (NOES Survey 1981-83) has statistically estimated that 15,140 workers are exposed to di-n-butylamine in the United States [17].

**Body Burdens:** Di-n-butylamine was found in the expired air of 6.5% of a sample of 54 carefully selected normal, health, nonsmoking adults who resided in urban areas [13]. The geometric mean concn level was 0.218 ng/L [13].

# Di-n-butylamine

## REFERENCES

1. Atkinson R; Chem Rev. 85: 69-201 (1985)
2. Calamari D et al; Chemosphere 9: 753-62 (1980)
3. Golovnya RV et al; Amines In Soil as Possible Precursors Of N-Nitroso Compounds USSR Acad Med Sci 1982 pp. 327-35 (1982)
4. Golovnya RV et al; USSR Acad Med Sci 1982: 327-35 (1982)
5. Graedel TE; Chemical Compounds in the Atmosphere. pp. 290 NY: Academic Press (1978)
6. Hamano T et al; Agric Biol Chem 45: 2237-43 (1981)
7. Hansch C, Leo AJ; Medchem Project Issue No.26 Pomona College, Claremont CA (1985)
8. Hanst PL et al; Environ Sci Technol 11: 403-5 (1977)
9. Hawley GG; pp. 327-8 in Condensed Chem Dictionary 10th ed NY: Von Nostrand Reinhold (1981)
10. Hine J, Mookerjee PK; J Org Chem 40: 292-8 (1975)
11. Irvine WJ, Saxby MJ; Phytochemistry 8: 473-6 (1969)
12. Kaplan DL et al; Gov Rep Announce Index 84: 67 (1984)
13. Krotoszynski BK et al; J Anal Toxicol 3: 225-34 (1979)
14. Lyman WJ et al; Handbook of Chem Property Estimation Methods NY: McGraw-Hill (1982)
15. Neurath GB et al; Food Cosmet Toxicol 15: 275-82 (1977)
16. NIOSH; National Occupational Health Survey (1975)
17. NIOSH; National Occupational Exposure Survey (1985)
18. Perrin DD; Dissociation Constants of Organic Bases in Aqueous Solution. IUPAC Chemical Data Series. Butterworth: London (1965)
19. Pitts JN Jr et al; Environ Sci Technol 12: 946-53 (19)
20. Riddick JA et al; Organic Solvents: Physical Properties and Methods of Purification. Techniques of Chemistry. 4th Ed. New York: Wiley-Interscience pp 1325 (1986)
21. Shackelford WM et al; Analyt Chim Acta 146: 15-27 (1983)
22. Singer GM, Lijinsky W; J Agric Food Chem 24: 553-5 (1976)
23. Sithole BB, Guy RD; Sci Total Environ 50: 227-35 (1986)
24. Sithole BB, Guy RD; Environ Internat 11: 499-504 (1986)
25. Thompson HC JR et al; J Toxicol Environ Health 13: 615-32 (1984)
26. Wszolek PC, Alexander M; J Agric Food Chem 27: 410-4 (1979)
27. Yoshimura K et al; J Amer Oil Chem Soc 57: 238-41 (1980)
28. Zahn R, Wellens H; Wasser Abwasser Forsch 13: 1-7 (1980)

# 1,1-Dichloroethane

## SUBSTANCE IDENTIFICATION

**Synonyms:**

**Structure:**

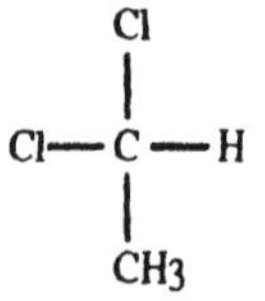

**CAS Registry Number:** 75-34-3

**Molecular Formula:** $C_2H_4Cl_2$

**Wiswesser Line Notation:** GYG1

## CHEMICAL AND PHYSICAL PROPERTIES

**Boiling Point:** 57.3 °C at 760 mm Hg

**Melting Point:** -96.98 °C

**Molecular Weight:** 98.96

**Dissociation Constants:**

**Log Octanol/Water Partition Coefficient:** 1.79 [17]

**Water Solubility:** 5060 mg/L at 25 °C [36]

**Vapor Pressure:** 227 mm Hg at 25 °C [2]

**Henry's Law Constant:** 5.87 x $10^{-3}$ atm-m³/mole [19]

## ENVIRONMENTAL FATE/EXPOSURE POTENTIAL

**Summary:** 1,1-Dichloroethane is released into the environment as fugitive emissions and in wastewater during its production and use as a chemical intermediate and solvent. If released on land, it will be

146

removed rapidly by volatilization, although it may also leach into ground water where its fate is unknown. Bioconcentration in aquatic organisms will not be important. If released in water it will be removed by volatilization with a half-life of 6-9 days, 5-8 days, and 24-32 hr, respectively in a typical pond, lake, or river. In the atmosphere, it will degrade (half-life 62 days) by reaction with photochemically produced hydroxyl radicals, and it will be scavenged by rain. Human exposure will be by inhalation to workers and those living or working near source areas.

**Natural Sources:** 1,1-Dichloroethane does not occur as a natural product [25].

**Artificial Sources:** 1,1-Dichloroethane is released as fugitive emissions during its production and use as a chemical intermediate, coupling agent in antiknock gasoline, paint and varnish remover, metal degreaser, and ore floatation agent [42]. Its largest industrial use is in the production of 1,1,1-trichloroethane [24].

**Terrestrial Fate:** If released on soil, 1,1-dichloroethane will be rapidly lost by evaporation. Due to its low adsorptivity, there is a possibility that it will leach into the ground water.

**Aquatic Fate:** The volatilization half-life of 1,1-dichloroethane from a typical pond, lake, or river is 6-9 days, 5-8 days, and 24-32 hr, respectively. This will be the principal removal mechanism; adsorption to sediment, biodegradation, and hydrolysis should be insignificant by comparison.

**Atmospheric Fate:** If released into the atmosphere, 1,1-dichloroethane will degrade by reaction with photochemically produced hydroxyl radicals (half-life 62 days). There will be considerable dispersal before it degrades. 1,1-Dichloroethane is moderately water soluble and will be washed out by rain.

**Biodegradation:** 50 and 29% degradation was reported when 5 and 10 ppm, respectively of 1,1-dichloroethane was incubated with sewage seed for 7 days of which 19 and 4%, respectively was lost by evaporation [39]. No degradation was detected when 1,1-dichloroethane was incubated with uncontaminated samples of subsurface material taken from positions immediately above and below the water table at Pickett, OK and Fort Polk, LA [44]. Halogenated aliphatic hydrocarbons are generally considered to be resistant to biodegradation [6]. Bank filtration, the passage of river water through earth, reduced the concn of dichloroethanes by 25% [38].

**Abiotic Degradation:** 1,1-Dichloroethane reacts with photochemically generated hydroxyl radicals in the atmosphere with a half-life of 62 days [20]. 1,1-Dichloroethane in a glass bulb with air had a half-life of 17 weeks when left outdoors [32]. HCl and $CO_2$ were formed in the degradation [32]. No information on the hydrolytic half-life of 1,1-dichloroethane could by found. However, since the half-life for ethyl chloride is 38 days [6], it is reasonable to expect that the hydrolytic half-life is somewhat longer because of the extra chloride substituent.

**Bioconcentration:** The bioconcentration factor estimated from 1,1-dichloroethane's water solubility is 1.2 [26], indicating insignificant bioconcentration in fish. All of the chloroethanes have an elimination half-life of <2 days as measured by whole body levels in exposed bluegills [41].

**Soil Adsorption/Mobility:** The Koc estimated from the water solubility is 40 [26] indicating little adsorption to soil organic matter. 1,1-Dichloroethane is readily leached from material representative of waste at land disposal sites [22] and was found in leachate from a simulated landfill lysimeter used to study the codisposal of metal plating sludge and municipal waste [14].

**Volatilization from Water/Soil:** In a laboratory experiment, the evaporative half-life of 1,1-dichloroethane from a stirred beaker filled to a depth of 6.5 cm was 32.2 min [9] which converts to a half-life of 8.3 hr at a 1 m depth. Three values of the overall liquid transfer coefficient of 1,1-dichloroethane relative to oxygen are 0.55 [5], 0.62 [27], and 0.71 [37]. Using values for the oxygen transfer coefficients in typical waters in conjunction with the experimental relative transfer coefficients, one estimates that the range of volatilization half-lives in a typical pond, lake, and river to be 6.4-9.4 day, 5.1-7.5 day, and 24-32 hr, respectively [37]. 1,1-Dichloroethane has a high vapor pressure and low adsorption to soil and would be therefore volatilize rapidly from soil.

**Water Concentrations:** DRINKING WATER: Not detected in drinking water of Love Canal residents [1]. US Ground Water Supply

148

## 1,1-Dichloroethane

Survey (954 supplies derived from ground water chosen both randomly and on the basis that they may contain VOC's) - 41 samples positive, 0.6 ppb median of positive, 4.2 ppb maximum [43]. Bank-filtered tap water from the Rhine River in the Netherlands 500 ppt, maximum [34]. 30 Canadian treatment plants serving all large centers of population 5 ppb, mean, 29 ppb, maximum, 11% detection frequency in Aug-Sept; <1 ppb mean, 10 ppb maximum 4% detection frequency in Nov-Dec [31]. UK: Detected in 1 of 14 water supplies tested - source of water was river/lowland reservoir [12]. According to federal studies, 18% of monitored drinking water wells contained 1,1-dichloroethane [10]. The highest reported concentration in wells was 11.3 ppm while the maximum surface concentration reported was 0.2 ppb [10]. Iowa (127 wells from 58 public water supplies) 5 wells (4 supplies contained residues) 1-24 ppb, and 2 supplies had positive values in finished water [23]. Polluted drinking water well in Maine 7 ppb [4]. GROUND WATER: Potomac-Raritan-Magothy aquifer system along the Delaware River in SW New Jersey (315 wells) 6.6% of wells positive [13]. Not detected (detection limit 1 ppb) in monitoring wells underlying the Amphenol metal plating plant in Broadview, IL [21]. Minnesota: found in ground water underlying 7 of 13 municipal landfills with suspected ground water contamination 0.5-1900 ppb but not in 7 others with no suspected contamination [35]. Ground water around Miami Drum Disposal site (Biscayne Aquifer - water supply for Dade County, FL) 2.6-14 ppb with the higher concn at 13-31 m depth and lower concentration at 3 m [28]. Detected in Love Canal water /sediment/ soil samples. SURFACE WATER: Lake Ontario Basin: Detected in Genesee River sample but absent in Niagara River and open water of Lake Ontario [15]. Ohio River System, 1980-81, 8 stations on mainstream and 3 on tributaries, 4972 samples 156 (3.1%) positive of which 122 were between 0.1 and 1.0 ppb, 33 between 1.0 and 10 ppb and 1 >10 ppb [30]. Ohio River System, 1978-79, 8 stations, 991 samples on mainstream; 2 stations, 359 samples on tributaries 3.9% of samples had detectable levels of 1,1-dichloroethane on mainstream that were <1.0 ppb and 5.3% of samples from the tributaries had detectable levels that were <1.0 ppb and 0.3% that were between 1.0 and 10.0 ppb [29]. Raw water for 30 Canadian water treatment plants - 2 ppb, mean, 33 ppb max, 19% detection frequency in Aug-Sept; <1.0 ppb mean, 11 ppb max, 13% detection frequency in Nov-Dec [31].

**Effluent Concentrations:** Municipal landfill leachate in Minnesota and Wisconsin - 9 of 13 positive, 0.5-6300 ppb [35]. National Urban Runoff Program in which 86 samples from 19 cities throughout the

US were analyzed: 4% of samples positive, 1.5-3 ppb, detected only in Long Island, NY and Eugene, OR [7]. Treated waste water from the following industries contained 1,1-dichloroethane (industry mean concn): coil coating (10 ppb); nonferrous metals manufacturing (0.6 ppb); organic chemical manufacturing/plastics (9.1 ppb); paint formulation (95 ppb); rubber processing (56 ppb) [40]. Additionally, untreated wastewater from the following industries contained 1,1-dichloroethane (industry mean concn): battery manufacturing (10 ppb); metal finishing (480 ppb); pharmaceutical manufacturing (13 ppb); and pulp and paperboard mills (12 ppb) [40]. Not detected in the final effluents of the Los Angeles City (Hyperion), Orange County, CA (OCSD), and San Diego City (Point Loma) treatment plants at a detection limit of 10 ppt [45]. Los Angeles County (JWPCP) municipal wastewater (final effluent) 3.5 ppb [45].

**Sediment/Soil Concentrations:** Not detected in bottom sediment of the submarine outfall of the Los Angeles County (JWPCP) municipal wastewater treatment plant at a detection limit of 0.5 ppb [45]. Detected in Love Canal sediment/soil/water samples [18].

**Atmospheric Concentrations:** RURAL/REMOTE: US (2 samples 0 ppt [3]. Southwest Washington <5 ppt [16]. URBAN/SUBURBAN US (455 samples) 61 ppt, median; 110 ppt, maximum [3]. SOURCE AREAS: US (101 samples) 11 ppt, median, 1400 ppt maximum [3]. Ambient air surrounding Kin-Buc Waste disposal site; Edison, NJ 5.5 ppb [33]. Not detected in air outside homes of Love Canal residents [1]. Detected, not quantified in Baton Rouge, LA where many organic chemicals and petroleum facilities are located [33].

**Food Survey Values:**

**Plant Concentrations:**

**Fish/Seafood Concentrations:** Not detected (detection limit 0.3 ppb) in whole inverts, fish liver, or shrimp muscle obtained from a station off the submarine outfall of the Los Angeles County (JWPCP) municipal treatment plant [45]. Lake Pontchartrain (New Orleans, Mississippi River delta) 33 ppt, mean in oysters; ND in clams [11].

**Animal Concentrations:**

**Milk Concentrations:**

# 1,1-Dichloroethane

## Other Environmental Concentrations:

**Probable Routes of Human Exposure:** Humans are exposed to 1,1-dichloroethane in ambient air and occupationally via inhalation.

**Average Daily Intake:** AIR INTAKE: (assume 61 ppt) 5.0 ug; WATER INTAKE: (assume 0-.6 ppb) 0-1.2 ug; FOOD INTAKE: (insufficient information but probably insignificant).

**Occupational Exposures:** NIOSH estimated that approximately 4600 workers in the US were exposed to 1,1-dichloroethane [25].

**Body Burdens:** Trace of 1,1-dichloroethane was found in exhaled air from 1 of 9 Love Canal residents tested [1]. Trace of 1,1-dichloroethane was found on breath of 8 healthy volunteers [8].

## REFERENCES

1. Barkley J et al; Biomed Mass Spectrom 7: 139-47 (1980)
2. Boublik T et al; The Vapor Pressure of Pure Substances. Vol 17: Amsterdam, Netherlands: Elsevier Sci Publ (1984)
3. Brodzinsky R, Singh HB; Volatile Organic Chem in the Atmos, An Assess of Available Data; Menlo Park, CA, Atmos Sci Centr, SRI Internatl pp 198 68-02-3452 (1982)
4. Burmaster DE; Environ 24: 6-13, 33-36 (1982)
5. Cadena F et al; J Water Pollut Control Fed 56: 460-3 (1984)
6. Callahan MA et al; Water-Related Environ Fate of 129 Priority Pollut Vol.II USEPA-440/4-79-029B (1979)
7. Cole RH et al; J Water Pollut Control Fed 56: 898-908 (1984)
8. Conkle JP et al; Arch Environ Health 30: 290-5 (1975)
9. Dilling WL; Environ Sci Technol 11: 405-9 (1977)
10. Dyksen JE, Hess AF III; J Amer Water Works Assoc 74: 394-403 (1982)
11. Ferrario JB et al; Bull Environ Contam Toxicol 34: 246-55 (1985)
12. Fielding M et al; Organic Micropollut in Drinking Water Medmenham, Eng Water Res Cent p 49 TR-159 (1981)
13. Fusillo TV et al; Ground Water 23: 354-60 (1985)
14. Gould JP et al; Water Chlorination Environ Impact Health Eff 4: 525-39 (1983)
15. Great Lakes Water Quality Board; An Inventory of Chem Substances Identified in the Great Lakes Ecosystem Vol.I pp 195 (1983)
16. Grimsrud EP, Rasmussen RA; Atmos Environ 9: 1014-7 (1975)
17. Hansch C, Leo AJ; Medchem Project, Claremont, CA: Pomona College (1985)
18. Hauser TR, Bromberg SM; Environ Monit Assess 2: 249-72 (1982)
19. Hine J, Mookerjee PK; J Org Chem 40: 292-8 (1975)
20. Howard CJ, Evenson KM; J Chem Phys 64: 4303-6 (1976)

# 1,1-Dichloroethane

21. IT Corporation; Preliminary Site Assess, Broadview, Illinois Plant, Amphenol Products Div TSCA Health and Safety Studies 8D Submissions No.878216382 (1985)
22. Jackson DR et al; Environ Sci Technol 18: 668-73 (1984)
23. Kelley RD; Synthetic Organic Compounds Sampling Survey of Public Water Supplies p 38 NTIS PB85-214427 (1985)
24. Kirk Othmer; Encycl Chem Tech 3rd New York Wiley InterScience 5: 722 (1979)
25. Konietzko H; Hazardous Assess Chem Curr Dev 3: 401-8 (1984)
26. Lyman WJ et al; Handbook of Chemical Property Estimation Methods. Environmental Behavior of Organic Compounds. McGraw-Hill NY (1982)
27. Matter-Mueller C et al; Water Res 15: 1271-9 (1981)
28. Myers VB; pp 354-7 in Natl Conf Manag Uncontrolled Hazard Waste Sites (1983)
29. Ohio River Valley Water Sanit Comm; Assess of Water Quality Conditions, Ohio River Mainstream 1978-79 (1980)
30. Ohio River Valley Water Sanit Comm; Assessment of Water Quality Conditions, Ohio River Mainstream 1980- 81 (1982)
31. Otson R et al; J Assoc Off Anal Chem 65: 1370-4 (1982)
32. Pearson CR, McConnell G; Proc Roy Soc London Ser B 189: 305-32 (1975)
33. Pellizzari ED; Environ Sci Technol 16: 781-5 (1982)
34. Piet GJ, Morra CF; pp 31-42 in Artificial Groundwater Recharge, Huisman L, Olsthorn TN eds. Pitman (1983)
35. Sabel GV, Clark TP; Waste Manag Res 2: 119-30 (1984)
36. Seidell A; Solubilities of Organic Compounds. New York, NY: D. Van Norstrand Co (1941)
37. Smith JH et al; Environ Sci Technol 14: 1332-7 (1980)
38. Sontheimer H; J Amer Water Works Assoc 72: 386-90 (1980)
39. Tabak HH et al; J Water Pollut Contr Fed 53: 1503-18 (1981)
40. USEPA; Treatability Manual I Treatability Data USEPA-600/2-82-001A (1981)
41. USEPA; Ambient Water Quality Criteria Document For Chlorinated Ethanes USEPA-440/5-80-029 (1980)
42. Verschueren K; Handbook of environmental data on organic chemicals. 2nd ed Von Nostrand Reinhold NY p 486 (1983)
43. Westrick JJ et al; J Amer Water Works Assoc 76: 52-9 (1984)
44. Wilson JR et al; Devel Indust Microbiol 24: 225-33 (1983)
45. Young DR et al; Water Chlorination; Environ Impact Health Eff 4(Book 2): 871-84 (1983)

# 1,2-Dichloroethane

## SUBSTANCE IDENTIFICATION

**Synonyms:** Ethylene dichloride

**Structure:**

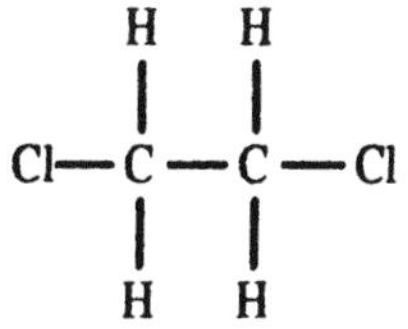

$$Cl-\underset{\underset{H}{|}}{\overset{\overset{H}{|}}{C}}-\underset{\underset{H}{|}}{\overset{\overset{H}{|}}{C}}-Cl$$

**CAS Registry Number:** 107-06-2

**Molecular Formula:** $C_2H_4Cl_2$

**Wiswesser Line Notation:** G2G

## CHEMICAL AND PHYSICAL PROPERTIES

**Boiling Point:** 83.47 °C at 760 mm Hg

**Melting Point:** -35.36 °C

**Molecular Weight:** 98.96

**Dissociation Constants:**

**Log Octanol/Water Partition Coefficient:** 1.48 [17]

**Water Solubility:** 8,524 mg/L at 25 °C [21]

**Vapor Pressure:** 78.7 mm Hg at 20 °C [3]

**Henry's Law Constant:** 9.77 x $10^{-4}$ atm-m³/mole [13]

## ENVIRONMENTAL FATE/EXPOSURE POTENTIAL

**Summary:** The majority of 1,2-dichloroethane released into the environment will enter the atmosphere mostly from its production and use as a chemical intermediate, solvent, and use as a lead scavenger in gasoline. Once in the atmosphere, it may be transported long distances and is primarily removed by photooxidation (half-life approx 1 month) with the products of photooxidation being $CO_2$ and HCl. Direct photolysis will not be a significant removal process. Releases to water will primarily be removed by evaporation (half-life several hours to 10 days). It will not significantly hydrolyze or sorb to sediment. Releases on land will dissipate by volatilization to air and by percolation into ground water where it is likely to persist for a very long time. 1,2-Dichloroethane is not expected to bioconcentrate in the food chain; its presence in some food products is probably due to its use as an extractant. Major human exposure is from urban air, drinking water from contaminated aquifers, and occupational atmospheres.

**Natural Sources:** None [24].

**Artificial Sources:** Atmospheric release may result from its production and use as a chemical intermediate, lead scavenger, extraction and cleaning solvent, diluent for pesticides, grain fumigant and in paint, coatings and adhesives [14,24,26,55,56]. Wastewater may contain 1,2-dichloroethane primarily from its use as a cleaning solvent and chemical intermediate [14,24,26,55,56]. Land release can occur primarily from its production and use as a cleaning solvent and diluent for pesticides [14,24,26,55,56]. One source of chloroethanes in the environment may be from "EDC-tars" (ethylene dichloride tars are by-products of vinyl chloride synthesis) [23].

**Terrestrial Fate:** Releases on land will evaporate fairly rapidly because of 1,2-dichlorethane's moderately high vapor pressure. It will leach rapidly through sandy soil into ground water. Significant biodegradation in soil or ground water is not expected based on laboratory experiments in aquatic systems. Hydrolysis will not be significant.

**Aquatic Fate:** When 1,2-dichloroethane is released to water, its primary loss will be by evaporation. The half-life for evaporation will depend on wind and mixing conditions and was of the order of hours in the laboratory. However, a modeling study using the EXAMS model for an eutrophic lake gave a half-life of 10 days [52]. The half-life for volatilization from a model river 1 m deep flowing 1 m/sec with a wind velocity of 3 m/sec is estimated to be 3.8 hr at 20 °C. Chemical and biological degradation is expected to be very slow. Adsorption to sediment, hydrolysis, and bioconcentration in aquatic organisms are not expected.

# 1,2-Dichloroethane

**Atmospheric Fate:** When released to the atmosphere, 1,2-dichloroethane will degrade by reaction with hydroxyl radicals which are formed photochemically in the atmosphere with a half-life of a little over a month. The products of this photooxidation are $CO_2$ and HCl. Direct photolysis will not be a significant removal process. One would expect the chemical to be transported long distances and be washed out in rain.

**Biodegradation:** Biodegradability tests with 1,2-dichloroethane resulted in no biodegradation to modest amounts in aerobic systems using sewage seed or activated sludge [20,30,38,47,50]. The one river die-away test reported no degradation [30]. The percent BOD produced in 5-10 days is 0-7% [20,38,47]. Another investigator reported slow to moderate biodegradation activity [50]. The extent of biodegradation was difficult to assess because loss also occurred due to volatilization. Of the 98.6% overall removal reported using a 6-day, complete mix, continuous-flow activated sludge system, 0% was attributed to biodegradation and 1% and 99% were attributed to sorption and stripping, respectively [27]. An experiment on the biodegradation of 1,2-dichloroethane in an acclimated anaerobic system reports no degradation in 4 months [4]. Greater than 99% removal was reported in experiments using continuous-flow laboratory methanogenic biofilm columns with a 2-day retention time [8].

**Abiotic Degradation:** The direct photolysis of 1,2-dichlorethane is not a significant loss process [58]. It is primarily degraded in the atmosphere by reaction with hydroxyl radicals, having a half-life of a little over a month with a 1.9% loss for a 12 hour sunlit day [22,44]. Indirect evidence for photooxidation of 1,2-dichloroethane comes from the observation that monitoring levels are highest during the night and early morning [43]. The products of photooxidation are $CO_2$ and HCl [36]. Although firm experimental data are lacking, the photooxidation of 1,2-dichloroethane in water is expected to be slow [14]. The rate of hydrolysis is not significant, being much slower than other pertinent environmental processes such as volatilization and photooxidation [14].

**Bioconcentration:** The measured log BCF in bluegill sunfish (Lepomis macrochirus) is 0.30 [2]. Based on the octanol/water partition coefficient, an estimated BCF of 8 can be calculated [22]. Based on the reported and estimated BCF values, 1,2-dichloroethane will not be expected to significantly bioconcentrate in aquatic organisms.

155

# 1,2-Dichloroethane

**Soil Adsorption/Mobility:** Little adsorption to soil is expected. An experimental Koc of 33 for silt loam [8] is in agreement with values calculated from the water solubility [25]. Based on the octanol/water partition coefficient, an estimated Koc of 152 was calculated [29]. Based on the reported and estimated Koc values, 1,2-dichloroethane will be expected to exhibit very high to high mobility in soil [48] and, therefore, may leach to the ground water. 1,2-Dichloroethane rapidly percolates through sandy soil [57].

**Volatilization from Water/Soil:** It rapidly evaporates from water in laboratory experiments (half-life 0.5-4 hours) [13,29,42]. It would be expected to evaporate rapidly from spills on land due to its high vapor pressure. The half-life for volatilization from a model river 1 m deep flowing 1 m/sec with a wind velocity of 3 m/sec is estimated to be 3.8 hr at 20 °C [29] based on the Henry's Law constant.

**Water Concentrations:** SURFACE WATER: US - 6 river basins 1-90 ppb, 53 of 204 sites pos, only 1 site above 15 ppb [16]; Ohio River basin (1977-78) 0.1-29 ppb, 39 of 243 samples pos [35]; Ohio River basin (1980-81, 4972 samples) 7% pos, 44 samples 1-10 ppb [34]; 105 US cities - raw drinking water 1-45 ppb, 0.55 ppb median, 9.5% pos [11]; 80 US municipal water systems - raw water 0-0.3 ppb, 14% pos [49]; Lake Erie - 2 sites, 4 ppb, 1 site pos [28]. USEPA STORET data base, 7,909 data points, 7.0% pos, <0.10 ppb median [46]. SEAWATER: Gulf of Mexico 0-210 ppt (anthropogenic influence) and not detected (unpolluted areas) [41]. GROUND WATER: 13 US cities - raw ground water 0.2 ppb, 7.7% pos [11]; State ground water survey - 2 states 400 ppb max, 7% pos [15], Aerojet General Rocket Plant - well water, Sacramento - up to 52 ppm [55]. New Jersey, 250 ppb [39]. DRINKING WATER: 133 US Cities - finished surface water 0.8-4.8 ppb, 1.8 ppb median, 4.5% pos [11]; 25 US cities - finished ground water - 0.2 ppb, 4.0% pos [11]. National Organic Monitoring Survey (1976-77) - 3 of 218 samples pos, limit of detection <0.2 ppb [14]. (Note: chlorination of water does not appear to contribute to 1,2-dichloroethane in drinking water) [14]. State data, 1212 sites sampled, 7.0% pos, 400 ppb max; NOMS, 113 cities, 1.8% pos, 0.1-1.8 ppb; National Screening Program, 1971-81, 142 supplies, 1.4% pos, 4.8 ppb max; Community Water Supply Survey, 1978, 451 supplies, 0.9% pos, 0.5-1.8 ppb; Ground Water Supply Survey of drinking water which used ground water as a source, random sample, 466 samples, 3.9% pos, 3.2 ppb max, 0.5 ppb median [12].

# 1,2-Dichloroethane

**Effluent Concentrations:** Industries whose wastewater exceed a mean of 1000 ppb include: photographic equipment/supplies, pharmaceutical mfg and organic chemicals/plastics mfg; max concn in wastewater was 14 ppm (pharmaceutical mfg) [51]. USEPA STORET data base, 97 data points, 2.1% pos, <10 ppb median [46]. Ground water at 178 CERCLA hazardous waste sites, 14.2% pos [37]. Minnesota municipal solid waste landfills, leachates, 6 sites, 33% pos, 5.5 ppb, contaminated ground water (by inorganic indices), 13 sites, 31% pos, other ground water (apparently not contaminated as indicated by inorganic indices), 7 sites, 29% pos, not quantified [40].

**Sediment/Soil Concentrations:** USEPA STORET data base, 40 data points, 0% pos [46].

**Atmospheric Concentrations:** RURAL/REMOTE - US (9 samples) - 0 ppt avg [7]; troposphere baseline concn (ppt), 1982-86, NORTHERN HEMISPHERE, 12 sites, 92% pos, 30 max, 12 avg [9], baseline level of 15-30 [10]; SOUTHERN HEMISPHERE, 2 sites, not detected, <4 [9,10]. URBAN/SUBURBAN - US (1230 samples) - 120 ppt avg [7]; 7 US cities - 110-1380 ppt avg, 7300 ppt max [43,45]. 3 western US cities - 83-519 ppt avg, 1450 ppt max [44]; 5 areas in New Jersey - 940-1500 ppt avg, 22-44% pos, 16,000 ppt max [6]; US - source dominated areas (436 samples) - 1200 ppt avg [7]; US - production and user sites - concn as high as 16, 38, and 45-113 ppb have been recorded at 3 sites [52].

**Food Survey Values:** Meat, oil and fats, tea, fruits and vegetables, 1-10 ppb, largest amount found in olive oil [54]. Not detected in wheat, flour, bran, middlings, and bread [54]. Spice oleoresins, 2-23 ppm, 11 of 17 spices pos [53,54]. Wheat, 10 samples, 20% pos, 110-180 ppb; not detected in corn, oats, corn meal or corn grits; bleached flour, 3 samples, 67% pos, 6.1-6.5 ppb [18]. Table-ready food items, 19, 5.3% pos (plain granola, 0.31 ppb); not detected in butter or margarine (7 samples each), ready-to-eat cereal products, 11 samples, 18% pos (plain granola, 12 ppb, shredded wheat cereal, 8.2 ppb) [19].

**Plant Concentrations:**

**Fish/Seafood Concentrations:** Liverpool Bay, England not detected in marine invertebrates and fish [36].

157

# 1,2-Dichloroethane

**Animal Concentrations:**

**Milk Concentrations:**

**Other Environmental Concentrations:**

**Probable Routes of Human Exposure:** Humans are primarily exposed to 1,2-dichloroethane from ambient air in urban and industrial areas. Only about 4-5% of the population is exposed from drinking water, but where drinking water comes from contaminated ground water sources, exposure can be considerable. Data on food is limited; it has been found in a variety of foods and spices, the latter being from its use as an extractant, grain fumigant, and pesticide diluent.

**Average Daily Intake:** AIR INTAKE: (assume 83-1500 ppt [6,44]) 7-133 ug; WATER INTAKE: (assume 0 ppb) 0 ug; FOOD INTAKE: - insufficient data.

**Occupational Exposures:** NIOSH estimates that 2 million workers are exposed to 1,2-dichloroethane in about 150,000 work places [31]. 12.5 million people are estimated to be exposed to avg annual concn of 0.009-9 ppb near production facilities [53]. Exposure estimates from filling tank with gasoline 0.1 ug/day (time-weighted avg) [53]. NIOSH (NOES Survey 1981-83) has statistically estimated that 81,147 workers are exposed to 1,2-dichloroethane in the United States [32]. NIOSH (NOHS Survey 1972-74) has statistically estimated that 1,351,190 workers are exposed to 1,2-dichloroethane in the United States [33].

**Body Burdens:** Human breath (Old Love Canal, Niagara Falls, NY) 0-54 ppt, 4 of 9 pos; Urine (Old Love Canal, Niagara Falls, NY) 0-140 ppt, 3 of 9 pos [1]. Human milk (women had occupational exposure of up to 14 ppm) 5.4-6.4 ppm immediately after exposure [53].

## REFERENCES

1. Barkley J et al; Biomed Mass Spectrom 7: 139-47 (1980)
2. Barrows ME et al; Dyn Exposure Hazard Asses Toxic Chem, Ann Arbor, MI, Ann Arbor Sci. p 379-92 (1980)
3. Boublik t et al; The Vapor Pressure of Pure Substances Vol 17: Amsterdam, Netherlands: Elsevier Sci Publ (1984)
4. Bouwer EJ, McCarty PL; App Environ Microbiol 45: 1286-94 (1983)
5. Bouwer EJ, McCarty PL; Ground Water 22: 433-40 (1984)

# 1,2-Dichloroethane

6. Bozzelli JW, Kebbekus BB; Analysis of selected volatile organic substances in ambient air, final report Apr-Nov 1978 Newark NJ, NJ Inst Tech 80 p (1979)
7. Brodzinsky R, Singh HB; Volatile organic chemicals in the atmosphere: an assessment of available data SRI contract 68-02-3452 Menlo Park, CA 198 p (1982)
8. Chiou CT et al; Science 206: 831-2 (1979)
9. Class T, Ballschmiter K; Z Anal Chem 327: 198-204 (1987)
10. Class T, Ballschmiter K; Chemosphere 15: 413-27 (1986)
11. Coniglio WA et al; Occurrence of volatile organics in drinking water. EPA exposure assessment project draft 47 p (1980)
12. Cotruvo JA et al; pp. 511-30 in: Org Carcinogens in Drinking Water (1986)
13. Dilling WL; Environ Sci Technol 11: 405-9 (1977)
14. Drury JS, Hammons AS; Investigations of Selected Environmental Pollutants 1,2-Dichloroethane. EPA-560/2-78-006. p 20-78 (1979)
15. Dyksen JE, Hess AF III; J Amer Water Works Assoc 1982: 394-403 (1982)
16. Ewing B et al; Monitoring to detect previously unrecognized pollutants in surface water. EPA-560/6-77-015 75 p (1977)
17. Hansch C, Leo AJ; Substituent constants for correlation analysis in chemistry and biology. NY NY, John Wiley and Sons 339 pp (1979)
18. Heikes DL, Hopper ML; J Assoc Off Anal Chem 69: 990-8 (1986)
19. Heikes DL; J Assoc Off Anal Chem 70: 215-26 (1987)
20. Heukelekian H, Rand MC; Water Pollut Control Assoc 29: 1040-53 (1955)
21. Horvath AL; Halogenated Hydrocarbons: Solubility-Miscibility with Water. New York, NY: Marcel Dekker, Inc pp 889 (1982)
22. Howard CJ, Evenson KM; J Chem Phys 64: 4303-6 (1976)
23. Jensen S, et al; Proc R Soc Lond B 189: 333-346 (1975)
24. Johns R; Air Pollution Assessment. 1,2-Dichloroethane. MTR-7164 The Mitre Corp, McLean, VA 34 pp (1976)
25. Kenaga EE; Ecotox Environ Safety 4: 26-38 (1980)
26. Khan ZS, Hughes TW; Source Assessment: Chlorinated Hydrocarbon Manufacture. EPA-600/2-79-019g. p 48-66 (1979)
27. Kincannon DF et al; J Water Poll Control Fed 55: 157-63 (1983)
28. Konasewich D et al; Status report on organic and heavy metal contaminants in lakes Erie, Michigan, Huron, Superior Basins. Great Lakes Water Qual Board 373 p (1978)
29. Lyman WJ et al; Handbook of Chemical Property Estimation Methods. Environmental Behavior of Organic Compounds. McGraw Hill, NY (1982)
30. Mudder TI; Amer Chem Soc Div Environ Chem Present. Kansas City Mo. Sept (1982)
31. NIOSH; Occupational Exposure to Ethylene Dichloride (1,2-dichloroethane) p 78-211 (1978)
32. NIOSH; National Occupational Exposure Survey (NOES) (1983)
33. NIOSH; National Occupational Hazard Survey (NOHS) (1974)
34. Ohio R Valley Water Sanit Comm; Assessment of water quality conditions, Ohio River Mainstream 1980-81 Cincinnati, OH Table 13 (1982)
35. Ohio R Valley Water Sanit Comm; Assessment of water quality conditions, Ohio River Mainstream 1978-79 Cincinnati, OH p T-45 (1980)
36. Pearson CR, McConnell G; Proc Roy Soc London B 189: 305-32 (1975)
37. Plumb RHJr; Ground Water Monit Rev 7: 94-100 (1987)
38. Price KS et al; J Water Pollut Control Fed 46: 63-77 (1974)
39. Rao PSC et al; Proc Soil Crop Sci Soc FL 44: 1-8 (1985)

# 1,2-Dichloroethane

40. Sabel GV, Clark TP; Waste Manag Res 2: 119-30 (1984)
41. Sauer TC Jr; Org Geochem 3: 91-101 (1981)
42. Scherb K; Muench Beitr Abwasser-Fisch-Flussbiol 30: 234-48 (1978)
43. Singh HB et al; Environ Sci Technol 16: 872-80 (1982)
44. Singh HB et al; Atmos Environ 15: 601-12 (1981)
45. Singh HB et al; Atmospheric measurements of selected hazards organic chemicals EPA-600/53-81-032 (1981)
46. Staples CA et al; Environ Toxicol Chem 4: 131-42 (1985)
47. Stover EL, Kincannon DF; J Water Pollut Control Fed 55: 97-109 (1983)
48. Swann RL et al; Res Rev 85: 17-28 (1983)
49. Symons JM et al; J Amer Water Works Assoc 67:634-47 (1975)
50. Tabak HH et al; J Water Pollut Control Fed 53: 1503-18 (1981)
51. Treatability Manual. EPA-600/2-82-001a page I.12.7-1 to I.12.7-4 (1981)
52. USEPA; An Exposure and Risk Assessment for Dichloroethanes. Draft Final Report p 4-14 to 4-24 (1980)
53. USEPA; An Exposure and Risk Assessment for Dichloroethanes. Draft Final Report. page 5-23 to 5-26 (1980)
54. USEPA; Ambient Water Quality Criteria for Chlorinated Ethanes. EPA-440/5-80-029 page C-1 to C12 (1980)
55. USEPA; An Exposure and Risk Assessment for Dichloroethanes Final Draft Report p 3-1 to 3-10, A-10 to D-3 (1980)
56. Verschueren K; Handbook of environmental data on organic chemicals. 2nd ed NY, NY Van Nostrand Reinhold Co, Inc. (1983)
57. Wilson JT et al; J Environ Qual 10: 501-6 (1981)
58. Yates WF, Hughes LJ; J Phys Chem 64: 672-3 (1960)

# cis-1,2-Dichloroethylene

## SUBSTANCE IDENTIFICATION

**Synonyms:** cis-1,2-Dichloroethene

**Structure:**

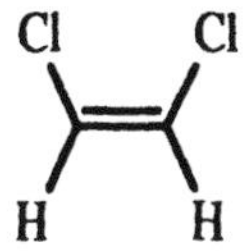

**CAS Registry Number:** 156-59-2

**Molecular Formula:** $C_2H_2Cl_2$

**Wiswesser Line Notation:** G1U1G-Z

## CHEMICAL AND PHYSICAL PROPERTIES

**Boiling Point:** 60.3 °C

**Melting Point:** -80.5 °C

**Molecular Weight:** 96.94

**Dissociation Constants:**

**Log Octanol/Water Partition Coefficient:** 1.86 [11]

**Water Solubility:** 3.5 g/L [18].

**Vapor Pressure:** 200 mm Hg at 35 °C [18]

**Henry's Law Constant:** 0.00337 atm-m³/mole [14]

## ENVIRONMENTAL FATE/EXPOSURE POTENTIAL

**Summary:** cis-1,2-Dichloroethylene may be released to the environment in emissions and wastewater during its production and use. Under anaerobic conditions that may exist in landfills or sediment, one is

161

likely to find 1,2-dichloroethylenes that are formed as breakdown products from the reductive dehalogenation of trichloroethylene and tetrachloroethylene. The cis-1,2-dichloroethylene is apparently the more common isomer found although it is mistakenly listed as the trans-isomer. The trans-isomer, being a priority pollutant, is more commonly analyzed for and the analytical procedures generally used do not distinguish the isomers. If cis-1,2-dichloroethylene is released on soil, it should evaporate and leach into the ground water where very slow biodegradation should occur. If released into water, cis-1,2-dichloroethylene will be lost mainly through volatilization (half-life 3 hr in a model river). Biodegradation, adsorption to sediment, and bioconcentration in aquatic organisms should not be significant. In the atmosphere, cis-1,2-dichloroethylene will be lost by reaction with photochemically produced hydroxyl radicals (half-life 8 days) and scavenged by rain. Because it is relatively long-lived in the atmosphere, considerable dispersal from source areas should occur. The general population is exposed to cis-1,2-dichloroethylene in urban air as well as in contaminated drinking water from ground water sources. Occupational exposure will be via dermal contact with the vapor and liquid or via inhalation.

**Natural Sources:**

**Artificial Sources:** cis-1,2-Dichloroethylene may be released to the environment in emissions and wastewater during its production and use as a solvent and extractant, in organic synthesis, and in the manufacture of perfumes, lacquers, and thermoplastics [13]. An assessment of the sources of cis-1,2-dichloroethylene is complicated by the fact that the trans-isomer is a priority pollutant while the cis-isomer is not and the standard EPA methods of analysis do not allow the isomers to be differentiated [5]. This has resulted in monitoring reports erroneously listing the trans-isomer when the cis-isomer is present [5]. The Michigan Department of Health has the capability to distinguish these isomers and claims that it frequently finds the cis-isomer and, if concentrations are high, occasionally finds traces of the trans-isomer [5]. In an anaerobic, high-organic matrix, such as a landfill site, one is likely to find 1,2-dichloroethylene as breakdown products from reductive dehalogenation [5]. Degradation products are found in increasing proportions further from a source and where there are high concentrations of other degradable organic compounds [5]. Under simulated landfill conditions, cis-1,2-dichloroethylene is formed from trichloroethylene, tetrachloroethylene, and 1,1,2,2-tetrachloroethane and therefore the common industrial solvents may be sources of dichloroethylenes in such environments [5,9]. Additionally, in muck microcosms, tetrachloroethylene is converted into dichloroethylene [17]. The preponderance of dichloroethylene so formed is the cis-isomer [17].

**Terrestrial Fate:** If cis-1,2-dichloroethylene is released on soil, it should evaporate and leach into the ground water where very slow biodegradation should occur.

**Aquatic Fate:** If released into water, cis-1,2-dichloroethylene will be lost mainly through volatilization (half-life 3 hr in a model river). Biodegradation and adsorption to sediment should not be significant.

**Atmospheric Fate:** In the atmosphere, cis-1,2-dichloroethylene will be lost by reaction with photochemically produced hydroxyl radicals (half-life 8 days). There is evidence that it will be scavenged by rain which is to be expected of a water soluble chemical.

**Biodegradation:** In a biodegradability screening test employing a wastewater inoculum and 5 ppm of cis-1,2-dichloroethylene, 54% of the chemical was lost in 7 days, whereas a 34% loss due to volatilization occurred in 10 days [20]. However, literature references to microbial degradation of low molecular weight chlorinated aliphatics generally find that they are not metabolized [4]. When cis-1,2-dichloroethylene was incubated with methanogenic aquifer material obtained adjacent to a landfill site in a serum bottle at 17 °C, the concn of the compound was reduced to <2% of the controls in 16 wk [24]. After 40 wk, only traces of the chemical remained but no vinyl chloride or other transformation product was detected [24]. Another investigator found that when cis-1,2-dichloroethylene was incubated anaerobically using an inoculum from a municipal waste digester in order to simulate conditions in a landfill, vinyl chloride appeared within 6 wk [9]. Biodegradation of trans-1,2-dichloroethylene was studied in microcosms prepared from uncontaminated organic sediment from the Everglades and allowed to sit to insure oxygen depletion. Under these anoxic conditions, 50% of the chemical was lost in 6 months [2]. The fact that ethyl chloride is produced as well as vinyl chloride indicates that there are different pathways in the sequential dechlorination of cis-1,2-dichloroethylene [2].

**Abiotic Degradation:** cis-1,2-Dichloroethylene in the atmosphere reacts with photochemically produced hydroxyl radicals resulting in a half-life

163

of 8 days [8]. The only product positively identified in this reaction was formyl chloride (89% yield) [8]. Chlorine substitution on alkenes markedly reduces their reactivity towards ozone and the half-life resulting from ozone attack of the double bond is 129 days [21]. cis-1,2-Dichloroethylene has a UV absorption band at 190 nm that extends to 240 nm [1,6]. However, minute light absorption was observed, up to 380 nm [1], but there is no evidence that this would contribute significantly to photolytic breakdown under environmental conditions.

**Bioconcentration:** The recommended log octanol/water partition coefficient for cis-1,2-dichloroethylene is 1.86 [11], from which one can estimate a BCF of 15 using a recommended regression equation [15]. Therefore, cis-1,2-dichloroethylene should not bioconcentrate significantly in aquatic organisms.

**Soil Adsorption/Mobility:** From the water solubility of cis-1,2-dichloroethylene, one can estimate a Koc of 49 using a recommended regression equation [15]. Therefore cis-1,2-dichloroethylene should not adsorb significantly to soil or sediment.

**Volatilization from Water/Soil:** From the Henry's Law constant, one can estimate that the half-life for volatilization from a model river 1 m deep with a 1 m/sec current and a 3 m/sec wind is 3.1 hr [15]. Transport through the liquid phase controls volatilization [15]. The mean volatilization half-life of cis-1,2-dichloroethylene from a slowly stirred beaker 6.5 cm deep was 19.4 minutes which is equivalent to a 5.0 hr half-life in a body of water 1 m deep [7]. 96.8% of the cis-1,2-dichloroethylene was removed from contaminated ground water in Wausau, WI by air stripping, a value that was within 1.5% of that predicted using the Henry's Law constant [10].

**Water Concentrations:** DRINKING WATER: cis-1,2-Dichloroethylene was found in Miami drinking water at 16 ppb and Cincinnati and Philadelphia drinking water at 0.1 ppb, but was absent from 7 other drinking waters surveyed [22]. In an EPA survey of finished drinking water from ground water sources, 1.1% of the 280 random sites serving fewer than 10,000 people contained 1,2-dichloroethylene whereas 7.0% of the 186 sites serving more than 10,000 people were positive [23]. The median of positive samples were 0.23 and 1.1 ppb, respectively. For state-selected sites, 3.4% of the 321 sites serving fewer than 10,000 people contained 1,2-dichloroethylene whereas 17.1% of the 158 sites serving more than 10,000 people were positive [23]. The median and maximum levels for these problem sites were 1.3 and 17 ppb for those serving <10,000 people and 2.7 and 120 ppb for those serving >10,000 people [23]. GROUND WATER: Raw water from a well in Wausau, WI contained 83.3 ppb of cis-1,2-dichloroethylene [10]. Studies of the contaminants in shallow ground water at the Miami Drum site, an inactive drum recycling facility, reported 839 and 13.3-17.9 ppb of cis-1,2-dichloroethylene [16]. The Biscayne aquifer, which supplies drinking water to residents of Dade County, contained 0-26 ppb of cis-1,2-dichloroethylene in the vicinity of the Miami Drum site [16]. SURFACE WATER: cis-1,2-Dichloroethylene was found along a 30 km stretch of the Glatt River in Switzerland at load levels of 1 g/hr [25].

**Effluent Concentrations:** In a comprehensive survey of wastewater from 4000 industrial and publicly owned treatment works (POTWs) sponsored by the Effluent Guidelines Division of the US EPA, cis-1,2-dichloroethylene was identified in discharges of the following industrial category (frequency of occurrence; median concn in ppb): steam electric (1; 1.6), leather tanning (1; 3.3), iron and steel mfg (2; 1400.8), nonferrous metals (1; 314.6), organics and plastics (2; 121.5), textile mills (1; 8.3), plastics and synthetics (3; 20.1), rubber processing (1; 712.0), explosives (1; 1.5) [19]. The highest effluent concn was 2059 ppb in the iron and steel mfg industry [19].

**Sediment/Soil Concentrations:** cis-1,2-Dichloroethylene has been detected, but not quantitated in sediment/soil/water samples at the Love Canal [12].

**Atmospheric Concentrations:** RURAL/REMOTE: Two site/samples in US contained no cis-1,2-dichloroethylene [3]. URBAN/SUBURBAN: US (669 site/samples) 68 ppt median, 3500 ppt maximum [3]. SOURCE AREAS: US (101 site/samples) 300 ppt median, 6700 ppt maximum [3].

**Food Survey Values:**

**Plant Concentrations:**

**Fish/Seafood Concentrations:**

**Animal Concentrations:**

# cis-1,2-Dichloroethylene

**Milk Concentrations:**

**Other Environmental Concentrations:**

**Probable Routes of Human Exposure:** Occupational exposure to cis-1,2-dichloroethylene will be via inhalation and dermal contact with the vapor as well as by dermal contact with the liquid during its use as a solvent. The general population is exposed to cis-1,2-dichloroethylene in urban air as well as in contaminated drinking water from ground water sources.

**Average Daily Intake:** AIR INTAKE: (assume air concn of 68 ppt) 5.4 ug. WATER INTAKE: (assume water concn from contaminated sources of 0.23-2.7 ppb) 0.5-5.4 ug when drinking water is contaminated. FOOD INTAKE: insufficient data.

**Occupational Exposures:**

**Body Burdens:**

## REFERENCES

1. Ausubel R, Wijnen MHJ; Int J Chem Kinetics 7: 739-51 (1975)
2. Barrio-Lage G et al; Environ Sci Technol 20: 96-9 (1986)
3. Brodzinsky R, Singh HB; Volatile Organic Chemicals in The Atmosphere Menlo Park,CA: SRI International Contract 68-02-3452 198 pp. (1982)
4. Callahan MA et al; Water-related Environmental Fate of 129 Priority Pollutants USEPA-440/4-79-029b (1979)
5. Cline PV, Viste DR; Waste Manag Res 3: 351-60 (1985)
6. Dahlberg JA; Acta Chemica Scandinavica 23: 3081-90 (1969)
7. Dilling WL; Environ Sci Technol 11: 405-9 (1977)
8. Goodman MA et al; ACS Div Environ Chem 192nd Natl Mtg 26: 169-71 (1986)
9. Hallen RT et al; ACS Div Environ Chem 192nd Natl Mtg 26: 344-6 (1986)
10. Hand DW et al; J Am Water Works Assoc 78: 87-97 (1986)
11. Hansch C, Leo AJ; MEDCHEM Project issue 26 Claremont College Pomona, CA (1986)
12. Hauser TR et al; Env Monit Assess 2: 249-72 (1982)
13. Hawley GG; Condensed Chemical Dict 10th ed NY: Von Nostrand Reinhold pp.335 (1981)
14. Hine J, Mookerjee PK; J Org Chem 40: 292-8 (1975)
15. Lyman WJ et al; Handbook of Chem Property Estimation Methods NY: McGraw-Hill (1982)
16. Myers VB; Remedial Activities at The Miami Drum Site, Florida Natl Conf Manage Uncontrolled Hazard Waste Sites pp. 354-7 (1983)

# cis-1,2-Dichloroethylene

17. Parsons F et al; J Amer Water Works Assoc 76: 56-9 (1984)
18. Riddick JA et al; Organic Solvents New York: Wiley-Interscience (1986)
19. Shackelford WM et al; Analyt Chim Acta 146: 15-27 [supplemental data] (1983)
20. Tabak HH et al; J Water Pollut Control Fed 53: 1503-18 (1981)
21. Tuazon EC et al; Arch Environ Contam Toxicol 13: 691-700 (1984)
22. USEPA; AWQC for Dichloroethylenes (NTIS PB81-117525) (1980)
23. Westrick JJ et al; J Amer Water Works Assoc 76: 52-9 (1984)
24. Wilson BH et al; Environ Sci Technol 20: 997-1002 (1986)
25. Zuercher F, Giger; Vom Waser 47: 37-55 (1976)

# trans-1,2-Dichloroethylene

## SUBSTANCE IDENTIFICATION

**Synonyms:** trans-1,2-Dichloroethene

**Structure:**

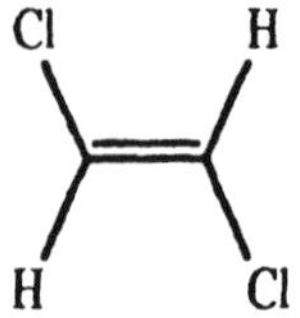

Cl      H
  \    /
   C==C
  /    \
 H      Cl

**CAS Registry Number:** 156-60-5

**Molecular Formula:** $C_2H_2Cl_2$

**Wiswesser Line Notation:** G1U1G-T

## CHEMICAL AND PHYSICAL PROPERTIES

**Boiling Point:** 48.0-48.5 °C

**Melting Point:** -50 °C

**Molecular Weight:** 96.94

**Dissociation Constants:**

**Log Octanol/Water Partition Coefficient:** 2.06 [13]

**Water Solubility:** 6.3 g/L at 25 °C [23]

**Vapor Pressure:** 340 mm Hg at 25 °C [23]

**Henry's Law Constant:** 0.00672 atm-m³/mole [16]

## ENVIRONMENTAL FATE/EXPOSURE POTENTIAL

**Summary:** trans-1,2-Dichloroethylene may be released to the environment in air emissions and wastewater during its production and use. Under anaerobic conditions that may exist in landfills, aquifiers,

168

or sediment one is likely to find 1,2-dichloroethylenes that are formed as breakdown products from the reductive dehalogenation of common industrial solvents trichloroethylene, tetrachloroethylene, and 1,1,2,2-tetrachloroethane. The cis-1,2-dichloroethylene is apparently the more common isomer found although it is mistakenly reported as the trans-isomer. The trans-isomer, being a priority pollutant, is more commonly analyzed for and the analytical procedures generally used do not distinguish between isomers. If trans-1,2-dichloroethylene is released on soil, it should evaporate and leach into the ground water where very slow biodegradation should occur. If released into water, trans-1,2-dichloroethylene will be lost mainly through volatilization (half-life 3 hr in a model river). Biodegradation, adsorption to sediment, and bioconcentration in aquatic organisms should not be significant. In the atmosphere, trans-1,2-dichloroethylene will be lost by reaction with photochemically produced hydroxyl radicals (half-life 3.6 days) and scavenged by rain. Because it is relatively long-lived in the atmosphere, considerable dispersal from source areas should occur. The general population is exposed to trans-1,2-dichloroethylene in urban air as well as in contaminated drinking water from ground water sources. Occupational exposure will be via dermal contact with the vapor and liquid or via inhalation.

**Natural Sources:**

**Artificial Sources:** trans-1,2-Dichloroethylene may be released to the environment in air emissions and wastewater during its production and use as a solvent and extractant, in organic synthesis, and in the manufacture of perfumes, lacquers, and thermoplastics [15]. An assessment of the sources of trans-1,2-dichloroethylene is complicated by the fact that it is a priority pollutant while the cis- isomer is not and the standard EPA methods of analysis do not allow the isomers to differentiated [5]. This has resulted in monitoring reports erroneously listing the trans-isomer when the cis-isomer is present [5]. The Michigan Department of Health has the capability of distinguishing these isomers and claims that it frequently finds the cis-isomer and, if concns are high, they occasionally find traces of the trans-isomer [5]. In an anaerobic, high-organic matrix, such as a landfill site, one is likely to find 1,2-dichloroethylene as a breakdown product due to reductive dehalogenation [5]. Degradation products are found in increasing proportions further from a source and where there are high concns of other degradable organic compounds [5]. Under simulated landfill conditions, it has been found that trans-1,2-dichlorothylene is

## trans-1,2-Dichloroethylene

formed from trichloroethylene, tetrachloroethylene, and 1,1,2,2-tetrachloroethane and therefore common industrial solvents may be sources of dichloroethylenes in such environments [5,12]. Additionally, in muck microcosms tetrachloroethylene is converted into dichloroethylene, although the relative amount of the trans-isomer produced is much less than the cis [22].

**Terrestrial Fate:** If trans-1,2-Dichloroethylene is released to soil, it should evaporate and leach into the ground water where it may very slowly biodegrade.

**Aquatic Fate:** If released into water, trans-1,2-dichloroethylene will be lost mainly through volatilization (half-life 3 hr in model river). Biodegradation and adsorption to sediment should not be significant.

**Atmospheric Fate:** In the atmosphere trans-1,2-dichloroethylene will be lost by reaction with photochemically produced hydroxyl radicals (half-life 3.6 days.) There is evidence that it will be scavenged by rain which is to be expected of a water soluble chemical.

**Biodegradation:** trans-1,2-Dichloroethylene was recalcitrant in shake flask tests modified to accommodate volatile chemicals [20,21]. The concns examined in these studies ranged from 0.80 to 25 ppm. A 21-day acclimation period and the addition of a lactose cometabolite did not alter the biodegradability. Similarly no biodegradation occurred in a river die-away test [21]. Contradictory results were obtained in a biodegradability screen test using a wastewater inoculum and 5 ppm, of trans-1,2-dichloroethylene [27]. 67% of the chemical was lost in 7 days, whereas 33% loss due to volatilization occurred in 10 days [27]. Literature references to microbial degradation of low molecular weight chlorinated aliphatics generally find that they are not metabolized [4]. When trans-1,2-dichloroethylene was incubated with aquifer material obtained adjacent to a landfill site in a serum bottle at 17 °C, at least 16 wk of incubation were required before disappearance began relative to autoclaved controls [33]. After 40 wk, the average concn was reduced to 18% of controls and vinyl chloride was identified as a degradation product [33]. Another investigator found that when trans-1,2-dichloroethylene was incubated anaerobically using an inoculum from a municipal waste digester in order to simulate conditions in a landfill, vinyl chloride appeared within 6 weeks [12]. Biodegradation of trans-1,2-dichloroethylene was studied in microcosms prepared from uncontaminated organic sediment from the Florida

Everglades and allowed to sit to insure oxygen depletion. Under these anoxic conditions, 73% of the chemical was lost in 6 months with the accompanying formation of vinyl chloride [1].

**Abiotic Degradation:** trans-1,2-Dichloroethylene in the atmosphere reacts with photochemically produced hydroxyl radicals resulting in a half-life of 3.6 days [11]. The only product positively identified in this reaction was formyl chloride (56% yield) [11]. Chlorine substitution on alkenes markedly reduces their reactivity towards ozone and the half-life resulting from ozone attack of the double bond is 44 days [29]. trans-1,2-Dichloroethylene has a UV absorption band that extends to about 240 nm [7]; therefore, direct photolysis would not be a significant degradative process.

**Bioconcentration:** The recommended octanol/water partition coefficient for trans-1,2-dichloroethylene is 2.06 [13], from which one estimates a BCF of 22 using a recommended regression equation [19]. Therefore, trans-1,2-dichloroethylene should not bioconcentrate significantly in aquatic organisms.

**Soil Adsorption/Mobility:** From the water solubility, one can estimate a Koc of 36 using a recommended regression equation [19]. Therefore, trans-1,2-dichloroethylene should not adsorb significantly to soil or sediment.

**Volatilization from Water/Soil:** From the Henry's Law constant, one can estimate that the half-life for volatilization from a model river 1 m deep with a 1 m/sec current and a 3 m/sec wind is 3.0 hr [19]. Transport through the liquid phase controls volatilization [19]. The mean volatilization half-life of trans-1,2-dichloroethylene from a slowly stirred beaker 6.5 cm deep was 24.0 minutes which is equivalent to a 6.2 hr half-life in a body of water 1 m deep [8].

**Water Concentrations:** DRINKING WATER: trans-1,2-Dichloroethylene was found in Miami drinking water at 1 ppb [30]. The concn of trans-1,2-dichloroethylene in private wells in 5 homes in northern Winnebago County, IL ranged from ND to 64 ppb, 8 ppb median [32]. The chemical was found in a ground water plume of predominantly trichloroethylene believed to originate from an old industrial source [32]. Two production wells belonging to the Lakewood Utility district near Tacoma, WA contained 200 ppb of trans-1,2-dichloroethylene from a nearby commercial facility [2]. In a survey of purgeable organics in 12 parts of the world outside of Europe and North America, only northern Egypts' contained trans-1,2-dichloroethylene [28]; measured concn 0.5 ppb [28]. GROUND WATER: 4.6% of the 315 wells sampled from the outcrop area of the Potomac-Raritan-Magothy aquifer system adjacent to the Delaware River contained trans-1,2-dichloroethylene [10]. The chemical was absent from wells downdip of the outcrop area [10]. A site study of a western Connecticut manufacturing plant that used large quantities of high quality trichloroethylene for degreasing found that 7 of 9 monitoring wells around the plant contained 1.2-320.9 ppb of trans-1,2-dichloroethylene [26].

**Effluent Concentrations:** In a comprehensive survey of wastewater from 4000 industrial and publicly owned treatment works (POTWs) sponsored by the Effluent Guidelines Division of the US EPA, trans-1,2-dichloroethylene was identified in discharges of the following industrial category (frequency of occurrence; median concn in ppb): iron and steel mfg (2; 2265.9), organics and plastics (3; 14.6), inorganic chemicals (2; 3.9), rubber processing (2; 19.0), auto and other laundries (1; 60.6), explosives (1; 3.9), electronics (7; 140.7), mechanical products (2; 13.7), transportation equipment (1; 29.3), publicly owned treatment works (63; 16.3) [24]. The highest effluent concn was 3013 ppb in the iron and steel mfg industry [24]. In another survey of the industrial occurrences of trans-1,2-dichloroethylene, 4 industries had wastewater discharges of >0.1 kg/day. These (industry, mean concn-ppb; max concn-ppb) were: metal finishing (260; 1700), photographic equipment/supplies (-; 2200), nonferrous metal mfg (75; 260), rubber processing (150; 290) [31]. The concn of trans-1,2-dichloroethylene in 3 sewage treatment effluents ranged from 31 to 43 ppb [17]. While effluent from the Los Angeles City, Orange County, and San Diego County, contained <10 ppb of trans-1,2-dichloroethylene, sludge from two of the plants contained 145 and 44 ppb of the chemical [34]. At the Valley of the Drums waste site near Louisville, KY, water samples contained trace amounts to 75 ppb of trans-1,2-dichloroethylene, while some sediment samples contained trace amounts of the chemical [25]. In the National Urban Runoff Program in which samples of runoff were collected from 19 cities (51 catchments) in the US, trans-1,2-dichloroethylene was detected in Eugene, OR and Little Rock, AR (5% of the samples) at levels of 1-3 ppb [6]. In a four-city study (Cincinnati, St. Louis, Atlanta, and Hartford) to determine the major source type of priority pollutants in tap water and publicly owned treatment work (POTW) influents, it was found that 43%, 38%, and 28% of commercial sources, industrial sources, and POTW influents contained trans-1,2-dichloroethylene [18]. The average level of the industrial sources was between 10 and 100 ppb while the others were <10 ppb [18].

**Sediment/Soil Concentrations:** trans-1,2-Dichloroethylene has been detected, but not quantitated in sediment/soil/water samples at the Love Canal [14].

**Atmospheric Concentrations:** SOURCE AREAS: Edison NJ - 930 ppt trans-1,2-dichloroethylene [3].

**Food Survey Values:**

**Plant Concentrations:**

**Fish/Seafood Concentrations:**

**Animal Concentrations:**

**Milk Concentrations:**

**Other Environmental Concentrations:** Primary sludges from three publicly owned treatment works treating municipal and industrial wastes contained 22, 1540, and 1317 ppb of trans-1,2-dichloroethylene [9].

**Probable Routes of Human Exposure:** Occupational exposure to trans-1,2-dichloroethylene will be via inhalation and dermal contact with the vapor as well as by dermal contact with the liquid during its use as a solvent. The general population is exposed to trans-1,2-dichloroethylene in urban air as well as in contaminated drinking water.

**Average Daily Intake:**

**Occupational Exposures:**

**Body Burdens:**

# REFERENCES

1. Barrio-Lage G et al; Environ Sci Technol 20: 96-9 (1986)
2. Boateng K et al; A Case History Ground Water Monit Rev 4:24-31 (1984)
3. Brodzinsky R, Singh HB; Volatile Organic Chemicals In The Atmosphere Menlo Park, CA: SRI International 198 pp. (1982)
4. Callahan MA et al; Water-related Environmental Fate of 129 Priority Pollutants USEPA-440/4-79-929b (1979)
5. Cline PV, Viste DR; Waste Manage Res 3: 351-60 (1985)
6. Cole RH et al; J Water Pollut Control Fed 56: 898-908 (1984)
7. Dahlberg JA; Acta Chemica Scandinavica 23: 3081-90 (1969)
8. Dilling WL; Environ Sci Technol 11: 405-9 (1977)
9. Feiler HD et al; pp.53-7 in Natl Conf Munic Ind Sludge Util Disposal, Silver Spring, MD (1980)
10. Fusillo TV et al; Ground Water 23:354-60 (1985)
11. Goodman MA et al; ACS Div Environ Chem 192nd Natl Mtg 26:169-71 (1986)
12. Hallen RT et al; ACS Div Environ Chem 192nd Natl Mtg 26: 344-6 (1986)
13. Hansch C, Leo AJ; MEDCHEM Project issue 26 Claremont College Pomona, CA (1986)
14. Hauser TR et al; EPA's Monitoring Program AT Love Canal 1980 Env Monit Assess 2:249-72 (1982)
15. Hawley GG; Condensed Chem Dictionary 10th ed pp. 335 Von Nostrand Reinhold NY (1981)
16. Hine J, Mookerjee PK; J Org Chem 40: 292-8 (1975)
17. Lao RC et al; pp.107-18 in Analytical Techniques in Environmental Chemistry II Albaiges J ed NY: Pergamon Press (1982)
18. Levins P et al; Sources of Toxic Pollutants in Influents to Sewage Treatment Plants p.118 USEPA-440/4-81-008 NTIS PB81-219685 (1981)
19. Lyman WJ et al; Handbook of Chem Property Estimation Methods. McGraw-Hill NY (1982)
20. Mudder TI; Diss Abstra Int B 42: 1804 (1981)
21. Mudder TI, Musterman JL; Development of Empirical Structure Biodegradability Relationships and Biodegradability Testing Protocol for Volatile and Slightly Soluble Priority Pollutants presented before the Div Environ Chem Amer Chem Soc Kansas City, MO pp. 52-3 (1982)
22. Parsons F et al; J Am Water Works Assoc 76: 56-9 (1984)
23. Riddick JA et al; Organic Solvents New York: Wiley Interscience (1986)
24. Shackelford WM et al; Analyt Chim Acta 146: 15-27 [supplemental data base] (1983]
25. Stonebraker RD, Smith AJ Jr; pp.1-10 in Control Hazard Mater Spills, Proc Natl Conf Nashville, TN (1980)
26. Stuart JD; Organics transported thru selected geological media Comm Univ Storrs Inst of Water Resources pp.37 NTIS PB83-224246 (1983)
27. Tabak HH et al: J Water Pollut Contr Fed 53: 1503-18 (1981)
28. Trussell AR et al; pp.39-53 in Water Chlorination Vol 4 Jolley RL ed Ann Arbor Sci Publ Ann Arbor, MI (1980)
29. Tuazon EC et al; Arch Environ Contam Toxicol 13: 691-700 (1984)
30. USEPA; AWQC for Dichloroethylenes (NTIS PB81-117525) (1980)
31. USEPA; Treatability Manual - Vol I USEPA-600/8-80-042 (1980)

32. Wehrmann HA; Investigation Of A Volatile Organic Chemical Plume In Northern Winnebago County, Illinois Ilenr/re-84/09 (NTIS PB85-114452/gar) Springfield, IL: IL Dept Energy Nat Res pp.95 (1985)
33. Wilson BH et al; Environ Sci Technol 20: 997-1002 (1986)
34. Young DR; Ann Rep South Calif Coastal Water Res Proj p.103-12 (1978)

# Dichloromethane

## SUBSTANCE IDENTIFICATION

**Synonyms:** Methylene chloride

**Structure:**

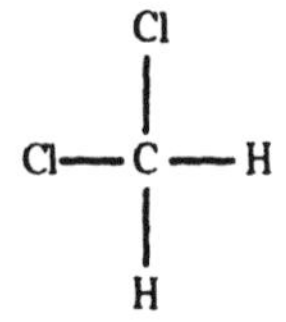

```
        Cl
        |
Cl —— C —— H
        |
        H
```

**CAS Registry Number:** 75-09-2

**Molecular Formula:** $CH_2Cl_2$

**Wiswesser Line Notation:** G1G

## CHEMICAL AND PHYSICAL PROPERTIES

**Boiling Point:** 39.75 °C at 760 mm Hg

**Melting Point:** -95.1 °C

**Molecular Weight:** 84.94

**Dissociation Constants:**

**Log Octanol/Water Partition Coefficient:** 1.25 [16]

**Water Solubility:** 13,000 mg/L at 25 °C [37]

**Vapor Pressure:** 434.9 mm Hg at 25 °C [2]

**Henry's Law Constant:** 2.68 x $10^{-3}$ atm-m³/mole [9]

## ENVIRONMENTAL FATE/EXPOSURE POTENTIAL

**Summary:** Large quantities of methylene chloride are used each year, primarily in aerosols, paint removers, and chemical processing. Most of the methylene chloride will be released to the atmosphere where it

176

will degrade by reaction with photochemically produced hydroxyl radicals with a half-life of a few months. It will not be subject to direct photolysis. Releases to water will primarily be removed by evaporation. Biodegradation is possible in natural waters but will probably be very slow compared with evaporation. It will not be expected to significantly adsorb to sediment or to bioconcentrate in aquatic organisms. Releases to soil will evaporate rapidly from near-surface soil and partially leach into ground water where its fate is unknown. Major route of human exposure is from air, which can be high near sources of emission, and contaminated drinking water.

**Natural Sources:** None.

**Artificial Sources:** Air emissions are likely to occur from methylene chloride use as an aerosol propellant, paint remover, metal degreaser, and a urethane foam blowing agent [4]. Wastewater may occur primarily from the following industries: paint and ink, aluminum forming, coal mining, photographic equipment and supplies, pharmaceutical, organic chemical/plastics, rubber processing, foundries, and laundries [50].

**Terrestrial Fate:** When spilled on land, methylene chloride is expected to evaporate from near-surface soil into the atmosphere because of its high vapor pressure. Although little work has been done on its adsorptivity, it is probable that it will leach through soil into ground water. Degradation in ground water is unknown. Hydrolysis in soil or ground water is not an important process under normal environmental conditions.

**Aquatic Fate:** Methylene chloride will be primarily lost by evaporation to the atmosphere which should take several hours depending on wind and mixing conditions. When released into a river, methylene chloride levels were non-detectable 3-15 miles from the source [7,20]. Biodegradation is possible in natural waters but will probably be very slow compared with evaporation. Little is known about adsorption to sediment or bioconcentration in aquatic organisms but these are not likely to be significant processes. Hydrolysis in water is not an important process under normal environmental conditions.

**Atmospheric Fate:** Methylene chloride released into the atmosphere will degrade by reaction with hydroxyl radicals with a half-life of several months. It will not be subject to direct photolysis. A small fraction of the chemical will diffuse to the stratosphere where it will rapidly degrade by photolysis and reaction with chlorine radicals. A moderately water-soluble chemical such as methylene chloride will be expected to partially return to earth in rain.

**Biodegradation:** Methylene chloride is reported to completely biodegrade under aerobic conditions with sewage seed or activated sludge between 6 hours to 7 days [6,24,38,48,49]. 86-92% conversion to $CO_2$ will occur after a varying acclimation period using anaerobic digestion in wastewater [14]. No information could be found concerning its biodegradability in natural bodies of water or in aquifers.

**Abiotic Degradation:** Hydrolysis is not an important degradation process under normal environmental conditions. The minimum reported half-life for hydrolysis is approximately 18 months [8]. Since methylene chloride does not absorb light >290 nm [21], it will not degrade by direct photolysis in the troposphere. It does not photodegrade when exposed to sunlight for 1 year in aerated water [8]. In the stratosphere, it would undergo photolysis and also degrade by reaction with Cl radicals [5,46]. Methylene chloride will degrade by reaction with hydroxyl radicals in the troposphere with a half-life of several months [5,15,41]. There is some disparity concerning the photooxidation of methylene chloride in the presence of nitrogen oxides. One investigator reported 11.5% degradation in 6 hours [53] and another claimed <5% degradation in the presence of much higher concentrations of nitrogen oxides [10]. The importance of photooxidation is supported by the fact that the highest concentrations of methylene chloride are observed at night or in the early morning [42].

**Bioconcentration:** Although experimental data is lacking, methylene chloride would not be expected to bioconcentrate due to its low octanol/water partition coefficient from which an estimated BCF of 5 can be calculated [26].

**Soil Adsorption/Mobility:** Little work has been done on the adsorption of methylene chloride to soil. A log Koc of 1.68 can be calculated [26] from a reported log Kom of 1.44 [40]. It is adsorbed strongly to peat moss, less strongly to clay, only slightly to dolomite limestone, and not at all to sand [8]. Its presence in ground water at high concentration levels is some confirmation of its low adsorptivity to subsurface soil which is of low organic content.

# Dichloromethane

**Volatilization from Water/Soil:** Methylene chloride has a high Henry's Law coefficient and will evaporate rapidly from water. Half-lives for the evaporation from water of 3-5.6 hours have been determined at moderate mixing conditions [26,36]. When released into an estuarine bay, all the chemical dissipated within 4 km of the release point in the spring and within 8 km in the winter under ice [20]. Due to its high vapor pressure, it will evaporate rapidly from near-surface soil.

**Water Concentrations:** SURFACE WATER: US - 14 heavily industrialized river basins (204 sites) - 16% positive, 1-30 ppb [12]; Ohio River Basin - 11 stations, 4972 samples - 33.1% positive, 219 samples between 1 and 10 ppb, 19 samples >10 ppb [31]; Lake Erie (Buffalo) - 5 ppb, Lake Michigan (9 sites) - 4 sites positive, 1-30 ppb [25]. Concentrations decrease rapidly away from sources in Delaware River and its tributaries [20] and the Black River estuary, Maryland [7]. USEPA STORET data base, 8,917 data points, 30.0% pos, 0.100 ppb median [47]. DRINKING WATER: 30 Canadian water treatment facilities - 50% positive - 10 ppb, avg, 50 ppb max. (summer), 30% pos, 3 ppb avg, 50 ppb max (winter) [32]; 10 state survey drinking water from ground water sources - 2% pos, 3600 ppb max, max surface water concn 13 ppb [11]; EPA Region V Survey (83 sites in 5 states: Minnesota, Wisconsin, Illinois, Indiana, Ohio) - 8% pos, 1-7 ppb [51], National Organics Monitoring Survey (1976) - 15 of 109 samples positive, 6.1 ppb, mean of positive samples [51]. GROUND WATER: The Netherlands 1976-78, 232 ground water pumping stations, max concn, 3000 ppb [55]. IA, 1984-85, 128 drinking water wells, 3.1% pos, 1.0-5.0 ppb [23]. SEAWATER: Surface seawater, eastern Pacific Ocean, 29.21 deg North to 28.95 deg South, 30 samples, 93.3% pos, 1.0-8.0 ppt, 0.68 ppt avg [45].

**Effluent Concentrations:** USEPA STORET data base, 1,480 data points, 38.8% pos, 10.0 ppb median [47]. Weser R, Germany - 72-179 ppb [52]. Industries in which wastewater exceeded an average of 1000 ppb: coal mining, aluminum forming, photographic equipment and supplies, pharmaceutical mfg, organic chemical/ plastics mfg, paint and ink formulation, rubber processing, foundries, and laundries [50]. Max concentration measured was 210,000 ppb in paint and ink industry and aluminum forming [50]. Outfalls from 4 municipal treatment plants in southern California with primary or secondary treatment - random samples - <10 to 400 ppb [54]. US, 178 CERCLA hazardous waste disposal sites, 19.2% pos [35]. Minnesota municipal solid waste landfills, leachates, 6 sites, 66.7% pos, 64-1300 ppb, contaminated ground water (by inorganic indices), 13 sites, 53.8% pos, 1-250 ppb, other ground water (apparently not contaminated as indicated by inorganic indices), 7 sites, 14.3% pos, 2.1-3.9 ppb [39].

**Sediment/Soil Concentrations:** SEDIMENT: Bottom sediment near sewage outfall - Southern California - <4 ppb. In Lake Pontchartrain, New Orleans: 1.5 ppb wet weight in sediment from Inner Harbor Navigation Canal; 3.2 ppb in sediment from Chef Menteur Pass; and not detected in sediment from Rigolets [13,54]. USEPA STORET data base, 338 data points, 20.0% pos, 13.0 ppb median [47].

**Atmospheric Concentrations:** BACKGROUND: Global 32 ppt, Northern Hemisphere - 44 ppt, Southern Hemisphere 20 ppt [44]. RURAL/REMOTE (5 samples) - 45 ppt [3]. Weighted avg (ppt), Eastern Pacific Ocean: Northern hemisphere, 38, Southern hemisphere, 21; global avg, 29 [45]. URBAN/SUBURBAN: 11 sites - 414-3751 ppt avg; 12,000 ppt max [41,42,43]; avg. of 718 samples - 630 ppt [3]. US SOURCE AREAS: Avg of 127 samples - 270 ppt [3], 11 - highly industrialized locations - 10-74,000 ppt [34]; Industrial sites in Newark, Elizabeth, and Camden, NJ (summer 1981) - 230-720 ppt geometric mean, 10.2 ppb max [17].

**Food Survey Values:** Intermediate grain-based foods (1984): 9 varieties, 77.8% pos, 1.9-30 ppb max concn in bleached flour, followed by a fudge brownie mix; wheat, corn, oats (1984), 10, 2, and 1 samples, respectively: not detected [18]. Table-ready foods: 19 varieties, 42% pos, 1.4-71 ppb; max concn in cheddar cheese; butter, 7 samples, 100% pos, 1.1-280 ppb; margarine, 7 samples, 100% pos, 1.2-81 ppb; cheese, 4 types, 8 samples, 100% pos, 3.9-98 ppb, max concn in Parmesan cheese [19].

**Plant Concentrations:**

**Fish/Seafood Concentrations:** Bottomfish, Commencement Bay and adjacent waterways, Tacoma, WA, 1982, highest avg level, 0.53 ppm, highest level, 0.7 ppm [27]. In Lake Pontchartrain, New Orleans: oysters from Inner Harbor Navigation Canal, 7.8 ng/g (ppb) wet weight; clams from Chef Menteur Pass, 27 ppb; clams from Rigolets, 4.5 ppb [13].

**Animal Concentrations:**

# Dichloromethane

**Milk Concentrations:**

**Other Environmental Concentrations:**

**Probable Routes of Human Exposure:** Large amounts (over 500 million lbs in 1983) of methylene chloride are used each year in aerosols, paint removers, metal degreasing, chemical processing, electronics, and as a urethane foam blowing agent [4]. Human exposure primarily will result from ambient air, particularly in the vicinity of these industries. Another source of exposure is from drinking water originating from contaminated ground water sources.

**Average Daily Intake:** AIR INTAKE: (assume 0.4-3.8 ppb intake [41,42,43]) - 28-268 ug; WATER INTAKE: (assume 0 ppb for 86% of people and 6.1 ppb average for 14% of people [51]) - 0 ug or 12.2 ug; FOOD INTAKE: insufficient data.

**Occupational Exposures:** NIOSH (NOES Survey 1981-83) has statistically estimated that 1,147,425 workers are exposed to dichloromethane in the United States [29]. NIOSH (NOHS Survey 1972-74) has statistically estimated that 2,175,499 workers are exposed to dichloromethane in the United States [30]. Casting room (1968-72) - 55-495 ppm; Plastic film factory - 3 year study, 318 samples - 30-5,000 ppm, 627 ppm avg; 1973-74 study of 7 jobs using methylene chloride: 6 of 7 jobs - 0-74 ppm, chemical plant 0-5,520 ppm [28].

**Body Burdens:** Detected in all 8 samples of human milk from 4 urban areas [33]. Mother's milk in Soviet women manufacturing rubber articles - 74 ppb mean in 17 of 28 samples approx 5 hours after start of work, level declined after termination of work [22]. Whole blood specimens, 250 subjects, not detected to 25 ppb, 0.7 ppb avg [1].

## REFERENCES

1.  Antoine SR et al; Bull Environ Contam Toxicol 36: 364-71 (1986)
2.  Boublik T et al; The Vapor Pressures of Pure Substances Vol 17 Amsterdam, Netherlands: Elsevier Science Publ (1984)
3.  Brodzinsky R, Singh HB; Volatile Organic Chemicals in the Atmosphere: An assessment of available data. SRI Contract 68-02-3452 (1982)
4.  Chemical Marketing Reporter; Chemical Profile Feb 28, 1983 (1983)
5.  Cox RA et al; Atmos Environ 10:305-8 (1976)
6.  Davis EM et al; Water Res 15:1125-7 (1981)
7.  DeWalle FB, Chain ESK; Proc Ind Waste Conf 32:908-19 (1978)

8.  Dilling WL et al; Environ Sci Technol 9:833-8 (1975)
9.  Dilling WL; Environ Sci Technol 11: 405-9 (1977)
10. Dilling WL et al; Environ Sci Technol 10:351-6 (1976)
11. Dyksen JE, Hess AF III; J Amer Water Works Assoc 74:394-403 (1982)
12. Ewing BB et al; Monitoring to detect previously unrecognized pollutants in surface waters. EPA-560/6-77-015, EPA-560/6-77-015A 79 pp (1977)
13. Ferrario JB et al; Bull Environ Contam Toxicol 34: 246-55 (1985)
14. Gossett JM; Anaerobic degradation of C1 and C2 chlorinated hydrocarbons. Air Force Eng Serv Cent, Eng Serv Lab. ESL-TR-85-38 153 pp (1985)
15. Hampson RF; Chemical Kinetic and Photochemical Data Sheets for Atmospheric Reactions. 1 Report FAA-EE-80-17 US Dept. of Transportation (1980)
16. Hansch C, Leo AJ; Medchem Project Issue No 26. Claremont CA: Pomona College (1985)
17. Harkov R et al; J Air Pollut Control Assoc 33:1177-83 (1983)
18. Heikes DL, Hopper ML; J Assoc Off Anal Chem 69: 990-8 (1986)
19. Heikes DL; J Assoc Off Anal Chem 70: 215-26 (1987)
20. Helz GR, Hsu RY; Limnol Oceanogr 23:858-69 (1978)
21. Hubrich C, Stuhl F; J Photochem 12:93-107 (1980)
22. Jensen AA; Res Rev 89: 1-128 (1983)
23. Kelley RD; Synthetic Organic Compounds Sampling Survey of Public Water Supplies. Iowa Dept Water, Air and Waste Management. (NTIS PB85-214427) (1985)
24. Klecka GM; Appl Environ Microbiol 44:701-7 (1982)
25. Konasewich D et al; Status Report on Organic and Heavy Metal Contaminants in the Lakes Erie, Michigan, Huron, and Superior Basins. Great Lake Water Quality Board 373 p (1978)
26. Lyman WJ et al; Handbook of Chemical Property Estimation Methods. Environmental Behavior of Organic Compounds. McGraw Hill, NY. (1982)
27. Nicola RM; J Environ Health 49: 342-7 (1987)
28. NIOSH; Criteria for a recommended standard. Occupational Exposure to Methylene Chloride. NIOSH 76-138. p 80-81, 152, 164 (1976)
29. NIOSH; The National Occupational Exposure Survey (NOES) (1983)
30. NIOSH; The National Occupational Hazard Survey (NOHS) (1974)
31. Ohio R Valley Water Sanit Comm; Assessment of Water Quality Conditions. Ohio River Mainstream 1980-1981 table 13 (1982)
32. Otson R et al; J Assoc Off Analyt Chem 65:1370-4 (1982)
33. Pellizzari ED et al; Bull Environ Contam Toxicol 28: 322-8 (1982)
34. Pellizzari ED; Quantification of Chlorinated Hydrocarbons in Previously Collected Air Samples. EPA-450/3-78-112 (1978)
35. Plumb RHJr; Ground Water Monit Rev 7: 94-100 (1987)
36. Rathbun RE, Tai DY; Water Res 15:243-50 (1981)
37. Riddick JA et al; Organic Solvents: Physical Properties and Methods of Purification, 4th Edit. New York: J Wiley & Sons (1986)
38. Rittmann BE, McCarty PL; Appl Environ B Microbiol 39:1225-6 (1980)
39. Sabel GV, Clark TP; Waste Manag Res 2: 119-30 (1984)
40. Sabljic A; J Agric Food Chem 32: 243-6 (1984)
41. Singh HB et al; Atmos Environ 15:601-12 (1981)
42. Singh HB et al; Environ Sci Technol 16:872-80 (1982)
43. Singh HB et al; Atmospheric Measurements of Selected Hazardous Organic Chemicals. EPA-600/5-3-81-032 (1981)

# Dichloromethane

44. Singh HB et al; Atmospheric Distributions, Sources and Sinks of Selected Halocarbons, Hydrocarbons, SF$_6$ and H$_2$O. EPA-600/3-79-107 p 4 (1979)
45. Singh HB et al; J Geophys Res 88: 3675-83 (1983)
46. Spence JW et al; J Air Pollut Control Assoc 76:994-6 (1976)
47. Staples CA et al; Environ Toxicol Chem 4: 131-42 (1985)
48. Stover EL, Kincannon DF; J Water Pollut Control Fed 55:97-109 (1983)
49. Tabak HH et al; J Water Pollut Control Assoc 53:1503-18 (1981)
50. USEPA; Treatability Manual. EPA-600/2-82-001A page I.12.2-1 to I.12.2-4 (1981)
51. USEPA; Ambient Water Quality Criteria for Halomethanes EPA-440/5-80-051 p. C-6 to C-17 (1980)
52. Von Dueszeln et al; Z Wasser Abwasser Forsch 15:272-6 (1982)
53. Yanagihara S et al; Photochemical Reactivities of Hydrocarbons. Proceedings of the 4th Clean Air Congress p 472-7 (1977)
54. Young DR et al; Water Chlorination: Environ Impact Health Effect 4(Book 2): 871-4 (1983)
55. Zoeteman BCJ; Sci Total Environ 21: 187-202 (1981)

# 1,2-Dichloropropane

## SUBSTANCE IDENTIFICATION

**Synonyms:** Propylene dichloride

**Structure:**

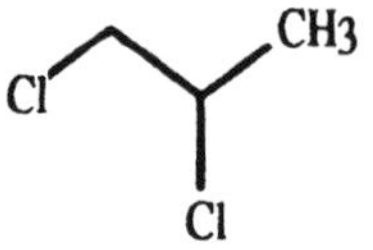

**CAS Registry Number:** 78-87-5

**Molecular Formula:** $C_3H_6Cl_2$

**Wiswesser Line Notation:** GY1&1G

## CHEMICAL AND PHYSICAL PROPERTIES

**Boiling Point:** 96.37

**Melting Point:** -100.44 °C

**Molecular Weight:** 112.99

**Dissociation Constants:**

**Log Octanol/Water Partition Coefficient:** 1.99 [44]

**Water Solubility:** 2,740 mg/L at 25 °C [34]

**Vapor Pressure:** 49.67 mm Hg at 25 °C [34]

**Henry's Law Constant:** 2.07 x $10^{-3}$ atm-mole/m³ [28]

## ENVIRONMENTAL FATE/EXPOSURE POTENTIAL

**Summary:** 1,2-Dichloropropane is released into soil during its use as a soil fumigant and into air as fugitive emissions and in wastewater during its production and use as a chemical intermediate, solvent, scouring, spotting, and metal degreasing agent. If injected into soil 1,2-dichloropropane will be primarily lost by volatilization. Some may leach into the ground water where its fate is unknown. If released into water, 1,2-dichloropropane will be lost by volatilization with half-lives ranging from approx 6 hr for a river to 10 days for a lake. Adsorption to soil and bioconcentration in fish will not be significant. In air it will react with photochemically generated hydroxyl radicals (half-life >23 days) and be washed out by rain. Therefore, there will be ample time for dispersal as is evidenced by its presence in ambient air. Human exposure is primarily due to inhalation. Occupational exposure, both dermal and via inhalation, will occur during and after its application as a soil fumigant as well as during its production and other uses.

**Natural Sources:**

**Artificial Sources:** 1,2-Dichloropropane may be released into the atmosphere or in wastewater during its production or use as a soil fumigant for nematodes, chemical intermediate in the production of carbon tetrachloride and perchloroethylene, lead scavenger for antiknock fluids, solvent, in ion exchange resin manufacture, paper coating, scouring, spotting, metal degreasing agent [1,17], and insecticide for stored grain. Municipal landfill leachates are an additional source [36]. The total estimated annual environmental releases from production and industrial use are 772,000 lb to water, and 176,000 lb to land disposal sites for a total of 1,146,000 lb [46]. These releases include process emissions to the air, secondary air emissions resulting from volatilization during freshwater treatment, releases to water in wastewater effluent, air release via incineration, and land disposal of solid waste residues, tar, and ash residues from incineration. One manufacturer of ion exchange resins annually discharges about 500,000 lb of 1,2-dichloropropane to the aquatic environment [45]. There are four ion exchange manufacturers in the US with potentially similar release patterns [45].

**Terrestrial Fate:** If released on soil, 1,2-dichloropropane will rapidly volatilize and readily leach into the ground especially in the sandy soils in areas of California and the coastal plains of Georgia, South Carolina, North Carolina, and Virginia where it is used as a nematicidal fumigant [11]. There is no evidence that it biodegrades in sandy soil although there is some evidence that a minor amount of degradation occurs in medium loam soil in 20 weeks in closed glass containers [35]. Under outdoor conditions >99% is lost through volatilization within 10 days

[35]. During its use as a fumigant for nematodes, 1,2-dichloropropane is injected into the root zone and then the soil is compacted to retain the fumigant longer. Leaching into ground water is possible especially if there is rain or the fields are irrigated as indicated by ground water contamination in agricultural areas.

**Aquatic Fate:** If released into water, 1,2-dichloropropane will be lost primarily by volatilization (half-life 5-8 hr in a typical river to 10 days in a lake). Adsorption to sediment, hydrolysis, and biodegradation will not be an important loss process.

**Atmospheric Fate:** If released in air, 1,2-dichloropropane will degrade by reaction with photochemically produced hydroxyl radicals (half-life >23 days). Due to its moderately high water solubility washout by rain should occur.

**Biodegradation:** 1,2-Dichloropropane is reported to be degraded during biological treatment provided suitable acclimation is achieved [18,41]. However, in a bench scale continuous flow activated sludge reactor with an 8 hr retention time, 1,2-dichloropropane was entirely removed by stripping [22]. It was reported to be degradation resistant in a 2 week screening test that utilizes a mixed inoculum of soil, surface water, and sludge [21]. 42% degradation was achieved in 7 days when incubated with sewage seed [40]. There was evidence of only slight biodegradation in 20 weeks (as evidenced by 5% of the radioactivity remaining unextracted from a medium loam soil) and it was stable in a sandy loam soil for 12 weeks when incubated in a closed container [35]. No volatile degradation products were detected from treated soil [35].

**Abiotic Degradation:** 1,2-Dichloropropane does not have any chromophores that absorb wavelengths >290 nm so direct photolysis will not be a significant fate process. Vapor phase photolysis under simulated sunlight did not occur after prolonged exposure [11]. Experimental determination of its rate of reaction with hydroxyl radicals yields a half-life of >23 days [2]. A computer estimate of its half-life due to H-atom abstraction by hydroxyl radical yields a conflicting half-life of 7.12 days [47]. The fact that levels of 1,2-dichloropropane were lowest in the afternoon is evidence of photooxidative degradation [38]. The hydrolytic half-life of 1,2-dichloropropane is expected to be 6 months to several years [8].

# 1,2-Dichloropropane

**Bioconcentration:** The log BCF in fish is <1 [21]. Based on the calculated log octanol/water partition coefficient, 1.99 [44], one would estimate a log BCF for 1,2-dichloropropane of 1.26 [43].

**Soil Adsorption/Mobility:** The Koc of 1,2-dichloropropane is 47 in a silt loam soil [10]. It sorbs to clay minerals in dry soil but desorbs as the soil absorbs moisture [11]. In the areas of the US where 1,2-dichloropropane is used as a fumigant, the soil is generally sandy with low organic carbon content and would probably have little impact on reducing mobility due to soil adsorption [11]. It was detected at concn up to 12.2 ppb throughout much of a 24-foot soil core collected from a field with recent documented use of Telone II where ground water contamination was found, thereby documenting vertical movement of the 1,2-dichloropropane after it was applied as a fumigant [1].

**Volatilization from Water/Soil:** 1,2-Dichloropropane has a high Henry's Law constant and will therefore volatilize rapidly, with diffusion through the liquid phase controlling the rate of evaporation. In a laboratory experiment, 1,2-dichloropropane volatilized with a half-life of 8 min from a stirred solution 1.6 cm deep [9] which is equivalent to an 8.3 hr half-life at a 1 m depth [27]. The half-life in a wind wave tank (wind speed 6 m/sec) is 6.7 hr [28]. The removal of 1,2-dichloropropane in a stream with 1 m/sec current and 1 m depth is estimated to be 5.5 hr based on laboratory determined relative transfer coefficients [7]. Greater than 99% removal from a wastewater treatment plant is attributable to stripping [22]. An EXAMS model of the fate of 1,2-dichloropropane in a pond, river, and two lakes yielded volatilization half-lives of 1.7-10 day [6]. 1,2-Dichloropropane has a high vapor pressure and low adsorptivity to soil and will therefore volatilize rapidly from soil [11].

**Water Concentrations:** DRINKING WATER: 1,2-Dichloropropane has been identified in drinking water in the US [23,24]: 11 water utilities along the Ohio River 0.1 ppb mean, 0.1 ppb max, 1.6% positive [31]. The Netherlands: Maximum concn in tap water derived from bank-filtered Rhine River water 300 ppt [33]. US Ground Water Supply Survey (945 supplies derived from ground water chosen both randomly and on the basis that they may contain VOC's) - 13 samples positive, median of positive samples, 0.9 ppb, maximum 21 ppb [48]. GROUND WATER: Found in 3 Maryland wells, 30 Long Island, NY wells, and over 60 California wells at levels ranging from 1-50 ppb [11]. Minnesota: Found in ground water under 8 of 13 municipal landfills with suspected ground water contamination - 0.5-43 ppb and 1 of 7 other municipal landfills - 1.1 ppb [36]. 1,2-Dichloropropane is the second most frequent ground water pesticidal contaminant in California [1]. It has been detected in 75 wells at concn ranging up to 1200 ppb in 9 countries [1]. Not detected in ground water underlying the Amphenol metal plating facility in Broadview, IL at a detection limit of 1 ppb [19]. Detected in ground water 83 days after application with 92%, 1,3-dichloropropene [26]. Identified in ground water under a leaky storage tank of a paint factory [4]. SURFACE WATER: Lake Ontario (95 stations) 4 stations have concentrations ranging from 210-440 ppt and 15 others have trace quantities [20]. Lower Niagara River (16 stations) 4 stations had concentrations ranging from 7-55 ppt and 5 other stations had trace quantities [20]. Ohio River 1977-78 (141 samples) 19 positive, 0.1 ppb max, Ohio River tributaries (95 samples) 5 positive, 0.8 ppb maximum [30]. Ohio River 1980-81 (11 stations 4972 samples) 8.8% of samples positive, 28 samples between 1-10 ppb and 1 sample >10 ppb [30]. 14 heavily industrialized river basins in US: 5 of 204 sites positive including those in the Illinois River Basin, Delaware River Basin, and Hudson River Basin 1-2 ppb [13]. Rhine River (km 865) <1 ppb [29]. Detected in the Niagara and Genesee Rivers but not in the open waters of Lake Ontario [14].

**Effluent Concentrations:** Municipal landfill leachate tested in Minnesota - 3 of 6 positive 2.0-81 ppb as was 1 of 5 in Wisconsin 54 ppb [36]. National Urban Runoff Program in which 86 samples from 19 cities throughout the US were analyzed: Detected only in Eugene, OR at 3 ppb, 1% frequency of detection nationwide [12]. Industries (mean concn) whose effluents contain 1,2-dichloropropane include photographic equipment/supplies (10 ppb), organic chemical manufacturing/plastics (25 ppb), and paint and ink formulation (210 ppb) [42].

**Sediment/Soil Concentrations:** Present in concn up to 12.2 ppb throughout much of soil core underlying recently fumigated field in California [1]. Not detected in boring in Anoka sandplain near Minneapolis, MN and upgradient of an asphalt lined municipal solid waste landfill which had polluted ground water, but 2.0 ppb was found in clay boring upgradient of landfill in SW Minnesota which had contaminated the ground water [36]. Detected in Love Canal sediment/soil/water samples [16]. Found in soil cores in California as far as 7 m down at 0.2-2.2 ppb [11].

# 1,2-Dichloropropane

**Atmospheric Concentrations:** RURAL: Island of Terschelling, The Netherlands 60 ppt mean, 1500 ppt, maximum [15]. URBAN/SUBURBAN: US (396 samples) 57 ppt, median, 110 ppt, maximum [5]. Seven US cities (round-the-clock sampling for 1-2 weeks) 21-78 ppt range of means [38]; Houston had highest mean value with highest values at night or early morning and low values in afternoon [38]. Delft and Vlaardingin, The Netherlands 140 ppt mean, 3000 ppt, maximum [15]. California Air Monitoring Program (24-hr sampling at four sites throughout the year) 2% of samples above the quantitation limit (0.2 ppt), while most positive values were near the quantitation limit, one monthly mean in Riverside was 1.1 ppb [37]. Portland, OR (7 rain events) 4.4-8.4 ppt in gas phase [25]. SOURCE AREAS: US (26 samples) 120 ppt, median, 130 ppt, maximum [5]. Niagara Falls - trace, 22% of samples positive; Baton Rouge 0-40 ppb, 38% of samples positive [32]. Philadelphia (24-hr, 3 month survey of 10 source-oriented sites, 31 samples) 259 ppt mean [39]. Traces of 1,2-dichloropropane detected outside 2 of 9 homes of Love Canal residents [3].

**Food Survey Values:**

**Plant Concentrations:**

**Fish/Seafood Concentrations:**

**Animal Concentrations:**

**Milk Concentrations:**

**Other Environmental Concentrations:**

**Probable Routes of Human Exposure:** The general public is exposed to 1,2-dichloropropane from ambient air via inhalation and from contaminated drinking water. Workers may be exposed via inhalation and dermally during its application as a soil fumigant and by being near fields within several days after treatment. Occupational exposure will also occur via inhalation and dermally during its production and use.

**Average Daily Intake:** AIR INTAKE: (assume 57 ppt) 5.3 ug; WATER INTAKE: (assume 0 ppt) 0 ug; FOOD INTAKE: insufficient data.

**Occupational Exposures:** Occupational exposure to 1,2-dichloropropane involves approximately 500 workers exposed to direct inhalation (estimated to range from 31 to 410 g/person/yr) at 1,2-dichloropropane production and industrial use facilities [46]. An estimated 900 workers may be exposed to direct inhalation (0.020 to 0.27 g/person/yr) as a result of the volatilization of 1,2-dichloropropane from wastewater during treatment operations [46]. There is also potential for exposure (4.8 to 100 mg/yr) to 1,2-dichloropropane of non-production workers at 1,2-dichloropropane production and use facilities [46].

**Body Burdens:**

# REFERENCES

1. Ali SM et al; Am Chem Soc Div Environ Chem 191st Natl Meet 26: 41 (1986)
2. Atkinson R; Chem Rev 85: 69-201 (1985)
3. Barkley J et al; Biomed Mass Spectrom 7: 139-47 (1980)
4. Botta D et al; Comm Eur 8518 (Annal Org Micropollut Water): 261-75 (1984)
5. Brodzinsky R, Singh HB; Menlo Park, CA Atmospheric Sci Center, SRI International p 198 68-02-3452 (1982)
6. Burns LH; Exposure Analysis Modeling System Environ Res Lab USEPA (1981)
7. Cadena F et al; J Water Pollut Control Fed 56: 460-3 (1984)
8. Callahan MA et al; Water-Related Environ Fate of 129 Priority Pollut Vol.II USEPA-440/4-79-029B (1979)
9. Chiou CT et al; Environ Inter 3: 231-6 (1980)
10. Chiou CT et al; Science 206: 831-2 (1979)
11. Cohen SZ et al; ACS Symp Ser 259: 297-325 (1984)
12. Cole RH et al; J Water Pollut Control Fed 56: 898-908 (1984)
13. Ewing BB et al; Monitoring to Detect Previously Unrecognized Pollut in Surface Waters p 75 USEPA-560/6-77-015 (1977)
14. Great Lakes Water Quality Board; An Inventory of Chemical Substances Identified in the Great Lakes Ecosystem Vol.I (1983)
15. Guicherit R, Schulting FL; Sci Total Environ 43: 193-219 (1985)
16. Hauser TR, Bromberg SM; Environ Monit Assess 2: 249-72 (1982)
17. Hawley GG; Condensed Chemical Dictionary 10th ed Von Nostrand Reinhold NY p. 864 (1981)
18. Ilisescu A; pp 249-66 in Stud Prot Epurarea Apelor (1971)
19. IT Corporation; Preliminary Site Assessment, Broadview, Illinois Plant, Amphenol Products Div, Industrial and Technol Sector, TSCA Health and Safety Studies 8D Submission No. 878216382 (1985)
20. Kaiser KLE et al; J Great Lakes Res 9: 212-23 (1983)
21. Kawasaki M; Ecotox Environ Safety 4: 444-54 (1980)
22. Kincannon DF et al; p 641-50 in Proc 37th Industrial Waste Conf Bell JM ed,

# 1,2-Dichloropropane

Ann Arbor MI, Ann Arbor Sci Pub (1983)

23. Kincannon DF et al; pp 641-50 in Proc 37th Industrial Waste Conf, Bell JM ed, Ann Arbor, MI Ann Arbor Sci Pub (1983)
24. Kool HJ et al; Crit Rev Environ Control 12: 307-57 (1982)
25. Ligocki MP et al; Atmos Environ 19: 1609-17 (1985)
26. Loria R; Plant Dis 70: 42-5 (1986)
27. Lyman, WJ et al; Handbook of Chemical Property Estimation Methods McGraw-Hill, NY (1982)
28. Mackay D, Yeun ATK; Environ Sci Technol 17: 211-7 (1983)
29. Malle KG; Z Wasser-Abwasser Forsch 17: 75-81 (1984)
30. Ohio River Valley Sanit Comm; Assessment of Water Quality Conditions Ohio River Mainstream 1980-81 (1982)
31. Ohio River Valley Water Sanit Comm; Water Treatment Process Modifications for Trihalomethane Control and Organic Substances in the Ohio River 10/76-8/79 (1979)
32. Pellizzari ED et al; Formulation of Prelim Assess of Halogenated Org Compounds in Man and Environ Media p 469 USEPA-560/13-79-006 (1979)
33. Piet GJ, Morra CF; pp 31-42 in Artificial Groundwater Recharge, Huisman L, Olsthorn TN, Eds; Pitman Pub (1983)
34. Riddick JA et al; Organic Solvents New York: Wiley Interscience (1986)
35. Roberts TR, Stoydin G; Pest Sci 7: 325-35 (1976)
36. Sabel GV, Clark TP; Waste Manag Res 2: 119-30 (1984)
37. Shikiya J et al; p 21 in Proc. APCA Annu Meet 77th Vol I 84-1.2 (1984)
38. Singh HB et al; Environ Sci Technol 16: 872-80 (1982)
39. Sullivan DA et al; p 15 in Proc APCA Annu Meet 78th Vol 2, 85-17.5 (1985)
40. Tabak HH et al; J Water Pollut Contr Fed 53: 1503-18 (1981)
41. Thom NS, Agg AR; Proc R Soc Lond B 189: 347-57 (1975)
42. USEPA; Treatability Manual I Treatability Data USEPA-600/2-82-001A (1981)
43. USEPA; GEMS Graphical Exposure Modeling System. CHEMEST. (1986)
44. USEPA; GEMS Graphical exposure modeling system; CLOGP3 (1987)
45. USEPA; 49 FR 900 1/6/84 (1984)
46. USEPA; 51 FR 32083 9/9/86 (1986)
47. USEPA; GEMS; Graphical Exposure Modeling System. FAP. Fate of Atmospheric Pollutants (1986)
48. Westrick JJ et al; J Amer Water Works Assoc 76: 52-9 (1984)

# 1,2-Dichloro-1,1,2,2-Tetrafluoroethane

## SUBSTANCE IDENTIFICATION

**Synonyms:** Freon 114

**Structure:**

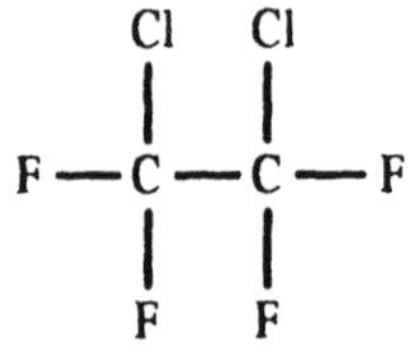

```
        Cl   Cl
        |    |
  F — C — C — F
        |    |
        F    F
```

**CAS Registry Number:** 76-14-2

**Molecular Formula:** $C_2Cl_2F_4$

**Wiswesser Line Notation:**

## CHEMICAL AND PHYSICAL PROPERTIES

**Boiling Point:** 4.1 °C at 760 mm Hg

**Melting Point:** -94 °C

**Molecular Weight:** 170.93

**Dissociation Constants:**

**Log Octanol/Water Partition Coefficient:** 2.82 [11]

**Water Solubility:** 130 mg/L at 25 °C [15]

**Vapor Pressure:** 2014 mm Hg at 25 °C [15]

**Henry's Law Constant:** 2.8 atm-m³/mole at 25 °C (calculated from vapor pressure and water solubility)

## ENVIRONMENTAL FATE/EXPOSURE POTENTIAL

**Summary:** 1,1-Dichloro-1,1,2,2-tetrafluoroethane (Freon 114) may be released to the environment as emissions from production, storage,

192

transport, or use as a foaming agent, refrigerant, and aerosol propellant, and it may be released to soil from the disposal of items which contain this compound (e.g., commercial/industrial refrigeration units). The global release rate of Freon 113 is estimated to be $2.1 \times 10^{+4}$ tons per year, which corresponds with a 6% annual increase in the abundance of Freon 114 in the atmosphere. If released to soil, Freon 114 would rapidly volatilize from soil surfaces or leach through soil possibly into ground water. If released to water, essentially all Freon 114 is expected to be lost by volatilization (half-life 4 hr from a model river). If released to the atmosphere, Freon 114 will not degrade in the troposphere. This compound will gradually diffuse into the stratosphere (half-life 20 years). In the stratosphere this compound will slowly photolyze or slowly react with singlet oxygen (stratospheric lifetime 126-310 years). Due to its stability, detection long distances from its sources of emissions has occurred. General population exposure occurs by inhalation of Freon 114 found in ambient air. Occupational exposure may occur by inhalation of contaminated air or dermal contact.

**Natural Sources:** No data are available which indicates that Freon 114 is a naturally occurring compound.

**Artificial Sources:** Freon 114 may be released to the environment as emissions from production, storage, transport, or use as a foaming agent, refrigerant, and aerosol propellant [10]. Freon 114 together with Freon 113, Freon 115, and Freon 13 contain about 3% of the organically bound chlorine present in the atmosphere [9]. The global release rate of Freon 113 is estimated to be $2.1 \times 10^{+4}$ tons per year, which corresponds with a 6% annual increase in the abundance of Freon 114 in the atmosphere [4,9]. This compound may be released to soil from the disposal of products containing this compound. These products include mobile air conditioners, retail food refrigeration units, and centrifugal and reciprocating chillers [5].

**Terrestrial Fate:** If released to soil, Freon 114 would rapidly volatilize from soil surfaces or leach through soil possibly into ground water.

**Aquatic Fate:** If released to water, essentially all Freon 114 is expected to be lost by volatilization (half-life 4 hr from a model river). Chemical hydrolysis, bioaccumulation, and adsorption to sediments would not be a significant fate processes in water.

# 1,2-Dichloro-1,1,2,2-Tetrafluoroethane

**Atmospheric Fate:** If released to the atmosphere, essentially all Freon 114 is expected to exist in the vapor phase due to its extremely high vapor pressure. The moderate water solubility of Freon 114 suggests that some loss by wet deposition occurs, but any loss by this mechanism is probably returned to the atmosphere by volatilization. Freon 114 will not degrade in the troposphere, thus diffusion from the troposphere to the stratosphere would be the sole removal mechanism (half-life 20 years [6]). As a result of its persistence in the atmosphere, this compound is transported long distances and its concentration should be fairly uniform throughout the globe away from known sources.

**Biodegradation:**

**Abiotic Degradation:** Chemical hydrolysis of Freon 114 is not an environmentally significant fate process [7]. Freon 114 is essentially inert to reaction with photochemically generated radicals and ozone molecules [1,2]. This compound will not undergo direct photolysis in the troposphere [13]. The stratospheric lifetime of Freon 114 has been estimated to range from 126 to 310 years with direct photolysis and reaction with singlet oxygen being the significant removal mechanisms [4]. In the stratosphere, this compound will slowly photolyze to release chlorine atoms, which in turn participate in the catalytic removal of stratospheric ozone [13].

**Bioconcentration:** Based on the water solubility and Kow value, bioconcentration factors (BCF) of 10 and 82, respectively, were estimated for Freon 114 [12]. These BCF values suggest that Freon 114 would not bioaccumulate significantly in aquatic organisms.

**Soil Adsorption/Mobility:** Soil adsorption coefficients (Koc) of 815 and 300 were estimated using linear regression equations based on the Kow and the water solubility, respectively [12]. These Koc values suggest that Freon 114 would have low to moderate mobility in soil and that adsorption to suspended solids and sediments in water would be moderate to high [16].

**Volatilization from Water/Soil:** The value of the Henry's Law constant suggests that Freon 114 would volatilize rapidly from all bodies of water and from soil surfaces [12]. Based on this value the volatilization half-life of Freon 114 from a model river 1 m deep flowing 1 m/sec with a wind velocity of 3 m/sec has been estimated to be 4 hours [12].

**Water Concentrations:**

**Effluent Concentrations:**

**Sediment/Soil Concentrations:**

**Atmospheric Concentrations:** Freon 114 was detected in air samples collected in ambient air over France between 1982 and 1984, avg concn 10.5 ppt/volume [8]. During spring 1983 in the Norwegian Arctic, avg concn 10.9 ppt/volume [8]. During 1979-81, the average concentration of Freon 114 in ambient air of the Northern and Southern hemispheres was 14 and 13 ppt/volume, respectively [8]. Freon 114 was detected in air samples collected throughout the U.S. between 1976 and 1980: RURAL/REMOTE: 5 data points, median concn 12 ppt/volume, mean concn 11 ppt/volume; URBAN/SUBURBAN: 267 data points, median concn 32 ppt/volume, mean concn 32 ppt/volume [3].

**Food Survey Values:**

**Plant Concentrations:**

**Fish/Seafood Concentrations:**

**Animal Concentrations:**

**Milk Concentrations:**

**Other Environmental Concentrations:**

**Probable Routes of Human Exposure:** The general population is exposed to Freon 114 in ambient air. In occupational settings, it is expected that exposure occurs by inhalation of contaminated air and dermal contact with this compound.

**Average Daily Intake:** AIR: (assume 10.5 - 32 ppt [3,8]) 1.5 - 4.5 ug/day.

# 1,2-Dichloro-1,1,2,2-Tetrafluoroethane

**Occupational Exposure:** NIOSH has estimated that 38 workers are potentially exposed to Freon 114 based on a survey conducted in 1972-74 in the United States [14].

**Body Burdens:**

## REFERENCES

1. Atkinson R; Internat J Chem Kinetics 19: 799-828 (1987)
2. Atkinson R; Chem Rev 85: 69-201 (1985)
3. Brodzinsky R, Singh HB; pp. 23 and 184 in Volatile Organic Chemicals in the Atmosphere: An Assessment of Available Data Menlo Park, CA: SRI International (1982)
4. Chou CC et al; J Phys Chem 82: 1-7 (1978)
5. Clayton GD, Clayton FE eds; p. 3102-2 in Patty's Industrial Hygiene and Toxicology Vol IIB 3rd ed NY: Wiley and Sons (1981)
6. Dilling WL; Environmental Risk Analysis for Chemicals; Conway RA ed NY: Van Nostrand Reinhold Co pp. 154-97 (1982)
7. Du Pont de Nemours Co; Freon Products Information B-2; A98825 12/80 (1980)
8. Fabian P et al; J Geophys Res 90: 13091-3 (1985)
9. Fabian P; pp. 23-51 in The Handbook of Environmental Chemistry, Vol 4/Part A; Hutzinger O ed NY: Springer-Veralg (1986)
10. Graedel TE; Chemical Compounds in the Atmosphere NY: Academic Press p. 327 (1978)
11. Hansch C, Leo AJ; Medchem Project Issue No.26 Pomona College, Claremont CA (1985)
12. Lyman WJ et al; Handbook of Chemical Property Estimation Methods. NY: McGraw-Hill (1982)
13. Makide T et al; Chem Lett 4: 355-8 (1979)
14. NIOSH; National Occupational Hazard Survey (NOHS) (1974)
15. Riddick JA et al; Organic Solvents: Physical Properties and Methods of Purification. Techniques of Chemistry. 4th Ed. New York: Wiley-Interscience pp 1325 (1986)
16. Swann RL et al; Res Rev 85: 17-28 (1983)

# Diethanolamine

## SUBSTANCE IDENTIFICATION

**Synonyms:**

**Structure:**

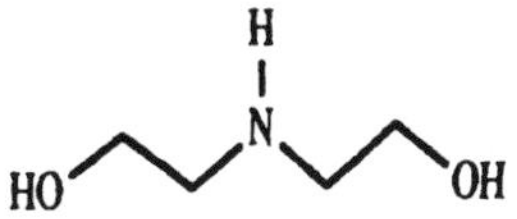

**CAS Registry Number:** 111-42-2

**Molecular Formula:** $C_4H_{11}NO_2$

**Wiswesser Line Notation:** Q2M2Q

## CHEMICAL AND PHYSICAL PROPERTIES

**Boiling Point:** 268.8 °C at 760 mm Hg

**Melting Point:** 28 °C

**Molecular Weight:** 105.14

**Dissociation Constants:** pKa = 8.97 [19]

**Log Octanol/Water Partition Coefficient:** -1.43 [9]

**Water Solubility:** Miscible [5]

**Vapor Pressure:** 0.00028 mm Hg at 25 °C [5]

**Henry's Law Constant:** 5.35 x $10^{-14}$ atm-m³/mole at 25 °C [11]

## ENVIRONMENTAL FATE/EXPOSURE POTENTIAL

**Summary:** Diethanolamine may be released to the environment in emissions or effluents from sites of its manufacture or industrial use,

197

from disposal of consumer products which contain this compound, and from application of agricultural chemicals in which this compound is used as a dispersing agent. In soil and water, diethanolamine is expected to biodegrade fairly rapidly following acclimation (half-life on the order of days to weeks). N-Nitrosodiethanolamine is a metabolite of diethanolamine. In soil, residual diethanolamine may leach into ground water. In the atmosphere, diethanolamine is expected to exist almost entirely in the vapor phase. Reaction with photochemically generated hydroxyl radicals is expected to be the dominant removal mechanism (half-life 4 hr). This compound may also be removed from the atmosphere in precipitation. The most probable route of exposure to diethanolamine is dermal contact with personal care products (i.e., soaps, shampoos, cosmetics), detergents, and other surfactants containing this compound.

**Natural Sources:**

**Artificial Sources:** Diethanolamine may be released to the environment in emissions or effluents from sites of its manufacture or industrial use, from disposal of consumer products which contain this compound, and from application of agricultural chemicals in which this compound is used as a dispersing agent [14,23].

**Terrestrial Fate:** If released to soil, diethanolamine is expected to biodegrade fairly rapidly following acclimation (half-life on the order of days to weeks). Residual diethanolamine may leach into ground water. Volatilization from soil surfaces is not expected to be an important fate process.

**Aquatic Fate:** If released to water, diethanolamine should biodegrade. The half-life of this compound is expected to range from a couple of days to a few weeks depending, in large part, on the degree of acclimation of the system. N-Nitrosodiethanolamine is a metabolite of diethanolamine. Bioconcentration in aquatic organisms, adsorption to suspended solids and sediments, and volatilization are not expected to be important fate processes in water.

**Atmospheric Fate:** Based on the vapor pressure, diethanolamine is expected to exist almost entirely in the vapor phase in the atmosphere [6]. Reaction with photochemically generated hydroxyl radicals is expected to be the dominant removal mechanism (half-life 4 hr). The complete solubility of diethanolamine in water suggests that this compound may also be removed from the atmosphere in precipitation.

**Biodegradation:** Grab sample (stream water), initial concn 21 mg/L, 210 ug/L, and 21 ng/L, 4 days - 5, 55, and 32% mineralization, respectively [2]. Grab sample (lake water), initial concn 0.001 ppm, 14 day - 31 and 1.2% mineralization, Cayuga Lake water and North Lake (acidic) water, respectively [24]. Sewage die-away, initial concn 0.001 ppm, 20 days 53% mineralization, sewage inoculum [24]. N-Nitrosodiethanolamine has been identified as a metabolite of diethanolamine in natural water samples and sewage [24]. Die-away, initial concn 50 ppm, 10 day 90% BODT, acclimated Kanawha River water as seed, sewage inoculum [16]. BOD water, initial concn 2.5 ppm, 5, 10, 15, and 20 days - 0.9, 1.4, 3.5, and 6.8% BODT, respectively, sewage inoculum [13]. BOD water, 20 day 88% BODT, sewage inoculum [21]. Synthetic seawater 20 day 76% BODT, sewage inoculum [21]. BOD water, 5 day 2% BODT (unadapted) and 77% BODT (adapted), inoculum was effluent from a biological waste treatment plant [3]. BOD water, initial concn 500 ppm, 10 day 97% BODT, acclimated activated sludge inoculum [7]. BOD water, initial concn equivalent to 100 ppm C, 5 day 97% COD removal, activated sludge inoculum [20]. Zahn-Wellens, initial concn equivalent to 400 ppm C, 3 day 94% DOC removal, activated sludge inoculum [8]. Sturm - $CO_2$ evolution, initial concn equivalent to 10 ppm C, 28 days 91% $CO_2$ evolution and 100% DOC removal, acclimated sewage inoculum [8]. OECD, initial concn equivalent to 3-20 ppm C, 19 day 100% DOC removal, sewage inoculum [8]. Modified Closed Bottle, initial concn 2 ppm, 30 day 94% BODT, enriched sewage inoculum [8]. French AFNOR, initial concn equivalent to 40 ppm C, 28 and 42 days - 97% and 98% DOC removal, respectively, sewage inoculum [8]. Japanese MITI, initial concn equivalent to 50 ppm C, 14 days 3% BODT, activated sludge inoculum [8]. Japanese MITI, initial concn 100 ppm, 14 days 74.6% BODT ($NH_3$ end product) and 51.4% BODT ($NO_2$ end product), activated sludge inoculum [12].

**Abiotic Degradation:** The half-life for diethanolamine vapor reacting with photochemically generated hydroxyl radicals in the atmosphere has been estimated to be 4 hr based on an estimated reaction rate constant of 8.9 x $10^{-11}$ cm$^3$/molecules-sec at 25 °C and an average ambient hydroxyl concentration of 5 x $10^{+5}$ molecules/cm$^3$ [1].

## Diethanolamine

**Bioconcentration:** A bioconcentration factor (BCF) of <1 was estimated for diethanolamine based on the Kow value [15]. This BCF value and complete solubility of diethanolamine in water suggest that this compound does not bioconcentrate significantly in aquatic organisms.

**Soil Adsorption/Mobility:** A soil adsorption coefficient (Koc) of 4 was estimated for diethanolamine based on the Kow [15]. This Koc value and the complete solubility of diethanolamine in water suggests that this compound would be extremely mobile in soil and would not adsorb appreciably to suspended solids and sediments in water [22].

**Volatilization from Water/Soil:** The value of the Henry's Law constant suggests that volatilization of diethanolamine from water and moist soil surfaces would not be a significant fate process [15].

**Water Concentrations:**

**Effluent Concentrations:**

**Sediment/Soil Concentrations:**

**Atmospheric Concentrations:**

**Food Survey Values:**

**Plant Concentrations:**

**Fish/Seafood Concentrations:**

**Animal Concentrations:**

**Milk Concentrations:**

**Other Environmental Concentrations:**

**Probable Routes of Human Exposure:** The most probable route of exposure to diethanolamine is dermal contact with personal care products (i.e., soaps, shampoos, cosmetics), detergents and other surfactants which contain this compound [4,10,23].

**Average Daily Intake:**

# Diethanolamine

**Occupational Exposure:** NIOSH (NOHS Survey 1972-74) has statistically estimated that 1,284,534 workers are potentially exposed to diethanolamine in the United States [18]. NIOSH (NOES Survey 1981-83) has statistically estimated that 573,025 workers are potentially exposed to diethanolamine in the United States [17].

**Body Burdens:**

# REFERENCES

1.  Atkinson R; Inter J Chem Kinet 19: 799-828 (1987)
2.  Boethling RS, Alexander M; Environ Sci Tech 13: 989-91 (1979)
3.  Bridie AL et al; Water Res 13: 627-30 (1979)
4.  Chemical Marketing Reporter; Chemical Profile: Ethanolamine NY: Schnell Publishing Nov 10 (1986)
5.  Dow Chemical Co; The Alkanolamines Handbook. Midland, MI (1980)
6.  Eisenreich SJ et al; Environ Sci Tech 15: 30-8 (1981)
7.  Gannon JE et al; Microbios 23: 7-18 (1978)
8.  Gerike P, Fischer WK; Ecotox Environ Safety 3: 159-73 (1979)
9.  Hansch C, Leo AJ; Medchem Project Issue No.26 Pomona College, Claremont CA (1985)
10. Hawley GG; The Condensed Chemical Dictionary 10 th ed NY: Van Nostrand Reinhold p 342 (1981)
11. Hine J, Mookerjee PK; J Org Chem 40: 292-8 (1975)
12. Kitano M; OECD Tokyo Meeting Reference Book Tsu-No. 3 (1978)
13. Lamb CB, Jenkins GF; pp 326-39 in Proc 8th Ind Waste Conf Purdue Univ (1952)
14. Liepins R et al; Industrial Process Profiles for Environmental Use. USEPA-600/2-77-023f NTIS PB-281 478 pp. 6-386 to 6-387 (1977)
15. Lyman WJ et al; Handbook of Chemical Property Estimation Methods NY: McGraw-Hill (1982)
16. Mills EJ, Stack VT; Proc 9th Ind Waste Conf Eng Bull Purdue Univ: Ext Ser 9: 449-64 (1955)
17. NIOSH; National Occupational Exposure Survey (NOES) (1983)
18. NIOSH; National Occupational Hazard Survey (NOHS) (1974)
19. Perrin DD; Dissociation Constants of Organic Bases in Aqueous Solution. IUPAC Chemical Data Series 1972 Supplement Buttersworth: London (1972)
20. Pitter P; Water Res 10: 63-77 (1976)
21. Price KS et al; J Water Poll Contr Fed 46: 63-77 (1974)
22. Swann RL et al; Res Rev 85: 17-28 (1983)
23. Windholz M ed; The Merck Index 10th ed Rahway, NJ: Merck and Co p. 451 (1983)
24. Yordy JR, Alexander M; J Environ Qual 10: 266-70 (1981)

# Diethylamine

**Synonyms:**

**Structure:**

$$H_3C \diagup \diagdown NH \diagup \diagdown CH_3$$

**CAS Registry Number:** 109-89-7

**Molecular Formula:** $C_4H_{11}N$

**Wiswesser Line Notation:** 2 M 2

## CHEMICAL AND PHYSICAL PROPERTIES

**Boiling Point:** 55.5 °C

**Melting Point:** -50 °C

**Molecular Weight:** 73.14

**Dissociation Constants:** pKa = 10.98 at 25 °C [22]

**Log Octanol/Water Partition Coefficient:** log Kow= 0.58 [11]

**Water Solubility:** Miscible [24]

**Vapor Pressure:** 233.5 mm Hg at 25 °C [24]

**Henry's Law Constant:** $2.57 \times 10^{-5}$ atm-m$^3$/mole at 25 °C [13]

## ENVIRONMENTAL FATE/EXPOSURE POTENTIAL

**Summary:** Diethylamine is produced in large quantities and large amounts of the chemical will be released, primarily as emissions during its production and use as a chemical intermediate and in the petroleum and rubber industries. It also occurs naturally in food and as a metabolic product. If released on land, diethylamine should volatilize and leach into the soil. The chemical would be expected to biodegrade; however, no estimate of degradative rates in soil are available. If released into water, diethylamine will readily biodegrade (half-life 0.9 days at 10 ppm and longer at lower concentrations) and also be removed by volatilization (half-life 31.6 hr in a model river). Adsorption to sediment and bioconcentration in aquatic organisms will not be appreciable. In the atmosphere, diethylamine will react with photochemically produced hydroxyl radicals with an estimated half-life in clean atmospheres of 0.21 days; in polluted atmospheres, diethylamine degrades in 2 hr. It will also be scavenged by rain. Diethylamine is a precursor of dimethylnitrosamine. The latter is formed in the atmosphere in the presence of nitrogen oxides but is destroyed by sunlight. Human occupational exposure to diethylamine is via inhalation of the gas. The general public is exposed primarily by ingesting food in which it occurs naturally.

**Natural Sources:**

**Artificial Sources:** Diethylamine is produced in large quantities, 19.7 million lbs in 1985 [29], and will be released in emissions and wastewater during its production and use as a rubber chemical, solvent, floatation agent, polymerization inhibitor, corrosion inhibitor, electroplating, and in the manufacture of dyes, pesticides, and pharmaceuticals [12]. It is found in emissions from fish processing as well as in tobacco smoke [9].

**Terrestrial Fate:** If released on soil, diethylamine will probably volatilize rapidly and leach readily into the soil. Based on screening studies, biodegradation should be the most important degradative process, but no experimental rates for soil are available.

**Aquatic Fate:** If released into water, diethylamine will volatilize slowly (estimated half-life 31.6 hr in a model river). Biodegradation will be rapid and is the most important removal process in most natural waters. The half-life is inversely proportional to concentration from the ng/L range to 10 mg/L. At the highest concentration, the half-life is 0.9 days. Direct photolysis and adsorption to sediment will not be significant.

# Diethylamine

**Atmospheric Fate:** Diethylamine, released into the atmosphere, should react rapidly with photochemically produced hydroxyl radicals with an estimated half-life of 0.21 days in clean atmospheres. Under polluted atmospheric conditions when nitrogen oxide concentrations are high, diethylamine degrades within 2 hr. Due to its high water solubility, washout by rain will also be an important removal process.

**Biodegradation:** In a screening study, diethylamine degraded at 10 ppm with both an activated sludge and freshwater/sediment inoculum [4]. 59% and 38% of theoretical BOD was obtained after 12 days incubation [4]. Inhibition was noted at moderate concentrations and sizeable reductions in BOD were noted at 50 ppm [4]. Another study concluded that diethylamine was degraded slowly by activated sludge even when acclimatized (53% of theoretical BOD after 13 days) [5]. However, the concentration levels used in this study could not be ascertained. When added to stream water, the maximum rate of biodegradation of diethylamine was proportional to initial amine concentration over a concentration range from several nanograms to several milligrams per liter [3]. At the highest concentration studied, 10 mg/L, the half-life of diethylamine was only 0.9 days [3].

**Abiotic Degradation:** Diethylamine is a strong base and undergoes the typical reactions of primary amines [25]. Many of these reactions may occur in the environment, but there is little documentation as to what reactions take place in the environment and at what rates. Reaction of aliphatic amines with photochemically produced hydroxyl radicals are rapid; the estimated reaction rate for diethylamine is 77.1 x $10^{-12}$ cm$^3$/molecule-sec which would give a half-life of 0.21 days at 5 x $10^5$ radicals/cm$^3$ [1]. Experiments show that diethylamine reacts with NO-NO$_2$-H$_2$0 mixtures to form diethylnitrosamine in the dark and ozone, PAN, acetaldehyde, diethylnitramine, diethylformamide, ethylacetamide, and diethylacetamide upon illumination [23]. These experiments were performed in large outdoor chambers under natural conditions of temperature, humidity, and illumination. Initially the mixture was allowed to react for two hours in the dark, forming the maximum diethylnitrosamine (2.8% conversion) after 10 minutes. The diethylamine had completely disappeared after two hours of illumination [23]. In another experiment, when a bulb containing ppm quantities of NO and NO$_2$, diethylamine, and air (45-50% relative humidity) was exposed to June sunlight for 95 min, most of the original diethylamine disappeared and diethylnitroamine and diethylnitrosamine, in addition to carbon monoxide and formaldehyde, were formed [28]. Based on data obtained from smog chambers, diethylamine has been classified as having negligible reactivity based on its rate of photochemical reaction or ozone-forming potential [6]. Diethylamine is known to form the nitrosamine in waters containing nitrite ions [16,21]. A similar reaction is possible for diethylamine; however, experimental data are lacking. The ability of amines to form complexes with metallic ions is well known [27]. Metallic ions in soils or natural waters may therefore combine with diethylamine but no information could be found on reactions with soil components. Humic acids that occur in natural water contain carbonyl groups that could also potentially react with the amine groups to form adducts, but again data in natural systems are lacking [25]. Diethylamine does not contain any chromophores which absorb radiation >290 nm so direct photolysis will not be significant.

**Bioconcentration:** Using the octanol/water partition coefficient, an estimated BCF of 1.62 was calculated for diethylamine using a recommended regression equation [15]. Diethylamine would therefore not be expected to bioconcentrate in aquatic organisms.

**Soil Adsorption/Mobility:** Using the reported octanol water partition coefficient, an estimated Koc of 50 was obtained using a recommended regression equation [15]. Based on this Koc diethylamine will not absorb appreciably to soils and sediments.

**Volatilization from Water/Soil:** Using the Henry's Law constant, a half-life of 31.6 hr was estimated for a model river 1 m deep flowing at 1 m/sec with a wind velocity of 3 m/sec [15]. Diethylamine is highly volatile and adsorbs poorly to soil, so it should readily volatilize from soil surface.

**Water Concentrations:** SURFACE WATER: 8 rivers in Germany not detected to 14 ppb [18].

**Effluent Concentrations:** Identified, not quantified, in volatiles from cattle feedyard [17]. Diethylamine has been reported in the exhaust from a gasoline engine [10].

**Sediment/Soil Concentrations:** Identified in uncultivated loamy soil from the Moscow region [8]. Since this soil is uncultivated, it is possible that the amines are formed naturally rather than being a contaminant or a metabolite of a fertilizer or pesticide [8].

# Diethylamine

**Atmospheric Concentrations:** INDOOR AIR: Problem house in Sweden in which casein and other proteins were used in a concrete matrix; 2, 2, and 0.2 ppb in kitchen, bedroom, and living room, respectively [2].

**Food Survey Values:** Diethylamine has been found in the following fresh food, concn in ppm: Spinach, 15; Apple, 3 [18]. Concentrations in preserved food items are: Broken beans, 0.1; Broken butterbeans, 2.4; Shelled peas, 0.1; Bean salad, 1.5; Red cabbage, 2.4; Paprika, 0.5; Cornichons, 0.1 [18]. Various samples of pickled vegetables contained 0-3.2 ppm of diethylamine [18]. Some other food products contained the following amounts of diethylamine: Herring, ND-5.2 ppm; Cod roe, 5.2 ppm; Cheese, ND; Brown bread, ND; Coffee, Tea, and Cocoa, ND; Barley, 5.7; Hops, 3.1; Malt, 0.6 [18]. Diethylamine has been identified as a volatile component of boiled beef [7]. The interest for the presence of amines in food arises in part because they are regarded as possible precursors of carcinogenic N-nitroso compounds [7].

**Plant Concentrations:**

**Fish/Seafood Concentrations:**

**Animal Concentrations:**

**Milk Concentrations:**

**Other Environmental Concentrations:** Diethylamine has been identified in tobacco leaf (0.1-35 ppm) and cigarette smoke concentrate (ND-0.4 ppm) [14,26].

**Probable Routes of Human Exposure:** The general population will be primarily exposed to diethylamine from ingesting food in which it occurs naturally and from tobacco smoke. Occupational exposure will occur primarily via inhalation of the vapor.

**Average Daily Intake:**

**Occupational Exposure:** NIOSH estimated that 26,640 workers are potentially exposed to diethylamine [19] from a survey conducted during 1981-83. NIOSH estimated that 58,989 workers are exposed to diethylamine [20] based upon a survey conducted during 1972-74.

# Diethylamine

## Body Burdens:

## REFERENCES

1. Atkinson R; Internat J Chem Kinetics 19: 799-828 (1987)
2. Audunsson G et al; Int J Environ Anal Chem 20: 85-100 (1985)
3. Boethling RS, Alexander M; Environ Sci Technol 13: 989-91 (1979)
4. Calamari D et al; Chemosphere 9: 753-62 (1980)
5. Chudoba J et al; Chem Prum 19: 76-80 (1969)
6. Farley FF; Photochemical Reactivity Classification of Hydrocarbons and Other Organic Compounds, EPA-600/3-77-001B pp. 713-27 (1977)
7. Golovnya RV et al; Chem Senses Flavour 4: 97-105 (1979)
8. Golovnya RV et al; Amines in Soil as Possible Precursors of N-Nitroso Compounds. USSR Acad Med Sci pp. 327-35 (1982)
9. Graedel TE; Chemical Compounds in the Atmosphere Academic Press NY p. 289 (1978)
10. Hampton CV et al; Environ Sci Technol 16: 287-98 (1982)
11. Hansch C, Leo AJ; Medchem Project Issue No.26 Pomona College, Claremont CA (1985)
12. Hawley GG; Condensed Chem Dictionary 10th ed Von Nostrand Reinhold NY (1981)
13. Hine J, Mookerjee PK; J Org Chem 40: 292-8 (1975)
14. Irvine WJ, Saxby MJ; Phytochemistry 8: 473-6 (1969)
15. Lyman WJ et al; Handbook of Chem Property Estimation Methods. Environ Behavior of Organic Compounds. McGraw-Hill NY (1982)
16. Mills AL, Alexander M; J Environ Qual 5: 437-40 (1976)
17. Mosier AR et al; Environ Sci Technol 7: 642-4 (1973)
18. Neurath GB et al; Food Cosmet Toxicol 15: 275-82 (1977)
19. NIOSH; National Occupational Exposure Survey (1985)
20. NIOSH; National Occupational Hazard Survey (1984)
21. Ohto T et al; Chemosphere 11: 979 (1982)
22. Perrin DD; Dissociation Constants of Organic Bases in Aqueous Solution. IUPAC Chemical Data Series. Buttersworth: London (1965)
23. Pitts JN et al; Environ Sci Technol 12: 946-53 (1978)
24. Riddick JA et al; Organic Solvents: Physical Properties and Methods of Purification. Techniques of Chemistry. 4th Ed. New York: Wiley-Interscience pp 1325 (1986)
25. Schweizer AE et al; Amines, Lower Aliphatic Encyclopedia of Chemical Technology 2: 272-83 (1978)
26. Singer GM, Lijinsky W; J Agric Food Chem 24: 553-5 (1976)
27. Stumm W, Morgan JJ; Aquatic Chemistry 2nd ed pp. 356-63 (1981)
28. Tuazon EC et al; Environ Sci Technol 12: 954-8 (1978)
29. USITC; United States Internal Trade Commission, USITC 1892 (1986)

# Dimethylamine

**Synonyms:**

**Structure:**

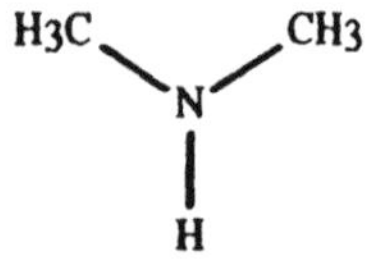

**CAS Registry Number:** 124-40-3

**Molecular Formula:** $C_2H_7N$

**Wiswesser Line Notation:** 1M1

## CHEMICAL AND PHYSICAL PROPERTIES

**Boiling Point:** 7.4 °C

**Melting Point:** -93 °C

**Molecular Weight:** 45.08

**Dissociation Constants:** pKa = 10.732 at 25 °C [30]

**Log Octanol/Water Partition Coefficient:** -0.38 [17]

**Water Solubility:** 1,630,000 mg/L at 40 °C [34]

**Vapor Pressure:** 1520 mm Hg at 25 °C [9]

**Henry's Law Constant:** 1.77 x $10^{-5}$ atm-m³/mole at 25 °C [20]

## ENVIRONMENTAL FATE/EXPOSURE POTENTIAL

**Summary:** Dimethylamine is produced in large quantities, and large amounts of the chemical will be released primarily as emissions during its production and use as a chemical intermediate. It also occurs

208

naturally in food and as a metabolic product. If released on land, dimethylamine should volatilize and leach into the soil. The chemical would be expected to biodegrade in several weeks. If released into water, dimethylamine will readily biodegrade (half-life 1.5 days) and also be removed by volatilization (half-life 35 days in a model river). Adsorption to sediment and bioconcentration in aquatic organisms will not be appreciable. In the atmosphere, dimethylamine will react with photochemically produced hydroxyl radicals and degrade with a 5.9 hr half-life. Degradation will be faster in polluted atmospheres. It will also be scavenged by rain. Dimethylamine is a precursor of dimethylnitrosamine. The latter is formed in the atmosphere in the presence of nitrogen oxides and in lake water, sewage, and soil in the presence of nitrite ions. Human occupational exposure to dimethylamine is via inhalation of the gas. The general public is exposed primarily by ingesting food in which it occurs naturally.

**Natural Sources:** Dimethylamine naturally occurs in many foods [26]. It is also formed as a volatile in cattle manure [25].

**Artificial Sources:** Dimethylamine is a gas which is produced in large quantities (65.9 million lb in 1985 [40]) and will be released to the atmosphere during its production and use (% of production) in the manufacture of the solvents dimethylformamide and dimethylacetamide (50%), water treatment (15%), surfactants (10%), pesticides (10%), and other uses (24%) [7]. The other uses include, as an antioxidant, dyes, floatation agent, gasoline stabilizer, pharmaceuticals, textile chemicals, rubber accelerators, electroplating, dehairing agent, rocket fuel, and reagent to Mg [19]. Based on its 1978 production of 71.8 million lb, it is estimated that 143,000 lb of dimethylamine was released into the atmosphere as process, storage, and fugitive emissions and another 71,800 lb were released associated with its use as a chemical intermediate [1]. Additionally, dimethylamine will be released in wastewater during its production and use. It is found in emissions from fish processing as well as in tobacco smoke [14].

**Terrestrial Fate:** Dimethylamine is a gas at room temperature with a low adsorptivity to soil. If released on soil, it will probably volatilize rapidly and leach readily through the soil. Biodegradation should be the most important degradative process and removal of the chemical should occur in several weeks.

# Dimethylamine

**Aquatic Fate:** If released into water, dimethylamine will volatilize slowly (estimated half-life 35 hr in a model river). Biodegradation will be rapid (half-life 1.5 days) and is the most important removal process in most natural waters. Direct photolysis and adsorption to sediment will not be significant. However, in water containing nitrite, dimethylnitrosamine is formed photochemically from dimethylamine especially at higher pH levels.

**Atmospheric Fate:** Dimethylamine, released into the atmosphere, will react with photochemically produced hydroxyl radicals (half-life 5.9 hr). Dimethylamine will disappear more rapidly under polluted atmospheric conditions when nitrogen oxide concentrations are high. Due to its high water solubility, washout by rain will also be an important removal process.

**Biodegradation:** In a screening study, dimethylamine completely degraded at 10 ppm with both an activated sludge and freshwater/sediment inoculum [6]. 70% and 80% of theoretical BOD was consumed, respectively, after 5 days incubation [6]. Inhibition was noted at 50 ppm with the sediment inoculum and 100 ppm with the sludge inoculum [6]. Another screening study that employed an activated sludge inoculum reported 100% degradation in 6 and 12 days when the concentration was 20 mg/L and 135 mg/L, respectively [11]. Other screening studies give similar results and dimethylamine is confirmed to be biodegradable according to the standard test of the Japanese Ministry of Industry and Trade (MITI) that employs a mixed inoculum obtained from freshwater, soil, and sludge [8,22,33]. In a laboratory activated sludge unit, dimethylamine was completely removed from inflows of up to 135 mg/L with retention times of 4 hr indicating that it should be readily degraded in biological treatment plants [11]. When dimethylamine was added to stream water, the maximum rate of biodegradation was proportional to the initial amine concentration over a concentration range from several nanograms to several milligrams per liter [5]. At the highest concentration studied, 10 mg/L, the half-life of dimethylamine was 1.5 days [5]. In 4 days all of the dimethylamine was mineralized when the initial concentration ranged from 2.8 ppt to 90 ppb, while 90% was mineralized when the concentration was 18 ppm [10]. The half-life in Vistula River water (Warsaw, Poland) was 1.6 days after a 0.3 day lag [11]. When 250 ppm dimethylamine was added to a fine sand loam and sandy soil amended with sewage and nitrite-N, 50% degradation occurred in 2 days in the sand loam, while 20% degradation occurred in the sandy

210

soil [15]. N-nitrosodimethylamine was formed in the degradation [15]. 50% to >90% degradation occurred in four silt loam or loam soils within 14 days [38]. Under anaerobic conditions, one of the silt loam soils tested required 35 days to achieve 86% removal of dimethylamine [38].

**Abiotic Degradation:** Dimethylamine is a strong base and undergoes the typical reactions of primary amines [34]. Many of these reactions may occur in the environment, but there is little documentation as to what reactions take place in the environment and at what rates. Dimethylamine does not contain chromophores which absorb radiation >290 nm so direct photolysis will not be significant [24]. It reacts with photochemically produced hydroxyl radicals by H-atom abstraction from both the C and N atoms, forming formaldehyde as one of the products [2,3]. Assuming a seasonally and diurnally averaged hydroxyl radical concentration of $5 \times 10^{+5}$ molecules/cm$^3$ [2], the half-life of dimethylamine would be 5.9 hr [3]. Under polluted atmospheric conditions reaction with nitrogen oxides will be important. In the presence of air, water vapor, NO, and $NO_2$, dimethylamine reacts to form dimethylnitrosamine, a product which is destroyed by sunlight [18]. When a bulb containing ppm quantities of NO and $NO_2$, dimethylamine, and air (45-50% relative humidity) was exposed to June sunlight for 2-4 hr, only 10% of the original dimethylamine remained and dimethylnitroamine, dimethylnitrosamine, formaldehyde, carbon monoxide, nitrate aerosol, small quantities of dimethylformamide, and possibly tetramethylhydrazine were formed [39]. It has been shown that in water containing nitrite, dimethylamine is photochemically converted to dimethylnitrosamine [29]. The conversion is facilitated by higher pH levels and nitrite ion concentrations [29]. Dimethylnitrosamines also can form in sewage, soils, and lake water when dimethylamine and nitrite-N are present [24]. Since it has been demonstrated that this occurs as readily in sterilized samples as unsterilized samples, this reaction apparently can occur by a nonenzymatic route; the presence of organic matter and acid conditions promotes its formation [24]. The ability of amines to form complexes with metallic ions is well known [37]. Metallic ions in soils or natural waters may therefore combine with dimethylamine but no information could be found on reactions with soil components. Humic acids that occur in natural waters contain carbonyl groups that could also potentially react with the amine groups to form adducts but again data in natural systems are lacking [34].

## Dimethylamine

**Bioconcentration:** Using the octanol/water partition coefficient, an estimated BCF of 0.30 was calculated for dimethylamine using recommended regression equations [23]. Dimethylamine would therefore not be expected to bioconcentrate in aquatic organisms.

**Soil Adsorption/Mobility:** The adsorption isotherm for dimethylamine in 5 soils was linear and resulted in a mean Koc of 434.9 [31]. Based on its extremely high water solubility or low octanol water partition coefficient, one can estimate much lower Koc values of 0.3 to 15 using recommended regression equation [23]. This discrepancy may indicate that there is some binding of the dimethylamine to a soil component.

**Volatilization from Water/Soil:** Using the Henry's Law constant, a half-life of 35.1 hr was estimated for dimethylamine in a model river 1 m deep flowing at 1 m/sec with a wind velocity of 3 m/sec [23].

**Water Concentrations:** SURFACE WATER: 8 rivers in Germany not detected to 11.9 ppb [26].

**Effluent Concentrations:** Dimethylamine was emitted into the atmosphere from an RDX munitions plant in the Houston area [32]. The concn of dimethylamine 40 m downwind from a fishmeal plant was 0.53 ppb [41].

**Sediment/Soil Concentrations:** Identified in uncultivated loamy soil from the Moscow region [13]. Since this soil is uncultivated, it is possible that the amines are formed naturally rather than resulting from contamination or as a metabolite of a fertilizer or a pesticide [13].

**Atmospheric Concentrations:** INDOOR AIR: Problem house in Sweden in which casein and other proteins were used in a concrete matrix; 2, 2, and 0.2 ppb in kitchen, bedroom, and living room, respectively [4].

**Food Survey Values:** Dimethylamine has been found in the following fresh food, concn in ppm: Red cabbage, 2.8; Cabbage, 2; Cauliflower, 14; Kale, 5.5; Red radish, 1.1; Celery, 5.1; Maize, 3.5; Green salad, 7.2 [26]. Concentrations in preserved food items are: Broken beans, 0.6; Broken butterbeans, <0.1; Shelled peas, 2.2; Bean salad, 0.2; Kale, 4.5; Red cabbage, 0.1; Paprika, 1; Cornichons, 0.5 [26]. Various samples of pickled vegetables contained 0-15.4 ppm of dimethylamine [26]. Some other food products contained the following amounts of dimethylamine,

# Dimethylamine

concn in ppm: Herring, 3.4-45; Cod roe, 6.3; Cheese, ND; Brown bread, 3.1; Coffee, 3-6; Barley, 1.6; Hops, 1.4; Malt, 0.5 [26]. Four samples of animal feed contained 0-8 ppm of dimethylamine [26]. Average levels of dimethylamine in food items, ppm: fish sausage, 0.902; baked ham, 1.310; Cod roe, 54.76; Spinach, 0.120; Miso, 0.055 [16]. Dimethylamine has been identified as a volatile component of boiled beef [12]. The interest for the presence of amines in food arises in part because they are regarded as possible precursors of carcinogenic N-nitroso compounds [12].

**Plant Concentrations:** 28 Marine algae (3 classes): Detected in 2 and occasionally in 3 other algae, all of the class Phaeophyceae [36]

**Fish/Seafood Concentrations:**

**Animal Concentrations:**

**Milk Concentrations:**

**Other Environmental Concentrations:** Dimethylamine has been identified in tobacco leaf (4-75 ppm) and cigarette smoke concentrate (110 ppm or 1.8 ug/cigarette) [21,35].

**Probable Routes of Human Exposure:** The general population will be primarily exposed to dimethylamine from ingesting food in which it occurs naturally. Occupational exposure will occur primarily via inhalation of the vapor.

**Average Daily Intake:**

**Occupational Exposure:** NIOSH in its 1981-83 National Occupational Exposure Survey estimates that 27,777 workers are potentially exposed to dimethylamine [28]. NIOSH in its 1972-74 National Occupational Hazard Survey estimates that 27,374 workers are potentially exposed to dimethylamine [27].

**Body Burdens:**

## REFERENCES

1.  Anderson GE; Human Exposure to Atmospheric Concentrations of Selected Chemicals p.230 NTIS PB84-102540 (1986)
2.  Atkinson R; Chem Rev 85: 69-201 (1985)

# Dimethylamine

3. Atkinson R et al; J Chem Phys 68: 1850-3 (1978)
4. Audunsson G et al; Intern J Environ Analyt Chem 20: 85-100 (1985)
5. Boethling RS, Alexander M; Environ Sci Technol 13: 989-91 (1979)
6. Calamari D et al; Chemosphere 9: 753-62 (1980)
7. Chem Marketing Reporter; Chemical Profiles 2/11/85 (1985)
8. Chudoba J et al; Chem Prum 19: 76-80 (1969)
9. Daubert TE, Danner RP; Data Compilation Tables of Properties of Pure Compounds. Amer Inst Chem Engr pp 450 (1985)
10. Digeronimo MJ et al; Effect of Chemical Structure and Concentration on Microbial Degradation in Model Ecosystems, EPA-600/9-79-012 pp.154-66 (1979)
11. Dojlido JR; Investigations of Biodegradability and Toxicity of Organic Compounds EPA-600/2-79-163 pp.118 (1979)
12. Golovnya RV et al; Chem Senses Flavour 4: 97-105 (1979)
13. Golovnya RV et al; USSR Acad Med Sci pp.327-35 (1982)
14. Graedel TE; Chem Compounds in the Atmosphere, Academic Press NY p.289 (1978)
15. Greene S et al; J Environ Qual 10: 416-21 (1981)
16. Hamano T et al; Agric Biol Chem 45: 2237-43 (1981)
17. Hansch C, Leo AJ; Medchem Project Issue No 26 Pomona CA Claremont College (1985)
18. Hanst PL et al; Environ Sci Technol 11: 403-5 (1977)
19. Hawley GG; Condensed Chemical Dictionary 10th ed Von Nostrand Reinhold NY (1981)
20. Hine J, Mookerjee PK; J Org Chem 40: 292-8 (1975)
21. Irvine WJ, Saxby MJ; Phytochemistry 8: 473-6 (1969)
22. Kitano M; Biodeg and Bioaccum Test on Chem Substances, OECD Tokyo Meeting, Reference Book TSU-No. 3 (1978)
23. Lyman WJ et al; Handbook of Chemical Property Estimation Methods. Environmental Behavior of Organic Compounds. McGraw-Hill NY (1982)
24. Mills AL, Alexander M; J Environ Qual 5: 437-40 (1976)
25. Mosier AR et al; Environ Sci Technol 4: 742-4 (1973)
26. Neurath GB et al; Food Cosmet Toxicol 15: 275-82 (1977)
27. NIOSH; National Occupation Hazard Survey (1975)
28. NIOSH; National Occupational Exposure Survey (1985)
29. Ohta T et al; Chemosphere 11: 979 (1982)
30. Perrin DD; Dissociation Constants of Organic Bases in Aqueous Solution. IUPAC Chemical Data Series. Buttersworth: London (1965)
31. Rao PSC, Davidson JM; Retention and Transformation of Selected Pesticides and Phosphorus in Soil-Water Systems, A Critical Review EPA-600/S3-82-060 (1982)
32. Ryon MG et al; Database Assessment of the Health and Environmental Effects of Munitions Production Waste Products NTIS DE84-016512 pp.217 (1984)
33. Sasaki S; pp.283-98 in Aquatic Pollutants: Transformation and Biological Effects, Hutzinger, O et al eds, Pergamon Press (1978)
34. Schweizer AE et al; Kirk-Othmer Encycl Chem Technol 3rd Ed 2: 272-83 (1978)
35. Singer GM, Lijinsky W; J Agric Food Chem 24: 553-5 (1976)
36. Steiner M, Hartmann T; Planta 79: 113-21 (1968)
37. Stumm W, Morgan JJ; Aquatic Chemistry 2nd ed pp. 356-63 (1981)
38. Tate RL, Alexander M; Appl Environ Microbiol 31: 399-403 (1976)

# Dimethylamine

39. Tuazon EC et al; Environ Sci Technol 12: 954-8 (1978)
40. USITC; US Internat Trade Commission 1985 (1986)
41. Verschueren K; Handbook of Environmental Data on Organic Chemicals. 2nd ed Von Nostrand Reinhold NY pp.545-6 (1983)

# 1,4-Dioxane

## SUBSTANCE IDENTIFICATION

**Synonyms:**

**Structure:**

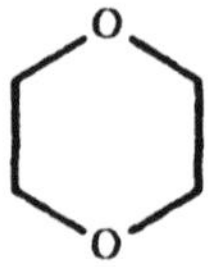

**CAS Registry Number:** 123-91-1

**Molecular Formula:** $C_4H_8O_2$

**Wiswesser Line Notation:** DOTJ

## CHEMICAL AND PHYSICAL PROPERTIES

**Boiling Point:** 101.1 °C at 760 mm Hg

**Melting Point:** 11.80 °C

**Molecular Weight:** 88.10

**Dissociation Constants:**

**Log Octanol/Water Partition Coefficient:** -0.27 [13]

**Water Solubility:** Miscible [23,29]

**Vapor Pressure:** 38.0 mm Hg at 25 °C [3]

**Henry's Law Constant:** 4.88 x $10^{-6}$ atm-m³/mole [17]

## ENVIRONMENTAL FATE/EXPOSURE POTENTIAL

**Summary:** 1,4-Dioxane is used primarily as a solvent in such widely used products as paints, varnishes, lacquers, cosmetics, and deodorants. When released to water, 1,4-dioxane is not expected to hydrolyze and volatilization should be very slow based upon the estimated Henry's Law constant and its miscibility in water. Based on its infinite water solubility and low estimated soil sorption partition coefficient, 1,4-dioxane released to soil is expected to leach to ground water. 1,4-Dioxane is not expected to bioconcentrate in fish or biodegrade in soil or water. 1,4-Dioxane which enters the atmosphere is expected to degrade fairly quickly. After 3.4 hr, 50% of the dioxane mixed with NO and subjected to environmental UV radiation had degraded. A half-life of 6.69 hr was estimated for the reaction of 1,4-dioxane with atmospheric hydroxyl radicals. The expected products of this reaction are aldehydes and ketones. Human exposure is expected to result primarily from contact with products containing 1,4-dioxane.

**Natural Sources:**

**Artificial Sources:** 1,4-Dioxane is produced in millions of pounds per year and used as a solvent for lacquers, paints, varnishes, paint and varnish removers, wetting and dispersing agents in textile processing, dye baths, stain and printing compositions, cleaning and detergent preparations, cements, cosmetics, deodorants, fumigants, emulsions, polishing compositions, stabilizer for chlorinated solvents, and in scintillation fluids [15]. 1,4-Dioxane is used as a solvent for cellulose acetate, ethyl cellulose, benzyl cellulose, resins, oils, waxes, oil, and spirit-soluble dyes [26].

**Terrestrial Fate:** Using a measured log octanol/water partition coefficient of -0.27 [13], a log soil-sorption coefficient (Koc) of 1.23 was estimated for 1,4-dioxane [25]. Compounds with a Koc of this magnitude are mobile in soil [20] so 1,4-dioxane should, therefore, readily leach to ground water. The estimated Henry's Law constant suggests that volatilization from moist soils will be slow, although the volatilization from dry soil should be fast based on its moderate vapor pressure (37 mm Hg at 25 °C [30]).

**Aquatic Fate:** No hydrolysis data were available for 1,4-dioxane. Ethers have been classified as generally resistant to hydrolysis; therefore, 1,4-dioxane is not expected to hydrolyze significantly. No volatilization data for 1,4-dioxane were available, but the estimated Henry's Law constant would suggest that volatilization will be slow. With an estimated Koc of 1.23, 1,4-dioxane is not expected to significantly adsorb on suspended sediments. 1,4-Dioxane exhibited a negligible biological oxygen demand in two activated sludge experiments and the compound has been classified as relatively undegradable. It is expected, therefore, that 1,4-dioxane will not biodegrade extensively in the aquatic environment.

**Atmospheric Fate:** The half-life of the reaction of 1,4-dioxane with photochemically produced hydroxyl radicals in the atmosphere was estimated to be 6.69 to 9.6 hr. Experimental results of sunlight irradiated mixtures of dioxane/NO suggest similar half-lives. The products of the reaction of ethers with hydroxyl radicals are likely to be aldehydes and ketones.

**Biodegradation:** 1,4-Dioxane has been found to be resistant to biodegradation [1,9,16,28] and has been classified as relatively undegradable [19,25]. 1,4-Dioxane, therefore, is not expected to biodegrade rapidly in the environment.

**Abiotic Degradation:** 1,4-Dioxane was mixed with NO at 27 °C and subjected to UV radiation equal to about 2.6 times the intensity of natural sunlight on a summer day in Freeport, TX. After 3.4 hr, 50% of the 1,4-dioxane had degraded [7]. The half-life of the reaction of 1,4-dioxane with hydroxyl radicals in the atmosphere was estimated to be 6.69 hr [11]. A half-life of 9.6 hr was estimated for the reaction of 1,4-dioxane with hydroxyl radicals in the atmosphere [4]. The products of the reaction of ethers with hydroxyl radicals are likely to be aldehydes and ketones [12]. 1,4-Dioxane is photooxidized by aqueous hydroxyl radicals with a half-life of 336 days at pH 7 [2].

**Bioconcentration:** No bioconcentration data for 1,4-dioxane were available. The log octanol/water partition coefficient (Kow) of 1,4-dioxane is -0.27 [13], however, and this very low Kow suggests that 1,4-dioxane will not bioconcentrate significantly in aquatic organisms.

**Soil Adsorption/Mobility:** No adsorption data for 1,4-dioxane were available. Using the octanol/water partition coefficient, a log soil-sorption coefficient (Koc) of 1.23 was estimated for 1,4-dioxane [25]. Compounds with a Koc of this magnitude are mobile in the soil [20] and 1,4-dioxane may leach to ground water.

**Volatilization from Water/Soil:** No data concerning the volatilization of 1,4-dioxane from water or soil were available. However, the estimated Henry's Law constant suggests that 1,4-dioxane volatilization from water and moist soil should be slow. 1,4-Dioxane has a moderate vapor pressure [3,30], however, so volatilization from dry soil is possible.

**Water Concentrations:** SURFACE WATER: Raw water collected from an unspecified river in the United Kingdom contained 1,4-dioxane, but no quantitative data were presented [8]. 1,4-Dioxane at 1 ug/L was detected in the Chicago Sanitary and Ship Channel in the Lake Michigan basin [21]. DRINKING WATER: 1,4-Dioxane at 1 ug/L was detected in United States drinking water [22]. A Massachusetts drinking water well contained 1,4-dioxane at 2100 ppb [5]. GROUND WATER: 1,4-Dioxane was detected in 37% of the samples of well water collected near a solid waste landfill located 60 miles southwest of Wilmington, DE [6]. No concentrations were presented. Leachates from wells located near low level radioactive waste disposal sites contained 1,4-dioxane, but no quantitative data were presented [10].

**Effluent Concentrations:** 1,4-Dioxane was detected at 1 ug/L in effluents from the North Side and Calumet sewage treatment plants on the Lake Michigan basin [21].

**Sediment/Soil Concentrations:**

**Atmospheric Concentrations:** Air samples at three urban sites in New Jersey were collected from July 6-August 16, 1981. The geometric mean 1,4-dioxane concentrations ranged from 0.01-0.02 ppb. Fifty-one percent of the samples were positive for 1,4-dioxane [14]. The same three sites were also sampled from January 18-February 26, 1982. The geometric means of these samples ranged from 0-0.01 ppb and 20% of samples were positive [14]. 1,4-Dioxane was detected but not quantified in the air of New Jersey [24] and Leningrad [18].

**Food Survey Values:**

**Plant Concentrations:**

**Fish/Seafood Concentrations:**

**Animal Concentrations:**

**Milk Concentrations:**

219

# 1,4-Dioxane

**Other Environmental Concentrations:**

**Probable Routes of Human Exposure:** Exposure to 1,4-dioxane in humans is most likely to result from contact with the wide variety of products containing the compound as a solvent.

**Average Daily Intake:**

**Occupational Exposures:** NIOSH (NOES Survey 1981-1983) has statistically estimated that 136,090 workers are exposed to 1,4-dioxane in the United States [27].

**Body Burdens:**

## REFERENCES

1. Alexander M; Biotechnol Bioeng 15: 611-47 (1973)
2. Anbar M, Neta P; Int J Appl Radiation and Isotopes 18: 493-523 (1967)
3. Boublik T et al; The Vapor Pressures of Pure Substances Vol 17 Amsterdam, Netherlands: Elsevier Science Publ (1984)
4. Brown et al; Research Program on Hazard Priority Ranking of Manufactured Chemicals NTIS PB84-263164 (1975)
5. Burmaster DE; Environ 24: 6-13, 33-6 (1982)
6. DeWalle FB, Chian ESK; J Am Water Works Assoc 73: 206-11 (1981)
7. Dilling WL et al; Environ Sci Technol 10: 351-6 (1976)
8. Fielding M et al; Formulation of Preliminary Assessment of Halogenated Organic Compounds in Man and Environmental Media. USEPA-560/13-79-006 (1979)
9. Fincher EL, Payne WJ; Appl Microbiol 10: 542-7 (1962)
10. Francis AJ et al; Nucl Technol 50: 158-63 (1980)
11. GEMS; Graphical Exposure Modeling System. Fate of atmospheric pollutants (FAP) data base. Office of Toxic Substances. USEPA (1986)
12. Graedel TE; Chemical Compounds in the Atmosphere p 267 (1978)
13. Hansch C, Leo AJ; Medchem Project Issue No 26 Claremont CA Pomona College (1985)
14. Harkov R et al; Sci Total Environ 38: 259-74 (1984)
15. Hawley GG; Condensed Chemical Dictionary 10th ed Van Nostrand Reinhold NY p 377 (1981)
16. Heukelekian H, Rand MC; J Water Pollut Control Assoc 27: 1040-53 (1955)
17. Hine J, Mookerjee PK; J Org Chem 40: 292-8 (1975)
18. Ioffe BV et al; J Chromatog 142: 787-95 (1977)
19. Kawasaki M; Ecotox Environ Safety 4: 444-54 (1980)
20. Kenaga EE; Ecotox Env Safety 4: 26-38 (1980)
21. Konasewich D et al; Status Report on Organic and Heavy Metal Contaminants in the Lakes Erie, Michigan, Huron and Superior Basins. Great Lakes Water Quality Board (1978)

# 1,4-Dioxane

22. Kraybill HF; NY Acad Sci Annals 298: 80-9 (1977)
23. Lange NA; Handbook of Chemistry 10th ed McGraw-Hill NY p 523 (1967)
24. Lioy PJ et al; J Water Pollut Cont Fed 33: 649-57 (1983)
25. Lyman WJ et al; Handbook of Chemical Property Estimation Methods. Environmental Behavior of Organic Compounds. McGraw-Hill NY (1982)
26. Merck Index; An Encyclopedia of Chemicals, Drugs and Biologicals 10th ed p 481 (1983)
27. Mills EJ, Stack VT; Proc 8th Ind Waste Conf Ext Ser 83: 492-517 (1954)
27. NIOSH; National Occupational Exposure Survey (1983)
28. Riddick JA et al; Organic Solvents: Physical Properties and Methods of Purification, 4th Edit. New York: J Wiley & Sons (1986)
29. Verschueren K; Handbook of environmental data on organic chemicals. 2nd ed Van Nostrand Reinhold NY p 579 (1983)

# Ethanol

## SUBSTANCE IDENTIFICATION

**Synonyms:**

**Structure:**

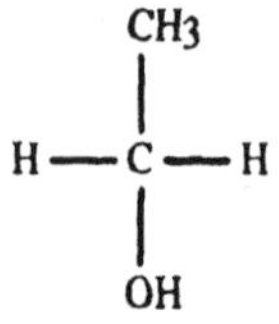

$$CH_3$$
$$|$$
$$H — C — H$$
$$|$$
$$OH$$

**CAS Registry Number:** 64-17-5

**Molecular Formula:** $C_2H_6O$

**Wiswesser Line Notation:** Q2

## CHEMICAL AND PHYSICAL PROPERTIES

**Boiling Point:** 78.5 °C

**Melting Point:** -114.1 °C

**Molecular Weight:** 46.07

**Dissociation Constants:** 15.9 [29]

**Log Octanol/Water Partition Coefficient:** -0.31 (recommended value) [14]

**Water Solubility:** Infinite [29]

**Vapor Pressure:** 59.03 mm Hg at 25 °C [29]

**Henry's Law Constant:** $6.29 \times 10^{-6}$ atm-m$^3$/mole [16]

## ENVIRONMENTAL FATE/EXPOSURE POTENTIAL

**Summary:** Ethanol will enter the environment as emissions from its manufacture, use as a solvent and chemical intermediate, and release

222

in fermentation and alcoholic beverage preparation. It naturally occurs as a plant volatile, microbial degradation product of animal wastes, and in natural fermentation of carbohydrates. When spilled on land it is apt to volatilize, biodegrade, and leach into ground water, but no data on the rates of these processes could be found. Its fate in ground water is unknown. When released into water it will volatilize and probably biodegrade. It would not be expected to adsorb to sediment or bioconcentrate in fish. Although no data on its biodegradation in natural waters could be found, laboratory tests suggest that it may readily biodegrade and its detection in water systems may be due in part to its extensive use in industry with possible relatively steady and large levels of discharges. When released to the atmosphere it will photodegrade in hours (polluted urban atmosphere) to an estimated range of 4 to 6 days in less polluted areas. Rainout should be significant. Human exposure will be primarily in occupational atmospheres and consumption of products containing ethanol. Exposure will also occur from other contaminated atmospheres especially in proximity to industries and cities and ingestion of contaminated drinking water, as well as proximity to sources of natural release.

**Natural Sources:** Emissions from animal wastes, plants, insects, forest fires, microbes, and volcanoes [12]. Emissions from natural fermentation of carbohydrates [24].

**Artificial Sources:** Emissions from petroleum manufacture and storage, plastics, printing, refuse combustion, tobacco smoke, wood pulping, and whiskey manufacture [12]. Leachate from landfills [30]. Emissions and wastewater from its manufacture and use as a solvent and chemical intermediate [24].

**Terrestrial Fate:** When spilled on soil, ethanol will both evaporate and leach into the ground due to the relatively high vapor pressure and low adsorption in soil. It will biodegrade in soil, probably to acetic acid and formaldehyde [13]. If degradation is not rapid, it will leach into ground water.

**Aquatic Fate:** When released into water, ethanol will volatilize (estimated half-life is 6 days) and biodegrade. It will not sorb to sediment or bioconcentrate in aquatic organisms. Although it readily biodegrades in laboratory tests, no data on its rate of degradation in natural waters could be found.

# Ethanol

**Atmospheric Fate:** When released into the atmosphere, ethanol will photodegrade with a half-life ranging from hours in polluted urban atmospheres to approximately 6 days in the atmosphere. Due to its solubility in water, rainout may be an important process.

**Biodegradation:** Ethanol is biodegraded in aerobic systems using activated sludge, sewage (including filtered and settled), wastewater, and soil inocula [9,11,13,15,27,40]. 5 day theoretical BOD values range from 37%-86% [11,27]. Biodegradation of 3, 7, and 10 mg/L with filtered sewage seed in fresh water resulted in 74% theoretical BOD in 5 days and 84% in 20 days; in salt water 45% theoretical BOD in 5 days and 75% in 20 days were observed [27]. Formaldehyde and acetic acid are products of biodegradation by a soil inoculum [13]. Anaerobic degradation (thermophilic digestion, 54 °C) of ethanol (5 mL of a 5% aqueous ethanol solution) produced approx 1000 mL gas/g sample using seed which had been prepared in a synthetic medium [34].

**Abiotic Degradation:** The estimated half-life of ethanol in the atmosphere ranges from 5.9 days [12] to 4 days (based on a hydroxyl radical concentration of $0.8 \times 10^{+6}$ molecules/cm$^3$ [4,23]). The half-life for ethanol in $H_2O_2$/$NO_2$/CO mixtures (total pressure 100 torr; typical sunlit atmosphere) is 10 hr at 19 degrees C [3]. Photochemical smog chamber tests with 500 ppm ethanol and 500 ppm $NO_2$, $SO_2$, and/or $H_2O$ resulted in varying amounts of degradation: 50% degradation in 0.7 hr ($NO_2$/$SO_2$/$H_2O$), 50% in 2.8 hr ($NO_2$), and 25% in 6.3 hr ($SO_2$) [18]. A smog chamber test with 2 ppm ethanol and 1 ppm $NO_x$ resulted in 20% degradation in 5 hr [41]. Ethanol is considered to have low reactivity (class 2 in a 5-class system (5 high)) in photochemical smog situations having ozone-forming potential slightly higher than that of toluene [10]. Reaction with hydroxyl radicals in aquatic media will not likely be a significant process [1,7]. Alcohols are known to be resistant to hydrolysis [23].

**Bioconcentration:** No information on the bioconcentration factor for ethanol could be found in the literature. However, its low octanol/water partition coefficient indicates that it will not bioconcentrate in fish.

**Soil Adsorption/Mobility:** No information on the adsorption of ethanol could be found in the literature. Its low octanol/water partition coefficient indicates that its adsorption to soil will be low.

# Ethanol

**Volatilization from Water/Soil:** The estimated half-life for evaporation of ethanol from water 1 m deep with a 1 m/sec current and 3 m/sec wind is 4 days and the gas exchange rate plays a more dominant role than the liquid exchange rate [23]. Ethanol is relatively volatile and would therefore readily evaporate from soil at the soil/air interface and solid surfaces.

**Water Concentrations:** DRINKING WATER: Detected (not quantified) in 5 city public supplies [6]; detected (not quantified) in city public supplies [21,36,38]. Philadelphia (1975-76), identified, not quantified, in 1 of 3 water treatment plants and in drinking water of 1 of 1 hotel [35]. GROUND WATER: Ethanol was found in ground water suspected of leachate contamination (based on levels of inorganics) 190 ppb (1/13 sites pos), and detected at 58 ppb in landfill ground water where inorganic levels indicated good or unknown water quality [30]. Not detected in Miami, FL [37]. SURFACE WATER: Detected (not quantified) in 4 raw water sources - uncontaminated and contaminated with agricultural runoff, municipal or industrial wastes [37]; Hayashida River (Japan) highly polluted by leather industry, 4020 ppb [42]. Detected at 58 ppb in landfill ground water where inorganic levels indicated good or unknown water quality [30]. RAIN/SNOW: Santa Rita, AZ (rural), concn in precipitation, 15 ppb (by mass), ratio of concn in precipitation/condensate 0.31 [33].

**Effluent Concentrations:** Ethanol was detected in leachate from Minnesota landfills in the range of 23,000 ppb to 110,000 ppb (2/6 sites pos) [30]. Traces found in 1 of 11 domestic wells near Granby, CT landfill, 1984 [31]. Concn in exhaust from simple hydrocarbon fuels (e.g. benzene, isooctane) <0.1-0.6 ppm [32].

**Sediment/Soil Concentrations:**

**Atmospheric Concentrations:** RURAL/REMOTE: Pt. Barrow, AK, 1967, 24 hr avg is 0.77 ppb (upper limit-methanol has same retention time; 17 of 25 samples pos) [5]. URBAN: Chicago, IL - detected near 9% and in 46% homes tested (min 0.5 ppb; max <100 ppb) [20]. Urban - Leningrad, USSR 1976 - detected [19]. Air pollution peak - Japan 29-57 ppb [2]. Concn mean atmospheric condensate/atmospheric (ppb by mass/volume): Tucson, AZ, Feb-Sept 1982, 150/3.3 ppb (17 samples each); Santa Rita and Mt. Lemmon (rural), Aug-Sept 1982, 19/0.40 ppb (18 samples each) [33].

225

## Ethanol

**Food Survey Values:** Identified, not quantified, as a volatile plant isolate in soy beans [25]. Present at various concn in many beverages [39]. Concn (ppb) in lima, common, mung, and soy beans (7, 5, 1, and 1 samples respectively): 1500-7900, 4200 avg; split peas, 3600; lentils 4400 [22]. Identified, not quantified, as a volatile flavor component in fried bacon [17] and mountain Beaufort cheese (French Alps, summer and winter) [8].

**Plant Concentrations:** Plant volatile [12].

**Fish/Seafood Concentrations:**

**Animal Concentrations:**

**Milk Concentrations:**

**Other Environmental Concentrations:**

**Probable Routes of Human Exposure:** Humans will be exposed to ethanol by ingestion of foods, flavorings, beverages, and pharmaceuticals. Workers will be exposed to ethanol in occupational settings associated with its manufacture, use as a solvent or use in synthesis [24], or when released as a product of fermentation, decomposition, or combustion (including cigarette smoke) [12].

**Average Daily Intake:**

**Occupational Exposures:** NIOSH (NOHS Survey 1972-74) has statistically estimated that 3,240,470 workers are exposed to ethanol in the United States [26]. NIOSH (NOES Survey 1981-83) has statistically estimated that 1,470,804 workers are exposed to ethanol in the United States [26]. Finnish furniture factory, 1975-84, 394 samples, 70% pos, 32 ppm avg of pos [28].

**Body Burdens:**

## REFERENCES

1. Anbar M, Netta P; Int J Appl Rad Isot 18: 493-523 (1969)
2. Anonymous; Kanagawa-Ken Taiki Osen Chosn Kenkyu Hokoku 20: 86-90 (1978)
3. Atkinson R et al; Adv Photochem 11: 375-488 (1979)
4. Campbell IM et al; Chem Phys Lett 38: 362-4 (1976)

# Ethanol

5. Cavanagh LA et al; Environ Sci Technol 3: 251-7 (1969)
6. Coleman WE et al; pp. 305-27 in Analysis and Identification of Organic Substances in Water Keith L, Ed (1976)
7. Dorfman LM, Adams GE; Reactivity of the Hydroxyl Radical in Aqueous Solution pp.51 NSRD-NBS-46 (1973)
8. Dumont J, Adda J; J Agric Food Chem 26: 364-7 (1978)
9. Ettinger MB; Ind Eng Chem 48: 256-9 (1956)
10. Farley FF; Int Conf on Photochemical Oxidant Pollution and its Control pp.713-27 USEPA-600/3-77-001b (1977)
11. Gerhold RM, Malaney GW; J Water Pollut Control Fed 38: 562-79 (1966)
12. Graedel TE; Chemical Compounds in the Atmosphere Academic Press New York (1978)
13. Griebel GE, Owens LD; Soil Biol Biochem 4: 1-8 (1972)
14. Hansch C, Leo AJ; Medchem Project Issue No 26. Claremont CA: Pomona College (1985)
15. Heukelekian H, Rand MC; J Water Pollut Control Assoc 29: 1040-53 (1955)
16. Hine J, Mookerjee PK; J Org Chem 40: 292-8 (1975)
17. Ho C et al; J Agric Food Chem 31: 336-42 (1983)
18. Hustert K et al; Chemosphere 7: 35-50 (1978)
19. Ioffe BV; J Chromatogr 142: 787-95 (1977)
20. Jarke FH et al; ASHRAE Trans 87: 153-66 (1981)
21. Kool HJ et al; Crit Rev Env Control 12: 307-57 (1982)
22. Lovegren NV et al; J Agric Food Chem 27: 851-3 (1979)
23. Lyman WJ et al; Handbook of Chemical Property Estimation Methods. Environmental Behavior of Organic Compounds McGraw Hill New York (1982)
24. Merck Index; An Encyclopedia of Chemicals, Drugs, and Biologicals, 10th Ed pp.34 (1983)
25. Nicholas HJ; Phytochem 2: 381 (1973)
26. NIOSH; National Occupational Hazard Survey (1974): NIOSH; National Occupational Exposure Survey (1983)
27. Price KS et al; J Water Pollut Control Fed 46: 63-77 (1974)
28. Priha E et al; Ann Occup Hyg 30: 289-94 (1986)
29. Riddick JA et al; Organic Solvents: Physical Properties and Methods of Purification, 4th Edit. New York: J Wiley & Sons (1986)
30. Sabel GV, Clark TP; pp. 108-25 in Ann Madison Conf Appl Res Pract Munic Ind Waste 6th (1983)
31. Sawhney BL, Raabe JA; Groundwater Contamination: Movement of Organic Pollutants in the Granby Landfill. Bull 833. Conn Exp Sta. New Haven CT (1986)
32. Seizinger DE, Dimitriades B; J Air Pollut Control Assoc 22: 47-51 (1972)
33. Snider JR, Dawson GA; J Geophys Res 90: 3797-805 (1985)
34. Sonoda Y, Seiko Y; J Ferment Technol 46: 796-801 (1968)
35. Suffet IH et al; Water Res 14: 853-7 (1980)
36. Suffet IH et al; pp. 375-97 in Identification and Analysis of Organic Pollutants in Water Keith L, Ed (1976)
37. USEPA; Preliminary Assessment of Suspected Carcinogens in Drinking Water. Interim Report to Congress pp. 9 June (1975)
38. USEPA; New Orleans Area Water Supply Study Draft Analytical Report by the Lower Mississippi River Facility (1974)
39. Verschueren K; Handbook of Environmental Data on Organic Chemicals. 2nd ed Van Nostrand Reinhold NY pp 616-9 (1983)

227

# Ethanol

40. Wagner R; Vom Wasser 42: 271-305 (1974)
41. Yanagihara et al; pp.472-7 in Proc Int Clean Air Congr 4th (1977)
42. Yasuhara A et al; Environ Sci Technol 15: 570-3 (1981)

# Ethanolamine

**Synonyms:** 2-Aminoethanol

**Structure:**

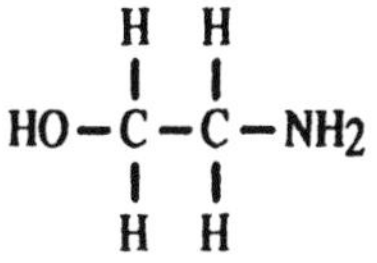

$$HO-CH_2-CH_2-NH_2$$

**CAS Registry Number:** 141-43-5

**Molecular Formula:** $C_2H_7NO$

**Wiswesser Line Notation:** Z2Q

## CHEMICAL AND PHYSICAL PROPERTIES

**Boiling Point:** 170.8 °C at 760 mm Hg

**Melting Point:** 10.3 °C

**Molecular Weight:** 61.08

**Dissociation Constants:** pKa = 9.48 at 25 °C [20]

**Log Octanol/Water Partition Coefficient:** -1.31 [8]

**Water Solubility:** Miscible [21]

**Vapor Pressure:** 0.26 mm Hg at 25 °C [6]

**Henry's Law Constant:** 4 x $10^{-8}$ atm-m³/mole at 25 °C (calculated using the bond method) [11]

## ENVIRONMENTAL FATE/EXPOSURE POTENTIAL

**Summary:** Ethanolamine may be released to the environment in emissions or effluents from sites of its manufacture or use, in urine,

229

from disposal of consumer products containing this compound (i.e. cleaning products), and use of agricultural chemicals in which this compound is used as a dispersing agent. In soil and water, ethanolamine is expected to biodegrade fairly rapidly following acclimation (half-life on the order of days to weeks). In soil, residual ethanolamine may leach into ground water. In the atmosphere, ethanolamine is expected to exist almost entirely in the vapor phase. The dominant removal mechanism is expected to be reaction with photochemically generated hydroxyl radicals (half-life 4 hr). This compound may also be removed from the atmosphere in precipitation. The most probable route of exposure to ethanolamine is dermal contact with personal care products (i.e., soaps and hair waving solutions), detergents, and other surfactants containing this compound.

**Natural Sources:** Ethanolamine is a normal metabolic intermediate in some animals species, having a part in the formation of phospholipids and choline [1] and is a constituent of urine [1].

**Artificial Sources:** Ethanolamine may be released to the environment in emissions or effluents from sites of its manufacture or industrial use, from disposal of consumer products which contain this compound, and from application of agricultural chemicals in which this compound is used as a dispersing agent [9,15].

**Terrestrial Fate:** If released to soil, ethanolamine is expected to biodegrade fairly rapidly following acclimation (half-life on the order of days to weeks). Residual ethanolamine may leach into ground water. Volatilization from soil surfaces is not expected to be an important fate process.

**Aquatic Fate:** If released to water, ethanolamine should biodegrade. The half-life of this compound is expected to range from a few days to a few weeks depending, in large part, on the degree of acclimation of the system. Bioconcentration in aquatic organisms, adsorption to suspended solids and sediments, and volatilization are not expected to be important fate processes in water.

**Atmospheric Fate:** Based on the vapor pressure, ethanolamine is expected to exist almost entirely in the vapor phase in the atmosphere [7]. The dominant removal mechanism is expected to be reaction with photochemically generated hydroxyl radicals (half-life 11 hr). The complete solubility of ethanolamine in water suggests that this compound may also be removed from the atmosphere in precipitation.

**Biodegradation:** BOD water - electrolytic respirometer, initial concn 100 ppm, 5 day 50% BODT, sewage inoculum [23,24]. BOD water, initial concn 10 ppm, 5 day - 34% BODT, 20 day - 40% BODT, sewage inoculum [25]. BOD water, initial concn 2.5 ppm, 5 day 61-84% BODT, sewage inoculum [10]. BOD water, 10 day 65% BODT, sewage inoculum [17]. BOD water, initial concn 2.5 ppm, 5, 10, 20, and 50 days - 0, 58.4, 64.0, and 75.0% BODT, respectively, sewage inoculum [14]. BOD water, 5 day 71% BODT and 98% COD removal, sewage inoculum [3]. Japanese MITI, initial concn 100 ppm, 14 days 49.2% BODT ($NO_2$ endproduct) and 93.6% BODT ($NH_3$ endproduct), activated sludge inoculum [12].

**Abiotic Degradation:** The half-life for vapor phase ethanolamine reacting with photochemically generated hydroxyl radicals in the atmosphere has been estimated to be 11 hours based on an estimated reaction rate constant of $3.5 \times 10^{-11}$ cm$^3$/molecules-sec at 25 °C and an average ambient hydroxyl concentration of $5 \times 10^{+5}$ molecules/cm$^3$ [2].

**Bioconcentration:** A bioconcentration factor (BCF) of <1 was estimated for ethanolamine based on the Kow [16]. This BCF value and complete solubility of ethanolamine in water suggest that this compound does not bioconcentrate significantly in aquatic organisms.

**Soil Adsorption/Mobility:** A soil adsorption coefficient (Koc) of 5 was estimated for ethanolamine based on the Kow [16]. This Koc value and the complete solubility of ethanolamine in water suggests that this compound would be extremely mobile in soil and would not adsorb appreciably to suspended solids and sediments in water [22].

**Volatilization from Water/Soil:** The Henry's Law constant suggests that volatilization of ethanolamine from water and moist soil surfaces would not be a significant fate process [16].

**Water Concentrations:**

**Effluent Concentrations:**

**Sediment/Soil Concentrations:**

# Ethanolamine

**Atmospheric Concentrations:**

**Food Survey Values:**

**Plant Concentrations:**

**Fish/Seafood Concentrations:** Ethanolamine has been identified in marine and freshwater algae [13].

**Animal Concentrations:**

**Milk Concentrations:**

**Other Environmental Concentrations:**

**Probable Routes of Human Exposure:** The most probable route of exposure to ethanolamine is dermal contact with personal care products (i.e., soaps and hair waving solutions), detergents, and other surfactants in which this compound is an ingredient [4,9].

**Average Daily Intake:**

**Occupational Exposure:** NIOSH (NOES Survey 1972-74) has statistically estimated that 1,754,175 workers are potentially exposed to ethanolamine in the United States [18]. NIOSH (NOES Survey 1981-83) has statistically estimated that 934,804 workers are potentially exposed to ethanolamine in the United States [19].

**Body Burdens:** Ethanolamine is a normal constituent of human urine. Average excretion rate: men - 0.162 mg/kg per day; women - 0.492 mg/kg per day; cats 0.47 mg/kg per day; rats - 01.46 mg/kg per day; and rabbits - 1.0 mg/kg per day [5].

## REFERENCES

1. Am Conf Ind Hyg; Appendix: Documentation of the Threshold Limit Values and Biological Exposure Indices 5th ed p. 235 Cincinnati, OH (1986)
2. Atkinson R; Inter J Chem Kinet 19: 799-828 (1987)
3. Bridie AL et al; Water Res 13: 627- 30 (1979)
4. Chemical Marketing Reporter; Chemical Profile: Ethanolamine NY: Schnell Publishing Nov 10 (1986)
5. Clayton GD, Clayton FE; Patty's Industrial Hygiene and Toxicology 3rd ed Vol 2B p. 3168 NY: Wiley-Interscience (1981)

6.  Dow Chemical; The Alkanolamine Handbook Midland, MI: Dow Chemical (1980)
7.  Eisenreich SJ et al; Environ Sci Tech 15: 30-8 (1981)
8.  Hansch C, Leo AJ; Medchem Project Issue No.26 Pomona College, Claremont CA (1985)
9.  Hawley GG; The Condensed Chemical Dictionary 10th ed NY: Van Nostrand Reinhold p. 420 (1981)
10. Heukulekian H, Rand MC; J Water Pollut Contr Assoc 29: 1040-53 (1955)
11. Hine J, Mookerjee PK; J Org Chem 40: 292-8 (1975)
12. Kitano M; OECD Tokyo Meeting Reference Book Tsu-No. 3 (1978)
13. Kneifel H et al; J Phycol 13: 36 (1977)
14. Lamb CB, Jenkins GF; pp 326-39 in Proc 8th Ind Waste Conf Purdue Univ (1952)
15. Liepins R et al; Industrial Process Profiles for Environmental Use. USEPA-600/2-77-023f NTIS PB-281 478 pp. 6-386-6-387 (1977)
16. Lyman WJ et al; Handbook of Chemical Property Estimation Methods NY: McGraw-Hill (1982)
17. Mills EJ, Stack VT; Proc 8th Ind Waste Conf Eng Bull Purdue Univ: Ext Ser 83: 492-517 (1954)
18. NIOSH; National Occupational Exposure Survey (NOES) (1983)
19. NIOSH; National Occupational Hazard Survey (NOHS) (1974)
20. Perrin DD; Dissociation Constants of Organic Bases in Aqueous Solution. IUPAC Chemical Data Series. Supplement 1972 Buttersworth: London (1972)
21. Riddick JA et al; Organic Solvents: Physical Properties and Methods of Purification. Techniques of Chemistry. 4th Ed. New York: Wiley-Interscience pp 1325 (1986)
22. Swann RL et al; Res Rev 85: 17-28 (1983)
23. Urano K, Kato Z; J Hazard Mater 13: 135-45 (1986)
24. Urano K, Kato Z; J Hazard Mater 13: 147-59 (1986)
25. Young RHF et al; J Water Pollut Contr Fed 40: 354-68 (1968)

# Ethylamine

## SUBSTANCE IDENTIFICATION

**Synonyms:**

**Structure:**

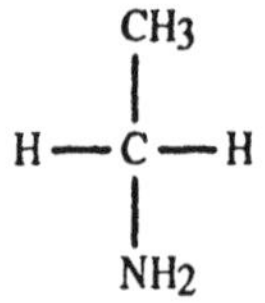

$$CH_3$$
$$|$$
$$H — C — H$$
$$|$$
$$NH_2$$

**CAS Registry Number:** 75-04-7

**Molecular Formula:** $C_2H_7N$

**Wiswesser Line Notation:** Z2

## CHEMICAL AND PHYSICAL PROPERTIES

**Boiling Point:** 16.6 °C

**Melting Point:** -81.1 °C

**Molecular Weight:** 45.08

**Dissociation Constants:** pKa = 10.87 at 20 °C [26]

**Log Octanol/Water Partition Coefficient:** -0.13 [13]

**Water Solubility:** Miscible [19]

**Vapor Pressure:** 1048 mm Hg at 25 °C [7]

**Henry's Law Constant:** $1.23 \times 10^{-5}$ at 25 °C [14]

## ENVIRONMENTAL FATE/EXPOSURE POTENTIAL

**Summary:** Ethylamine occurs widely in the environment as a decomposition product of amino acids. It forms during biological degradation of sewage, animal waste, and solid waste and is also found in a variety of foods. Major sources of anthropogenic release of ethylamine appear to be effluents and emissions from manufacturing plants and facilities at which it is used as an intermediate. If released to moist soil, ethylamine may readily biodegrade (half-life <2 months). Chemical hydrolysis is not expected to be a significant removal mechanism. It is not certain whether ethylamine would absorb strongly to soil; however, cationic compounds have been found strongly associated with humates and clay soil. If released to dry soil, this compound should volatilize rapidly. If released to water, ethylamine may biodegrade or volatilize (half-life >2 days from a model river). Adsorption to humate and clays in sediments may also be important. Bioaccumulation, chemical hydrolysis, and reaction with photochemically generated hydroxyl radicals are not expected to be significant fate processes in water. If released to air, ethylamine should exist almost entirely in the vapor phase. Reaction with photochemically generated hydroxyl radicals is expected to be the primary removal mechanism (estimated half-life 8.6 hr). The most probable routes of exposure to the general population are inhalation of tobacco smoke and ingestion of food in which this compound is found. Worker exposure may occur by dermal contact and/or inhalation.

**Natural Sources:** Ethylamine occurs naturally as a decomposition product of amino acids [1]. It is formed during biological degradation of sewage and solid waste and it has been found in emissions from animal waste and sewage treatment facilities [1,11]. Ethylamine is also found in a wide variety of foods (i.e., freeze-dried coffee, barley, camembert cheese, herring, and spinach to name a few) [22].

**Artificial Sources:** Ethylamine may be released to the environment in the effluent or emissions from its manufacturing or use facilities. Ethylamine is used mainly as an intermediate in the manufacture of triazine herbicides (particularly atrazine and cyanazine), 1,3-diethylthiourea (a corrosion inhibitor), ethylaminoethanol, 4-ethylmorpholine (urethane foam catalyst), ethyl cyanate, and dimethylolethyltriazone (agent used in wash-and-wear fabrics [28,32]).

**Terrestrial Fate:** If released to moist soil, ethylamine may biodegrade (half-life <2 months). Chemical hydrolysis is not expected to be an important fate process. If released to dry soil, ethylamine is expected to volatilize rapidly. Data on adsorption to soil were not available, although cations have been found to strongly adsorb to humic materials and clays in soil.

# Ethylamine

**Aquatic Fate:** If released to water, ethylamine may biodegrade or volatilize (half-life >2 days from a model river). Bioaccumulation in aquatic organisms, chemical hydrolysis, and reaction with photochemically generated hydroxyl radicals in water are not expected to be important fate processes. Adsorption to humic materials and clays in sediments may be an important fate process.

**Atmospheric Fate:** Based on the vapor pressure, ethylamine should exist almost entirely in the vapor phase in the atmosphere [8]. Reaction with photochemically generated hydroxyl radicals in the atmosphere is expected to be the dominant removal mechanism (half-life 8.6 hr). Its complete water solubility suggests that ethylamine also has the potential to be removed by wet deposition.

**Biodegradation:** Ethylamine should be degraded by biological sewage treatment provided suitable acclimation has been achieved [31]. Activated sludge acclimated to aniline removed 34% BODT in 130 hr, initial concn 500 mg/L ethylamine [18]. Activated and non-activated sludge cultures were observed rapidly degrading ethylamine [6,29]. Extract from a methylotrophic bacterium, Arthrobacter sp, oxidized ethylamine to hydrogen peroxide [16]. Ethylamine is readily biodegraded in soil (lifetime 1-2 months) [1].

**Abiotic Degradation:** The half-life for the reaction of ethylamine with photochemically generated hydroxyl radicals in water has been estimated to be 321 days based on a reaction rate constant of $2.5 \times 10^{+9}$ 1/mole-sec and an ambient hydroxyl concentration of $1 \times 10^{-17}$ mole/L [12,20]. Based on the molecular structure of this compound, ethylamine is not expected to be susceptible to chemical hydrolysis under environmental conditions. The principal chemical loss mechanism for amines in the atmosphere is expected to be reaction with photochemically generated hydroxyl radicals [11]. The half-life for the reaction of ethylamine with hydroxyl radicals has been estimated to be approximately 8.6 hr based on a reaction rate constant of $6.54 \times 10^{-11}$ cm$^3$/molecule-sec at 25.5 °C and assuming an average ambient hydroxyl radical concentration of $5.0 \times 10^{+5}$ molecules/cm$^3$ [3]. Reaction with ozone (estimated half-life 1.3 years based of a measured reaction rate constant of $2.76 \times 10^{-20}$ cm$^3$/molecule-sec at 23 °C and an ambient ozone concentration of $6.0 \times 10^{+11}$ molecules/cm$^3$) and reaction with atomic oxygen (estimated half-life 241 days based on a measured reaction rate constant of $1.33 \times 10^{-12}$ cm$^3$/molecule-sec at 25.6 °C and an ambient O(3P) concentration of $2.5 \times 10^{-4}$ molecules/cm$^3$) are not expected to be important fate processes in the atmosphere [2,3,4,11].

**Bioconcentration:** A bioconcentration factor (BCF) of <1 was estimated for ethylamine using a linear regression equation based on the Kow [17]. This BCF value and the complete water solubility of ethylamine suggest that bioaccumulation in aquatic organisms would not be significant.

**Soil Adsorption/Mobility:** The complete water solubility of ethylamine and its relatively low octanol water partition coefficient suggest that this compound would be extremely mobile in soil and that adsorption to suspended solids and sediments would not be significant. However, ethylamine is a base and should exist predominantly in the protonated form under environmental conditions (pH 5-9). Cations have been found to strongly bind with humates and clays [5,25]. Therefore, protonation may result in greater adsorption and less mobility than its water solubility or Kow indicate.

**Volatilization from Water/Soil:** Based on the Henry's Law constant, the volatilization half-life for unprotonated ethylamine from a model river 1 m deep flowing 1 m/sec has been estimated to be approximately 2 days [17]. However, ethylamine will exist primarily in the protonated form under environmental conditions and the protonated form is expected to volatilize more slowly than the unprotonated form. The high vapor pressure suggests that ethylamine would volatilize rapidly from dry soil surfaces.

**Water Concentrations:** SURFACE WATER: Detected in surface water in Europe: River Elbe, 16.2 ug/L; River Alster, 2 ug/L; River Stor, ug/L; River Au near Heligen, approx 1 ug/L; River Kruckau, 5 ug/L; River Pinnau, 5 ug/L; River Ammersbek, 37.1 ug/L; Timmermoor Swamp. 0.6 ug/L [22]. DRINKING WATER: Detected in drinking water from District of Columbia [27].

**Effluent Concentrations:**

**Sediment/Soil Concentrations:** Ethylamine has been identified as a volatile component of uncultivated soil [9].

**Atmospheric Concentrations:**

# Ethylamine

**Food Survey Values:** Ethylamine has been identified in: spinach, 8.4 mg/kg; red cabbage (fresh), 1.3 mg/kg; carrots, 1 mg/kg; white beet, 4.3. mg/kg; large radish, 10 mg/kg; red radish, 40 mg/kg; broken butter beans, 0.8 mg/kg; shelled peas, 0.1 mg/kg; kale, 0.3 mg/kg; red cabbage (preserved), 0.1 mg/kg; red paprika, 0.1 mg/kg; maize grains, 2.4 mg/kg; green salad, 3.3 mg/kg; apple flesh, 3 mg/kg; herring, 0.1-0.4 mg/kg; Camembert cheese, 4 mg/kg; Limburger cheese, 1 mg/kg; freeze dried coffee, 1.5-2 mg/kg; barley, 3.4 mg/kg; hops, 5.2 mg/kg; and malt, 0.3 mg/kg [22]. Ethylamine has been identified as a volatile component of boiled beef [10].

**Plant Concentrations:** Ethylamine occurs in tobacco leaves [15]. It has also been found to be a volatile constituent of marine algae [30].

**Fish/Seafood Concentrations:**

**Animal Concentrations:**

**Milk Concentrations:**

**Other Environmental Concentrations:** Ethylamine has been determined to be a volatile component of cattle feed lots, probably from the decomposition of manure [21]. Ethylamine also is a constituent of tobacco smoke [15].

**Probable Routes of Human Exposure:** The most probable routes of human exposure to ethylamine are inhalation of tobacco smoke and ingestion of foods in which this compound occurs. Limited data suggest that exposure by ingestion of drinking water may also occur, but this route of exposure is expected to be minor. Worker exposure may occur by dermal contact and/or inhalation.

**Average Daily Intake:**

**Occupational Exposure:** NIOSH (NOES Survey 1981-83) has statistically estimated that 1797 workers are exposed to ethylamine in the United States [23]. NIOSH (NOHS Survey 1972-74) has statistically estimated that 5066 workers are exposed to ethylamine in the United States [24].

**Body Burdens:**

# Ethylamine

## REFERENCES

1. Abrams EF et al; Identification of Organic Compounds in Effluents from Industrial Sources p. 77 USEPA 560/3-75-002 (1975)
2. Atkinson R, Pitts JN; J Chem Phys 68: 911-15 (1978)
3. Atkinson R; Chem Rev 85: 69-201 (1985)
4. Atkinson R, Carter WP; Chem Rev 84: 437-70 (1984)
5. Callahan MA et al; Water Related Fate Of 129 Priority Pollutants pp. 99-1 to 105-8 USEPA 440/4-79-029A (1979)
6. Chudoba J et al; Chem Prum 19: 76-80 (1969)
7. Daubert TE, Danner RP; Data Compilation Tables of Properties of Pure Compounds. Amer Inst Chem Engn pp 450 (1985)
8. Eisenreich SJ et al; Environ Sci Tech 15: 30-8 (1981)
9. Golovnya RV et al; USSR Acad Med Sci pp. 327-35 (1982)
10. Golovnya RV et al; Chem Sense Flavor 4: 97-105 (1974)
11. Graedel TE; Chemical Compounds in the Atmosphere Academic Press New York NY p. 287 (1978)
12. Guesten H et al; Atmos Environ 15: 1763-5 (1981)
13. Hansch C, Leo AJ; Medchem Project Issue No.26 Pomona College, Claremont CA (1985)
14. Hine J, Mookerjee PK; J Org Chem 40: 292-8 (1975)
15. Irvine WJ, Saxby MJ; Phytochemistry 8: 473-6 (1969)
16. Levering PR et al; Arch Microbiol 129: 72-80 (1981)
17. Lyman WJ et al; Handbook of Chemical Property Estimation Methods. Environment Behavior of Organic Compounds. McGraw-Hill NY (1982)
18. Malaney GW; J Water Pollut Control Fed 32: 1300-11 (1960)
19. Merck Index; The Merck Index An Encyclopedia Chem Drugs. 9th Ed. Rahway, NJ: Merck and Co (1976)
20. Mill T et al; Science 207: 886-7 (1980)
21. Mosier AR et al; Environ Sci Tech 7: 642-4 (1973)
22. Neurath GB et al; Food Cosmet Toxicol 15: 275-82 (1977)
23. NIOSH, National Occupational Exposure Survey (NOES) (1983)
24. NIOSH; National Occupational Hazard Survey (NOHS) (1984)
25. Parris GE; Environ Sci Tech 14: 1099-106 (1980)
26. Perrin DD; Dissociation Constants of Organic Bases in Aqueous Solution. IUPAC Chemical Data Series. Buttersworth: London (1965)
27. Scheiman MA et al; Biomed Mass Spec 4: 209-11 (1974)
28. Schweitzer AE et al; Kirk-Othmer Encycl Chem Tech 3rd ed NY Wiley 2: 274 (1978)
29. Slave T et al; Rev Chim 25: 666-70 (1974)
30. Steiner M, Hartmann T; Planta 79: 113-21 (1968)
31. Thom NS, Agg AR; Proc R Soc Lond B 189: 347-57 (1975)
32. USEPA; Chemical Hazard Information Profile p. 127 USEPA 560/11-80-011 (1980)

# Ethyl Acetate

## SUBSTANCE IDENTIFICATION

**Synonyms:**

**Structure:**

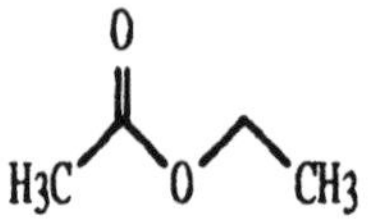

**CAS Registry Number:** 141-78-6

**Molecular Formula:** $C_4H_8O_2$

**Wiswesser Line Notation:** 2OV1

## CHEMICAL AND PHYSICAL PROPERTIES

**Boiling Point:** 77 °C

**Melting Point:** -83 °C

**Molecular Weight:** 88.10

**Dissociation Constants:**

**Log Octanol/Water Partition Coefficient:** 0.73 [14]

**Water Solubility:** 64,000 mg/L [38]

**Vapor Pressure:** 69 mm Hg at 19 °C [1]

**Henry's Law Constant:** $1.2 \times 10^{-4}$ atm-m³/mole [3]

## ENVIRONMENTAL FATE/EXPOSURE POTENTIAL

**Summary:** Ethyl acetate is emitted to the air and discharged into wastewater during its production and use as an industrial solvent and in organic synthesis. It is also released into the air during the formation of whiskey and beer. If released into water, ethyl acetate will be lost primarily by evaporation (half-life 10 hr in a typical river) and biodegradation. While ethyl acetate is readily biodegraded in most oceanic tests, the rate in natural waters is unknown. Bioconcentration in fish will be insignificant. If released on land, ethyl acetate will partially evaporate and partially leach into the ground. Biodegradation will probably occur both in soil and ground water, however experimental data are lacking. In the atmosphere, ethyl acetate will react with photochemically produced hydroxyl radicals (half-life 8.3 days). A few percent an hour will disappear under photochemical smog situations. Humans will be exposed to ethyl acetate in the workplace, by ingesting certain food items (e.g., beer) of which it is a natural component, and while using consumer products such as airplane dope and nail polish remover.

**Natural Sources:** Natural product of fermentation [13,22]. Plant volatile [13]. Animal waste, microbes [13].

**Artificial Sources:** Wastewater, emissions, and spills from its production, transport, storage, and use as a solvent for such substances as coatings and plastics, and organic synthesis [6,15]. Nail polish remover, airplane dopes, artificial fruit essence [13,15,21]. Whiskey fermentation vats [37].

**Terrestrial Fate:** If released on land, ethyl acetate will be lost by evaporation and leaching into ground water. Biodegradation should also occur because it is readily biodegradable in both aerobic and anaerobic systems, but rates for these processes in natural systems are lacking.

**Aquatic Fate:** If released in water, ethyl acetate will primarily be lost by volatilization (half-life 10 hr from a typical river). Biodegradation should also occur, but rates of biodegradation relevant to natural waters are lacking. Ethyl acetate would not be expected to adsorb to sediment or particulate matter.

**Atmospheric Fate:** Ethyl acetate will react with photochemically produced hydroxyl radicals (half-life 8.3 days). Under photochemical smog conditions 1.9-3.4% of the ethyl acetate is lost per hour.

**Biodegradation:** Ethyl acetate is normally easily biodegraded. Reported 5 day BOD values using a sewage inoculum range from 36 to 68% of theoretical [10,16,29,41] with the value being somewhat reduced in salt water [29]. One investigator reported that ethyl acetate was completely degraded in 20 hr using activated sludge [33] and in a bench-scale continuous-flow activated sludge reactor with an 8 hr retention time; 99.9% removal (including 17% volatilization loss) was obtained and the BOD was 80% of theoretical [35]. A review concluded that ethyl acetate is easily removed by biological treatment [36]. No data could be found to assess the biodegradation in natural waters or soil. Ethyl acetate has also been demonstrated to be amenable to anaerobic biodegradation [34]. It was mineralized (>75% of theoretical methane production) in 10% sludge from a secondary digester within 8 weeks [31]. 96% utilization occurred in an anaerobic reactor with a 20-day retention time [7].

**Abiotic Degradation:** Ethyl acetate reacts with photochemically produced hydroxyl radicals in the atmosphere by H-atom abstraction from the $OCH_2$ entity with a measured half-life of 8.3 days (12 hr, sunlit day, clean atmosphere) and 2.1 days (12 hr, sunlit day, moderately polluted atmosphere [5]. This value is about a quarter of the calculated value [12] leading some investigators to question the data and our understanding of the atmospheric chemistry of ethyl acetate and other members of its class [2]. Under simulated photochemical smog conditions, 1.9-3.4% of the ethyl acetate is lost per hour [8,17,38]. High ratios of ethyl acetate to nitrogen oxides are necessary to produce significant quantities of ozone [32]. The presence of $SO_2$ and water appear to increase the rate of loss [17]. Ethyl acetate is resistant to hydrolysis under neutral conditions. The base catalyzed process is dominant and leads to a half-life of 2 yr at pH 7 and 25 °C [26].

**Bioconcentration:** The recommended value for the octanol/water partition coefficient of ethyl acetate suggests that its bioconcentration in fish would be insignificant [25].

**Soil Adsorption/Mobility:** Ethyl acetate is very soluble in water and therefore would not be expected to adsorb significantly to soil [25].

**Volatilization from Water/Soil:** From the Henry's Law constant, one can calculate a half-life for volatilization from a river 1 m deep with a 1 m/sec current and 3 m/sec wind of 10.1 hr [25]. Diffusion through the liquid and vapor phases are important elements in the volatilization process so changes in current and wind will affect the rate [25].

# Ethyl Acetate

Because of its high vapor pressure and low adsorption to soil, ethyl acetate would be expected to volatilize rapidly from soil and other surfaces.

**Water Concentrations:** DRINKING WATER: Detected, not quantified, in 3 New Orleans drinking water plants [19]. Detected in drinking water - no levels given [24]. GROUND WATER: Not detected in aquifer polluted by a paint factory even though ethyl acetate was stored in buried tanks [4]. SURFACE WATER: US - 14 heavily industrialized river basins in US (204 sites) - only 1 site (Delaware River Basin) positive - 1 ppb [11]. Hayashida River in Tatsumo City, Japan (site of leather industry) 585 ppb [40].

**Effluent Concentrations:** Oil refinery final effluent - detected, not quantified [9]. Detected, not quantified, in effluent from sewage treatment plants [30].

**Sediment/Soil Concentrations:**

**Atmospheric Concentrations:** URBAN/SUBURBAN: Leningrad, USSR detected, not quantified [18]. SOURCE AREAS: Ambient air surrounding Kin Buc waste disposal site (4 sites) 0, trace, 4.1, 230 ug/m$^3$ [28].

**Food Survey Values:** U.S. lager beer 25-50 ppm [22]. Component of Orleans vinegar [20] and artificial grape flavoring [23].

**Plant Concentrations:**

**Fish/Seafood Concentrations:**

**Animal Concentrations:**

**Milk Concentrations:**

**Other Environmental Concentrations:** Nail polish remover typically 40 wt% ethyl acetate [21].

**Probable Routes of Human Exposure:** Exposure to ethyl acetate is primarily occupational including fermentation industries which produce whisky and beer. The general public will be exposed to ethyl acetate

# Ethyl Acetate

in products such as airplane dope and nail polish remover as well as in foods such as beer and those containing artificial fruit flavoring.

**Average Daily Intake:**

**Occupational Exposures:** NIOSH (NOES Survey 1981-83) has statistically estimated that 375,906 workers are exposed to ethyl acetate in the United States [27].

**Body Burdens:** Human milk from 4 urban areas in the US - 1 of 8 samples positive [28].

## REFERENCES

1. Ambrose D et al; J Chem Thermodyn 13: 795-802 (1981)
2. Atkinson R; Kinetics and mechanisms of the gas phase reactions of the hydroxyl radicals with organic compounds with atmospheric conditions Chemical Reviews (in press 1985) (1977)
3. Bocek K; Experientia Suppl 23: 231-40 (1976)
4. Botta D et al; Comm Eur Com Eur 851 (Anal Org Micropollut Water): 261-75 (1984)
5. Campbell IM, Parkinson PE; Chem Phys Lett 53: 385-9 (1978)
6. Chemical Profiles; Chemical Marketing Reporter 17 Jan (1983)
7. Chou WL et al; Biotech Bioeng Symp 8: 391-414 (1979)
8. Dilling WL et al; Environ Sci Technol 10: 351-6 (1976)
9. Ellis DD et al; Arch Environ Contam Toxicol 11: 373-82 (1982)
10. Ettinger MB; Ind Eng Chem 48: 256-9 (1956)
11. Ewing BB et al; Monitoring to detect previously unrecognized pollutants in surface waters p75 USEPA-560/6-77-015 (1977)
12. GEMS; Graphical Exposure Modeling System. Fate of atmospheric pollutants (FAP) data base; Office of Toxic Substances USEPA (1986)
13. Graedel TE; Chemical compounds in the atmosphere Academic Press New York p 223-34 (1978)
14. Hansch C, Leo AJ; Substituent constants for correlation analysis in chemistry and biology. John Wiley & Sons New York p.339 (1979)
15. Hawley GG ed; Condensed Chemical Dictionary 10th Van Nostrand Reinhold New York p 422 (1981)
16. Heukelekian H, Rand MC; J Water Pollut Control Assoc 29: 1040-53 (1955)
17. Hustert K et al; Chemosphere 7: 35-50 (1978)
18. Ioffe BV et al; J Chromatogr 142: 787-95 (1977)
19. Keith LH et al; pp.329-73 in Identification and analysis of organic pollutants in water Keith LH ed Ann Arbor Press Ann Arbor MI (1976)
20. Kirk-Othmer Encyclopedia of Chemical Technology; 3rd ed Wiley Interscience New York 23: 757 (1983)
21. Kirk-Othmer Encyclopedia of Chemical Technology; 3rd ed. Wiley Interscience New York 7: 163 (1979)

22. Kirk-Othmer Encyclopedia of Chemical Technology; 3rd ed Wiley Interscience New York 24: 789 (1983)
23. Kirk-Othmer Encyclopedia of Chemical Technology 3rd ed Wiley Interscience New York 4: 712 (1978)
24. Kool HJ et al; Crit Rev Environ Control 12: 307-57 (1982)
25. Lyman WJ et al; Handbook of property estimation methods. Environmental behavior of organic compounds. McGraw-Hill New York (1982)
26. Mabey W, Mill T; J Phys Chem Ref Data 7: 383-415 (1978)
27. NIOSH; National Occupational Exposure Survey (1983)
28. Pellizarri ED et al; Environ Sci Technol 16: 781-5 (1982)
29. Price KS et al; J Water Pollut Control Fed 46: 63-77 (1974)
30. Shackelford WM, Keith LH; Frequency of organic compounds identified in water USEPA-600/4-76-062 (1976)
31. Shelton DR, Tiedje JM; Appl Environ Microbiol 47: 850-7 (1984)
32. Singh HB et al; Reactivity/volatility classification of selected organic chemicals: existing data p190 USEPA-600/3-84-082 (1984)
33. Slave T et al; Rev Chim 25: 666-70 (1974)
34. Speece RE; Environ Sci Tech 17: 416A-27A (1983)
35. Stover EL, Kincannon DF; J Water Pollut Control Fed 55: 97-109 (1983)
36. Thom NS, Agg AR; Proc Roy Soc London B 189: 347-57 (1975)
37. Verscheuren K; Handbook of environmental data on organic chemicals 2nd ed Van Nostrand Reinhold New York p 623-4 (1983)
38. Wasik SP et al; Octanol/water partition coefficients and aqueous solubilities of organic compounds p66 NBS TR81-2406 (1981)
39. Yanagihara S et al; pp 472-7 in Proc Int Clean Air Congress 4th
40. Yasuhara A et al; Environ Sci Technol 15: 570-3 (1981)
41. Young RHF et al; J Water Pollut Control Fed 40: 354-68 (1968)

# Ethyl Chloride

**Synonyms:**

**Structure:**

$$CH_3$$

$$H—C—H$$

$$Cl$$

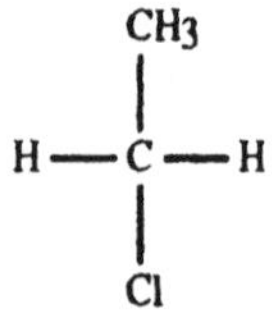

**CAS Registry Number:** 75-00-3

**Molecular Formula:** $C_2H_5Cl$

**Wiswesser Line Notation:** G2

## CHEMICAL AND PHYSICAL PROPERTIES

**Boiling Point:** 12.3 °C at 760 mm Hg

**Melting Point:** -138.7 °C

**Molecular Weight:** 64.52

**Dissociation Constants:**

**Log Octanol/Water Partition Coefficient:** 1.43 [13]

**Water Solubility:** 5710 mg/L at 20 °C [21]

**Vapor Pressure:** 766 mm Hg at 12.5 °C [3]

**Henry's Law Constant:** 8.48 x $10^{-3}$ atm-m$^3$/mole [15]

## ENVIRONMENTAL FATE/EXPOSURE POTENTIAL

**Summary:** Environmental emission sources of ethyl chloride include process and fugitive emissions from its production and use as a chemical intermediate, evaporation from solvent, aerosol, and anesthetic applications, stack emissions from plastics and refuse combustion, inadvertent formation during chlorination treatment, leaching from landfills, and formation via microbial degradation of other chlorinated solvents. Most releases of ethyl chloride will eventually reach the atmosphere since it is a gas at ordinary conditions. If released to the atmosphere, the dominant environmental fate process will be reaction with hydroxyl radicals which has a half-life of about 40 days in typical air. If released to surface water, volatilization will be the dominant process as half-lives ranging from 1.1-5.6 days have been predicted for representative bodies of water. In ground water, where volatilization may not be able to occur, hydrolysis may be the most important removal mechanism. The hydrolysis half-life has been estimated to be 38 days at 25 °C. Very limited biodegradation data suggest that ethyl chloride may be biodegradable, but insufficient data are available to estimate the relative importance of biodegradation in the environment. Aquatic bioconcentration, adsorption, direct photolysis, and oxidation are not important. If released to soil, ethyl chloride will evaporate rapidly where release to air is possible. Ethyl chloride is susceptible to significant leaching. General population exposure to ethyl chloride can occur through inhalation of contaminated ambient air and may occur through oral consumption of contaminated drinking water. Direct dermal exposure occurs through use of ethyl chloride as a topical anesthetic cooling agent. Probable routes of occupational exposure are inhalation and dermal contact.

**Natural Sources:** Ethyl chloride does not occur as a natural product of nature [17].

**Artificial Sources:** Ethyl chloride can be released to the environment through process and fugitive emissions involved with its production and primary use as a chemical intermediate, through evaporation from solvent, aerosol, and anesthetic uses, and through stack emissions from plastics and refuse combustion [10]. Ethyl chloride has been shown to form as a metabolite of the microbial degradation of various chlorinated solvents in soil systems [2,5]. Standard chlorination treatment of landfill leachate can inadvertently yield small amounts of ethyl chloride [9]. Wastewater effluents from treatment facilities and direct leaching from landfills can also release ethyl chloride to the environment.

**Terrestrial Fate:** Ethyl chloride is a gas at room temperature indicating that evaporation from soil surfaces will be a rapid and major removal process. Estimated Koc values of 33 and 143 indicate that ethyl chloride is highly mobile in soil and susceptible to significant leaching. In moist soil systems or ground water where evaporation can not occur, ethyl chloride will react with water via hydrolysis; the estimated hydrolysis half-life in pure water at 25 °C is 38 days. Very limited biodegradation data suggest that ethyl chloride may be biodegradable, but insufficient data are available to estimate the relative importance of biodegradation in the environment.

**Aquatic Fate:** The dominant environmental fate process for ethyl chloride in surface waters is probably volatilization. Volatilization half-lives from a representative pond, river, and lake have been estimated to 5.6, 1.1, and 4.5 days, respectively. Ethyl chloride hydrolyzes in water; however, the estimated hydrolysis half-life of 38 days at 25 °C is not competitive with volatilization. Very limited biodegradation data suggest that ethyl chloride may be biodegradable, but insufficient data are available to estimate the relative importance of biodegradation in the environment. Aquatic bioconcentration, adsorption, direct photolysis, and oxidation are not important.

**Atmospheric Fate:** The dominant atmospheric degradation process for ethyl chloride is probably the vapor-phase reaction with hydroxyl radicals which has a half-life of about 40 days in a typical ambient atmosphere. This tropospheric half-life suggests that <1% of the ethyl chloride will eventually diffuse above the ozone layer where it will be destroyed by photolysis [4]. Ethyl chloride will not directly photolyze below the ozone layer. Atmospheric removal via washout may be possible; however, any ethyl chloride which is removed in this manner will probably revolatilize into the air.

**Biodegradation:** Chlorinated ethanes and methanes were found to release 50-70% of the organically bound chlorine when incubated under anaerobic laboratory conditions [12]. Although ethyl chloride was not tested, the dichloroethanes were found to be potentially biodegradable (53-91% degradation in 28 days of incubation) using a static flask screening procedure [30].

**Abiotic Degradation:** The recommended rate constant for the vapor phase reaction of ethyl chloride with photochemically produced hydroxyl radicals in the atmosphere at 25 °C is $4.0 \times 10^{-13}$ $m^3$/molecule-sec [1]; assuming a typical atmospheric hydroxyl radical concn of $5 \times 10^{+5}$ molecules/$m^3$ yields a half-life of 40 days [1]. Ethyl chloride does not absorb UV light above 290 nm indicating that direct photolysis will not occur in the troposphere [16,19]. Oxidation of ethyl chloride in water via singlet oxygen or peroxy radicals is too slow to be environmentally important [19]. The hydrolysis half-life of ethyl chloride in water at 25 °C has been estimated to be 38 days based on an experimental half-life of 1.68 hr at 100 °C with ethanol and HCl being the hydrolysis products [20].

**Bioconcentration:** Based on the octanol/water partition coefficient and water solubility, the log BCF for ethyl chloride can be estimated to be 0.86 and 0.67, respectively, from recommended regression equations [18]. Measured steady-state log BCFs of di- and trichloroethane in bluegill fish have been reported to be 0.30-0.95 [32].

**Soil Adsorption/Mobility:** Based on the octanol/water partition coefficient and water solubility, the Koc for ethyl chloride can be estimated to be 143 and 33, respectively, from appropriate regression equations [18]; these estimated Koc values indicate high mobility in soil [29].

**Volatilization from Water/Soil:** Laboratory studies concerning the evaporation of chlorinated compounds from dilute aqueous solutions found that 50% of added levels of ethyl chloride evaporated from beakers (65 mm water depth, stirred) in 21-26 minutes [6,7]. Based on the Henry's Law constant, the volatilization half-life of ethyl chloride from a model river (1 m deep) can be estimated to be about 2.5 hr [18]. Based on an estimated oxygen reaeration rate ratio of 0.645, the volatilization half-lives from a representative environmental pond, river, and lake can be estimated to be 5.6 days, 1.1 days, and 4.5 days, respectively [19].

**Water Concentrations:** DRINKING WATER: Ethyl chloride was qualitatively identified in drinking water from Miami, FL, Philadelphia, PA and Cincinnati, OH [31]. GROUND WATER: Ethyl chloride was detected in ground water in Wisconsin at a level of 90 ug/L 80 m downgradient from a contamination source of chlorinated solvents; the presence of the ethyl chloride was attributed to breakdown of other chlorinated compounds [5]. Ethyl chloride levels of 4.3-136 ppb were identified in ground water beneath the Miami Drum waste site in Florida [22]. SURFACE WATER: An analysis of the USEPA STORET data base has reported positive detection of ethyl chloride in 6.0% of 994 ambient water observation stations at a median concn below 10 ppb [28].

**Effluent Concentrations:** An ethyl chloride concn of 1.5 ppb was detected in the final wastewater effluent from a Los Angeles County treatment facility in 1981 [35]. An analysis of the USEPA STORET data base has reported positive detection of ethyl chloride in 2.6% of 1323 effluent observation stations at a median concn below 10 ppb [28]. Raw wastewaters from 9 organic chemical manufacturing facilities contained a mean ethyl chloride concn of 240 ppb [33]. Ethyl chloride was qualitatively detected in leachates and associated ground waters from various municipal landfills in Minnesota [26]. Ethyl chloride has been identified as a gas emitted from landfill simulators [34].

**Sediment/Soil Concentrations:** Ethyl chloride was qualitatively identified in the soil-sediment-water matrix of the Love Canal near Niagara Falls, NY in 1980 [14]. An analysis of the USEPA STORET data base has reported positive detection of ethyl chloride in 0.3% of 354 sediment observation stations [28]. Mean ethyl chloride levels of 0.2 ppb (wet wt) were detected in sediment taken from one location of Lake Pontchartrain in LA in 1980 [8].

**Atmospheric Concentrations:** Atmospheric levels of ethyl chloride in the rural air of Pullman, WA were below 5 ppt during monitoring between Dec 1974 and Feb 1975 [11]. Mean ambient air concentrations of 227, 46, 41, and 87 ppt ethyl chloride were detected in Houston, St. Louis, Denver, and Riverside, respectively, in 1980 [27].

**Food Survey Values:**

**Plant Concentrations:**

**Fish/Seafood Concentrations:** Mean ethyl chloride levels of 7.6 ppb (wet wt) were detected in oysters taken from one location of Lake Pontchartrain in Louisiana in 1980 [8].

**Animal Concentrations:**

**Milk Concentrations:**

**Other Environmental Concentrations:**

**Probable Routes of Human Exposure:** General population exposure to ethyl chloride can occur through inhalation of contaminated ambient

# Ethyl Chloride

air and may occur through oral consumption of contaminated drinking water. Direct dermal exposure occurs through use of ethyl chloride as a topical anesthetic cooling agent. Probable routes of occupational exposure are inhalation and dermal contact.

**Average Daily Intake:** AIR INTAKE: 2.4-13.5 ug/day (average) based on monitoring of four US cities in 1980 [27]; WATER INTAKE: insufficient data; FOOD INTAKE: insufficient data.

**Occupational Exposures:** 142,416 workers are potentially exposed to ethyl chloride based on statistical estimates derived from the NIOSH survey conducted from 1972-74 in the United States [24]. 47,230 workers are potentially exposed to ethyl chloride based on preliminary statistical estimates derived from the NIOSH survey conducted from 1981-83 in the United States [23]. Human exposure risk to ethyl chloride from production, storage, and transport is slight in comparison to exposure risk from medical use as a refrigeration anesthetic [17].

**Body Burdens:** Ethyl chloride was qualitatively identified in 2 of 12 human milk samples collected from women at four US locations [25].

## REFERENCES

1.  Atkinson R; Chem Rev 85: 109-193 (1985)
2.  Barrio-Lage G et al; Environ Sci Technol 20: 96-99 (1985)
3.  Boublik T et al; The Vapor Pressures of Pure Substances Vol 17 Amsterdam, Netherlands: Elsevier Science Publ (1984)
4.  Callahan MA et al; Water-Related Environmental Fate of 129 Priority Pollutants-Volume II EPA-440/4-79-029b Chapter 42 (1979)
5.  Cline PV, Viste DR; Waste Manage Res 3: 351-60 (1985)
6.  Dilling WL et al; Environ Sci Technol 9: 833-8 (1975)
7.  Dilling WL; Environ Sci Technol 11: 405-9 (1977)
8.  Ferrario JB et al; Bull Environ Contamin Toxicol 34: 246-55 (1985)
9.  Gould JP et al; p. 525-39 in Water Chlorination: Environ Impact Health Eff Vol 4 (1983)
10. Graedel TE; Chemical Compounds in the Atmosphere NY: Academic Press p. 326 (1978)
11. Grimsrud EP, Rasmussen RA; Atmos Environ 9: 1014-17 (1975)
12. Haider K; p.200-4 in Comm. Eur. Communities, EUR 1980 EUR 6388, Environ Res Programme (1980)
13. Hansch C, Leo AJ; Medchem Project Issue No 26. Claremont CA: Pomona College (1985)
14. Hauser TR, Bromberg SM; Environ Monitor Assess 2: 249-72 (1982)
15. Hine J, Mookerjee PK; J Org Chem 40: 292-8 (1975)
16. Hubrich C, Stuhl F; J Photochem 12: 93-107 (1980)

17. Konietzko H; Hazard Assess Chem:Curr Dev 3: 407-10 (1984)
18. Lyman WJ et al; Handbook of Chemical Property Estimation Methods NY: McGraw-Hill (1982)
19. Mabey WR et al; Aquatic Fate Process Data for Organic Priority Pollutants EPA-440/4-81-014 p.141, 142, 428 (1981)
20. Mabey W, Mill T; J Phys Chem Ref Data 7: 383-415 (1978)
21. Mackay D, Shiu WY; J Phys Chem Ref Data 19: 1175-99 (1981)
22. Myers VB; Natl Conf Manage Uncontrolled Hazard Waste Sites p.354-7 (1983)
23. NIOSH; National Occupational Exposure Survey (NOES) (1983)
24. NIOSH; National Occupational Hazard Survey (NOHS) (1974)
25. Pellizzari ED et al; Bull Environ Contamin Toxicol 28: 322-8 (1982)
26. Sabel GV, Clark TP; Waste Manage Res 2: 119-30 (1984)
27. Singh HB et al; Atmospheric Measurements of Selected Hazardous Organic Chemicals EPA-600/S3-81-032 p. 4-5 (1981)
28. Staples CA et al; Environ Toxicol Chem 4: 131-42 (1985)
29. Swann RL et al; Res Rev 85:23 (1983)
30. Tabak HH et al; J Water Pollut Control Fed 53: 1503- 18 (1981)
31. USEPA; Preliminary Assessment of Suspected Carcinogens in Drinking Water. Interim Report to Congress, June, 1975. Washington, DC p.9 (1975)
32. USEPA; Ambient Water Quality Criteria for Chlorinated Ethanes EPA-440/5-80-028 p.C-11 (1980)
33. USEPA; Treatability Manual EPA-660/2-82-001a p.I.12.5-2 (1981)
34. Vogt WG, Walsh JJ; p.9 in Proc APCA Annual Meet 78th(Vol 6) (1983)
35. Young DR et al; p.871-84 in Water Chlorination: Environ Impact Health Eff 4(Book2) (1983)

# Ethylene Glycol

## SUBSTANCE IDENTIFICATION

**Synonyms:** 2-Hydroxyethanol

**Structure:**

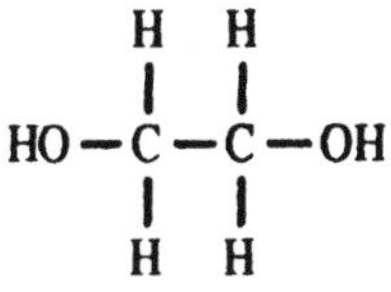

$$HO-\underset{\underset{H}{|}}{\overset{\overset{H}{|}}{C}}-\underset{\underset{H}{|}}{\overset{\overset{H}{|}}{C}}-OH$$

**CAS Registry Number:** 107-21-1

**Molecular Formula:** $C_2H_6O_2$

**Wiswesser Line Notation:** Q2Q

## CHEMICAL AND PHYSICAL PROPERTIES

**Boiling Point:** 197.6 °C at 760 mm Hg

**Melting Point:** -13 °C

**Molecular Weight:** 62.07

**Dissociation Constants:**

**Log Octanol/Water Partition Coefficient:** -1.36 [12]

**Water Solubility:** Miscible

**Vapor Pressure:** 0.0878 mm Hg at 25 °C [22]

**Henry's Law Constant:** 6 X $10^{-8}$ atm-m³/mole [3]

## ENVIRONMENTAL FATE/EXPOSURE POTENTIAL

**Summary:** Ethylene glycol is produced in large quantities primarily for the manufacture of polyester fiber and film and for use in antifreeze. It will enter the environment when released in wastewater

253

or from disposal on land. It will readily biodegrade in water (half-life ca. 3 days). No data are available that report its fate in soils, however, by analogy to its fate in water, biodegradation is probably fast and the dominate removal mechanism. Should ethylene glycol leach into the ground water, biodegradation may occur. Ethylene glycol is not expected to bioconcentrate in aquatic organisms. If ethylene glycol volatilizes, it will react in the atmosphere with hydroxyl radicals with a half-life of about 1 day. Human exposure appears to be primarily from contact with antifreeze solutions, although industrial exposure during manufacture and use may also occur.

**Natural Sources:**

**Artificial Sources:** Ethylene glycol is discharged into wastewater from its production and use as an antifreeze agent, heat transfer agents and in polyester fiber and film manufacture. It also enters the environment from its uses in de-icing runways and from spills and improper disposal of antifreeze, coolant, and solvents containing ethylene glycol [4,15].

**Terrestrial Fate:** When released on land, ethylene glycol, because it is completely soluble in water, may leach to ground water. Its fate in soil has not been studied, although it is easily biodegraded in water, which suggests that it will biodegrade rapidly in the soil.

**Aquatic Fate:** When released into water, ethylene glycol will readily biodegrade (half-life several days). Ethylene glycol is not expected to adsorb to sediment.

**Atmospheric Fate:** Ethylene glycol will react with hydroxyl radicals in the atmosphere. Based on a hydroxyl radical concentration of $10^{+6}$ molecules/cm$^3$, ethylene glycol is predicted to have a half-life of about 1 day in the atmosphere.

**Biodegradation:** There is a large body of information confirming the biodegradability of ethylene glycol in aerobic systems using activated sludge, sewage, and soil inocula. Degradation was essentially complete in <1-4 days although 100% theoretical biological oxygen demand may not be realized for several weeks [2,5,11,13,17,18,20,21,23,25]. In a river die-away test, degradation was completed in 3 days at 20 °C and 5-14 days at 8 °C [8]. Data are scant for anaerobic systems, but the evidence indicates that it readily biodegrades in these systems also [7].

# Ethylene Glycol

**Abiotic Degradation:** Photooxidation in aqueous systems will not be significant [6,14]. Glycols have no hydrolyzable groups and are not susceptible to hydrolysis [16]. The hydroxyl radical rate constant for ethylene glycol is $7.7 \times 10^{-12}$ cm$^3$/molecule-sec [4]. Based on a hydroxyl radical concentration of $10^{+6}$ molecules/cm$^3$, ethylene glycol is predicted to have a half-life of about 1 day in the atmosphere.

**Bioconcentration:** The bioconcentration factor for ethylene glycol in fish (Golden ide) was reported to be 10 after 3 days of exposure [9]. In algae (Chlorella fusca), the bioconcentration factor was 190 after 1 day [9]. Its extremely low octanol/water partition coefficient suggests that it will not bioconcentrate in fish.

**Soil Adsorption/Mobility:** No information concerning the adsorption of ethylene glycol could be found in the literature. Its low octanol/water partition coefficient indicates that its adsorption to soil will be low.

**Volatilization from Water/Soil:** No information could be found concerning the rate of evaporation of ethylene glycol from water. Its relatively low Henry's Law constant suggests that it will not evaporate rapidly from either water or soil.

**Water Concentrations:**

**Effluent Concentrations:** Detected, not quantified, in chemical effluent in Brandenburg, KY [24].

**Sediment/Soil Concentrations:**

**Atmospheric Concentrations:**

**Food Survey Values:**

**Plant Concentrations:**

**Fish/Seafood Concentrations:**

**Animal Concentrations:**

**Milk Concentrations:**

**Other Environmental Concentrations:** Detected in tobacco smoke [10].

**Probable Routes of Human Exposure:** Humans are propably exposed to ethylene glycol primarily from contact with antifreeze and coolants containing ethylene glycol.

**Average Daily Intake:**

**Occupational Exposures:** NIOSH (NOES Survey 1981-83) has statistically estimated that 1,133,792 workers are exposed to ethylene glycol in the United States [19].

**Body Burdens:**

## REFERENCES

1. Atkinson R; Chem Rev 85:69-201 (1985)
2. Bridie AL et al; Water Res 13: 627-30 (1979)
3. Butler JAU, Ramchandani CN; J Chem Soc 952-955 (1935)
4. Chemical Profile; Chemical Marketing Reporter 2/13 (1984)
5. Conway RA et al; Environ Sci Technol 17: 107-12 (1983)
6. Dorfman LM, Adams GE; Reactivity of the hydroxyl radical in aqueous solutions 51 p NSRD-NBS-46 (1973)
7. Dwyer DF, Tiedje JM; Appl Environ Microbiol 46: 185-90 (1983)
8. Evans WH, David EJ; Water Res 8: 97-100 (1974)
9. Freitag D et al; Chemosphere 14:1589-1616 (1985)
10. Graedel TE; Chemical compounds in the atmosphere p 250 New York, NY Academic Press (1978)
11. Haines JR, Alexander M; Appl Microbiol 29: 621-5 (1975)
12. Hansch C, Leo AJ; Medchem Project Issue No 26. Claremont CA: Pomona College (1985)
13. Helfgott TB et al; An index of refractory organics USEPA 600/2-77-174 (1977)
14. Hendry DG, Kenley RA; Atmospheric reaction products of organic compounds. 80 p USEPA 560/12-79-001 (1979)
15. Lovell R; Organic Chemical Manufacturing Volume 9: Selected Processes. Ethylene Glycol EPA-450/3-80-028d (1980)
16. Lyman WJ et al; Handbook of property estimation methods. Environmental Behavior of Organic Compounds, p.7-1 to 7-48 New York, NY McGraw Hill Book Co. (1982)
17. Matsui S et al; Prog Water Technol 7: 645-59 (1975)
18. Means JL, Anderson SJ; Water Air Soil Poll 16: 301-15 (1981)
19. NIOSH; National Occupational Exposure Survey (1983)
20. Pitter P; Water Res 10: 231-5 (1976)
21. Price KS et al; J Water Pollut Control Fed 46: 63-77 (1974)

22. Riddick JA et al; Organic Solvents: Physical Properties and Methods of Purification, 4th Edit. New York: J Wiley & Sons p 263 (1986)
23. Schefer W, Waelchli O; Z Wasser Abwasser Forsch 13: 205-9 (1980)
24. Shakelford WM, Keith LH; Frequence of Org Compounds Identified in Water pp 447 USEPA 600/4-76-062 (1976)
25. Zahn R, Wellens H; Z Wasser Abwasser Forsch 13: 1-7 (1980)

# Ethylene Glycol Monoethyl Ether

## SUBSTANCE IDENTIFICATION

**Synonyms:**

**Structure:**

**CAS Registry Number:** 110-80-5

**Molecular Formula:** $C_4H_{10}O_2$

**Wiswesser Line Notation:** Q2O2

## CHEMICAL AND PHYSICAL PROPERTIES

**Boiling Point:** 135 °C at 760 mm Hg

**Melting Point:**

**Molecular Weight:** 90.12

**Dissociation Constants:**

**Log Octanol/Water Partition Coefficient:** -0.10 [7]

**Water Solubility:** Miscible with water [5]

**Vapor Pressure:** 5.3 mm Hg at 25 °C [5]

**Henry's Law Constant:** $5.13 \times 10^{-2}$ atm-m$^3$/mole [5]

## ENVIRONMENTAL FATE/EXPOSURE POTENTIAL

**Summary:** Release of ethylene glycol monoethyl ether (EGM) to the environment is expected to result from its use as a solvent in a variety of products. Release to the soil is expected to result in volatilization

258

from the soil surface and leaching to ground water. Biodegradation is expected to be significant. Release to water will result in volatilization from the water surface and biodegradation. Minimal adsorption to sediments is expected and bioconcentration is not expected to be significant. Release to the atmosphere is expected to result in rapid degradation by nitrogen dioxides. The estimated half-life for the reaction between vapor phase EGM and photochemically generated hydroxyl radicals is 11.41 hr. Human exposure to EGM is expected to result primarily from industrial use of the compound and the presence of EGM in a variety of consumer and industrial products as a solvent.

**Natural Sources:**

**Artificial Sources:** Ethylene glycol monoethyl ether is used as a solvent for nitrocellulose, lacquers and dopes, in varnish removers, cleansing solutions, and dye baths, in leather finishing with water pigments and dye solutions, to increase the stability of emulsions [14] and in dry cleaning compounds and enamels, and in textiles to prevent spotting in printing or dyeing [9].

**Terrestrial Fate:** Ethylene glycol monoethyl ether is not expected to tightly bind to soils and so may leach to ground water and volatilization from the soil surface may occur. Biodegradation may also be significant.

**Aquatic Fate:** Ethylene glycol monoethyl ether is expected to volatilize readily from the water surface and biodegradation is expected to be significant. Ethylene glycol monoethyl ether is not expected to bind tightly to sediments.

**Atmospheric Fate:** Ethylene glycol monoethyl ether is expected to react rapidly with nitrogen dioxides in the atmosphere and with hydroxyl radicals with an estimated half-life of 11.41 hr [6].

**Biodegradation:** Incubation of EGM in 500 mL with 10 mL of effluent from a biological sanitary waste treatment plant at 20 °C for 5 days resulted in a biological oxygen demand (BOD) 65% of the theoretical oxygen demand (ThOD) when the seeding was adapted, and 53% of ThOD when unadapted [2]. Five-day, 20 °C ThOD values of 7.6% and 54.3% were observed for EGM upon incubation with sewage seed and acclimated activated sludge seed, respectively [3]. Upon incubation of 5-10 ppm EGM with activated sludge for 5 days at 20

°C, 81% of the theoretical BOD was achieved [10]. After 5 days incubation of 50 mg/L EGM in a Warburg respirometer at 20 °C with acclimated activated sludge, 54% of the theoretical BOD was achieved [12]. Ethanol and acetate ion were the products of the incubation of EGM with <u>Pelobacter</u> <u>venetianus</u> sp nov, a strict anaerobe of marine and limnic origin [17]. Ethoxyacetic acid was obtained from a culture of the soil bacterium, <u>Alcaligenes</u> MC11, supplemented with EGM [8].

**Abiotic Degradation:** When EGM was mixed with nitrogen dioxides at 20:1 and 2:1 (EGM:nitrogen dioxides) in a 440 L glass-Teflon smog chamber at relative humidities of 30-60% and 25 °C, the half-lives observed were 9.8 and >7.5 hr, respectively [11]. The half-life for the reaction between vapor phase EGM and photochemically generated atmospheric hydroxyl radicals was estimated to be 11.41 hr [6]. A half-life for the reaction of hydroxyl radicals with EGM in water at pH 9 of about 2.2 yr was obtained from a rate constant of $1.0 \times 10^{+9}$ 1/mol-sec and a hydroxyl radical concentration in water of $1 \times 10^{-17}$ M [1].

**Bioconcentration:**

**Soil Adsorption/Mobility:** A measured log Kow was used to estimate a log Koc of 1.32 for EGM [13]. With a log Koc of this magnitude and due to its miscibility with water, EGM is not expected to strongly bind to soil and may, therefore, leach to ground water.

**Volatilization from Water/Soil:** The Henry's Law constant indicates that EGM will rapidly volatilize from water [13]. EGM may also volatilize from moisture soil, based on its weak absorption to soils and high Henry's Law constant.

**Water Concentrations:** Hayashida River water (Japan) - 250-1200 ppb [18].

**Effluent Concentrations:**

**Sediment/Soil Concentrations:**

**Atmospheric Concentrations:**

**Food Survey Values:**

# Ethylene Glycol Monoethyl Ether

**Plant Concentrations:**

**Fish/Seafood Concentrations:**

**Animal Concentrations:**

**Milk Concentrations:**

**Other Environmental Concentrations:**

**Probable Routes of Human Exposure:** No data were available but it is expected that human exposure to EGM will result primarily from inhalation and dermal contact from industrial use of the compound and through contact with consumer and industrial products which contain EGM as a solvent.

**Average Daily Intake:**

**Occupational Exposures:** In a manufacturing plant, personal and area monitoring were ND-0.6 ppm and 0.5-1.5 ppm EGM, respectively [4]. EGM was 0.6, 1.1, and 1.5 ppm in the areas MICC (undefined), switchbank and lab sink [4]. No EGM was detected in personal monitoring samples collected in the drumming areas [4]. NIOSH (NOES Survey 1981-83) has statistically estimated that 119,247 workers are exposed to EGM in the United States [16]. NIOSH (NOHS Survey 1972-74) has statistically estimated that 411,270 workers are exposed to EGM in the United States [15].

**Body Burdens:**

## REFERENCES

1. Anbar M, Neta P; Int J Appl Rad Isot 18: 493-523 (1967)
2. Birdie Al et al; Water Res 13: 627-30 (1979)
3. Bogan RH, Sawyer CN; Sewage Ind Waste 27: 917-28 (1955)
4. Clapp DE et al; Environ Health Perspec 57: 91-5 (1984)
5. Dow Chemical Company; The Glycol Ethers Handbook (1981)
6. GEMS; Graphical Exposure Modeling System. FAP. Fate of Atmos Pollut (1986)
7. Hansch C, Leo AJ; Medchem Project Issue No.19 Pomona College (1985)
8. Harada T, Nagashima Y; J Ferment Technol 53: 218-22 (1975)
9. Hawley GG; Condensed Chemical Dictionary 10th ed Van Nostrand Reinhold NY p 433 (1981)
10. Heukelekian H, Rand MC; J Water Pollut Cont Fed 30: 1040-53 (1955)

# Ethylene Glycol Monoethyl Ether

11. Joshi SB et al; Atmos Environ 16: 1301-10 (1982)
12. Ludzack FJ, Ettinger MB; J Water Pollut Cont 30: 1173-200 (1960)
13. Lyman WJ et al; Handbook of Chemical Property Estimation Methods. Environmental Behavior of Organic Compounds. McGraw-Hill NY (1982)
14. Merck Index; An Encyclopedia of Chemicals, Drugs and Biologicals 10th ed p 545 (1983)
15. NIOSH; National Occupational Hazard Survey (1972)
16. NIOSH; National Occup Exposure Survey (1984)
17. Schink B, Stieb M; Appl Environ Microbiol 45: 1905-13 (1983)
18. Yasuhara A et al; Environ Sci Technol 15: 570-3 (1981)

# Ethyl Formate

## SUBSTANCE IDENTIFICATION

**Synonyms:**

**Structure:**

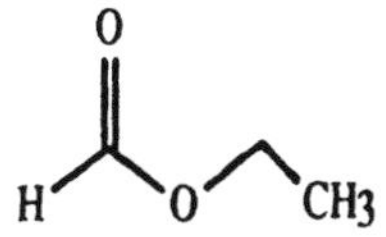

**CAS Registry Number:** 109-94-4

**Molecular Formula:** $C_3H_6O_2$

**Wiswesser Line Notation:** VHO2

## CHEMICAL AND PHYSICAL PROPERTIES

**Boiling Point:** 54.5 °C at 760 mm Hg

**Melting Point:** -80.5 °C

**Molecular Weight:** 74.09

**Dissociation Constants:**

**Log Octanol/Water Partition Coefficient:** 0.23 [4]

**Water Solubility:** 88,250 at 25 °C [4]

**Vapor Pressure:** 244.6 mm Hg at 20.6 °C [3]

**Henry's Law Constant:** 3.85 x $10^{-4}$ atm-m$^3$/mole at 25 °C [2]

## ENVIRONMENTAL FATE/EXPOSURE POTENTIAL

**Summary:** Release of ethyl formate to the environment will result from its use in synthetic flavors for lemonades, essences, artificial rum, and arrac and as a fungicide and larvicide for tobacco, cereals, dried

263

fruit, and other crops. It is also used as a solvent for cellulose nitrate and acetate, as an acetone substitute and fumigant, and in organic synthesis. If released to soil, it should not sorb strongly to the soil and should exhibit high mobility. No data were found concerning biodegradation or hydrolysis in soils. Evaporation from the surface of soils and other surfaces may be an important transport process. If released to water, it will not sorb to sediment or bioconcentrate in aquatic organisms. Hydrolysis may be an important fate process in water. No information was found about biodegradation in natural water; however, based on limited data from a laboratory screening test, it may be susceptible to biodegradation in natural waters. A half-life of 4.5 hr was predicted for evaporation from a model river 1 m deep flowing at 1 m/sec with a wind velocity of 3 m/sec. The estimated vapor phase half-life in the atmosphere is 11.1 days as a result of hydrogen abstraction reaction by photochemically produced hydroxyl radicals. Exposure will occur as a result of ingestion of foods containing ethyl formate and through occupational contact.

**Natural Sources:**

**Artificial Sources:** Ethyl formate is used in synthetic flavors for lemonades, essences, artificial rum, and arrack and as a fungicide and larvicide for tobacco, cereals, dried fruit, and other crops [9]. It is also used as a solvent for cellulose nitrate and acetate, as an acetone substitute and fumigant, and in organic synthesis [5].

**Terrestrial Fate:** If released to soil, ethyl formate should not sorb strongly to the soil and should exhibit high mobility. No data were found concerning biodegradation or hydrolysis in soils. Evaporation from the surface of soils and other surfaces may be an important transport process based on the reported vapor pressure.

**Aquatic Fate:** If released to water, ethyl formate will not sorb to sediment or bioconcentrate in aquatic organisms. Hydrolysis may be an important fate process in water based on a report that ethyl formate gradually decomposes to formic acid and ethanol in water solutions. No information was found about biodegradation in natural waters. Based on limited data from a laboratory screening test, ethyl formate may be susceptible to biodegradation in natural waters. A half-life of 4.5 hr was predicted for evaporation from a model river 1 m deep flowing at 1 m/sec with a wind velocity of 3 m/sec.

# Ethyl Formate

**Atmospheric Fate:** The estimated vapor phase half-life in the atmosphere is 11.1 days as a result of hydrogen abstraction by photochemically produced hydroxyl radicals.

**Biodegradation:** No information was found about biodegradation of ethyl formate in natural media. Theoretical BOD was 33% after 10 days with a dispersed seed aeration laboratory screening test using a settled sewage culture [10].

**Abiotic Degradation:** Ethyl formate gradually decomposes in water to formic acid and ethanol [9]. The estimated vapor phase half-life in the atmosphere is 11.1 days as a result of hydrogen abstraction reaction by photochemically produced hydroxyl radicals [1].

**Bioconcentration:** Ethyl formate did not bioconcentrate in rainbow trout [8]. Using the octanol/water partition coefficient, a BCF of 0.88 was calculated [7]. Based on this estimated and experimental BCF, ethyl formate will not be expected to bioconcentrate in aquatic organisms.

**Soil Adsorption/Mobility:** Using the octanol/water partition coefficient, an estimated Koc of 32 was calculated [7]. Based on this estimated Koc, ethyl formate will not adsorb to soils or sediments.

**Volatilization from Water/Soil:** Using the Henry's Law constant, a half-life of 4.5 hr was predicted for evaporation from a model river 1 m deep flowing at 1 m/sec with a wind velocity of 3 m/sec [7].

**Water Concentrations:**

**Effluent Concentrations:**

**Sediment/Soil Concentrations:**

**Atmospheric Concentrations:**

**Food Survey Values:** Identified, not quantified, in roasted filberts [6].

**Plant Concentrations:**

**Fish/Seafood Concentrations:**

# Ethyl Formate

**Animal Concentrations:**

**Milk Concentrations:**

**Other Environmental Concentrations:**

**Probable Routes of Human Exposure:** Ethyl formate is used in flavorings for lemonades and essences, in artificial rum and arrack, and as a fungicide and larvicide for tobacco, cereals, dried fruits, and other crops [9]. Oral ingestion may occur as a result of the above reported uses of ethyl formate.

**Average Daily Intake:**

**Occupational Exposure:** NIOSH estimates that 8149 employees are exposed according to a survey conducted during 1981-83 [12]. A National Occupational Hazard Survey (NOHS) from 1972-74 estimated that 264 workers are exposed to ethyl formate [11].

**Body Burdens:**

## REFERENCES

1. Atkinson R; Internat J Chem Kinetics 19: 799-828 (1987)
2. Bocek K; Experientia Suppl 23: 231-40 (1976)
3. Daubert TE, Danner RP; Data Compilation Tables of Properties of Pure Compounds. Amer Inst Chem Engn pp 450 (1985)
4. Hansch C et at; J Org Chem 33: 347 (1968)
5. Hawley GG; Condensed Chemical dictionary 10th ed Von Nostrand Reinhold NY p 436 (1981)
6. Kinlin TE et al; J Agric Food Chem 20: 1021-8 (1972)
7. Lyman WJ et al; Handbook of Chemical Property Estimation Methods Environ Behavior of Org Compounds McGraw-Hill NY (1982)
8. McKim J et al; Absorption Dynamics of Organic Chemical Transport Across Trout Gills as Related to Octanol-Water Partition Coefficient NTIS PB-198315 (1985)
9. Merck Index; An Encyclopedia of Chemicals, Drugs and Biologicals 10th ed p 551 (1983)
10. Mills EJ, Stack VT Jr; Proc 8th Industrial Waste Conf Eng Bull Purdue Univ, Eng Ext Ser 8: 492-517 (1954)
11. NIOSH; NOHS (National Occupational Hazard Survey (1974)
12. NIOSH; NOES (National Occupational Exposure Survey) (1988)

# Furan

## SUBSTANCE IDENTIFICATION

**Synonyms:**

**Structure:**

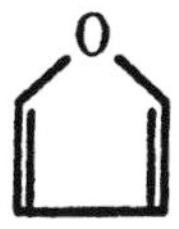

**CAS Registry Number:** 110-00-9

**Molecular Formula:** $C_4H_4O$

**Wiswesser Line Notation:**

## CHEMICAL AND PHYSICAL PROPERTIES

**Boiling Point:** 32 °C at 758 mm Hg

**Melting Point:** -85.65 °C

**Molecular Weight:** 68.08

**Dissociation Constants:**

**Log Octanol/Water Partition Coefficient:** 1.34 [11]

**Water Solubility:** 10,000 mg/L at 25 °C [28]

**Vapor Pressure:** 602.3 mm Hg at 25 °C [5]

**Henry's Law Constant:** $5.4 \times 10^{-3}$ atm-m$^3$/mole at 25 °C (calculated from vapor pressure and water solubility)

## ENVIRONMENTAL FATE/EXPOSURE POTENTIAL

**Summary:** Furan is both a natural and synthetic compound. It occurs in oil obtained by the distillation of pine wood containing rosin and has been identified as a volatile component of sorb trees. Furan is released to air in cigarette smoke, wood smoke, and engine exhaust gas. Major releases to the environment may also occur in effluents and emissions from its one manufacturing plant or sites at which it is used. If released to land, furan will be susceptible to rapid volatilization and significant leaching, possibly into ground water. Chemical hydrolysis is not expected to be an important fate process. If released to water, volatilization (half-life of 2.5 hr from a shallow model river) and reaction with singlet oxygen (half-life 1 hr) are expected to be the dominant fate processes. This compound is not expected to chemically hydrolyze, bioaccumulate significantly in aquatic organisms, or adsorb significantly to suspended solids and sediments. If released to air, furan is expected to exist almost entirely in the vapor phase. During daylight hours reaction with photochemically generated hydroxyl radicals is expected to be the dominant removal process (half-life 9.5 hr) and during night time hours reaction with nitrate radicals is expected to be the dominant removal process (half-life 0.5 hr). Removal from the atmosphere by reaction with ozone or physical processes is not expected to be significant. The most probable route of human exposure is by inhalation. A small number of breastfed infants may be exposed by ingestion since furan has been detected in human milk. Furan has been detected in the expired air of both smokers and nonsmokers. The range of expiration rates were from 0-98 ug/hr and 0-28 ug/hr, respectively.

**Natural Sources:** Furan occurs in oils obtained by the distillation of pine wood containing rosin [29]. Furan has been identified in volatile emissions from sorb trees [13].

**Artificial Sources:** Furan is released to air as a gas phase component of cigarette smoke [25], wood smoke [15], and exhaust gas from diesel and gasoline engines [10]. Major releases to the environment may also occur in effluents and emissions from manufacturing and use facilities. Furan is produced in commercial quantities by only one manufacturer in the US [26], and is an intermediate in the manufacture of a number of other industrial chemicals, especially pyrrole, tetrahydrofuran, and thiophene [12,18].

**Terrestrial Fate:** If released to soil, furan will be susceptible to rapid volatilization and significant leaching, possibly into ground water. Furan should be resistant to chemical hydrolysis.

# Furan

**Aquatic Fate:** If released to water, volatilization (half-life of 2.5 hr from a shallow model river 1 m deep flowing 1 m/sec with a wind speed of 3 m/sec) and reaction with singlet oxygen (half-life of 1 hr) are expected to be the dominant removal mechanisms. Chemical hydrolysis, bioaccumulation in aquatic organisms, and physical adsorption to suspended solids or sediments are not expected to be significant fate processes.

**Atmospheric Fate:** Based on the high vapor pressure, furan is expected to exist almost entirely in the vapor phase in the atmosphere [7]. During daylight hours, reaction with photochemically generated hydroxyl radicals is predicted to be the dominant removal mechanism (half-life 9.5 hr). During night time hours, reaction with nitrate radicals is predicted to be the dominant removal mechanism (half-life 1/2 hr). Removal from the atmosphere by reaction with ozone or physical processes is not expected to be significant.

**Biodegradation:**

**Abiotic Degradation:** Furan is expected to be resistant to chemical hydrolysis under environmental conditions [17]. The half-life for the reaction of furan with singlet oxygen in water has been estimated to be 1 hr based on a measured reaction rate constant of $1.4 \times 10^{+8}$ 1/mole-sec and assuming an ambient singlet oxygen concentration of $1 \times 10^{-12}$ mole/L [19]. The half-life for the reaction of furan vapor with photochemically generated hydroxyl radicals in the atmosphere has been estimated to be 9.5 hr using an experimental rate constant of $4.0 \times 10^{-11}$ cm$^3$/molecule-sec [1] and an average atmospheric hydroxyl radical concentration of $5.0 \times 10^5$ molecules/cm$^3$ [3]. The half-life for furan reacting with ozone in the atmosphere has been estimated to be approximately 4.7 days using an experimentally derived reaction rate constant of $2.4 \times 10^{-18}$ cm$^3$/molecule-sec at room temperature and an average ambient ozone concentration of $7.0 \times 10^{+11}$ molecules/cm$^3$ [2]. The half-life for the nighttime reaction of furan with nitrate radicals has been estimated to be 34 minutes based on an experimentally derived reaction rate constant of $1.4 \times 10^{-2}$ cm$^3$/molecule-sec at room temperature and an average night time nitrate radical concentration of $2.4 \times 10^{+8}$ molecules/cm$^3$ (nitrate radicals are unstable in sunlight) [4].

**Bioconcentration:** Bioconcentration factors (BCF) of 3-6 were estimated for furan using linear regression equations based on the Kow

269

and the water solubility [17]. These BCF values suggest that furan will not bioaccumulate significantly in aquatic organisms.

**Soil Adsorption/Mobility:** Soil adsorption coefficients of 27-128 were estimated for furan using linear regression equations based on the Kow value and the water solubility [17]. These Koc values suggest that physical adsorption of furan to suspended solids and sediments in water would not be significant and that furan would be highly mobile in soil [27].

**Volatilization from Water/Soil:** Based on the value of Henry's Law Constant, the half-life for furan volatilizing from a shallow river 1 m deep flowing 1 m/sec with a wind speed of 3 m/sec was estimated to be 2.5 hr [17]. The relatively high vapor pressure of furan suggests that this compound would volatilize rapidly from dry soil surfaces. It appears that furan would also volatilize fairly rapidly from moist soil surfaces since this compound does not adsorb strongly to soil and is predicted to volatilize rapidly from water.

**Water Concentrations:** SURFACE WATER: Furan has been qualitatively identified in the Niagara River and 2 creeks in the Niagara River watershed [8,9].

**Effluent Concentrations:** Furan was detected in 1/63 industrial effluents at a concentration of <10 ug/L [24]. Furan was detected in aqueous condensate samples from low-Btu gasification of rosebud coal at a concentration of $7 \pm 4$ ppb [23]. It was not detected (detection limit 0.1 ppb) in ground water or coal water prior to in situ coal gasification, product water samples obtained during in situ coal gasification, retort water from in situ oil shale processing, or boiler blowdown water from in situ oil shale processing [23].

**Sediment/Soil Concentrations:**

**Atmospheric Concentrations:**

**Food Survey Values:** Furan has been identified as a volatile component of roasted filberts [14].

**Plant Concentrations:**

**Fish/Seafood Concentrations:**

# Furan

**Animal Concentrations:**

**Milk Concentrations:**

**Other Environmental Concentrations:**

**Probable Routes of Human Exposure:** The most probable route of exposure to furan by the general population and workers is inhalation. Detection of furan in 1 out of 12 samples of human milk suggests that a small number of breastfed infants may be exposed to this compound by ingestion [22].

**Average Daily Intake:**

**Occupational Exposure:** NIOSH estimates that 35 workers are potentially exposed to furan based upon the NIOSH survey conducted in 1981-83 in the United States [21]. NIOSH estimates that 6804 workers are potentially exposed to furan based upon a survey conducted in 1972-74 in the United States [20].

**Body Burdens:** Furan was detected in the expired air of 2 of 3 male smokers and 4 of 5 male nonsmokers at the Brooks Air Force Base, TX with the rate of expiration ranging from 0.25-98 ug/hr for smokers and from 0.33-28 ug/hr for nonsmokers [6]. This compound was also identified in 15 of 387 samples of expired air taken from 54 male and female nonsmokers from Chicago, IL with a mean concn 0.547 ng/L [16]. Furan was qualitatively identified in 1 of 12 samples of human milk obtained from woman in 4 different urban areas [22].

## REFERENCES

1.  Atkinson R; Chem Rev 85: 69-201 (1985)
2.  Atkinson R, Carter WPL; Chem Rev 84: 437-470 (1984)
3.  Atkinson R; Internat J Chem Kinetics 19: 799-828 (1987)
4.  Atkinson R et al; Environ Sci Technol 19: 87-9 (1985)
5.  Chao J et al; J Phys Chem Ref Data 12: 1033-63 (1983)
6.  Conkle JP et al; Arch Environ Health 30: 290-5 (1975)
7.  Eisenreich SJ; Env Sci Tech 15: 30-8 (1981)
8.  Elder VA et al; Environ Sci Tech 15: 1237-43 (1981)
9.  Great Lakes Water Quality Board; An Inventory of Chemical Substances Identified in the Great Lake Ecosystem. Vol 1. p. 195 (1983)
10. Hampton CV et al; Environ Sci Tech 16: 287-98 (1982)

# Furan

11. Hansch C, Leo AJ; Medchem Project Issue No.26 Pomona College, Claremont CA (1985)
12. Hawley GG; Condensed Chem Dictionary 10th ed Von Nostrand Reinhold NY p. 483 (1981)
13. Isidorov VA et al; Atmos Environ 19: 1-8 (1985)
14. Kinlin TE et al; J Agric Food Chem 20: 1021 (1972)
15. Kleindienst TE et al; Environ Sci Tech 20: 493-501 (1986)
16. Krotoszynski BK et al; J Anal Toxicol 3: 225-34 (1979)
17. Lyman WJ et al; Handbook of Chem Property Estimation Methods McGraw-Hill NY (1982)
18. McKillip WJ. Sherman E; in Kirk-Othmer Encycl Chem Tech 3rd ed. Wiley NY 11: 516-20 (1980)
19. Mill T, Mabey W; p. 221 in Environmental Exposure from Chemicals Vol 1 Neely WB, Blau GE eds CRC Press Boca Raton FL (1985)
20. NIOSH; National Occupational Hazard Survey (1974)
21. NIOSH; National Occupational Exposure Survey (NOES)(1983)
22. Pellizzari ED et al; Bull Environ Contam Toxicol 28: 322-8 (1982)
23. Pellizzari ED et al; ASTM Spec Tech Publ STP 686: 256-74 (1979)
24. Perry DL et al; Identification of Organic Compounds in Industrial Effluent Discharges USEPA 600/4-79-016 NTIS PB-294794 (1979)
25. Sakuma H et al; Nippon Sembai Kosha Chuo Kenkyusho Kenkyu Hokoku 117: 47-54 (1975)
26. SRI International; 1987 Directory of Chemical Producers, US Menlo Park CA p.682 (1987)
27. Swann RL et al; Res Rev 85: 17-28 (1983)
28. Valvani SC et al; J Pharm Sci 70: 502-7 (1981)
29. Windholz M ed; The Merck Index 10th ed. Merck Co. Rahway, NJ p. 613 (1983)

# Isoamyl Acetate

## SUBSTANCE IDENTIFICATION

**Synonyms:** 3-Methylbutyl acetate

**Structure:**

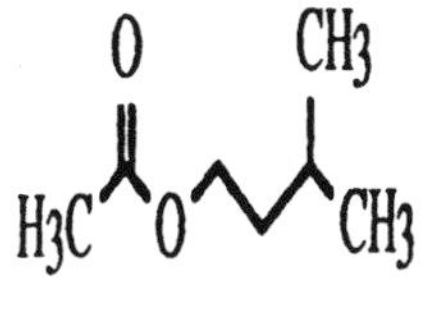

H3C—C(=O)—O—CH2—CH2—CH(CH3)—CH3

**CAS Registry Number:** 123-92-2

**Molecular Formula:** $C_7H_{14}O_2$

**Wiswesser Line Notation:**

## CHEMICAL AND PHYSICAL PROPERTIES

**Boiling Point:** 142 °C

**Melting Point:** -78.5 °C

**Molecular Weight:** 130.18

**Dissociation Constants:**

**Log Octanol/Water Partition Coefficient:** 2.13 [17]

**Water Solubility:** 2000 mg/L at 25 °C [28]

**Vapor Pressure:** 4.5 mm Hg at 20 °C [20]

**Henry's Law Constant:** 5.87 x $10^{-4}$ at 25 °C [8]

## ENVIRONMENTAL FATE/EXPOSURE POTENTIAL

**Summary:** Isoamyl acetate will be released to the atmosphere and in wastewater during its use as a solvent as well as from foods in which it is used as a flavoring agent. It is also released naturally from plants,

273

as a pheromone from insects, and during some fermentations. If released on land or in water, volatilization would be important (half-life 5 hr in a model river) and biodegradation, while not demonstrated experimentally, may be a dominant process. Adsorption to soil or sediment would not occur to any significant extent, so leaching into ground water may occur. Some chemical hydrolysis may occur but only under fairly alkaline conditions. Isoamyl acetate would not be expected to bioconcentrate in aquatic organisms. In air, isoamyl acetate will be scavenged by rain and degrade by reaction with photochemically produced hydroxyl radicals (estimated half-life 2.6 days). Occupational exposures, both dermal and inhalation, would occur primarily from its use as a solvent. The general public will be exposed from ingesting foods in which it occurs naturally or is a flavoring, or from contact with products which contain the chemical as a solvent.

**Natural Sources:** Isoamyl acetate is a plant volatile [5]. It is also released during the fermentation process in making beer and whiskey [1]. It is a sting pheromone of the honey bee (Apis mellifera) [1,22], and a pheromone for Lobesia botrana and Manduca sexta [21].

**Artificial Sources:** Isoamyl acetate will be released to the atmosphere or in wastewater during its use as a flavoring agent commonly used in soft drinks, chewing gum, and candies, in perfumes, for masking undesirable odors, and as a solvent for tannins, nitrocellulose, lacquers, celluloid, and camphor [16,21]. Releases may also occur during its use in manufacturing artificial silk, leather, or pearls, photographic films, celluloid cements, waterproof varnish, bronzing liquids, and metallic paints, dyeing, and finishing textiles [7,27]. It is also released to the environment during the fermentation process in making beer and whiskey and is found in gaseous emissions from whiskey fermentation units [3,21]. Because of its intense odor, isoamyl acetate is used as an agent in respirator fit tests and may be released to the environment during this procedure [25].

**Terrestrial Fate:** If released on land, isoamyl acetate would volatilize and leach into the soil. Chemical hydrolysis would not be important except possibly under fairly alkaline conditions (half-life 14 days at pH 9). It is probable that biodegradation will be the most important chemical fate process but experimental data in soil systems are lacking.

**Aquatic Fate:** If released into water, isoamyl acetate will be lost by volatilization (estimated half-life 5 hr in a model river). Results of screening studies on similar chemicals suggest that biodegradation may be significant in natural waters. Hydrolysis and adsorption to sediment are not expected to be significant.

**Atmospheric Fate:** Due to its moderately high vapor pressure, isoamyl acetate will exist primarily as the vapor if released into air. It should react with photochemically produced hydroxyl radicals (estimated half-life 2.7 days [2]). It is soluble in water and would be scavenged by rain.

**Biodegradation:** No information could be found in the literature on the biodegradation of isoamyl acetate. An analogous compound, isobutyl acetate, has been shown to be readily biodegradable (60% and 81% of theoretical BOD in 5 and 20 days, respectively [18]) in a screening test using a filtered sewage seed inoculum. The same investigators found that for n-butyl acetate and n-amyl acetate 58% and 64% of the theoretical BOD was respectively consumed after 5 days, and 83% and 72% after 20 days [18]. Since longer chained groups tend to biodegrade more rapidly than shorter chained groups and straight-chain compounds tend to degrade more rapidly than branched ones [12], one might also expect isoamyl acetate to be somewhat more biodegradable than isobutyl acetate and somewhat less biodegradable than n-amyl acetate.

**Abiotic Degradation:** No information on the hydrolysis of isoamyl acetate could be found in the literature. The base catalyzed process is the dominant hydrolytic process for aliphatic esters. However, since simple esters are resistant to hydrolysis, hydrolysis would not be expected to be a significant degradative under environmental conditions [13]. The estimated basic hydrolysis constant (pH > 8) at 25 °C is $5.723 \times 10^{-2}$ 1/mole-sec [26]. At pH 9, the hydrolytic half-life would be 14.0 days. At pH 8, the half-life would be 140 days. There are only limited data on acid catalyzed hydrolysis of aliphatic esters in dilute acid. However, the available rates are not greatly influenced by steric and electronic factors and are all roughly $1 \times 10^{-4}$ 1/mole-sec at 25 °C [13]. This would indicate a half-life of 2.2 yr at pH 4. Isoamyl acetate does not contain any chromophores which absorb light >290 nm and therefore it will not be subject to direct photolysis. It should react with photochemically produced hydroxyl radicals by H-atom abstraction, having an estimated half-life of 2.7 days [2].

## Isoamyl Acetate

**Bioconcentration:** Using the estimated octanol/water partition coefficient, one can estimate a bioconcentration factor of 24 for isoamyl acetate using recommended regression equations [12]. Bioconcentration would therefore not be expected to be a significant transport process.

**Soil Adsorption/Mobility:** The Koc for isoamyl acetate calculated from its water solubility is 66 [12]. This indicates that isoamyl acetate should not adsorb significantly to soil or sediments.

**Volatilization from Water/Soil:** Using the Henry's Law constant, the estimated volatilization half-life of isoamyl acetate in a model river 1 m deep with a 1 m/s current and a 3 m/s wind speed is 5.0 hr [12]. Volatilization will be controlled by resistance in both the liquid and gas phases. The half-life in a pond or lake would be much longer. The moderate vapor pressure and low soil adsorption constant would suggest moderate evaporation from soil and other surfaces.

**Water Concentrations:** DRINKING WATER: Detected, not quantitated, in drinking water from Miami, Florida and Seattle, Washington [11].

**Effluent Concentrations:**

**Sediment/Soil Concentrations:**

**Atmospheric Concentrations:** Isoamyl acetate was identified, but not quantitated, in samples of urban air from Tuscaloosa, AL and rural air from the Talladega National Forest, AL [10].

**Food Survey Values:** Isoamyl acetate is a plant volatile [5] and has been identified as a volatile component of ripening bananas [14] and nectarines [23]. It has also been found in alcoholic beverages such as whiskey, beer, and cognac [5,6,24]. Other food products in which isoamyl acetate has been identified include fried bacon [9] and Beaufort cheese, a Gruyere type cheese manufactured in the French Alps [4]. The concentration of esters, including isoamyl acetate, in US lager beer is 25-50 ppm and is responsible for giving beer a fruity flavor [19].

**Plant Concentrations:**

**Fish/Seafood Concentrations:**

# Isoamyl Acetate

**Animal Concentrations:**

**Milk Concentrations:**

**Other Environmental Concentrations:**

**Probable Routes of Human Exposure:** Workers will be exposed to isoamyl acetate primarily during its use as a solvent. The general public will be exposed to it in food in which it occurs naturally or to which it is added as a flavoring agent and possibly in consumer products which contain isoamyl acetate as a solvent.

**Average Daily Intake:**

**Occupational Exposure:** NIOSH (NOES Survey 1981-83) has statistically estimated that 12,528 workers are exposed to isoamyl acetate in the United States [15].

**Body Burdens:**

## REFERENCES

1.  Amer Chem Soc; Chemcyclopedia 5: 284 (1986)
2.  Atkinson R; Int J Chem Kinet 19: 799-828 (1987)
3.  Carter RV, Linsky B; Atmos Environ, 8: 57-62 (1974)
4.  Dumont JP, Adda J; J Agric Food Chem 26: 364-7 (1978)
5.  Graedel TE; Chemical Compounds in the Atmosphere NY: Academic Press (1978)
6.  Hashimoto H, Kuroiwa Y; J Inst Brewing 72: 151-62 (1966)
7.  Hawley GG; Condensed Chem Dictionary 10th ed Von Nostrand Reinhold NY pp. 573 (1981)
8.  Hine J, Mookerjee PK; J Org Chem 40: 292-8 (1975)
9.  Ho CT et al; J Agric Food Chem 31: 336-42 (1983)
10. Holzer G et al; J Chromatography 142: 755-64 (1977)
11. Lucas SV; GC/MS Analysis of Organics in Drinking Water Concentrates and Advanced Waste Treatment Concentrates USEPA-600/1-84-020A (1984)
12. Lyman WJ et al; Handbook of Chem Property Estimation Methods. Environ Behavior of Organic Compounds. McGraw-Hill NY pp 9-1 to 9-85 (1982)
13. Mabey W, Mill T; J Phys Chem Ref Data 7: 383-415 (1978)
14. Macku C, Jennings WG; J Agric Food Chem 35: 845-8 (1987)
15. NIOSH; National Occupational Exposure Survey (NOES) (1988)
16. Parrish CF; Kirk-Othmer Encycl Chem Tech 3rd Ed. 21: 377-401 (1983)

17. PCGEMS; Graphical Exposure Modelling System, CLOGP3 Program, Office of Toxic Substances, U.S. EPA (1989)
18. Price KS et al; J Water Pollut Contr Fed 46: 63 (1974)
19. Reed G; Kirk-Othmer Encycl Chem Tech 3rd NY: Wiley-Interscience 24: 771-806 (1984)
20. Riddick JA et al; Organic Solvents: Physical Properties and Methods of Purification. Techniques of Chemistry. 4th Ed. New York: Wiley-Interscience pp 1325 (1986)
21. Rogers JA Jr, Fischette F Jr; Kirk Othmer Encyclopedia of Organic Compounds 3rd ed. 10: 456-88 (1980)
22. Sigma Chem Co; Biochemicals, Organic Compounds for Research and Diagnostic Reagents. MO: St. Louis. Sigma Chemical Co (1989)
23. Takeoka GR et al; J Agric Food Chem 36: 553-60 (1988)
24. TerHeide R et al; pp. 249-81 in Anal Food Bev Proc Symp 1977. Chavalambous G Ed. NY: Academic Press (1978)
25. U.S. OSHA. Occupational Safety and Health Administration. Occupational exposure to lead; respirator fit testing. Fed Reg. 47(219): 5110-19 Nov 12, (1982)
26. USEPA; Environmental Protection Agency. GEMS PCHYDRO (1989)
27. Windholz M; Merck Index; An Encyclopedia of Chemicals, Drugs and Biologicals 10th ed pp. 4960 (1983)
28. Yalkowski SH et al; ARIZONA dATABASE of Aqueous Solubility, U. Arizona, Tucson, AZ (1987)

# Isobutanol

SUBSTANCE IDENTIFICATION

**Synonyms:** 2-Methyl-1-propanol

**Structure:**

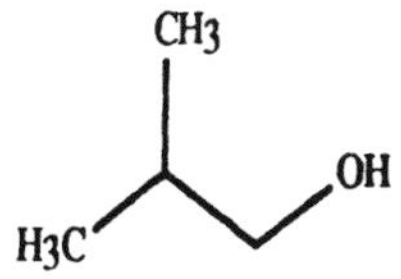

$CH_3$

$H_3C$ OH

**CAS Registry Number:** 78-83-1

**Molecular Formula:** $C_4H_{10}O$

**Wiswesser Line Notation:** Q1Y1&1

## CHEMICAL AND PHYSICAL PROPERTIES

**Boiling Point:** 108.1 °C

**Melting Point:** -108 °C

**Molecular Weight:** 74.14

**Dissociation Constants:**

**Log Octanol/Water Partition Coefficient:** 0.76 [11]

**Water Solubility:** 76.27 g/L [1]

**Vapor Pressure:** 10.41 mm Hg at 25 °C [4]

**Henry's Law Constant:** $4.0 \times 10^{-4}$ atm-m$^3$/mole [30]

## ENVIRONMENTAL FATE/EXPOSURE POTENTIAL

**Summary:** Isobutanol will enter the environment as emissions from its manufacture and use as a solvent and release in fermentation. It naturally occurs as a plant volatile and is released during the microbial degradation of animal wastes. When released into water, isobutanol will be lost by evaporation and biodegradation. It would not be expected to adsorb to sediment or bioconcentrate in fish. No data on its rate of degradation in natural waters could be found. When spilled on land it is apt to volatilize, biodegrade, and leach into ground water but no data on the rates of these processes could be found. Its fate in ground water is unknown. In the atmosphere it will photodegrade primarily by reaction with hydroxyl radicals in hours to days, with the half-life being shorter in more polluted atmospheres. Human exposure will be primarily in occupational atmospheres, ingesting food in which it is used as a fruit flavoring, or being in proximity to sites where it is released naturally (some plants, decay of animal wastes, etc.).

**Natural Sources:** Emissions from animal wastes and microbes; plant volatiles [8]. Emissions from natural fermentation of carbohydrates [25]. Product of atmospheric isobutene photo-oxidation [8].

**Artificial Sources:** Emissions from petroleum manufacture and storage, whiskey manufacturing, and wood pulping [8]. Emissions and wastewater from its manufacture and use as a chemical intermediate (principally in lacquer production) and solvent [8,16]. Emissions from wastewater treatment plants [10].

**Terrestrial Fate:** When spilled on soil, isobutanol will both evaporate and leach into the ground due to its relatively high vapor pressure and low adsorption to soil. Although it readily degrades in laboratory tests, its degradation in soil has not been determined. If degradation is not rapid, it is apt to leach into ground water.

**Aquatic Fate:** When released into water, isobutanol will volatilize (half-life in a river approximately 4 days) and biodegrade. Although it is readily degradable in laboratory tests and is reported to degrade in natural waters, no data on its rate of degradation in surface waters could be found. Its degradation in ground water is unknown.

**Atmospheric Fate:** When released into the atmosphere, isobutanol will photodegrade with a half-life ranging from hours in polluted urban atmospheres, to days in cleaner atmospheres.

**Biodegradation:** Isobutanol is readily degradable in laboratory tests using sewage [2,5,13,20,28,33], acclimated sludge [7,35] or mixed inocula containing sewage sludge, soil, and surface water [15]. The 5 day BOD values reported are generally 65% of theoretical [2,5,7,28,33] with most of the degradation occurring in 4-6 hr [13,20]. The half-life in an activated sludge die-away system was 2.5 days [35]. Degradation is also reported to occur in river water although no rates were given [9,31].

**Abiotic Degradation:** Isobutanol is considered moderately reactive (class 3 in a 5 class system (5 high)) in photochemical smog situations [6] having an ozone-forming potential slightly greater than toluene [19]. The photodegradation of aliphatic alcohols in the atmosphere is primarily due to their reaction with hydroxyl radicals [3,22]. The half-life of n-butanol in a typical sunlit atmosphere is 6 hr and several days in a clean atmosphere [3,22]. The reactivity of isobutanol would not be expected to be appreciably different. Alcohols are generally resistant to hydrolysis [22].

**Bioconcentration:** Isobutanol has a very low octanol/water partition coefficient and would therefore not be expected to bioconcentrate in fish.

**Soil Adsorption/Mobility:** Isobutanol has a very low octanol/water partition coefficient and therefore would not be expected to significantly adsorb to soil.

**Volatilization from Water/Soil:** The half-life for evaporation of isobutanol from water 1 m deep with a wind speed of 7 m/sec is 50.6 hr as measured in a wind-wave tank [23]. Using the Henry's Law constant, a half-life of 79.7 hr was calculated for evaporation of isobutanol from a river 1 m deep with a current of 3 m/sec and with a wind velocity of 3 m/sec [22]. It would readily evaporate from soil or surfaces due to its relatively high vapor pressure [4].

**Water Concentrations:** SURFACE WATER: Hayashida River, Japan-site of leather industry 685 ppb [34]. DRINKING WATER: Detected, not quantified, in Shreveport, LA and Grand Forks, ND [29].

**Effluent Concentrations:** Detected, not quantified, in air over sedimentation tank of water treatment plant [10].

**Sediment/Soil Concentrations:**

# Isobutanol

**Atmospheric Concentrations:** Measurements were conducted on volatile components employed in the production of paints at the Hiltrup plant of BASF Farben und Fasein AG.; isobutanol was detected at 1,781 ppm [17] Identified, not quantified, in the forest air of the Southern Black Forest [14].

**Food Survey Values:** Concn in dried legumes (ppb): beans, 22-300, 72 avg; split peas, 140 avg; lentils, 100 avg [21]. Identified, not quantified, in volatiles from ripening bananas [24]. Identified in Japanese beers, >1.5 ppm [12].

**Plant Concentrations:**

**Fish/Seafood Concentrations:**

**Animal Concentrations:**

**Milk Concentrations:**

**Other Environmental Concentrations:**

**Probable Routes of Human Exposure:** Humans may be exposed to isobutanol in occupational settings associated with its manufacture, use as a solvent, or when released as a product of fermentation or decomposition as in whiskey manufacture or sewage treatment plants [25]. In addition, the general public may ingest isobutanol since it is used in fruit flavorings [25].

**Average Daily Intake:**

**Occupational Exposures:** Half-shift personal samples for female operators of silk-screening printing installations for the surface marking of plastic containers, 111 half-shift personal samples, 58.6% pos, 1.5 mg/m³ avg, 3.4 mg/m³ max [32]. NIOSH (NOES Survey 1981-83) has statistically estimated that 192,949 workers are exposed to isobutanol in the United States [26]. NIOSH (NOHS Survey 1972-74) has statistically estimated that 386,225 workers are exposed to isobutanol in the United States [27].

**Body Burdens:** Expired air from normal, healthy, non-smoking human subjects, 54 subjects, 387 samples, 13.9% pos, 0.328 ng/L [18].

# Isobutanol

## REFERENCES

1.  Barton AFM; Alcohols with water. IUPAC Solubility Data Series Vol 15:438 (1984)
2.  Bridie AL et al; Water Res 13: 627-30 (1979)
3.  Campbell IM et al; Chem. Phys. Lett. 38: 362-4 (1976)
4.  Daubert TE, Danner RP; Data compilation tables of properties of pure compounds; pp 450 American Institute of Chemical Engineers (1985)
5.  Dias FF, Alexander M; Appl Microbiol 22: 1114-8 (1971)
6.  Farley FF; pp.713-27 in Inter. conf. on photochemical oxidant pollution and its control; USEPA 600/3-77-001b (1977)
7.  Gerhold RM, Malaney GW; J Water Pollut Control Fed 38: 562-79 (1966)
8.  Graedel TE; Chemical Compounds in the Atmosphere; p.245 Academic Press New York (1978)
9.  Hammerton C; J Appl Chem 5: 517-24 (1955)
10. Hangartner M; Inter J Environ Anal Chem 6: 161-9 (1979)
11. Hansch C, Leo AJ; Medchem Project Issue No 26. Claremont CA: Pomona College (1985)
12. Hashimoto H, Kuroiwa Y; J Inst Brewing 72: 151-62 (1966)
13. Hatfield R; Ind Eng Chem 49: 192-6 (1957)
14. Juttner F; Chemosphere 15: 985-92 (1986)
15. Kawasaki M; Ecotox Environ Safety 4: 444-54 (1980)
16. Kirk-Othmer; Encyclopedia of Chemical Technology; 3rd ed. John Wiley & Sons New York 4: 338-45 (1978)
17. Knappe E et al; Collect Med Leg Toxicol Med 125(Toxicovigilance Ind Part 1): 185-191 (1984)
18. Krotosynski BK et al; J Anal Toxicol 3: 225-34 (1979)
19. Laity JL et al; Adv Chem Ser 124: 95-112 (1973)
20. Langley WD; Intermediate products in the bacterial decomposition of hexadecanol and octadecanol; TR-29 NTIS PB194237 (1970)
21. Lovegren NV et al; J Agric Food Chem 27: 851-3 (1979)
22. Lyman WJ et al; Handbook of Chemical Property Estimation Methods McGraw Hill New York (1982)
23. Mackay D, Yeun ATK; Environ Sci Technol 17: 211-7 (1983)
24. Macku C, Jennings WG; J Agric Food Chem 35: 845-8 (1987)
25. Merck Index: An Encyclopedia of Chemicals and Drugs; 10th ed (1983)
26. NIOSH; The National Occupational Exposure Survey (NOES) (1983)
27. NIOSH; The National Occupational Hazard Survey (NOHS) (1974)
28. Price KS et al; J Water Pollut Control Fed 46: 63-77 (1974)
29. Shakelford WM, Reith LH; Frequency of organic compounds identified in water; p.90 USEPA 600/4-76-062 (1976)
30. Snider JR, Dawson GA; J Geophys Res 90: 3797-3805 (1985)
31. Sokolova LP, Kaplin VT; Gidrokhim Mater 51: 186-91 (1969)
32. Veulemans H et al; Scand J Work Environ Health 13: 239-42(1987)
33. Wagner R; Vom Wasser 42: 271-305 (1974)
34. Yasuhara A et al; Environ Sci Technol 15: 570-3 (1981)
35. Yonezawa Y, Urushigawa Y; Chemosphere 8: 139-42 (1979)

# Isobutylamine

## SUBSTANCE IDENTIFICATION

**Synonyms:**

**Structure:**

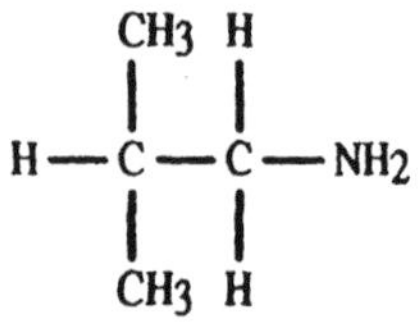

$$
\begin{array}{c}
\quad\ \ CH_3\ \ H \\
\quad\ \ |\quad\ | \\
H-C-C-NH_2 \\
\quad\ \ |\quad\ | \\
\quad\ \ CH_3\ \ H
\end{array}
$$

**CAS Registry Number:** 78-81-9

**Molecular Formula:** $C_4H_{11}N$

**Wiswesser Line Notation:** Z1Y1&1

## CHEMICAL AND PHYSICAL PROPERTIES

**Boiling Point:** 68-69 °C

**Melting Point:** -85 °C

**Molecular Weight:** 73.14

**Dissociation Constants:** pKa = 10.685 at 25 °C [20]

**Log Octanol/Water Partition Coefficient:** 0.73 [7]

**Water Solubility:** Miscible [21]

**Vapor Pressure:** 140.7 mm Hg at 25 °C [21]

**Henry's Law Constant:** 1.69 x $10^{-5}$ atm-m³/mole at 25 °C [10]

## ENVIRONMENTAL FATE/EXPOSURE POTENTIAL

**Summary:** Isobutylamine has been produced in relatively small annual quantities (ca 20,000 lb/yr). Therefore, only minor amounts may be expected to be released to the environment from its commercial production and use in organic synthesis. If released to soil, isobutylamine may be susceptible to significant leaching; it has been detected in leachate from a municipal waste disposal site. Isobutylamine may be expected to evaporate quite rapidly from dry surfaces. Sufficient data are not available to predict the significance of biodegradation in soil or natural water; the results of one screening study suggests that isobutylamine may undergo biodegradation under properly acclimated conditions. If released to water, isobutylamine is not expected to significantly adsorb to sediment or to bioconcentrate. The volatilization half-life from a model river 1 m deep flowing 1 m/sec with a wind velocity of 3 m/sec is estimated to be 1.95 days; the volatilization half-life from a similar river 10 m deep is estimated to be 23.5 days. If released to air, isobutylamine should exist almost entirely in the vapor phase due to its high vapor pressure. The half-life for the vapor-phase reaction with photochemically produced hydroxyl radicals has been estimated to be 0.478 days in a typical atmosphere. Due to the complete water solubility of isobutylamine, physical removal from the atmosphere by washout or by dissolution into clouds with subsequent rainfall should occur. General population exposure to isobutylamine may occur through oral consumption of food and beverages (such as cheese, wine, beer, coffee, smoked herring, and cooked beef) which may contain isobutylamine and through inhalation of tobacco smoke. Occupational exposure by inhalation or dermal routes may occur.

**Natural Sources:** Isobutylamine has been found to occur in various species of marine algae [24]. It has also been found to occur naturally in Latakia tobacco leaves [11]. Proteus mirabilis was also found to produce isobutylamine [13].

**Artificial Sources:** Isobutylamine has been produced on a small commercial-scale of about 20,000 lb/yr [22] for use in organic synthesis (such as in insecticides) [9]. Therefore, only minor amounts of isobutylamine may be expected to be released via wastewater and atmospheric effluents generated during production and use.

**Terrestrial Fate:** When released to soil, isobutylamine may be susceptible to significant leaching due to its complete water solubility and low Kow value. It has been detected in a leachate from a municipal waste disposal site. Sufficient data are not available to predict the significance of biodegradation in soil; The results of one screening study suggest that isobutylamine may undergo biodegradation under properly acclimated conditions. Isobutylamine may be expected to evaporate quite rapidly from dry surfaces because of its high vapor pressure.

**Aquatic Fate:** Isobutylamine is not expected to significantly adsorb to sediment or to bioconcentrate in aquatic organisms. The volatilization half-life from a model river 1 m deep flowing 1 m/sec with a wind velocity of 3 m/sec is estimated to be 1.95 days; the volatilization half-life from a similar model river 10 m deep is estimated to be 23.5 days. Sufficient data are not available to predict the significance of biodegradation in natural water; the results of one screening study suggest that isobutylamine may undergo biodegradation under properly acclimated conditions.

**Atmospheric Fate:** Isobutylamine should exist almost entirely in the vapor phase in the atmosphere due to its relatively high vapor pressure. The half-life for the vapor phase reaction of isobutylamine with photochemically produced hydroxyl radicals has been estimated to be 0.478 days in a typical atmosphere. Due to the complete water solubility of isobutylamine, physical removal from the atmosphere by dissolution into clouds with subsequent rainfall or by washout may be possible.

**Biodegradation:** Isobutylamine was biologically-oxidized by aniline-acclimated activated sludge using a Warburg respirometer [15]. The bacteria <u>Alcaligenes</u> <u>faecalis</u>, isolated from activated sludge, was able to biodegrade isobutylamine in a Warburg respirometer [16].

**Abiotic Degradation:** The half-life for the vapor phase reaction of isobutylamine with photochemically produced hydroxyl radicals has been estimated to be 0.478 days at 25 °C assuming an average atmospheric hydroxyl radical concn of $5 \times 10^{+5}$ molecules/cm$^3$ [1].

**Bioconcentration:** Based on the Kow, the BCF value for isobutylamine can be estimated to be 2.1 from a recommended regression-derived equation [14]. The complete water solubility of isobutylamine also suggests that isobutylamine will not bioconcentrate significantly.

**Soil Adsorption/Mobility:** Based on the Kow and a complete solubility in water, isobutylamine is not expected to adsorb significantly to soil or sediments. Therefore, significant leaching may be possible.

# Isobutylamine

**Volatilization from Water/Soil:** The value of the Henry's Law constant suggests that volatilization may be significant from shallow rivers [14]. The volatilization half-life from a model river 1 m deep flowing 1 m/sec with a wind velocity of 3 m/sec is estimated to be 1.95 days; the volatilization half-life from a similar river 10 m deep is estimated to be 23.5 days [14]. Based on the high vapor pressure, isobutylamine may be expected to evaporate quite rapidly from dry surfaces.

**Water Concentrations:**

**Effluent Concentrations:** Isobutylamine was detected in the leachate from a municipal refuse waste disposal site in the Netherlands at a concn of 32 ppm [8].

**Sediment/Soil Concentrations:** Isobutylamine was qualitatively detected in a loam soil sample, which was not under cultivation, from Moscow, USSR [4].

**Atmospheric Concentrations:**

**Food Survey Values:** Isobutylamine has been detected in smoked herring (0.3 mg/kg), cod (0.3 mg/kg), Camembert and Limburger cheese (0.2 mg/kg), coffee extract and freeze-dried coffee (1 mg/kg), and defatted cocoa (6 mg/kg) [17]. It has been qualitatively detected in various Italian cheeses, in German wine, in American beef, in Valencia oranges and lemons from California, and in cooked and boiled beef [3,5,6,12,19,23].

**Plant Concentrations:** Isobutylamine has been detected in various species of marine algae [24]. It has been found to occur naturally in Latakia tobacco leaves [11].

**Fish/Seafood Concentrations:**

**Animal Concentrations:**

**Milk Concentrations:**

**Other Environmental Concentrations:** Isobutylamine was qualitatively detected in tobacco smoke [18]. It was detected at a concn of 1.06 ppm in a commercially fermented egg product set to attract coyotes [2].

287

# Isobutylamine

**Probable Routes of Human Exposure:** General population exposure to isobutylamine may occur through oral consumption of food and beverages (such as cheese, wine, beer, coffee, smoked herring, and cooked beef) which may contain isobutylamine. General population exposure may also occur through inhalation of tobacco smoke. Occupational exposure may occur through inhalation of contaminated air or dermal contact.

**Average Daily Intake:**

**Occupational Exposure:**

**Body Burdens:**

## REFERENCES

1.  Atkinson R; Internat J Chem Kinetics 19: 799-828 (1987)
2.  Bullard RW et al; J Agric Food Chem 26: 155 (1978)
3.  Drawert F; Vitis 5: 127 (1965)
4.  Golovnya RV et al; USSR Acad Med Sci 1982, pp 327-35 (1982)
5.  Golovnya RV et al; Prikl Biokhim Mikrobiol 15: 295 (1979)
6.  Golovnya RV et al; Chem Senses Flavour 4: 97 (1979)
7.  Hansch C, Leo AJ; Medchem Project Issue No.26 Pomona College, Claremont CA (1985)
8.  Harmsen J; Water Res 17: 699 (1983)
9.  Hawley GG; Condensed Chemical Dictionary 10th ed Von Nostrand Reinhold NY p 576 (1981)
10. Hine J, Mookerjee PK; J Org Chem 40: 292-8 (1975)
11. Irvine WP, Saxby MJ; Phytochem 8: 473 (1969)
12. Kolbezen MJ et al; Proc Int Citrus Symp, 1st pp 1077 (1969)
13. Larsson L et al; Acta Pathol Microbiol Scand Ser B 86: 207-13 (1978)
14. Lyman WJ et al; Handbook of Chemical Property Estimation Methods. Environmental Behavior of Organic Compounds. McGraw-Hill NY p 5-4 (1982)
15. Malaney GW; J Water Pollut Control Fed 35: 1300 (1960)
16. Marion CV, Malaney GW; J Water Pollut Control Fed 35: 1269 (1963)
17. Neurath GB et al; Food Cosmet Toxicol 15: 275 (1977)
18. Pailer M et al; Monatsh Chem 97: 1448 (1966)
19. Palamand SR et al; Amer Soc Brew Chem Proc p 54 (1969)
20. Perrin DD; Dissociation Constants of Organic Bases in Aqueous Solution. 1972 Supplement. IUPAC Chemical Data Series. Buttersworth: London (1972)
21. Riddick JA et al; Organic Solvents: Physical Properties and Methods of Purification. Techniques of Chemistry. 4th Ed. New York: Wiley-Interscience pp 1325 (1986)

# Isobutylamine

22. Schweizer AE et al; Kirk Othmer Encycl Chem Technol 3rd ed. Wiley NY 21: 381 (1978)
23. Spettoli P; Ind Agric 9: 42 (1971)
24. Steiner M, Hartmann T; Planta 79: 113 (1968)

# Isobutyl Acetate

## SUBSTANCE IDENTIFICATION

**Synonyms:**

**Structure:**

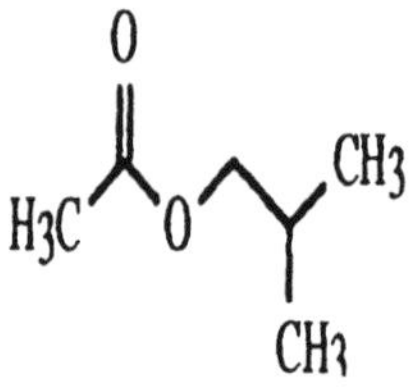

**CAS Registry Number:** 110-19-0

**Molecular Formula:** $C_6H_{12}O_2$

**Wiswesser Line Notation:** 1Y1&1OV1

## CHEMICAL AND PHYSICAL PROPERTIES

**Boiling Point:** 117.2 °C

**Melting Point:** -98.58 °C

**Molecular Weight:** 116.16

**Dissociation Constants:**

**Log Octanol/Water Partition Coefficient:** 1.60 [12]

**Water Solubility:** 6700 mg/L at 20 °C [14]

**Vapor Pressure:** 17.92 mm Hg at 25 °C [14]

**Henry's Law Constant:** 4.47 x $10^{-4}$ atm-m$^3$/mole at 25 °C [8]

## ENVIRONMENTAL FATE/EXPOSURE POTENTIAL

**Summary:** Evaporation of isobutyl acetate solvent from lacquers, painting, and other coatings, is the dominant anthropogenic emission source of isobutyl acetate into the environment. If released to soil, isobutyl acetate may be susceptible to biodegradation based on its demonstrated biodegradability in a single BOD study. Chemical hydrolysis in moist alkaline soils (pH approaching 9 or higher) may be important, but not in neutral or acidic soils. Isobutyl acetate should be subject to moderate-to-high leaching based on estimated Koc values of 36 and 177. Volatilization from dry soil surfaces is likely to be rapid. If released to water, volatilization may be expected to be an important removal mechanism. The volatilization half-life from a river 1 m deep flowing 1 m/sec with a wind velocity of 3 m/sec has been estimated to be 6.1 hr. The hydrolysis half-lives of isobutyl acetate are probably similar to the half-lives of sec-butyl acetate which are about 12.6 years, 1.26 years, and 46 days at pH's 7.0, 8.0, and 9.0, respectively, at 25 °C indicating that hydrolysis will be important only in very alkaline environmental waters. Biodegradation in natural water may be possible based on a demonstrated biodegradability in a single BOD study. Aquatic adsorption and bioconcentration are not expected to be significant. If released to air, isobutyl acetate will exist almost entirely in the vapor phase in the ambient atmosphere. The dominant removal mechanism in the atmosphere will be the vapor-phase reaction with photochemically produced hydroxyl radicals which has an estimated half-life of about 6 hr in an average atmosphere. General population exposure to isobutyl acetate can occur through consumption of food (in which it has been added as a flavoring agent) and by inhalation of contaminated air, especially in the vicinity of usage of lacquers, paints, or coatings containing isobutyl acetate solvent. Occupational exposure by inhalation and dermal routes may be significant.

**Natural Sources:** Isobutyl acetate has been identified as a major aroma-producing compound emitted from potato broth by the fungus Chalaropsis thielaviodis [4]. Animal waste (cattle) has been identified as a source [6].

**Artificial Sources:** Isobutyl acetate is used as a solvent for nitrocellulose, in topcoat lacquers, thinners, and sealants, in perfumery, and as a flavoring agent [7]; its use in many lacquer formulations is similar to n-butyl acetate except that it dries slightly faster [1]. With the exception of the flavoring agent use (which probably comprises only a small fraction of the total use volume), isobutyl acetate will evaporate directly into the surrounding air from its uses, thereby becoming the dominant anthropogenic emission source of isobutyl acetate to the environment.

# Isobutyl Acetate

**Terrestrial Fate:** Isobutyl acetate was shown to be significantly biodegradable in one BOD study which may suggest that soil microbes may be able to biodegrade the compound. Chemical hydrolysis of isobutyl acetate in moist alkaline soils (pH approaching 9 or higher) may be important, but hydrolysis in neutral or acidic soils is not expected to be significant. Based on estimated Koc values of 36 and 177, isobutyl acetate may be subject to moderate-to-high leaching. Volatilization from dry soil surfaces is likely to be rapid.

**Aquatic Fate:** Isobutyl acetate was shown to be significantly biodegradable in one BOD study which may suggest that biodegradation in natural water is possible. The volatilization half-life from a river 1 m deep flowing 1 m/sec with a wind velocity of 3 m/sec has been estimated to be 5.3 hours; the volatilization half-life from a similar river 10 m deep has been estimated to be 7.1 days. The hydrolysis half-lives of isobutyl acetate are probably similar to the half-lives of sec-butyl acetate which are about 12.6 years, 1.26 years, and 46 days at pH's 7.0, 8.0, and 9.0, respectively at 25 °C; this indicates that hydrolysis will be important only in very alkaline environmental waters. Aquatic adsorption to sediments and bioconcentration are not expected to be significant.

**Atmospheric Fate:** Isobutyl acetate will exist almost entirely in the vapor phase in the ambient atmosphere due to its relatively high vapor pressure. The half-life for the vapor-phase reaction of isobutyl acetate with photochemically produced hydroxyl radicals has been estimated to be about 6 hr in an average atmosphere indicating that this reaction will be the dominant removal mechanism.

**Biodegradation:** Using a filtered sewage seed, 5-day and 20-day theoretical BOD's of 60% and 81% were measured in freshwater dilution tests; 5-day and 20-day theoretical BOD's of 23% and 37% were measured in salt water [13].

**Abiotic Degradation:** Based on an experimentally derived base-catalyzed hydrolysis rate constant at 25 °C, the hydrolysis half-lives of sec-butyl acetate at pH's 7.0, 8.0, and 9.0 are estimated to be 12.6 years, 1.26 years, and 46 days, respectively [10]; similar hydrolysis rates should be expected for isobutyl acetate. The half-life for the vapor phase reaction of isobutyl acetate with photochemically produced hydroxyl radicals in the atmosphere has been estimated to be about 6 hr assuming an average atmospheric hydroxyl radical concn of 5 x $10^{+5}$ molecules/cm$^3$ [2].

**Bioconcentration:** Based on the Kow and the water solubility, the BCF value for isobutyl acetate can be estimated to be 9.7 and 4, respectively, by regression derived equations [9]. These BCF values suggest that bioconcentration is not significant.

**Soil Adsorption/Mobility:** Based on the Kow and the water solubility, the Koc value for isobutyl acetate can be estimated to be 177 and 36, respectively, by regression derived equations [9]. These Koc values indicate a high-to-medium soil mobility [16].

**Volatilization from Water/Soil:** The value of Henry's Law constant suggests that volatilization is probably significant from environmental bodies of water [9]. The volatilization half-life from a river 1 m deep flowing 1 m/sec with a wind velocity of 3 m/sec is estimated to be 5.3 hours [9]; the volatilization half-life from a similar river 10 meters deep is estimated to be 7.1 days [9]. Isobutyl acetate evaporates relatively rapidly from dry surfaces; the evaporative half-life of isobutyl acetate at 25 °C was found to be 2.67 min according to a standardized solvent evaporation test which classified isobutyl acetate as a medium rate evaporating solvent [5].

**Water Concentrations:** GROUND WATER: Isobutyl acetate was qualitatively detected in ground water beneath underground storage tanks of solvents at a paint factory in Italy [3].

**Effluent Concentrations:**

**Sediment/Soil Concentrations:**

**Atmospheric Concentrations:** Isobutyl acetate was detected at 4 monitoring locations in the Netherlands during 1979-81 monitoring with mean concentrations of 1.8-10 ppb found in the ambient air of Delft [15].

**Food Survey Values:**

**Plant Concentrations:**

**Fish/Seafood Concentrations:**

# Isobutyl Acetate

**Animal Concentrations:**

**Milk Concentrations:**

**Other Environmental Concentrations:**

**Probable Routes of Human Exposure:** General population exposure to isobutyl acetate can occur through oral consumption of food (to which it has been added as a flavoring agent) and inhalation of contaminated air, particularly in the vicinity of usage of lacquers, paints, and other coatings containing isobutyl acetate solvent. Dermal exposure may occur upon contact with lacquers, paints, or coatings containing isobutyl acetate. Occupational exposure by inhalation and dermal routes related to the use of isobutyl acetate as a solvent may be significant.

**Average Daily Intake:**

**Occupational Exposure:** NIOSH (NOES Survey 1981-1983) has statistically estimated that 121,883 workers are exposed to isobutyl acetate in the United States [11]. A National Occupational Hazard Survey (NOHS) conducted between 1972 and 1974 estimated that 901,213 US workers are potentially exposed to isobutyl acetate [11]. A mean concn of 0.1 ppm was detected in the workplace air of a company performing spray painting and gluing [17].

**Body Burdens:**

## REFERENCES

1. American Chemical Society; Chemcyclopedia 4: 88 (1985)
2. Atkinson R; Internat J Chem Kinetics 19: 799-828 (1987)
3. Botta D et al; Comm Eur Comm Eur 8518 (Anal Org Micropollut Water): 261 (1984)
4. Collins RP, Morgan ME; Sci 131: 933 (1960)
5. Davis DS; Am Perfumer Cosmet 81: 32 (1966)
6. Graedel TE; Chemical Compounds in the Atmosphere NY: Academica Press (1978)
7. Hawley GG; The Condensed Chemical Dictionary 10th ed (1981)
8. Hine J, Mookerjee PK; J Org Chem 40: 292-8 (1975)
9. Lyman WJ et al; Handbook of Chemical Property Estimation Methods. Environmental Behavior of Organic Compounds. McGraw-Hill NY (1982)
10. Mabey W, Mill T; J Phys Chem Ref Data 7: 383 (1978)

# Isobutyl Acetate

11. NIOSH; National Occupational Hazard Survey (1974): NIOSH; National Occupational Exposure Survey (1983)
12. PCGEMS; Graphical Exposure Modelling System, CLOGP3 Program, Office of Toxic Substances, U.S. EPA (1989)
13. Price KS et al; J Water Pollut Contr Fed 46: 63 (1974)
14. Riddick JA et al; Organic Solvents: Physical Properties and Methods of Purification. Techniques of Chemistry. 4th Ed. New York: Wiley-Interscience pp 1325 (1986)
15. Smeyers-Verbeke J et al; Atmos Environ 18: 2471 (1984)
16. Swann RL et al; Res Rev 85: 17 (1983)
17. Whitehead LW et al; Am Ind Hyg Assoc J 45: 767 (1984)

# Isophorone

## SUBSTANCE IDENTIFICATION

**Synonyms:**

**Structure:**

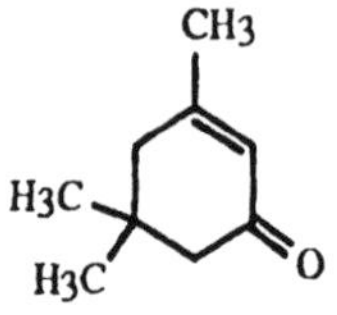

**CAS Registry Number:** 78-59-1

**Molecular Formula:** $C_9H_{14}O$

**Wiswesser Line Notation:** L6V BUTJ C1 D1 D1

## CHEMICAL AND PHYSICAL PROPERTIES

**Boiling Point:** 214 °C at 754 mm Hg

**Melting Point:** -8.1 °C

**Molecular Weight:** 138.23

**Dissociation Constants:**

**Log Octanol/Water Partition Coefficient:** 2.22 (estimated from structure) [6]

**Water Solubility:** 12,000 mg/L at 25 °C [24]

**Vapor Pressure:** 0.38 mm Hg at 20 °C [39]

**Henry's Law Constant:** $5.8 \times 10^{-6}$ atm-m$^3$/mole (approximate - calculated from the water solubility and vapor pressure)

# Isophorone

## ENVIRONMENTAL FATE/EXPOSURE POTENTIAL

**Summary:** Isophorone is used as a solvent for a large number of natural and synthetic polymers, resins, waxes, fats, oils, and pesticides, in addition to being used as a chemical intermediate. As a result this compound may be released to the environment from a wide variety of industries, from the disposal of many different products, and during the application of some pesticides. If released to soil or water, isophorone is predicted to be removed partially by volatilization (half-life 7.5 days from a model river) and partially by biodegradation. Potential biodegradation products include: 3,5,5-trimethyl-2-cyclohexene-1,4-dione, 3,5,5-trimethylcyclohexane-1,4-dione, (S)-4-hydroxy-3,5,5-trimethyl-2-cyclohexen-1-one, and 3-hydroxymethyl-5,5-dimethyl-2-cyclohexen-1-one. Potential exists for contamination of ground water by leaching through soil. Isophorone is not expected to absorb significantly to suspended solids and sediments in water, bioaccumulate significantly in aquatic organisms, photolyze, oxidize by reaction with singlet oxygen or alkylperoxy radicals in water, or undergo chemical hydrolysis. If released to air, isophorone is expected to exist primarily in the vapor phase. Reaction with ozone is expected to be the dominant removal process and reaction with photochemically generated hydroxyl radicals is expected to be of minor importance (overall reaction half-life 32 min). Isophorone emitted to the atmosphere in particulate form may be removed by wet or dry deposition. The most probable route of human exposure to isophorone in the ambient environment is by ingestion of drinking water contaminated with this compound. Worker exposure may occur by inhalation or dermal contact.

**Natural Sources:** There are no known natural sources of isophorone [1].

**Artificial Sources:** Isophorone is used extensively as a solvent for lacquers, inks, vinyl resins, copolymers, coatings and finishes, ink thinners, and pesticides [23,27]. In addition it is used as an intermediate in the manufacture of 3,5-xylenol, 3,3,5-trimethylcyclohexanol and plant growth retardants [37]. Since this compound has many different applications, release to the environment may originate from a wide variety of industrial sources including iron and steel manufacturers, photographic equipment and supplies manufacturers, automobile tire plants, and many others. Release may also result from its use in pesticide formulations or disposal of lacquers, inks, etc., containing this compound. Detection of isophorone in coal fly ash indicates that coal-fired power plants may be a source of atmospheric emissions in particulate form [9].

**Terrestrial Fate:** If released to soil, isophorone may be removed partially by volatilization and partially by biodegradation. Leaching into ground water may occur.

**Aquatic Fate:** If released to water, isophorone may biodegrade or it may volatilize (half-life 7.5 days from a shallow river). Results of biodegradation screening studies are conflicting; however, it has been shown that mixed microbial populations are capable of degrading isophorone fairly rapidly. Potential biodegradation products include 3,5,5-trimethyl-2-cyclohexene-1,4-dione, 3,5,5-trimethylcyclohexane-1,4-dione, (S)-4-hydroxy-3,5,5-trimethyl-2-cyclohexen-1-one, and 3-hydroxymethyl-5,5-dimethyl-2-cyclohexen-1-one. Adsorption to suspended solids and sediments, bioaccumulation in aquatic organisms, photolysis, oxidation by reaction with singlet oxygen or alkylperoxy radicals, and chemical hydrolysis are not expected to be important fate processes in water.

**Atmospheric Fate:** Based on the vapor pressure, isophorone is expected to exist primarily in the vapor phase in the atmosphere [4]. Isophorone is predicted to be removed form the atmosphere by reaction with ozone (half-life 39 min) and reaction with photochemically generated hydroxyl radicals (half-life 3 hr). The overall half-life for isophorone has been estimated to be 32 min [5]. Direct photolysis is not expected to occur. Isophorone emitted to the atmosphere in particulate form may be removed by wet or dry deposition.

**Biodegradation:** <30% degradation occurred after 2 weeks incubation of 100 ppm isophorone in 30 ppm activated sludge under aerobic conditions (Japanese MITI Test) [13,29]. 100% loss was observed when 5 and 10 mg/L isophorone underwent a 7-day static incubation in the dark at 25 °C under aerobic conditions using settled domestic wastewater as inoculum [33]. Removal of isophorone from unacclimated fresh and salt water seeded with settled domestic wastewater was 42 and 9%, respectively, after 20 days [25]. Removal of isophorone from wastewater treated by various different biological treatment processes (trickling filter, activated sludge, aerated lagoon, and facultative lagoon) was 19%, 98%, 24%, and 30%, respectively [8]. Metabolites identified in the degradation of isophorone by a pure culture of <u>Aspergillus niger</u> were 3,5,5-trimethyl-2-cyclohexene-1,4-

dione, 3,5,5-trimethylcyclohexane-1,4-dione, (S)-4-hydroxy-3,5,5-trimethyl-2-cyclohexen-1-one, and 3-hydroxymethyl-5,5-dimethyl-2-cyclohexen-1-one [20].

**Abiotic Degradation:** Isophorone contains no functional groups which would chemically hydrolyze under environmental conditions [2,18]. Oxidation by reaction with alkylperoxy radicals or singlet oxygen in water is not expected to be environmentally significant [18]. Isophorone in methanol exhibits almost no adsorption of UV light wavelengths >290 nm [26], which suggests that this compound does not have the potential to undergo direct photolysis under environmental conditions. The half-life for isophorone vapor reacting with photochemically generated hydroxyl radicals in the atmosphere has been estimated to be 3 hr based on a reaction rate constant of $8.14 \times 10^{-11}$ cm$^3$/molecule-sec at 25 °C and an ambient hydroxyl radical concentration of $8.0 \times 10^{+5}$ molecules/cm$^3$ [5]. The half-life for isophorone vapor reacting with ozone has been estimated to be 39 min based on a reaction rate constant of $5 \times 10^{-16}$ cm$^3$/molecule-sec at 25 °C and an ambient ozone concentration of $6.0 \times 10^{+11}$ molecules/cm$^3$ [5].

**Bioconcentration:** A bioconcentration factor (BCF) of 7 was measured for isophorone in bluegill sunfish [38]. The half-life of isophorone in fish tissue was found to be 1 day [38]. These data suggest that isophorone will not bioaccumulate significantly in aquatic organisms.

**Soil Adsorption/Mobility:** Soil adsorption coefficients (Koc) of 25 and 384 have been estimated for isophorone using linear regression equations with the water solubility and octanol/water partition coefficient, respectively [17]. These Koc values suggest that this compound would be extremely mobile in soil and that adsorption to suspended solids and sediment in water would be insignificant [32].

**Volatilization from Water/Soil:** Based on the value for the Henry's Law constant, the volatilization half-life of isophorone in a model river 1 m deep flowing 1 m/sec can be estimated to be approximately 7.5 days [17]. Based on the vapor pressure and Henry's Law constant of isophorone, this compound is expected to volatilize slowly from both wet and dry soil surfaces.

**Water Concentrations:** SURFACE WATER: USEPA STORET data base - 795 water samples, 1% pos, median concn <10.00 ug/L [10]. During Aug 1977 in Delaware River at river mile 106, 108, and 110 isophorone concentration was found to be 3, 0.6, and <0.01 ug/L, respectively [30]. Isophorone was detected in the St. Joseph River and not detected in the Cuyahoga River [7]. USEPA National Urban Runoff Program results as of July 1982 indicate that isophorone was found in runoff in 1/19 cities across US [34]. 10 ug/L was detected in urban runoff of Washington, DC [34]. DRINKING WATER: Isophorone identified in finished drinking water from: Cincinnati, OH - Oct 1978 and Jan 1980; Philadelphia, PA - Feb 1976; Ottumwa, IA - Sept 1976; and Seattle, WA - Nov 1976 [16]. During the USEPA 1974 National Organics Reconnaissance Survey isophorone was detected in 1/10 finished drinking water supplies [37]. Approximately 0.2 ug/L was found in drinking water from Cincinnati, OH [37]. Trace levels detected in Philadelphia, PA drinking water during Aug 1977 [30]. Detected in drinking water from New Orleans, LA during 1974, max concn 2.9 ug/L [3]. GROUND WATER: Identified in ground water in the Netherlands, max concn 10 ug/L [40].

**Effluent Concentrations:** USEPA STORET data base - 1272 effluent samples, 1.6% pos., median concn <10.0 ug/L [1]. Isophorone has been found in the treated wastewater from the following industries: iron and steel mfg., 1/5 samples pos., concn 170 ug/L; coil coating, 5/31 samples pos., max concn 560 ug/L, mean concn 120 ug/L; foundries, 7/7 samples pos., max concn 28 ug/L, mean concn 12 ug/L; photographic equipment and supplies, 2/4 samples pos., max concn 10 ug/L, mean concn 10 ug/L; paint and ink formulation, 1/1 samples pos., concn <7 ug/L; automobile tire plant, 40 ug/L; and oil shale retorting, 340-5800 ug/L [11,12,36]. The influent and effluent of the Philadelphia, PA Northeast Sewage Treatment plant during Aug 1977 were 100 and 10 ug/L, respectively [30].

**Sediment/Soil Concentrations:** Isophorone was qualitatively identified in sediment/soil/water samples taken from Love Canal in Niagara Falls, NY during 1980 [10]. USEPA STORET data base - 318 sediment samples, 0% pos. [31]. Detected in sediments taken from Lake Pontchartrain, LA, concn range 0.98-12 ng/g (ppb) dry wt. [19].

**Atmospheric Concentrations:** Isophorone has been detected in coal fly ash at a concentration of 490 ug/g [9].

**Food Survey Values:**

**Plant Concentrations:**

300

# Isophorone

**Fish/Seafood Concentrations:** USEPA STORET data base - 123 samples of biota, 0% pos [10].

**Animal Concentrations:**

**Milk Concentrations:**

**Other Environmental Concentrations:**

**Probable Routes of Human Exposure:** The most probable route of exposure to isophorone by the general population is ingestion of contaminated drinking water. Worker exposure may occur by inhalation or dermal contact.

**Average Daily Intake:**

**Occupational Exposures:** Concentration of isophorone in breathing zone samples from an isophorone manufacturing plant (Exxon Chemical) were reported to range from 0.01-0.63 ppm, mean concn 0.07 ppm [35]. Time-weighted average (TWA) concentration in breathing zones and workplace air of a screen printing plant ranged from 8.3-23 ppm and 3.5-14.5 ppm, respectively [28]. Up to 25.7 ppm was detected in air of a silk screen printing plant in Pittsburgh, PA [14]. Concentration in breathing zone samples from a decal manufacturing plant in Ridgefield, NJ was 0.7-14 ppm [15]. 40,007 workers are potentially exposed to isophorone based on statistical estimates derived from the NIOSH Survey conducted 1981-83 in the United States [21]. 1,004,686 workers are potentially exposed to isophorone based on statistical estimates derived from the NIOSH Survey conducted 1972-74 in the United States [22].

**Body Burdens:**

## REFERENCES

1.  Abrams EF et al; Identification of organic compounds in effluents from industrial sources. USEPA-560/3-75-002 (1975)

2. Callahan MA et al; Water-related environmental fate of 129 priority pollutants. USEPA-440-4-79-029a (1979)
3. Cole RH et al; J Water Pollut Control Fed 56: 898-908 (1984)
4. Eisenreich SJ et al; Environ Sci Tech 15: 30-8 (1981)
5. GEMS; Graphical Exposure Modeling System. FAP. Fate of Atmos Pollut (1987)
6. GEMS; Graphical Exposure Modeling System, CLOGP Program (1987)
7. Great Lakes Quality Board; An inventory of chemical substances identified in the Great Lakes ecosystem Vol 1 summary. p.195 (1983)
8. Hannah SA et al; J Water Pollut Control Fed 58:27 (1986)
9. Harrison FL et al; Environ Sci Tech 19: 186-93 (1985)
10. Hauser TR, Bromberg SM; Environ Monit Assess 2: 249-72 (1982)
11. Hawthorne SB, Sievers RE; Environ Sci Tech 18: 483- 90 (1984)
12. Jungclaus GA et al; Anal Chem 48: 1894-6 (1976)
13. Kawasaki M; Ecotox Environ Safety 4: 44-54 (1980)
14. Kominsky JR; National Institute of Occupational Safety and Health (NIOSH) Health Hazard Report No. HE78-107-563 NTIS PB 81-14371 (1981)
15. Lee SA, Frederick L; NIOSH Health Hazard Report No. HHE80-103-827 NTIS PB82-189226 (1982)
16. Lucas SV; GC/MS analysis of organics in drinking water concentrates and advanced water treatment concentrates Vol 3. p.186 USEPA-600/1-84-020 NTIS PB85 128221 (1984)
17. Lyman WJ et al; Handbook of Chem Property Estimation Methods NY: McGraw-Hill (1982)
18. Mabey WR et al; Aquatic fate process data for organic priority pollutants p.434 USEPA-440/4-81-014 (1981)
19. McFall JA et al; Chemosphere 14: 1561-9 (1985)
20. Mikami Y et al; Agric Biol Chem 45: 791-3 (1981)
21. NIOSH; National Occupational Exposure Survey (NOES) (1983)
22. NIOSH; National Occupational Hazard Survey (NOHS) (1974)
23. Papa AJ, Sherman PD; Kirk-Othmer Encycl of Chem Tech 3rd ed NY: Wiley 13:921 (1981)
24. Parrish CF; Kirk-othmer Encycl Chem Tech 3rd ed NY: Wiley 21: 385 (1983)
25. Price KS et al; J Water Pollut Control Fed 46: 63-77 (1974)
26. Sadtler; Sadtler Standard Spectra UV
27. Samimi B; Amer Ind Hyg Assoc J 43: 43-8 (1982)
28. Samimi B; Amer Ind Hyg Assoc J 43: 858-62 (1982)
29. Sasaki S; pp.283-91 in Aquatic Pollutants: Transformation and Biological Effects. Hutzinger O et al eds Oxford: Pergamon Press (1978)
30. Sheldon LS, Hites RA; Environ Sci Tech 13: 574-9 (1979)
31. Staples CA et al. Environ Toxicol Chem 4: 131-42 (1985)
32. Swann RL et al; Res Rev 85: 17-28 (1983)
33. Tabak HH et al; J Water Pollut Control Fed 53: 1503-18 (1981)
34. USEPA; New Orleans area water supply study draft (1979)
35. USEPA; Exxon Chemical Americas. Office of Toxic Substances Microfiche No. 206267 (1982)
36. USEPA; Treatability Manual I. USEPA-600/2-82-001a (1981)

# Isophorone

37. USEPA; Ambient Water Quality Criteria for Isophorone. USEPA-440/5-80-056 NTIS PB81 11767j (1980)
38. Veith GD et al; pp.116-29 in Aquatic Toxicology Easton JG et al eds; ASTM STP 707 (1980)
39. Verschueren K; Handbook of Environmental Data on Organic Chemicals 2nd ed NY: Van Nostrand Reinhold p.773 (1983)
40. Zoeteman BCJ et al; Sci Total Environ 21: 187-202 (1981)

# Isopropanol

## SUBSTANCE IDENTIFICATION

**Synonyms:** Isopropyl alcohol

**Structure:**

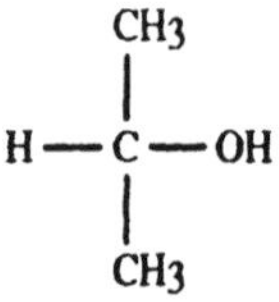

$$H - \overset{\overset{\displaystyle CH_3}{|}}{\underset{\underset{\displaystyle CH_3}{|}}{C}} - OH$$

**CAS Registry Number:** 67-63-0

**Molecular Formula:** $C_3H_8O$

**Wiswesser Line Notation:** QY1&1

## CHEMICAL AND PHYSICAL PROPERTIES

**Boiling Point:** 82.5 °C at 760 mm Hg

**Melting Point:** -88.5 °C

**Molecular Weight:** 60.09

**Dissociation Constants:** pKa = 17.1 at 25 °C [29]

**Log Octanol/Water Partition Coefficient:** 0.05 [11]

**Water Solubility:** Miscible [26]

**Vapor Pressure:** 43 mm Hg at 25 °C [26]

**Henry's Law Constant:** $8.07 \times 10^{-6}$ atm-m³/mole [13]

## ENVIRONMENTAL FATE/EXPOSURE POTENTIAL

**Summary:** Isopropanol will enter the environment as emissions from its manufacture and use as a solvent. It naturally occurs as a plant volatile and is released during the microbial degradation of animal wastes. When released on land it is apt to volatilize and leach into ground water and possibly biodegrade but no data on the rates of these processes were found. Its fate in ground water is unknown. When released into water, isopropanol will volatilize and biodegrade. It would not be expected to adsorb to sediment or bioconcentrate in fish. Although no data on its rate of degradation in natural waters could be found, laboratory tests suggests that it is not very long lived in water. In the atmosphere it will photodegrade primarily by reaction with hydroxyl radicals with a half-life of one to several days. Human exposure will be both in occupational atmospheres and from use of consumer products containing isopropanol as a volatile solvent.

**Natural Sources:** Plant volatile, animal waste, and emissions from volcanoes and microbes [9].

**Artificial Sources:** Emissions from petroleum storage, auto exhaust, plastics combustion, printing, sewage treatment, wood pulping, and steel manufacturing, as well as use as a solvent including consumer products [9,16]. Leachate from landfills [27,28]; wastewater from shale tar distillation [15].

**Terrestrial Fate:** When spilled on soil, isopropanol will both evaporate quickly and leach into the ground due to its high vapor pressure and low adsorption to soil. Degradation in soil and ground water has not been determined. If soil degradation is not rapid, it is apt to leach into the ground water.

**Aquatic Fate:** When released into water, isopropanol will volatilize (estimated half-life approximately 5.4 days) and may biodegrade. Although it is readily degradable in laboratory tests, no data on its degradability in natural waters could be found.

**Atmospheric Fate:** When released into the atmosphere, isopropanol will photodegrade with an estimated half-life ranging from one to several days. Due to its solubility in water, rainout may be significant.

**Biodegradation:** Degradation with sewage at 20 °C for 5 days resulted in 58% theoretical BOD (average of 4 results) [12]. Filtered sewage seed resulted in 49% theoretical BOD and acclimated sewage seed resulted in 72% of theoretical BOD after 5 days [4]. Concentrations of 3, 7, and 10 mg/L with filtered sewage seed in fresh water resulted in 28% theoretical BOD in 5 days and 78% in 20 days. In salt water 13%

theoretical BOD in 5 days and 72% in 20 days was observed [25]. Biodegradation with activated sludge and wastewater gave similar results [8,24,32]. Isopropanol was 99% degraded with acclimated activated sludge at 20 °C (52 mg COD/g-hr rate) [24]. Available information suggests that isopropanol is readily degraded in these systems also [5]. Anaerobic degradation (thermophilic digestion, 54 °C) of isopropanol (5 mL of a 5% aqueous isopropanol solution) produced approx 60 mL gas/g sample using seed which had been prepared in a synthetic medium [30]. Percent removal in semi-pilot scale anaerobic lagoons: dilute wastes (includes 60 ppm isopropanol), 50% avg removal in 7.5-10 days; concentrated wastes (includes 175 ppm isopropanol), 69-74% removal in 20-40 days [14].

**Abiotic Degradation:** The estimated half-life of isopropanol in the atmosphere ranges from 3.5 days [9] to 33 hr [3] based on a hydroxyl ion concentration of $0.8 \times 10^{+6}$ molecules/cm$^3$. Isopropanol is considered to have low reactivity (class 2 in a 5 class system (5 high) [7]) in photochemical smog situations [7,19,20] having an ozone forming potential 68% that of toluene [7]. A 20% decrease in isopropanol was observed after 5 hr in a smog chamber at 30 °C containing 2 ppm isopropanol and 1 ppm NO$_x$ at 55% relative humidity [34]. Reaction with hydroxyl radicals in aquatic media will not likely be a significant process [1,6]. Alcohols are known to be resistant to hydrolysis [20].

**Bioconcentration:** No information on the bioconcentration factor for isopropanol could be found in the literature. Its low octanol/water partition coefficient indicates that it will not bioconcentrate in aquatic organisms.

**Soil Adsorption/Mobility:** No information on the adsorption of isopropanol on soils and sediments could be found in the literature. Its low octanol/water partition coefficient indicates that its adsorption will be low.

**Volatilization from Water/Soil:** The estimated half-life for evaporation of isopropanol from water 1 m deep with a 1 m/sec current and 3 m/sec wind is 3.6 days; the gas exchange rate plays a more dominant role than the liquid exchange rate [20]. Isopropanol is relatively volatile and would therefore readily evaporate from dry soil and surfaces.

**Water Concentrations:** DRINKING WATER: detected in trace quantities in some samples of unspecified sources [15]. SURFACE

# Isopropanol

**WATER:** Identified, not quantified, in Cuyahoga River (tributary of Lake Erie) [10].

**Effluent Concentrations:** Leachate from 5 Connecticut town landfill sites, 1 of 5 pos, 3.4 ppm [28]. Tar water from distillation of shale tar [15]. Gas emission streptomycin production, 2.7% of total [15]. Isopropanol was found in leachate from Minnesota landfills in the range of 94-41,000 mg/L (6/6 samples pos) [27]. It was also found in ground water suspected of leachate contamination (based on levels of inorganics) - 8.6-2600 mg/L (6 of 13 samples pos), but not detected in ground water where inorganics levels indicted good or unknown water quality [27].

**Sediment/Soil Concentrations:**

**Atmospheric Concentrations:** Industrial - Linden, NJ 4-59 ng/m³ (23 ng/m³ avg, 4 of 12 samples pos) [23]. Air pollution peak - Japan, 13-24 ppb [2]. Industrial - polypropylene plant, up to 340 mg/m³ [15]. Identified, not quantified, in forest air of the Southern Black Forest, W. Germany, Nov 1984 and Jan 1985 [18]. Stockholm, Sweden, Aug 1982 - Jun 1983 (ppb): 2 busy streets, 56 and 96 samples, 14.3 and 5.66 avg all samples, ranges 2.84-44.0 and 0.28-33.3, respectively; 3 locations with moderate to little traffic: 56, 24, and 56 samples, 3.13, 0.62, and 0.30 avg, ranges 0.34-10.4, 0.25-1.59, and 0.05-1.19, respectively [17].

**Food Survey Values:** Roasted filbert nuts and milk products, detected in volatiles [15].

**Plant Concentrations:** Plant volatile.

**Fish/Seafood Concentrations:**

**Animal Concentrations:**

**Milk Concentrations:**

**Other Environmental Concentrations:**

**Probable Routes of Human Exposure:** Humans are exposed to isopropanol through contact with many commercial products such as cosmetics, hair tonics, pharmaceuticals, and antifreezes [31]. Rubbing alcohol is 70% isopropanol and 30% water [31]. Occupational exposure may also be significant.

**Average Daily Intake:**

**Occupational Exposures:** NIOSH (NOES Survey 1981-83) has statistically estimated that 4,122,567 workers are exposed to isopropanol in the United States [21]. NIOSH (NOHS Survey 1972-74) has statistically estimated that 2423 workers are exposed to isopropanol in the United States [21]. Working atmospheric concentrations of 206-771 ppm have been determined [15]. Exposure to isopropanol in spray painting averaged 1.9 ppm (TWA; highest sample 22 ppm) for higher-aromatic paints and averaged 4.3 ppm (TWA; highest sample 246 ppm) for lower aromatic paints; for solvent wiping exposure averaged 1.0 ppm(TWA; highest sample 5.1 ppm); for paint mixing exposure average 1.7 ppm(TWA; highest sample, 4.1 ppm) [33].

**Body Burdens:** Isopropanol detected in all 8 samples of human milk from 4 urban areas [22].

## REFERENCES

1.  Anbar M, Neta P; Int J Appl Radiat Isotopes 18: 493-523 (1967)
2.  Anonymous; Kanagawa-Ken Taiki Osen Chosa Kenkyu Hokoku 20: 86-90 (1978)
3.  Atkinson R et al; Adv Photochem 11: 375-488 (1979)
4.  Bridie AL et al; Water Res 13: 627-30 (1979)
5.  Chou WL et al; Biotech Bioeng Symp 8: 391-414 (1979)
6.  Dorfman LM, Adams LGE; Reactivity of the hydroxyl radical in aqueous solution p.51 NSRD-NBS-46 (1973)
7.  Farley FF; Int Conf Photochem Oxidant Pollut Control pp.713-27 USEPA-600/3-77-001b (1977)
8.  Gerhold RM, Malaney GW; J Water Pollut Control Fed 38: 562-79 (1966)
9.  Graedel TE; Chemical compounds in the atmosphere Academic Press New York p.244 (1978)
10. Great Lakes Quality Review Board; An Inventory of Chemical Substance Identified in the Great Lakes Ecosystem Vol 1 - Summary. Windsor Ontario, Canada (1983)
11. Hansch C, Leo AJ; Medchem Project Issue No 26. Claremont CA: Pomona College (1985)
12. Heukelekian H, Rand MC; J Water Pollut Control Assoc 29: 1040-53 (1955)
13. Hine J, Mookerjee PK; J Org Chem 40: 292-8 (1975)
14. Hovious JC et al; Anaerobic Treatment of Synthetic Organic Wastes USEPA-12020 DIS 01/72 p. 41 (1972)

# Isopropanol

15. IARC; Monograph. Some Fumigants, the Herbicides 2,4-D and 2,4,5-T, Chlorinated Dibenzodioxins and Miscellaneous Industrial Chemicals 15: 223-43 (1977)
16. Ishida T et al; Okayama-Ken Kankyo Hoken Senta Nempo 2: 265 (1978)
17. Jonsson A et al; Environ Inter 11: 383-92 (1985)
18. Juttner F; Chemosphere 15: 985-92 (1986)
19. Levy A; Amer Chem Soc Ser 124: 70-94 (1973)
20. Lyman WJ et al; Handbook of chemical property estimation methods Environmental behavior of organic compounds McGraw Hill New York (1982)
21. NIOSH; National Occupational Hazard Survey (1974): NIOSH; National Occupational Exposure Survey (1983)
22. Pellizzari ED et al; Bull Environ Contam Toxicol 28: 322-8 (1982)
23. Pellizzari ED; Quantification of Chlorinated Hydrocarbons in Previously Collected Air Samples USEPA 450/3-78-112 (1978)
24. Pitter P; Water Res 10: 231-5 (1976)
25. Price KS et al; J Water Pollut Control Fed 46: 63-77 (1974)
26. Riddick JA et al; Organic Solvents: Physical Properties and Methods of Purification, 4th Edit. New York: J Wiley & Sons (1986)
27. Sabel GV, Clark TP; pp 108-25 in Ann Madison Conf App Res Pract Munic Ind Waste 6th (1983)
28. Sawhney BL, Kozloski RP; J Environ Qual 13: 349-52 (1984)
29. Serjeant EP, Dempsey B; IUPAC Chem Data Series No. 23 (1979)
30. Sonoda Y, Seiko Y; J Ferment Technol 46: 796-801 (1968)
31. USEPA; CHIP Isopropyl Alcohol pp.159 (1979)
32. Wagner R; Vom Wasser 42: 271-305 (1974)
33. Whitehead LW et al; Am Ind Hyg Assoc J 45: 767-72 (1984)
34. Yanagihara S et al; pp.472-7 in Proc Int Clean Air Congr 4th (1977)

# Methanol

**Synonyms:**

**Structure:**

```
          H
          |
H ——— C ——— OH
          |
          H
```

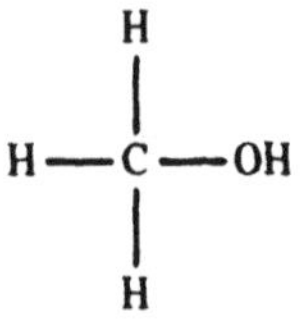

**CAS Registry Number:** 67-56-1

**Molecular Formula:** $CH_4O$

**Wiswesser Line Notation:** Q1

## CHEMICAL AND PHYSICAL PROPERTIES

**Boiling Point:** 64.7 °C at 760 mm Hg

**Melting Point:** -97.8 °C

**Molecular Weight:** 32.04

**Dissociation Constants:**

**Log Octanol/Water Partition Coefficient:** -0.77 [17]

**Water Solubility:** Miscible [32]

**Vapor Pressure:** 92 mm Hg at 20 °C [56]

**Henry's Law Constant:** $1.35 \times 10^{-4}$ atm-m³/mole at 25 °C [45]

## ENVIRONMENTAL FATE/EXPOSURE POTENTIAL

**Summary:** Methanol has been identified as a natural emission product from various plants and as a biological decomposition product of biological wastes and sewage. The largest anthropogenic source of

310

methanol release to the environment is evaporation from solvent uses (1.1 billion lb/yr). If released to the atmosphere, methanol degrades via reaction with photochemically produced hydroxyl radicals with an approximate half-life of 17.8 days. Physical removal from air can occur via rainfall. If released to water, significant decomposition via biodegradation is expected to occur in conjunction with volatilization. Volatilization half-lives of 5.3 hr and 2.6 days have been estimated for a model river (1 m deep) and an environmental pond, respectively. If released to soil, methanol is expected to degrade significantly via biodegradation and be susceptible to significant leaching. Relatively rapid evaporation from dry surfaces is likely to occur. Occupational and general exposure occurs through inhalation and dermal contact. Exposure also occurs through consumption of various foods and waters.

**Natural Sources:** Methanol has been identified as a volatile emission product from evergreen cypress trees [19]. Methanol is formed during biological decomposition of biological wastes, sewage, and sludges [1]. Natural emission sources include volcanic gases, vegetation, microbes, and insects [14].

**Artificial Sources:** The largest anthropogenic source of methanol release to the environment is evaporation from solvent uses which amounts to an estimated 1.1 billion lbs annually [50]. Annual emission releases from methanol production, end-product mfg, and storage/handling have been estimated to be 68, 49, and 12 million lb, respectively [50]. Methanol is emitted in exhaust from gasoline and diesel engines [21]. Other artificial sources include combustion of biomass, refuse, and plastics; manufacture of petroleum, charcoal, plastics, and starch; rendering; and wood pulping [14].

**Terrestrial Fate:** Methanol is expected to be significantly biodegradable in soil based on the results of a large number of biological screening studies, which include soil microcosm studies. Its miscibility in water and low octanol/water partition coefficient suggest high mobility in soil. The relatively high vapor pressure suggests significant evaporation from dry surfaces.

**Aquatic Fate:** The important environmental fate processes for methanol in water are biodegradation and volatilization. A large number of screening studies have found methanol to be rapidly biodegraded. Volatilization half-lives of 5.3 hr and 2.6 days have been estimated for a model river (1 m deep) and an environmental pond, respectively.

# Methanol

Aquatic hydrolysis, oxidation, photolysis, adsorption to sediment, and bioconcentration are not significant.

**Atmospheric Fate:** Methanol is expected to exist almost entirely in the vapor phase in the ambient atmosphere, based on its vapor pressure [9]. It is degraded by reaction with photochemically produced hydroxyl radicals with an estimated half-life of 17.8 days in a typical ambient atmosphere. Atmospheric methanol can also react with nitrogen dioxide in polluted air to yield methyl nitrite. Because of methanol's water solubility, rain would be expected to physically remove some from the air [50]; the detection of methanol in a thunderstorm water tends to confirm this supposition.

**Biodegradation:** Standard dilution BOD water, 5-day 48% BODT, sewage inocula [8]. Warburg respirometer, 2-day 93% BODT, activated sludge inocula [11]. Warburg respirometer, 1-day 21% BODT, activated sludge inocula [12]. Standard dilution BOD water, 5-day 53.4% BODT, 50-day 97.7% BODT, sewage inocula [27]. Warburg respirometer, 0.96-day 55% BODT, activated sludge inocula acclimated to methanol [31]. Standard dilution BOD water, 5-day 76% BODT, 20-day 97% BODT, sewage inocula [41]. Respirometric dilution, 5-day 82.9% BODT, sewage inocula [55]. Sewage die-away, 0.4 day half-life, sewage inocula [55]. Anaerobic-water, 75-80% degradation, sewage inocula [4]. Biological treatment simulation, 80% degradation, adapted activated sludge [46]. Anaerobic-water die-away, marinewater and sediment from the San Francisco Bay inocula, 3-day incubation, 83-91% degradation [38]. Standard dilution, 5-day 88.7% BODT; seawater dilution, 5-day 70.7% BODT [48]. Significant biodegradation of organic waste (methanol + acetic acid + formic acid) observed when injected into wells (850-1000 ft depth) as determined by concn monitoring and microbial population count [7]. Methanol found to be susceptible to biodegradation in subsurface regions in microcosm studies simulating subsurface conditions; complete degradation within one year or less [36]. Methanol degraded readily in test tube microcosms simulating subsurface soils and ground waters from sites in Virginia and New York [13]. Soil-sediment suspensions, aerobic conditions, 5-day $CO_2$ evolution ($^{14}C$) of 53.4% [42]; soil-sediment suspensions, anaerobic conditions, 5-day $CO_2$ evolution ($^{14}C$) of 46.3% [42].

**Abiotic Degradation:** The experimentally recommended rate constant for the vapor-phase reaction of methanol with photochemically produced hydroxyl radicals has been reported to be $0.9 \times 10^{-12}$ $cm^3$/molecule-sec at 25 °C [2]; the atmospheric half-life for this reaction can be estimated to be 17.8 days, assuming an average atmospheric hydroxyl radical concn of $5 \times 10^{+5}$ molecules/$cm^3$ [2]. Formaldehyde is formed from the reaction of methanol with hydroxyl radicals in the atmosphere [2]. The reaction of methanol with nitrogen dioxide may be the major source of methyl nitrite found in polluted atmospheres [47]. The rate constant for the reaction of methanol with hydroxyl radicals in aqueous solution is approximately $1 \times 10^{+9}$ 1/mol-sec [15]; if the hydroxyl radical concn of sunlit natural water is assumed to be $1 \times 10^{-17}$ mol/L [33], the half-life would be approximately 2.2 years. Methanol in aqueous solution exhibited no degradation when exposed to sunlight using an EPA test protocol [18]. Sediment an clay suspensions solution did not photocatalyze the degradation of methanol in aqueous solution during irradiation with UV light [37]. Alcohols are generally resistant to environmental aqueous hydrolysis [30].

**Bioconcentration:** The BCF of methanol experimentally measured in fish (golden ide) was less than 10 [10]. Based on the octanol/water partition coefficient, the BCF value for methanol can be estimated to be 0.2 from a recommended regression-derived equation [30].

**Soil Adsorption/Mobility:** Methanol is completely miscible in water and has a low octanol/water partition coefficient. These properties are indicative of high mobility in soil.

**Volatilization from Water/Soil:** The value of Henry's Law constant indicates that volatilization from environmental waters may be significant [30]. The volatilization half-life from a model river (1 meter deep flowing 1 m/sec with a wind speed of 3 m/sec) has been estimated to be 5.3 hr [30]. The volatilization half-life from an environmental pond has been estimated to be 2.6 days [53].

**Water Concentrations:** DRINKING WATER: Methanol has been qualitatively detected in drinking water from Miami, FL, Seattle, WA, Philadelphia, PA, Cincinnati, OH and New Orleans, LA [52,51]. As part of the USEPA National Organics Reconnaissance Survey (NORS), methanol was detected in 6 of 10 drinking waters from US cities [3]. RAIN WATER: Methanol was detected at a mean level of 22 ppb in thunderstorm water collected from Santa Rita, AZ in Sept, 1982 [45].

# Methanol

**Effluent Concentrations:** Methanol levels of 18-70 ppm were detected in wastewater effluents from a chemical manufacturing facility (near the Brackish River), but none were detected in associated river water or sediments [22]. Methanol has been identified in wastewater effluents from chemical, paper, and latex manufacturing plants and from sewage treatment plants [44]. Concn of 42.4 ppm detected in leachate from the Love Canal in Niagara Falls, NY [54]. Concn of 1050 ppm detected in condensate waters from a coal-gasification plant [34]. Levels of 0.1-0.6 ppm were found in exhausts from engines using simple hydrocarbon fuels [43]. Methanol has been identified in exhausts from both gasoline and diesel engines [21].

**Sediment/Soil Concentrations:**

**Atmospheric Concentrations:** Methanol was detected at mean ambient atmospheric concn of 7.9 and 2.6 ppb at Tucson, AZ and two remote Arizona locations, respectively, during 1982 monitoring [45]. Concn of 0.0-1.2 ppb (ave 0.77 ppb methanol + ethanol) were identified in arctic air from Point Barrow, Alaska in Sept 1967 [5]. Ave ambient concn of 3.83-26.7 ppb detected at 5 sites in and around Stockholm, Sweden [21]. Methanol has been detected (concn not reported) in indoor air of residential and office buildings [20,49].

**Food Survey Values:** Methanol has been identified as a volatile component of dried legumes (concn 1.5-7.9 ppm), baked potatoes, and roasted filbert nuts [6,23,29].

**Plant Concentrations:** Methanol has been identified as a volatile emission product from evergreen cypress trees [19] and alfalfa [39].

**Fish/Seafood Concentrations:**

**Animal Concentrations:**

**Milk Concentrations:**

**Other Environmental Concentrations:** Methanol was identified as a component of several industrial paint strippers [16]. Engine exhausts from both gasoline and diesel vehicles have been found to contain methanol [21]. Methanol has been identified as a constituent of tobacco smoke [14].

# Methanol

**Probable Routes of Human Exposure:** The general population is exposed to methanol through inhalation of air, through consumption of various drinking waters and foods, and through dermal contact of various consumer products such as paint thinners and strippers, adhesives, cleaners, and inks. Widespread occupational exposure occurs through inhalation and dermal contact.

**Average Daily Intake:** AIR INTAKE: assume 1.0-25.0 ppb (0.76-19 ug/m³): 15.2-380 ug; WATER INTAKE: insufficient data; FOOD INTAKE: insufficient data.

**Occupational Exposures:** NIOSH (NOES Survey 1981-83) has statistically estimated that 1,219,371 workers are exposed to methanol in the United States [35]. 2,062,431 workers are potentially exposed to methanol based on statistical estimates derived from the NIOSH Survey conducted between 1972-74 in the United States [35]. In a survey conducted between 1978-82 of solvent products used in industrial workplaces and having worker exposure, methanol was identified in 9.8% of the 275 solvent samples collected by factory inspectors [28]; the products represented solvent classes such as thinners, degreasers, paints, inks, and adhesives [28].

**Body Burdens:** Methanol was detected in 1 of 12 samples of human milk collected from volunteers in 4 US cities [40]. Methanol has been detected in expired human air [24,25,26]; in one study, it was detected in 3.6% of 387 expired air samples collected from 54 volunteers at a geometric mean concn of 0.549 ng/L [24].

## REFERENCES

1. Abrams EF et al; Identification of Organic Compounds in Effluents from Industrial Sources EPA- 560/3-75-002 p.102 (1975)
2. Atkinson RA; Chem Rev 85: 60-201 (1985)
3. Bedding ND et al; Sci Total Environ 25: 143-67 (1982)
4. Bekes J et al; Proc Hung 15th Annu Meet Biochem p.27-8 (1975)
5. Cavanagh LA et al; Environ Sci Technol 3: 251-7 (1969)
6. Coleman EC et al; J Agric Food Chem 29: 42-8 (1981)
7. Ditommaso A, Elkan GH; Underground Waste Manage Artif Recharge, Prepr Pap Int Symp, 2nd 1: 585-99 (1973)
8. Dore M et al; Trib Cebedeau 28: 3-11 (1975)
9. Eisenreich SJ et al; Environ Sci Technol 15: 30-8 (1981)
10. Freitag D et al; Chemosphere 14: 1589-1616 (1985)
11. Gellman I, Heukelekian H; Sew Indust Wastes 27: 793-801 (1955)

# Methanol

12. Gerhold RM, Malaney GW; J Water Pollut Control Fed 38: 562-79 (1966)
13. Goldsmith CD; Diss Abstr Int B 46: 3767 (1985)
14. Graedel TE et al; Atmospheric Chemical Compounds. Sources, Occurrence, and Bioassay. Orlando,FL: Academic Press p.232 (1986)
15. Guesten H et al; Atmos Environ 15: 1763-5 (1981)
16. Hahn WJ, Werschulz PO; Evaluation of Alternatives to Toxic Organic Paint Strippers. USEPA-600/S2-86-063 (1986)
17. Hansch C, Leo AJ; Medchem Project Issue No 26. Claremont CA: Pomona College (1985)
18. Hustert K et al; Chemosphere 10: 995-8 (1981)
19. Isidorov VA et al; Atmos Environ 19: 1-8 (1985)
20. Jarke FH et al; ASHRAE Trans 87: 153-66 (1981)
21. Jonsson A et al; Environ International 11: 383-92 (1985)
22. Jungclaus GA et al; Environ Sci Technol 12: 88-96 (1978)
23. Kinlin TE et al; J Agric Food Chem 20: 1021-8 (1972)
24. Krotosynski BK et al; J Anal Toxicol 3: 255-43 (1979)
25. Krotosynski BK, O'Neill HJ; J Environ Sci Health Part A-Environ Sci Eng 17: 855-83 (1982)
26. Krotosynski B; J Chromat Sci 15: 239-44 (1977)
27. Lamb CB, Jenkins, GF; Proc 8th Industrial Waste Conf,. Purdue Univ p.326-9 (1952)
28. Lehmann E et al; pp.31-41 in Safety Health Aspects Org Solvents. Alan R Liss Inc (1986)
29. Lovegren NV et al; J Agric Food Chem 27: 851-3 (1979)
30. Lyman WJ et al; Handbook of Chemical Property Estimation Methods NY:McGraw-Hill (1982)
31. McKinney RE, Jeris JS; Sew Indust Wastes 27: 728-35 (1955)
32. Merck Index; An Encyclopedia of Chemicals, Drugs and Biologicals 10th ed p.853 (1983)
33. Mill T et al; Science 207: 886-7 (1980)
34. Mohr DH and King J; Environ Sci Technol 19: 929-35 (1985)
35. NIOSH; National Occupational Hazard Survey (1974): NIOSH; National Occupational Exposure Survey (1983)
36. Novak JT et al; Wat Sci Tech 17: 71-85 (1985)
37. Oliver BG et al; Environ Sci Technol 13: 1075-7 (1979)
38. Oremland RL et al; Nature 296: 143-5 (1982)
39. Owens LD et al; Phytopathology 59: 1468-72 (1969)
40. Pellizzari ED et al; Bull Environ Contam Toxicol 28: 322-8 (1982)
41. Price KS et al; J Water Pollut Control Fed 46: 63-77 (1974)
42. Scheunert I et al; Chemosphere 16: 1031-41 (1987)
43. Seizinger DE, Dimitriades B; J Air Pollut Control Assoc 22: 47-51 (1972)
44. Shackelford WM, Keith LH; Frequency of Organic Compounds Identified in Water EPA-600/4-76- 062 p.169 (1976)
45. Snider JR, Dawson GA; J Geophys Res, D2, 90: 3797-3805 (1985)
46. Swain HM, Somerville HJ; J Appl Bacteriol 45: 147- 51 (1978)
47. Takagi H et al; Environ Sci Technol 20: 387-93 (1986)
48. Takemoto S et al; Suishitsu Odaku Kenkyu 4: 80-90 (1981)
49. Tsuchiya Y; Volatile Organic Compounds in Indoor Air. Am Chem Soc Div Environ Chem Preprint, New Orleans, LA 27: 183-5 (1987)
50. USEPA; Chemical Hazard Information Profiles (CHIPS). USEPA-560/11-80-011 p.196-7 (1980)

# Methanol

51. USEPA; New Orleans Area Water Supply Draft Analytical Report by the Lower Mississippi River Facility, Sliddell, LA. Dallas, TX (1974)
52. USEPA; Preliminary Assessment of Suspected Carcinogens in Water. Interim Report to Congress, June, 1975. Washington DC (1975)
53. USEPA; EXAMS II Computer Simulation (1987)
54. Venkataramani ES et al; CRC Crit Rev Environ Control 14: 333-76 (1984)
55. Wagner R; Vom Wasser 47: 241-65 (1976)
56. Weber RC et al; Vapor Pressure Distribution of Selected Organic Chemicals. USEPA-600/2-81-021 p.24 (1981)

# 2-Methoxyethanol

## SUBSTANCE IDENTIFICATION

**Synonyms:** Ethylene glycol monomethyl ether

**Structure:**

**CAS Registry Number:** 109-86-4

**Molecular Formula:** $C_3H_8O_2$

**Wiswesser Line Notation:** Q2O1

## CHEMICAL AND PHYSICAL PROPERTIES

**Boiling Point:** 125 °C at 768 mm Hg

**Melting Point:** -85.1 °C

**Molecular Weight:** 76.09

**Dissociation Constants:**

**Log Octanol/Water Partition Coefficient:** -0.77 [5]

**Water Solubility:** Miscible [13]

**Vapor Pressure:** 4.7 mm Hg at 25 °C [13]

**Henry's Law Constant:** $2.9 \times 10^{-3}$ atm-m$^3$/mole [3]

## ENVIRONMENTAL FATE/EXPOSURE POTENTIAL

**Summary:** 2-Methoxyethanol is released to the environment as emissions from manufacture and use as a solvent, including volatilization from consumer products. In the atmosphere, it will

318

photodegrade primarily by reaction with hydroxyl radicals in hours and rainout may be significant. Although no data on its rate of degradation in natural waters could be found, laboratory tests suggest that it will degrade in water. It will slowly volatilize, but will not sorb to sediment or bioconcentrate in aquatic organisms. Its fate when spilled on land is unknown, although volatilization from near-surface soil and leaching to ground water are possible. Human exposure will be primarily in occupational atmospheres and from use of certain consumer products.

**Natural Sources:**

**Artificial Sources:** Volatilization losses and wastewaters from its use as a solvent for resins, dyes, use in nail polishes, and quick drying varnishes, enamels, and woodstains [8] are possible sources of environmental release or consumer exposure.

**Terrestrial Fate:** When released on land 2-methoxyethanol is expected to volatilize from soil as well as leach rapidly into the ground. Its biodegradation in soils is unknown.

**Aquatic Fate:** When released into water, 2-methoxyethanol will slowly volatilize and will not be expected to readily adsorb to sediment or bioconcentrate in fish. Although it is degradable in laboratory tests, no data on its rate of degradation in surface waters could be found. Its degradation in ground water is unknown.

**Atmospheric Fate:** When released into the atmosphere, 2-methoxyethanol will photodegrade with an estimated half-life of less than 1 day. Based on its miscibility with water, washout by rain may be significant.

**Biodegradation:** Biodegradation of 100-1000 mg/L 2-methoxyethanol with activated sludge at 20 degrees C for 10 days resulted in 64.7% theoretical BOD [9]. Biodegradation of 2-methoxyethanol at concentrations of 3, 7, and 10 mg/L with filtered sewage seed in fresh water resulted in 30% theoretical BOD in 5 days and 88% theoretical BOD in 20 days [12]; in salt water 6% theoretical BOD in 5 days and 39% theoretical BOD in 20 days were observed [12]. Sewage seed degraded 2-methoxyethanol over 5 days resulting in 7% theoretical BOD using unadapted seed and 30% theoretical BOD using adapted seed [1]. No information is available on soil biodegradation.

# 2-Methoxyethanol

**Abiotic Degradation:** Ethers and alcohols are known to be resistant to hydrolysis [7]. The estimated half-life of 2-methoxyethanol in the atmosphere is 17.54 hr as a result of H atom abstraction by photochemically produced hydroxyl radicals [4].

**Bioconcentration:** No information on the bioconcentration factor for 2-methoxyethanol could be found in the literature; however, its low octanol/water partition coefficient indicates that it should not bioconcentrate in fish.

**Soil Adsorption/Mobility:** No information concerning the adsorption of 2-methoxyethanol could be found in the literature. Its low octanol/water partition coefficient indicates that its adsorption to soil will be low.

**Volatilization from Water/Soil:** The high Henry's Law constant would suggest that evaporation from water or moist soil is quite rapid. The calculated half-life of evaporation from a model river 1 m deep flowing at 1 m/s and with a wind speed of 3 m/s is 2.8 hours [7].

**Water Concentrations:** Identified, not quantified, in drinking water concentrates from New Orleans, LA, Jan 1976 [6].

**Effluent Concentrations:**

**Sediment/Soil Concentrations:**

**Atmospheric Concentrations:** Indoor air in Italy, 6 samples, 16.7% pos, 70 ug/m$^3$ [2].

**Food Survey Values:**

**Plant Concentrations:**

**Fish/Seafood Concentrations:**

**Animal Concentrations:**

**Milk Concentrations:**

**Other Environmental Concentrations:**

# 2-Methoxyethanol

**Probable Routes of Human Exposure:** Exposure to 2-methoxyethanol would be primarily occupational.

**Average Daily Intake:**

**Occupational Exposures:** NIOSH (NOES Survey 1981-83) has statistically estimated that 119,247 workers are exposed to 2-methoxyethanol in the United States [11]. NIOSH (NOHS Survey 1972-74) has statistically estimated that 103,459 workers are exposed to 2-methoxyethanol in the United States [10].

**Body Burdens:**

## REFERENCES

1.  Bridie AL et al; Water Res 13: 627-30 (1979)
2.  DeBortoli M et al; Environ Internat 12: 343-50 (1986)
3.  Dow Chemical Co; The Glycol Ethers Handbook Midland, MI: Dow Chemical Co (1981)
4.  GEMS. Graphical Environmental Modeling System. Fate of Atmospheric Pollutants (FAP) Data Base; USEPA Office of Toxic Substances (1985)
5.  Hansch C, Leo AJ; Medchem Project Pomona College Claremont, CA Issue No.26 (1985)
6.  Lucas SV; GC/MS (Gas Chromatography-Mass Spectrometry) Analysis of Organics in Drinking Water Concentrates and Advanced Waste Treatment Concentrates. Vol 1. p 144 USEPA-600/1-84-020a (1984)
7.  Lyman WJ et al; Handbook of Property Estimation Methods Environmental Behavior of Organic Compounds, New York, NY McGraw Hill pp.7-1 to 7-48 (1982)
8.  Merck Index; An Encyclopedia of Chemicals, Drugs, and Biologicals 10th Ed. pp.5905 (1983)
9.  Mills EJ JR, Stack VT JR; pp.492-517 in Proc Indust Waste Conf 8th Eng Bull Purdue Univ Eng Ext Ser (1954)
10. NIOSH; National Occupational Hazard Survey (NOHS) (1974)
11. NIOSH; National Occupational Exposure Survey (NOES) (1983)
12. Price KS et al; J Water Pollut Contr Fed 46: 63-77 (1974)
13. Riddick JA et al; Organic Solvents: Physical Properties and Methods of Purification, 4th Edit. New York: J Wiley & Sons (1986)

# Methylamine

## SUBSTANCE IDENTIFICATION

**Synonyms:**

**Structure:**

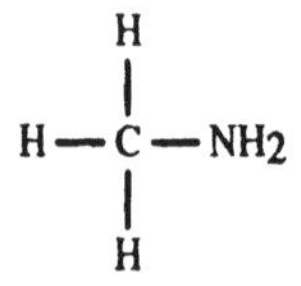

**CAS Registry Number:** 74-89-5

**Molecular Formula:** CH₅N

**Wiswesser Line Notation:** Z1

## CHEMICAL AND PHYSICAL PROPERTIES

**Boiling Point:** -6.3 °C at 760 mm Hg

**Melting Point:** -93.5 °C

**Molecular Weight:** 31.07

**Dissociation Constants:** pKa = 10.657 [21]

**Log Octanol/Water Partition Coefficient:** -0.57 [11]

**Water Solubility:** 1,080,000 mg/L at 25 °C [23]

**Vapor Pressure:** 2650 mm Hg at 25 °C [7]

**Henry's Law Constant:** $1.11 \times 10^{-5}$ atm-m³/mole at 25 °C [13]

## ENVIRONMENTAL FATE/EXPOSURE POTENTIAL

**Summary:** Methylamine is used in tanning, organic synthesis, dyeing of acetate textiles, paint removers, as an intermediate for accelerators, dyes, pharmaceuticals, insecticides, fungicides, and surface active agents, and as a fuel additive, polymerization inhibitor, photographic developer, and rocket propellent. Methylamine occurs naturally in the urine of dogs after eating meat and in certain plants such as Mentha aquatica and has been detected in a variety of vegetables and other foods. Methylamine released to the surface of soils (dissolved in water) will probably rapidly volatilize. Methylamine that enters soil is not expected to sorb to organic matter and thus may leach rapidly through soil to ground water. Biodegradation will probably be the most important degradative process. Methylamine released to water will volatilize (half-life from a model river 1.9 days). Sorption to biota and sediment will not be significant. Biodegradation will probably be the most significant degradation process in water. Hydrolysis of methylamine will not be significant. No information was found on photolysis, but methylamine should not absorb sunlight. Reaction with hydroxyl radicals will be the fastest chemical removal process for methylamine in the atmosphere (half-life 3-22 hours). Dissolution into rain droplets may be the most important physical removal process for methylamine in the atmosphere. Methylamine has been found in various European rivers, in a swamp, and in soil. It has also been found naturally occurring in a variety of foods, and ingestion of food containing methylamine should be a significant exposure route for humans.

**Natural Sources:** Methylamine occurs in urine of dogs after eating meat and in certain plants such as Mentha aquatica [17]. It is also found in a variety of vegetables and other foods [19].

**Artificial Sources:** Methylamine is used in tanning, organic synthesis [17], dyeing of acetate textiles, paint removers, as an intermediate for accelerators, dyes, pharmaceuticals, insecticides, fungicides, and surface active agents, and as a fuel additive, polymerization inhibitor, photographic developer, and rocket propellent [12].

**Terrestrial Fate:** Methylamine released to the surface of soils (dissolved in water) will volatilize rapidly. Methylamine that enters soil is not expected to sorb strongly to organic matter (estimated $K_{oc}$ is 12) and thus is expected to leach rapidly through soil to ground water. Biodegradation will probably be the most important degradative process. Hydrolysis will not be a significant removal process.

**Aquatic Fate:** Methylamine released to water will volatilize rapidly (half-life from a model river 1.9 days). Sorption to biota and sediment will not be significant (estimated BCF is 0.22). Biodegradation will probably be the most significant degradation process. Hydrolysis will not be a significant removal mechanism for methylamine. No information was found on photolysis, but methylamine should not absorb sunlight.

**Atmospheric Fate:** Reaction with hydroxyl radicals will be the fastest chemical removal process for methylamine in the atmosphere (half-life 3-22 hours). Degradation of methylamine by reaction with oxygen atoms and ozone will not be significant when compared to the rapid degradation of methylamine by hydroxyl radicals (half-life of several years versus several hours, respectively). Dissolution into rain droplets may be the most important physical removal process for methylamine in the atmosphere.

**Biodegradation:** Methylamine biodegradation was 96% in the OECD screening test and 107% in the closed bottle test (aerobic mixed cultures) [22]. Pseudomonas sp MA (an aerobe) grew on methylamine as a sole carbon, nitrogen, and energy source [5]. Sixteen non-methylotrophic bacteria, including representatives of Arthrobacter, Bacillus, Pseudomonas, and Enterobacteriaceae, were able to grow with methylamine as sole nitrogen sources in the presence of a mixture of organic compounds, but were unable to grow with methylamine as sole carbon source [6]. Methylamine may also be biodegraded under anaerobic conditions. Mixed cultures from anaerobic marine sediments [15] and pure cultures of Methanosarcina barkeri [14] degraded methylamine under anaerobic conditions.

**Abiotic Degradation:** A rate constant for aqueous hydroxyl radical reaction with methylamine in water is $1.1 \times 10^{+7}$/M-sec at pH 5 [1]. At a typical aqueous hydroxyl radical concn of $10^{-7}$ M [18], methylamine half-life would be 199 years. Hydrolysis will not be a significant degradative process for methylamine (stable pH 3,5, and 9) [22]. Under atmospheric conditions, a rate constant for the reaction of ozone with methylamine is $2.13 \times 10^{-20}$ cm$^3$/molecule-sec (296 deg K) [4]. At a typical average ozone concentration of $1.0 \times 10^{+12}$ molecules/cm$^3$ [9], methylamine half-life would be 1.03 years. A rate constant for the reaction of oxygen atoms O(3P) with methylamine in the atmosphere is $0.565 \times 10^{-12}$ cm$^3$/molecule-sec (298.0 deg K) [3]. At a typical average oxygen atom O(3P) concentration of $2.5 \times 10^4$ molecules/cm$^3$ [9], methylamine half-life would be 1.56 years. Reaction with hydroxyl radicals will be the fastest chemical removal process for methylamine in the atmosphere. A rate constant for the reaction of hydroxyl radicals with methylamine at room temperature is 2.17 x 10$^-$$^{11}$ cm$^3$/molecule-sec [2,10], resulting in half-lives of 3-22 hours (using hydroxyl radical concentrations of 3 x 10$^{+6}$ molecules/cm$^3$ [2] and 4.1 x 10$^{+5}$ molecules/cm$^3$ [9]).

**Bioconcentration:** The accumulation ratio for methylamine in mature blade sections of <u>Macrocystis pyrifera</u> ranged from 141 to 183 (methylamine concentration in the medium ranged from 2.15-8.14 uM) [24]. An estimated bioconcentration factor (BCF) for methylamine, using the Kow and a recommended regression equation [16], is 0.22. This indicates that methylamine has an extremely low potential to bioconcentrate.

**Soil Adsorption/Mobility:** An estimated soil adsorption coefficient (Koc) for methylamine, using the Kow and a recommended regression equation [16], is 12. This indicates that methylamine will not strongly adsorb to organic matter in soil or sediment and is expected to readily leach through most soils.

**Volatilization from Water/Soil:** From the Henry's Law constant, one can estimate a half-life for volatilization from a model river 1 m deep with a 1 m/sec current and a 3 m/sec wind speed of 1.9 days [16]. This indicates that volatilization from surface waters and from the surface of moist soils should be a significant removal mechanism for methylamine.

**Water Concentrations:** Methylamine concentrations in various European rivers ranged from 1-20.6 ug/kg (ppb) and was 6.2 ug/kg in a swamp [19].

**Effluent Concentrations:**

**Sediment/Soil Concentrations:** Methylamine was found in the soil (loam) of the Moscow, USSR region at an unspecified concentration [8].

**Atmospheric Concentrations:**

**Food Survey Values:** Methylamine concentrations (mg/kg) in various West German fresh vegetables were: 12 (spinach), 22.7 (red cabbage), 3.4 (cabbage), 65 (cauliflower), 16.6 (kale), 17.6 (white beet), 3.8

(carrots), 30 (red beet), 42 (large radish), and 6.4 (celery) [19]. Other methylamine concentrations (mg/kg) in West German foods were: 26.8 (maize, grains), 37.5 (green salad), 5.6 (apple flesh), 4.5 (apple peel), 12 (Camembert cheese), 3 (Limburger cheese), 27 (coffee extract), 80 (freeze-dried coffee), 16 (freeze-dried coffee), 60 (defatted cocoa), 50 (black tea), 4.5 (barley), 3.7 (hops), and <0.1 (malt) [19].

**Plant Concentrations:** Methylamine occurs in certain plants such as Mentha aquatica [17].

**Fish/Seafood Concentrations:** Methylamine concentrations (mg/kg) in various types of herring were: 2 (Bismarck), 3.4 (salted), and 7 (in oil) [19]. Methylamine concentration in cod roe (egg-laden ovary) was 10.3 mg/kg [19].

**Animal Concentrations:**

**Milk Concentrations:**

**Other Environmental Concentrations:**

**Probable Routes of Human Exposure:** Monitoring data for methylamine is limited. However, ingestion of contaminated food containing methylamine appears to be a significant exposure route.

**Average Daily Intake:**

**Occupational Exposure:** NIOSH (NOES Survey 1981-1983) has statistically estimated that 10,891 workers are exposed to methylamine in the United States [20].

**Body Burdens:**

## REFERENCES

1. Anbar M, Neta P; Int J Appl Radiation Isotopes 18: 493-523 (1967)
2. Atkinson R et al; J Chem Phys 68: 1850-3 (1978)
3. Atkinson R, Pitts JN Jr; J Chem Phys 68: 911-5 (1978)
4. Atkinson R, Carter WPL; Chem Rev 84: 437-70 (1984)
5. Bellion E et al; J Bacteriol 142: 786-90 (1980)
6. Bicknell B, Owens JD; J Gen Microbiol 117: 89-96 (1980)
7. Daubert TE, Danner RP; Data Compilation Tables of Properties of Pure Compounds. Amer Inst Chem Engn pp 450 (1985)
8. Golovnya RV et al; USSR Acad Med Sci 1982 pp. 327-35 (1982)

# Methylamine

9. Graedel TE; Chemical Compounds in the Atmosphere; New York, Academic Press (1978)
10. Guesten H et al; J Atmos Chem 2: 83-94 (1984)
11. Hansch C, Leo AJ; Medchem Project Issue No.26 Pomona College, Claremont CA (1985)
12. Hawley GG; Condensed Chemical Dictionary 10th ed. Von Nostrand Reinhold NY p 668 (1981)
13. Hine J, Mookerjee PK; J Org Chem 40: 292-8 (1975)
14. Hippe H et al; Proc Natl Acad Sci USA 76: 494-8 (1979)
15. King GM et al; Appl Environ Microbiol 45: 1848-53 (1983)
16. Lyman WJ et al; Handbook of Chem Property Estimation Methods. Environ Behavior of Organic Compounds McGraw-Hill NY (1982)
17. Merck Index; An Encyclopedia of Chemicals, Drugs and Biologicals 10th ed. p 864 (1983)
18. Mill T et al; Science 207: 886-7 (1980)
19. Neurath GB et al; Food Cosmet Toxicol 15: 275-82 (1977)
20. NIOSH; National Occupational Exposure Survey (1983)
21. Perrin DD; Dissociation Constants of Organic Bases in Aqueous Solution. IUPAC Chemical Data Series. Buttersworth: London (1965)
22. Schmidt-Bleek F et al; Chemosphere 11: 383-415 (1982)
23. Schweizer AE et al; Kirk-Othmer Encycl Chem Technol 3rd Ed 2: 272-83 (1978)
24. Wheeler PA; J Phycol 15: 12-7 (1979)

# 2-Methylpyridine

## SUBSTANCE IDENTIFICATION

**Synonyms:**

**Structure:**

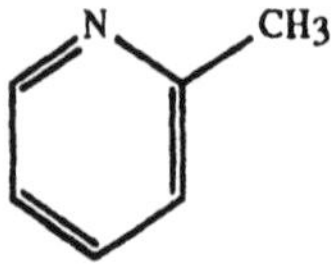

**CAS Registry Number:** 109-06-8

**Molecular Formula:** $C_6H_7N$

**Wiswesser Line Notation:**

## CHEMICAL AND PHYSICAL PROPERTIES

**Boiling Point:** 128-129 °C

**Melting Point:** -70 °C

**Molecular Weight:** 93.12

**Dissociation Constants:** pKa = 6.00 [25]

**Log Octanol/Water Partition Coefficient:** 1.11 [9]

**Water Solubility:** Miscible [25]

**Vapor Pressure:** 10.0 mm Hg at 22.8 °C [25]

**Henry's Law Constant:** 9.96 x $10^{-6}$ at 25 °C [13]

## ENVIRONMENTAL FATE/EXPOSURE POTENTIAL

**Summary:** 2-Methylpyridine is released to the environment in wastewater and as fugitive emissions during its production and use as a chemical intermediate and solvent. Energy-related processes such as coal and shale oil gasification are another important source of release. Several food items have been found to contain 2-methylpyridine which is either in the food naturally or formed during cooking. 2-Methylpyridine is contained in tobacco smoke and may contribute to its presence in indoor air. If released on land, 2-methylpyridine will leach into the ground and biodegrade (complete degradation in one soil in 16 days). If released into water, 2-methylpyridine may be lost through biodegradation, photooxidation, and volatilization (half-life 88 hr for a model river). No biodegradation and photooxidation rates in natural waters are available. Bioconcentration in aquatic organisms should not be significant. In the atmosphere, 2-methylpyridine should react with photochemically produced hydroxyl radicals (half-life 11.2 days) and be scavenged by rain. In polluted areas containing appreciable nitric acid vapor, reaction with nitric acid may be the major removal process. People are primarily exposed to 2-methylpyridine in occupational settings, although the general public will be exposed from tobacco smoke and some food items.

**Natural Sources:**

**Artificial Sources:** 2-Methylpyridine may be released to the environment in wastewater or in fugitive emissions during its production and use as a chemical intermediate in the synthesis of pharmaceuticals, dyes, and rubber chemicals, and as a solvent [10]. 2-Methylpyridine is produced during coal and shale oil gasification [22]. It is produced in coke ovens and is recovered from the resulting coal tar [5]. Therefore it may be released as fugitive emissions and in wastewater from coal and shale oil processing. It is also released to the air in tobacco smoke [8].

**Terrestrial Fate:** If released on land, 2-methylpyridine will leach into the ground and biodegrade. Complete degradation in one soil occurred in about 16 days.

**Aquatic Fate:** If released into water, 2-methylpyridine should biodegrade and be lost through volatilization (half-life 88 hr for a model river). No estimates of biodegradation rates in natural waters are available. It may also be lost by photooxidation but no data containing actual rates were available. Adsorption to sediment or particulate matter in the water column should not be important.

# 2-Methylpyridine

**Atmospheric Fate:** If released into the atmosphere, 2-methylpyridine should react with photochemically produced hydroxyl radicals (estimated half-life 11.2 days) and be scavenged by rain.

**Biodegradation:** 2-Methylpyridine is reported to be readily biodegradable in the MITI test, a biodegradability screening test used by the Japanese Ministry of International Trade and Industry [27]. In a screening test that employed enrichment cultures from 1 mM of 2-methylpyridine and a fertile garden soil suspension as an inoculum, 100% degradation was obtained in 14-32 days [20]. When this test was repeated under anaerobic conditions, degradation was much slower, taking >97 days [20]. When 2 micromole/g of 2-methylpyridine was incubated with a silt loam soil, 2.7% remained after 16 days [29].

**Abiotic Degradation:** 2-Methylpyridine should react with photochemically produced hydroxyl radicals in the atmosphere with a predicted half-life of 11.2 days [1]. The UV absorption of 2-methylpyridine should be similar to that of pyridine with no adsorption >290 nm [26], eliminating direct photolysis as a possible degradative process. It might react with alkoxy and hydroxyl radicals in natural waters as pyridine does [19] but data for the 2-methylpyridine is lacking. Another environmentally relevant reaction that may be important in polluted atmospheres is that with vapor phase nitric acid. This reaction may be the dominant one for pyridine in some situations [2] and may similarly be important for 2-methylpyridine.

**Bioconcentration:** From the octanol/water partition coefficient for 2-methylpyridine, one can predict a BCF of 4 using a recommended regression equation [18]. Such a low value would indicate that bioconcentration of 2-methylpyridine is not significant in aquatic organisms.

**Soil Adsorption/Mobility:** 2-Methylpyridine is miscible in water and therefore would not be likely to adsorb appreciably to soil or sediment [18]. The pKa of 2-methylpyridine indicates that in acidic soils it will be largely in an ionic form, suggesting that cationic adsorption is probable and adsorption to clayey soil is possible. Data for such an effect are lacking. In an experiment involving leaching of shale-oil process water through soil columns packed by horizons from Rock Springs, WY soil cores to the original 1016 mm depth, the 2-methylpyridine emerged in somewhat under 2 soil column void volumes [16]. This indicates that soil is an effective adsorbent only if less than this void volume of retort water is applied (small spills). Rainfall leaching after a spill also will probably enhance solute migration [16]. The pH and clay content of the soil was not specified [16].

**Volatilization from Water/Soil:** From the Henry's Law constant, one can estimate a half-life for volatilization from a model river 1 m deep with a 1 m/sec current and a 3 m/sec wind speed of 88 hr [18]. In mineral salts-soil suspensions incubated at 28 °C, 15% was volatilized in 24 days [30]. Volatilization from soil alone was only 2-3% after 60 days [29].

**Water Concentrations:** DRINKING WATER: 2-Methylpyridine was reported in drinking water in Cincinnati, OH [17]. GROUND WATER: Contaminated ground water from St. Louis Park, MN - site of a coal tar distillation and wood-preserving facility that operated from 1918-72 - contained 41 ppb of 2 methylpyridine [24]. Two aquifers under the Hoe creek coal gasification site contained 0.88-61 ppb of 2-methylpyridine 15 months after gasification was complete [31]. Not detected in wells in Hanna and Gillette, WY prior to coal gasification [22].

**Effluent Concentrations:** 2-Methylpyridine has been identified in effluents from the following industries: timber products, organic chemicals, pharmaceuticals, and publicly owned treatment works [28]. 2-Methylpyridine is contained in shale oil wastewater (5 ppm) and would be released to the atmosphere if the wastewater is heated as it would be when used to cool hot, retorted oil shale [3,11]. It was also found in the effluents from an advanced water treatment facility in Pomona, CA [17]. Wastewater from coal gasification contained an estimated 3.71 ppm of 2-methylpyridine [4].

**Sediment/Soil Concentrations:** 2-Methylpyridine was detected but not quantified in non-agricultural loamy soil from the Moscow region [6]. Less than 0.22 ppm of the chemical was found in Eagle Harbor sediment, an area of Puget Sound that is contaminated with creosote [15].

**Atmospheric Concentrations:** Indoor and outdoor air in and near the shale oil wastewater treatment facility of Occidental Oil Shale Inc at the Logan Wash site, CO contained 7 and 28 ug/m$^3$ of 2-methylpyridine, respectively [12]. Rural air in an undeveloped area of the oil shale region as well as urban air (Boulder, CO) contained no 2-methylpyridine.

**Food Survey Values:** 2-Methylpyridine has been identified as a volatile flavor compound in fried bacon [14] and boiled beef [7].

**Plant Concentrations:**

**Fish/Seafood Concentrations:**

**Animal Concentrations:**

**Milk Concentrations:**

**Other Environmental Concentrations:** 2-Methylpyridine has been identified in cigarette smoke [8].

**Probable Routes of Human Exposure:** People are primarily exposed to 2-methylpyridine in occupational settings via dermal contact or inhalation with the vapor or aerosols. The general public will be exposed from tobacco smoke and some food items.

**Average Daily Intake:**

**Occupational Exposure:** NIOSH (NOES Survey 1981-83) has statistically estimated that 9874 workers are exposed to 2-methylpyridine in the United States [21]. 2-Methylpyridine is formed in the thermal decomposition of amine-cured epoxy powder paint and this could lead to occupational exposures if the epoxy resin is deposited on a surface hot enough to degrade the polymer (350 °C) [23].

**Body Burdens:**

# REFERENCES

1. Atkinson R; Internat J Chem Kinetics 19: 799-828 (1987)
2. Atkinson R et al; Environ Sci Technol 21: 64-72 (1987)
3. Dobson KR et al; Water Res J 19: 849-56 (1985)
4. Giabbai MF et al; Int J Environ Anal Chem 20: 113-29 (1985)
5. Goe GL; Encyclopedia of Chemical Technology Wiley-Interscience 19: 454-83 (1978)
6. Golovnya RV et al; USSR Acad Med Sci pp. 325-35 (1982)

7. Golovnya RV et al; Chem Senses Flavour 4: 97-105 (1979)
8. Graedel TE; Chemical Compounds in the Atmosphere, Academic Press NY (1978)
9. Hansch C, Leo AJ; Medchem Project Issue No.26 Pomona College, Claremont CA (1985)
10. Hawley GG; Condensed Chem Dictionary 10th ed Van Nostrand Reinhold NY p. 815 (1981)
11. Hawthorne SB et al; Environ Sci Tech 19: 922-7 (1985)
12. Hawthorne SB, Sievers RE; Environ Sci Technol 18: 483-90 (1984)
13. Hine J, Mookerjee PK; J Org Chem 40: 292-8 (1975)
14. Ho CT et al; J Agric Food Chem 31: 336-42 (1983)
15. Krone CA et al; Environ Sci Technol 20: 1144 (1986)
16. Leenheer JA, Stuber HA; Environ Sci Technol 15: 1467-75 (1981)
17. Lucas SV; GC/MS Analysis of Organics in Drinking Water Concentrations and Advanced Waste Treatment Concentrates Vol 1, EPA-600/1-84-020A pp. 321 (1984)
18. Lyman WJ et al; Handbook of Chem Property Estimation Methods. Environ Behavior of Organic Compounds McGraw-Hill NY (1982)
19. Mill T et al; Science 207: 886-7 (1980)
20. Naik MN et al; Soil Biol Biochem 4: 313-23 (1972)
21. NIOSH; National Occupational Exposure Survey (1985)
22. Pellizzari ED et al; ASTM Spec Tech Publ, STP 686: 256-74 (1979)
23. Peltonen K; J Anal Appl Pyrolysis 10: 51-7 (1986)
24. Pereira WE et al; Environ Toxicol Chem 2: 283-94 (1983)
25. Riddick JA et al; Organic Solvents: Physical Properties and Methods of Purification. Techniques of Chemistry. 4th Ed. New York: Wiley-Interscience pp 1325 (1986)
26. Sadlter; Sadlter Standard Spectra, Sadlter Res Lab, 9 UV, Philadelphia, PA
27. Sasaki S; pp. 283-98 in Aquatic Pollutants, Transformation and Bio Effects, Hutzinger O, Von Letyoeld LH, Zoeteman BCJ, eds Oxford Pergamon Press (1978)
28. Shackelford WM et al; Analyt Chem Acta 146: 15-27 (1983)
29. Sims GK, Sommers LE; J Environ Qual 14: 480-4 (1985)
30. Sims GK, Sommers LE; Environ Toxicol Chem 5: 503-9 (1986)
31. Stuermer DH et al; Environ Sci Technol 16: 582-7 (1982)

# Methyl Ethyl Ketone

## SUBSTANCE IDENTIFICATION

**Synonyms:** 2-Butanone

**Structure:**

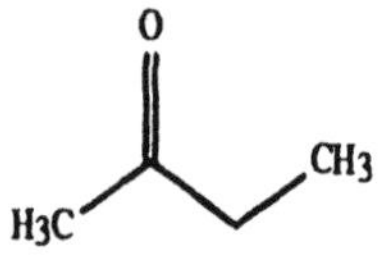

**CAS Registry Number:** 78-93-3

**Molecular Formula:** $C_4H_8O$

**Wiswesser Line Notation:** 2V1

## CHEMICAL AND PHYSICAL PROPERTIES

**Boiling Point:** 79.6 °C

**Melting Point:** -86.35 °C

**Molecular Weight:** 72.10

**Dissociation Constants:**

**Log Octanol/Water Partition Coefficient:** 0.29 [24]

**Water Solubility:** 239,000 mg/L [45]

**Vapor Pressure:** 90.6 mm Hg at 25 °C [1]

**Henry's Law Constant:** $1.05 \times 10^{-5}$ atm-m³/mole [42]

## ENVIRONMENTAL FATE/EXPOSURE POTENTIAL

**Summary:** Large quantities of methyl ethyl ketone (MEK) are used as a solvent especially in the coatings industry. MEK will be discharged into the atmosphere from this and other industrial uses. It

334

will also be discharged in wastewater. In addition, high atmospheric MEK levels are associated with photochemical smog episodes although it is generally absent from ambient air. It is formed as a result of the natural photooxidation of olefinic hydrocarbons which get in the air from automobiles and other sources. If MEK is released to soil, it will partially evaporate into the atmosphere from near-surface soil and may leach into the ground water. Biodegradability studies in anaerobic systems suggest that MEK present in ground water may degrade slowly after a long acclimation period. If released into water, MEK will be lost by evaporation (half-life 3-12 days) or be slowly biodegraded. It will not hydrolyze in water or soil under normal environmental conditions. It will not significantly indirectly photooxidize in surface waters, adsorb to sediment or bioconcentrate in aquatic organisms. It may be subject to direct photolysis. If released to the atmosphere, it will exist primarily in the gas phase. It will photodegrade at a moderate rate (half-life 2.3 days or less) and it may be subject to direct photolysis. It may be removed in rain since it has been detected in rainwater. Major human exposure is from occupational atmospheres or ambient air in the vicinity of industrial sources or during photochemical smog episodes. Although there is limited data, MEK is a natural component of some foods so ingestion is also a source of exposure.

**Natural Sources:** Volcanos, forest fires, products of biological degradation, natural component of food [21,30].

**Artificial Sources:** Emissions from its use as a solvent for lacquers, adhesives, rubber cement, printing inks, paint removers and cleaning solutions, catalyst, carrier, and wastewater from these uses [21,29,30,46]. Stack emission, fugitive emissions, and wastewater relating to its production, storage, transport, and disposal [30]. Photochemical air pollution [23]; combustion processes such as gasoline exhaust and cigarette smoke [21,30].

**Terrestrial Fate:** If MEK is released to soil, it will partially evaporate into the atmosphere from near-surface soil and may leach into ground water. It will not significantly hydrolyze in soil. It may be subject to slow biodegradation in soil and ground water based on results of biodegradation screening studies in aquatic media and a river die-away test.

**Aquatic Fate:** If MEK is released into water, it will evaporate into the atmosphere with estimated half-lives of 3 and 12 days in rivers and lakes, respectively. It will also biodegrade slowly in both fresh and salt water. No information is available concerning its fate in ground water but biodegradability studies in anaerobic systems suggest that it may degrade slowly after a long acclimation period. It will not significantly hydrolyze, photooxidize, adsorb to sediment, or bioconcentrate in aquatic organisms. It may be subject to direct photolysis.

**Atmospheric Fate:** If MEK is released into the atmosphere, it will exist primarily in the gas phase [16]. It will degrade principally by reaction with photochemically produced hydroxyl radicals (half-life 2.3 days) and the product of this reaction is acetaldehyde [10]. It may be subject to direct photolysis. It may be subject to removal in rain since it has been detected in rainwater.

**Biodegradation:** Complete removal (and 87% mineralization) in 5 days in screening tests using municipal wastewater inoculum; complete removal in 9 days using activated sludge treatment [13]. Percent theoretical BOD in 5 and 20 days were 76% and 89%, respectively, using nonacclimated settled domestic wastewater inoculum in fresh water and 32% and 69%, respectively, using settled raw wastewater seed developed in actual seawater, with the test being run in synthetic seawater [38]. Percent theoretical BOD and COD were 76% and 79%, respectively, after 5 days at 20 °C using a standard dilution method with filtered biological sanitary waste treatment plant seed [5]. 88% theoretical BOD in 5 days using sew seed in standard dilution screening tests [26]. Complete removal in 2 days in river die-away tests [13]. Degradation also occurs in anaerobic systems but time required for acclimating degrading microorganisms is long (ca 1 wk) [7].

**Abiotic Degradation:** MEK adsorbs radiation near the short wavelength cutoff of the solar spectrum at ground level; however, the reaction with photochemically produced hydroxyl radicals with a half-life of 2.3 days is the dominant atmospheric process [10]. Acetaldehyde is the primary product of this reaction [10]. Under simulated smog conditions, it is of intermediate reactivity with degradation rates ranging from 1.5%/hr to 33% in 6.5 hr [11,12,18,48]. In the surface layer of water, photolysis is possible because UV radiation reaching these layers can be absorbed by MEK. However, no determination of the actual photolysis rate could be found [32]. Indirect photooxidation in water [25] and hydrolysis will not be significant processes [32].

# Methyl Ethyl Ketone

**Bioconcentration:** Using the log Kow, a BCF of 1.0 was estimated [32]. Based on this estimated BCF, MEK will not be expected to significantly bioconcentrate in aquatic organisms.

**Soil Adsorption/Mobility:** Using the log Kow, a Koc value of 34 was estimated [32]. Based on this estimated Koc value, MEK will be expected to exhibit very high mobility in soil [43] and therefore may leach to the ground water.

**Volatilization from Water/Soil:** Based on laboratory data for the avg rate of evaporation of MEK from water relative to the reaeration rate (0.27) [33,39] and typical reaeration rates for rivers and lakes [34], the half-life for evaporation from a river and lake will be 3 and 12 days, respectively. Due to its high vapor pressure, volatilization from soil will be rapid.

**Water Concentrations:** DRINKING WATER: In a federal survey of ground water supplies, <5% occurrence [15]. Detected, not quantified, in drinking water supplies from 7 cities from varied sources and with different types of pollutant sources [8,41,44]. Concn (ppm) in drinking water at laboratory in Florida, 6 mo after PVC pipe installation (residence time in PVC pipe, hr): 0 ppm (0 hr), 0.4 (4), 1.8 (16), 3.9 (48), 4.5 (64), 4.5 (96); 8 mo after installation: 0 ppm (0 hr), 0.1 (4), 0.6 (16), 2.1 (48), 2.2 (72); source of contamination was PVC pipe cement [47]. SURFACE WATER: 14 heavily industrialized river basins in US - 1 of 204 sites pos, 23 ppb [17]; Detected, not quantified, in Black Warrior River in Tuscaloosa, AL [3]. Strait of Florida and Eastern Mediterranean Sea 7-16 ppb [9]. RAIN/SNOW: Detected in rain in Japan [28] but not at 5 sites in California [22]. In clouds in California 0-470 ppm and in fog or ice fog 0-trace [22].

**Effluent Concentrations:** Wastewater going into brackish US river 8-20 ppm [27]. Detected in trench leachate from 2 low-level radioactive disposal sites [20]. Gasoline engine exhaust <0.1-1.0 ppm [46]; Cigarette smoke 50 ppm [30,46]. Minnesota municipal solid waste landfills, leachates, 6 sites, 100% pos, 110-27,000, contaminated ground water (by inorganic indices), 13 sites, 54% pos, 6.8-6200, 7 sites, 28.6% pos, 5.1-1100 ppb [40].

**Sediment/Soil Concentrations:** Not detected in sediment of brackish industrial river in US receiving MEK effluent [27].

337

**Atmospheric Concentrations:** RURAL: Pine Barrens, NJ - not detected [4]. URBAN/SUBURBAN: US (181 samples) - all samples negative [6]; 5 cities in New Jersey - all samples below detection limit of 0.01 ppb [4]; Air pollution episode in Japan 4.3 ppb [2]. Los Angeles, CA area during moderate to severe photochemical pollution episodes (Sept-Oct, 1980) 0-14 ppb [23]. SOURCE DOMINATED AREAS: New Jersey - 33 samples, 64 ppt median, 10-1900 ppt [6]. Vicinity of chemical reclamation plant 94 ppm [30].

**Food Survey Values:** Swiss cheese 0.3 ppm, cream 0.154-0.177 ppm [30]. Detected in roasted barley, bread, honey, chicken, oranges, black tea, and rum [30]. Natural component of some foods [30]. Identified, not quantified, in: Mountain cheese [14]. Dried legumes: beans, 74-330 ppb, 148 ppb avg; split peas, 110 ppb; lentils, 50 ppb [31].

**Plant Concentrations:** 32 samples of southern pea seed were analyzed for volatiles. MEK was found in all samples and the mean concentration was 109 ppb (31 to 650 ppb) [19].

**Fish/Seafood Concentrations:**

**Animal Concentrations:**

**Milk Concentrations:** Milk 0.077-0.079 ppm, milk fat 8 ppm [30].

**Other Environmental Concentrations:**

**Probable Routes of Human Exposure:** The principal routes of exposure to MEK are from ingestion of contaminated drinking water and food (although exposure may be scattered and slight), and inhalation of contaminated air near industrial sources of emissions and during air pollution episodes.

**Average Daily Intake:**

**Occupational Exposures:** Occupational surveillance <25 ppm [30]. NIOSH (NOES Survey 1981-83) has statistically estimated that 1,221,857 workers are exposed to MEK in the United States [35]. NIOSH (NOHS Survey 1972-74) has statistically estimated that 2,431,748 workers are exposed to toluene in the United States [36].

# Methyl Ethyl Ketone

**Body Burdens:** Detected in 5 of 8 samples of human milk from 4 US urban areas [37].

## REFERENCES

1. Ambrose D et al; J Chem Therm 7: 453-72 (1975)
2. Anonymous; Kanagawa-Kan Taiki Osen Chosa Kenkyu Hokaku 20: 86-90 (1978)
3. Bertsch W et al; J Chromatogr 112: 701-18 (1975)
4. Bozzelli JW et al; Analysis of selected toxic and carcinogenic substances in ambient air in New Jersey (1980)
5. Bridie AL et al; Water Res 13: 627-30 (1979)
6. Brodzinsky R, Singh HB; Volatile organic chemicals in the atmosphere: an assessment of available data p 23 SRI 68-02-3452 (1982)
7. Chou WL et al; Biotech Bioeng Symp 8: 391-414 (1979)
8. Coleman WE et al; pp. 305-207 in Analysis and identification of organic substances in water. L Keith ed Ann Arbor, MI Ann Arbor Press (1976)
9. Corwin JF; Bull Mar Sci 19: 504-9 (1969)
10. Cox RA et al; Environ Sci Technol 15: 587-92 (1981)
11. Dilling WL et al; Environ Sci Technol 10: 351-6 (1976)
12. Dimitriade B, Joshi SB; p 705-11 in Int Conf on photochemical oxidant pollution and its control. B Dimitriade ed USEPA 600/3-77-001b (1977)
13. Dojlido JR; Investigation of biodegradability and toxicity of organic compounds: final report 1975-79 USEPA 600/2-79-163 (1979)
14. Dumont JP, Adda J; J Agric Food Chem 26: 364-7 (1978)
15. Dyksen JE, Hess AF III; J Am Water Works Assoc 74:394-403 (1982)
16. Eisenreich SJ et al; Environ Sci Technol 15: 30-8 (1981)
17. Ewing BB et al; Monitoring to detect previously unrecognized pollutants in surface waters 79 p USEPA 560/6-77-015, appendix USEPA 560/6-77-015a (1977)
18. Farley FF; p 713-27 in Int Conf on photochemical oxidant pollution and its control. B Dimitriade ed USEPA 600/3-77-001b (1977)
19. Fisher GS et al; J Agric Food Chem 27: 7-11 (1979)
20. Francis AJ et al; Nuclear Tech 50: 158-63 (1980)
21. Graedel TE; Chemical compounds in the atmosphere. p 182 New York, NY Academic Press (1978)
22. Grosjean D, Wright B; Atmos Environ 17: 2093-6 (1983)
23. Grosjean D; Environ Sci Technol 16: 254-62 (1982)
24. Hansch C, Leo AJ; Medchem Project Issue No 26. Claremont CA: Pomona College (1985)
25. Hendry DG et al; J Chem Phys Ref Data 3: 944-78 (1974)
26. Heukelekian H, Rand MC; J Water Pollut Control Assoc 29: 1040-53 (1955)
27. Jungclaus GA et al; Environ Sci Technol 12: 88-96 (1978)
28. Kato T et al; Yokohama Kokuritsu Daigaku Kankyo Kagaku Kenkyu Senta Kiyo 6: 11-20 (1980)
29. Kirk-Othmer Encyclopedia of Chemical Technology 3rd ed 13: 907 (1981)
30. Lande SS et al; Investigation of selected potential environmental contaminants: ketonic solvents p 43-128 USEPA 560/2-76-003 (1976)
31. Lovegren NV et al; J Agric Food Chem 27: 851-3 (1979)
32. Lyman WJ et al; Handbook of Chem Property Estimation Methods. McGraw-

Hill NY (1982)
33. Mackay D et al; Environ Sci Technol 16: 645-9 (1982)
34. Mill T et al; Laboratory protocols for evaluating the fate of organic chemicals in air and water p 255 USEPA 600/3-82-022 (1982)
35. NIOSH; The National Occupational Exposure Survey (NOES) (1983)
36. NIOSH; The National Occupational Hazard Survey (NOHS) 1974.
37. Pellizzari ED et al; Bull Environ Contam Toxicol 28: 322-8 (1982)
38. Price KS et al; J Water Pollut Control Fed 46: 63-77 (1974)
39. Rathbun RE, Tai DY; Water Air Soil Pollut 17: 281-93 (1982)
40. Sabel GV, Clark TP; Waste Manag Res 2: 119-30 (1984)
41. Scheeman MA et al; Biomed Mass Spectrom 4: 209-11 (1974)
42. Snider JR, Dawson GA; J Geophys Res D Atm 90: 3797-805 (1985)
43. Swann RL et al; Res Rev 85: 17-28 (1983)
44. USEPA; New Orleans area water supply study draft analytical report by the lower Mississippi River facility, Slidell LA (1974)
45. Valvani SC et al; J Pharm Sci 70: 502-7 (1981)
46. Verschueren K; Handbook on environmental data on organic chemicals 2nd ed p 850-1, New York, NY Van Nostrand Reinhold Co (1983)
47. Wang TC, Bricker JL; Bull Environ Contam Toxicol 23: 620-3 (1979)
48. Yanagihara S et al; 4th Int Clean Air Congr p 472-7 (1977)

# Methyl Isobutyl Ketone

## SUBSTANCE IDENTIFICATION

**Synonyms:**

**Structure:**

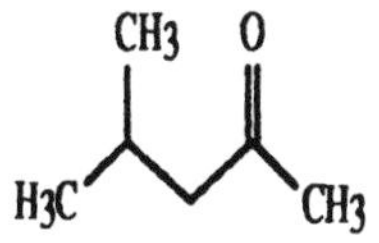

**CAS Registry Number:** 108-10-1

**Molecular Formula:** $C_6H_{12}O$

**Wiswesser Line Notation:** 1Y1&1V1

## CHEMICAL AND PHYSICAL PROPERTIES

**Boiling Point:** 116.8 °C at 760 mm Hg

**Melting Point:** -84.7 °C

**Molecular Weight:** 100.16

**Dissociation Constants:**

**Log Octanol/Water Partition Coefficient:** 1.19 (estimated) [28]

**Water Solubility:** 20,400 mg/L at 20 °C [22]

**Vapor Pressure:** 14.5 mm Hg at 20 °C [4]

**Henry's Law Constant:** 9.4 x $10^{-5}$ atm-m³/mole (calculated from water solubility and vapor pressure)

## ENVIRONMENTAL FATE/EXPOSURE POTENTIAL

**Summary:** Methyl isobutyl ketone (MIBK) is released to the environment in effluent and emissions from its manufacturing and use

341

facilities, in exhaust gas from vehicles, and from land disposal and ocean dumping of consumer products and industrial wastes which contain this compound. A large number of industries may release or dispose of this compound including: rare metal extracters and manufacturers of coatings (i.e., lacquers, varnishes, paints), pharmaceuticals, pesticides, rubber processing chemicals, and adhesives. If released to soil, MIBK may be removed by direct photolysis on soil surfaces, volatilization, or aerobic biodegradation. This compound is also susceptible to extensive leaching and has been detected in landfill leachate. Chemical hydrolysis is not expected to be environmentally significant. If released to water, the primary removal mechanisms for MIBK are expected to be volatilization (half-life 15-33 hr) and direct photolysis. Aerobic biodegradation may be of minor importance. MIBK is not expected to undergo chemical oxidation or chemical hydrolysis, bioaccumulate in aquatic organisms, or adsorb significantly to suspended solids or sediments in water. In the atmosphere, MIBK will be subject to direct photolysis (half-life 15 hours in sunlight) and reaction with hydroxyl radicals (half-life 16-17 hr). In photochemical smog situations, MIBK may also react with nitrogen oxides. Acetone is a major photooxidation product of MIBK, and in the presence of nitrogen oxides, peroxyacetylnitrate (PAN) and methyl nitrate will also be formed. The most probable routes of exposure to MIBK by general populations are inhalation and dermal contact during use of consumer products which contain this compound. Such products would include coatings, adhesives, rubber cements, pesticides, as well as a variety of other products. Some segments of the general population may also be exposed to MIBK by inhalation of contaminated air in source dominated areas or areas near landfills and by ingestion of contaminated drinking water.

**Natural Sources:**

**Artificial Sources:** MIBK is released to the environment in effluent and emissions from its manufacturing and use plants, in exhaust gas from vehicles, and from land disposal and ocean dumping of waste which contains this compound. Since MIBK is a solvent and denaturant with a wide variety of applications, a large number of industries could potentially release this compound. Such industries would include: rare metal extracters and manufacturers of coatings (i.e., lacquers, varnishes, paints), pharmaceuticals, pesticides, rubber processing chemicals, and adhesives. Likewise, MIBK occurs in a variety of consumer products which are ultimately disposed of in landfills.

**Terrestrial Fate:** If released to soil, MIBK may be removed by direct photolysis on soil surfaces, volatilization, or aerobic biodegradation. MIBK is also susceptible to extensive leaching and has been detected in landfill leachate. This compound is not expected to undergo chemical hydrolysis.

**Aquatic Fate:** If released to water, the primary removal mechanisms for MIBK are expected to be volatilization (half-life 15-33 hr) and direct photolysis. Aerobic biodegradation may be of minor importance. MIBK is not expected to undergo chemical oxidation or chemical hydrolysis, bioaccumulate in aquatic organisms, or adsorb significantly to suspended solids or sediments in water.

**Atmospheric Fate:** In the sunlit atmosphere, MIBK will be subject to direct photolysis (half-life 15 hr) and reaction with hydroxyl radicals (half-life 16-17 hr). In photochemical smog situations MIBK may also react with nitrogen oxides. Acetone is a major photooxidation produce of MIBK, and in the presence of nitrogen oxides, peroxyacetylnitrate (PAN) and methyl nitrate will also be formed.

**Biodegradation:** Biodegradation screening studies indicate that MIBK is susceptible to aerobic biodegradation by mixed populations of microorganisms [5,16,21,30,38,41,49]. 2.5 ppm MIBK incubated with settled domestic sewage as seed was found to have 5, 10, and 40 day BODT's of 4.4, 49.3, and 64.8, respectively [16]. Other studies indicate that freshwaters seeded with settled domestic sewage exerted 5 and 20 day BODT's of 4.4-76 and 69%, respectively [5,38,49]. In seeded synthetic seawater, 5 and 20 day BODTs of 15-31 and 53%, respectively, have been observed [38,49]. 500 ppm MIBK incubated in 3 different samples of activated sludge exerted an average BODT of 3% in 24 hours [21].

**Abiotic Degradation:** Ketones, in general, are resistant to chemical hydrolysis under environmental conditions; therefore, chemical hydrolysis is not expected to be an important fate process for MIBK [31]. MIBK is stable to molecular oxygen, and as a result it is not expected to react with dissolved oxygen in water [27]. Based on the lack of reactivity of methyl ethyl ketone, a structurally similar compound, towards hydroxyl radicals (half-life 2-4 years [1,14,32]) and alkyl peroxy radicals (half-life approximately 400 years [25,32]) in natural sunlit water, chemical oxidation of MIBK in the aquatic environment is not expected to be an important fate process. MIBK in cyclohexane exhibits strong absorption of UV light >290 nm [40] suggesting that MIBK has the potential to undergo direct photolysis in the environment. The half-life for direct photolysis of MIBK in the atmosphere is predicted to be on the order of 15 hr based on an overlap of the solar spectrum with the absorption spectrum at a solar zenith angle of 30 °C [12]. The half-life for the reaction of MIBK vapor with photochemically generated hydroxyl radicals in the atmosphere has been estimated to be 16-17 hr based on experimentally determined reaction rate constants ranging from $1.31 \times 10^{-11}$ to $1.45 \times 10^{-11}$ cm$^3$/molecule-sec at 22-27 °C and an average ambient hydroxyl radical concentration of $8.0 \times 10^{+5}$ molecules/cm$^3$ [2,3,20]. Smog chamber studies indicate that MIBK is moderately reactive with nitrogen oxides [11,12,13,26,29]. Acetone is a major photooxidation product of MIBK, and in the presence of nitrogen oxides, peroxyacetylnitrate (PAN) and methyl nitrate are also formed [2].

**Bioconcentration:** Bioconcentration factors (BCF) of 2-5 were estimated for MIBK using linear regression equations based on the estimated log Kow and the water solubility [31]. These BCF values suggest that MIBK will not bioaccumulate significantly in aquatic organisms.

**Soil Adsorption/Mobility:** Soil adsorption coefficients (Koc) of 19-106 were estimated for MIBK using linear regression equations based on the estimated log Kow and a water solubility [31]. These Koc values suggest that MIBK would be highly mobile in soil and would not adsorb significantly to suspended solids and sediments in soil [48].

**Volatilization from Water/Soil:** The overall mass-transfer coefficient for the volatilization of methyl isobutyl ketone from water in a stirred (557-2020 rpm) laboratory bath at 25 °C was found to range from 0.497-1.11 m/day [39]. Based on these values the volatilization half-life of MIBK from water 1 m deep can be estimated to be 15-33 hr. The value of the Henry's Law constant suggests that volatilization would be significant from all bodies of water [31]. Volatility from water and expected lack of adsorption to soil suggests that MIBK would be susceptible to volatilization from moist soil surfaces. The relatively high vapor pressure of MIBK suggests that this compound would volatilize rapidly from dry soil surfaces.

# Methyl Isobutyl Ketone

**Water Concentrations:** SURFACE WATER: Methyl isobutyl ketone was qualitatively identified in Cuyahoga River [23]. Identified, but not quantified, in 1 out of 204 samples of surface water collected near heavily industrialized areas across the US [17]. Qualitatively identified in 1 out of 17 samples of Delaware River water collected between Aug 1976 to March 1977 at river mile 78 and 132 [46]. DRINKING WATER: Methyl isobutyl ketone was detected in 4 out of 14 drinking water supplies sampled between 1977 and 1979 [18]. GROUND WATER: Leachate collected from the Southington, CT municipal landfill during 1982-83 contained MIBK at a concn ranging from 172-263 ug/L [45]. During 1981-82, MIBK was detected in leachate from a Granby, CT municipal landfill, concn range 25-150 ppb [44]. During 1984, after an attempt to abate ground water contamination by capping the landfill and diverting stormwater, MIBK was not detected in leachate; the detection limit was not reported [44]. Qualitatively identified in leachate from Maxey Flats, KY low-level radioactive waste disposal site [19].

**Effluent Concentrations:** MIBK detected in aqueous condensate from low-Btu gasification of rosebud coal at a concn of 78 ppm. Detected in Omega-9 retort water from in situ oil shale processing at a concn of 105 ppm [37]. MIBK has been identified in the final effluent from one plant in each of the following industries: printing and publishing, coal mining, electronic, and organic chemicals [8]. Detected at a concn of 190 ug/L in formation water discharged from an offshore (Shell Oil) production operation in the Gulf of Mexico [43]. MIBK was not detected in air samples taken from the Allegheny Tunnel during 1979; the detection limit was not reported [24].

**Sediment/Soil Concentrations:**

**Atmospheric Concentrations:** URBAN/SUBURBAN AREAS: During 1979 in 6 locations in New Jersey - 159 samples, median concn 0 ppt [6]. SOURCE DOMINATED AREA: During 1976 in Edison, NJ - 29 samples, median concn 270 ppt [6]. MIBK was found in air space of 3 municipal waste treatment plants in Cincinnati, OH during the summer of 1982, concn range < 0.5-13 ppm [15]. Estimated levels of MIBK in air surrounding the Kin-Buc Waste Disposal Site in Edison, NJ during June to July 1976 ranged from 2.1 to 6.0 ug/m$^3$ [36].

**Food Survey Values:** Methyl isobutyl ketone was identified as a volatile flavor component of baked potatoes [10].

**Plant Concentrations:**

**Fish/Seafood Concentrations:**

**Animal Concentrations:**

**Milk Concentrations:**

**Other Environmental Concentrations:** MIBK has been found in gasoline engine exhaust gas [24]. During 1979 methyl isobutyl ketone was found in waste material from pharmaceutical production, which was later disposed of by discharge into the ocean at a dumpsite north of Puerto Rico [7]. Identified in water from drum storage area in the "Valley of the Drums" hazardous waste site near Louisville, KY, concn range 880-1600 ug/L [47].

**Probable Routes of Human Exposure:** The most probable routes of exposure to MIBK by the general population are inhalation and dermal contact during use of consumer products which contain this compound. Such products would include coatings (i.e., paints), adhesives, rubber cements, and pesticides (i.e., pyrethrins) as well as a variety of other products [9,35]. Some segments of the general population may also be exposed to MIBK by inhalation of contaminated air in source dominated areas or areas near landfills and ingestion of contaminated drinking water (MIBK was detected in 4 out of 14 water supplies sampled between 1977 and 1979) [18]. The most probable routes of worker exposure to MIBK are dermal contact and inhalation.

**Average Daily Intake:**

**Occupational Exposures:** Level of worker exposure to MIBK during spray painting was found to be 0.6 ppm time-weighted average [50]. 467,763 workers are potentially exposed to MIBK based on statistical estimates derived from the NIOSH survey conducted 1981-83 in the United States [34]. 1,433,813 workers are exposed to MIBK based on statistical estimates derived from the NIOSH survey conducted 1972-74 in the United States [33].

# Methyl Isobutyl Ketone

**Body Burdens:** Methyl isobutyl ketone has been identified in the expired air from a non-smoking heterogenous study population [42].

## REFERENCES

1.   Anbar M, Neta P; J Appl Rad Iso 18: 493-523 (1967)
2.   Atkinson R et al; Inter J of Chem Kinet 14: 839-47 (1982)
3.   Atkinson R; Chem Rev 85: 69-201 (1985)
4.   Boublik T et al; The Vapor Pressures of Pure Substances Vol 17 Amsterdam, Netherlands: Elsevier Science Publ (1984)
5.   Bridie AL et al; Water Res 13: 627-30 (1974)
6.   Brodzinsky R, Singh HB; Volatile Organic Chemicals in the Atmosphere: An Assessment of Available Data Menlo Park, CA: SRI International p. 23, 187 (1982)
7.   Brooks JM et al; pp. 171-98 in Waste Ocean Vol.1 NY: Wiley (1983)
8.   Bursey JT, Pellizzari ED; Analysis of Industrial Wastewater of Organic Pollutants in Consent Survey USEPA Contract No. 68-03-2867 (1982)
9.   Chem Marketing Reporter; Chemical Profile on Methyl Isobutyl Ketone; NY: Schnell Publishing (1987)
10.   Coleman EC et al; J Agric Food Chem 29: 42-8 (1981)
11.   Cox RA et al; Environ Sci Tech 14: 57-61 (1980)
12.   Cox RA et al; Environ Sci Tech 15: 587-92 (1981)
13.   Dilling WL et al; Environ Sci Tech 10: 351-56 (1976)
14.   Dorfman LM, Adams GE; Reactivity of the Hydroxyl Radical in Aqueous Solution; Wash, DC: Nat Bureau of Standards NTIS COM-73-50623 (1973)
15.   Dunovant VS et al; J Water Pollut Control Fed 58: 886-95 (1986)
16.   Ettinger MB; Ind Eng Chem 48: 256-9 (1956)
17.   Ewing BB et al; Monitoring to Detect Previously Unrecognized Pollutants in Surface Waters. Appendix: Organic Analysis Data USEPA 560/6-77-015 (1977)
18.   Fielding M et al; Organic Pollutants in Drinking Water; Medmenham, England: Water Res Center pp. 17-28 (1981)
19.   Francis AJ et al; Nuclear Tech 50: 158-63 (1980)
20.   GEMS; Graphical Exposure Modeling System. FAP. Fate of Atmos Pollut (1986)
21.   Gerhold RM, Malaney GW; J Water Poll Control Fed 38: 562-79 (1966)
22.   Ginnings PM et al; J Am Chem Soc 62: 1923-4 (1940)
23.   Great Lakes Water Quality Board; An Inventory of Chemical Substances Identified in the Great Lakes Ecosystem Vol.1 - Summary; Windsor Ontario, Canada (1983)
24.   Hampton CV et al; Environ Sci Tech 16: 287-98 (1982)
25.   Hendry DG et al; J Phys Chem Ref Data 3: 944-47 (1974)
26.   Laity JL et al; Adv Chem. Ser No. 124: 95-112 (1973)
27.   Lande SS et al; Investigation of Selected Potential Environmental Contaminants; Ketonic Solvents p. 25, 151 USEPA 560/2-76-003 (1976)
28.   Leo A, Hansch, C; CLOGP Program, Pomona College CA
29.   Levy A; Adv Chem Ser 124: 70-94 (1973)
30.   Ludzack FJ, Ettinger MB; J Water Poll Control Fed 32: 1173-200 (1960)
31.   Lyman WJ et al; Handbook of Chemical Property Estimation Methods NY:McGraw-Hill (1982)
32.   Mill T et al; Science 207: 886-87 (1980)

# Methyl Isobutyl Ketone

33. NIOSH; National Occupational Hazard Survey (1974)
34. NIOSH; National Occupational Exposure Survey (1983)
35. Papa AJ, Sherman PD; Kirk-Othmer Encycl Chem Tech 3rd ed NY : Wiley 13: 912 (1981)
36. Pellizzari ED; Environ Sci Tech 16: 781-5 (1982)
37. Pellizzari ED et al; ASTM Spec Tech Publ 686: 256-74 (1979)
38. Price KS et al; J Water Poll Control Fed 46: 63-77 (1974)
39. Rathbun RE, Tai DY; Water Air Soil Poll 17: 281-93 (1982)
40. Sadlter; Standard UV Spectra No. 21; Philadelphia: Sadlter Res Lab (1961)
41. Sasaki S; pp.283-98 in Aquatic Pollutants: Transformation and Biological Effects; Hutzinger O et al eds Oxford: Pergamon Press (1978)
42. Sauer TC; Environ Sci Technol 15: 917-23 (1981)
43. Sauer TC; Environ Sci Tech 16: 287-98 (1982)
44. Sawhney BL, Raabe JA; The Connecticut Agric Experiment Bull 833: 1-9 (1986)
45. Sawhney BL, Kozlosku RP; J Environ Qual 13: 349-52 (1984)
46. Sheldon LS, Hites RA; Environ Sci Tech 12: 1188-94 (1978)
47. Stonebaker RD, Smith AJ; pp. 1-10 in Control Hazard Mater Spills Nashville, TN: Proc Natl Conf (1980)
48. Swann RL et al; Res Rev 85: 17-28 (1983)
49. Takemoto S et al; Suishitsu Odaku Kenkyu 4: 80-90 (1981)
50. Whitehead LW et al; Am Ind Hyg Assoc J 45: 767-72 (1984)

# Methyl n-Propyl Ketone

## SUBSTANCE IDENTIFICATION

**Synonyms:** 2-Pentanone

**Structure:**

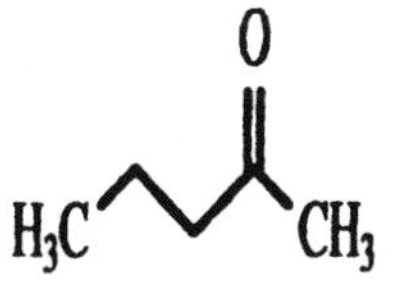

**CAS Registry Number:** 107-87-9

**Molecular Formula:** $C_5H_{10}O$

**Wiswesser Line Notation:**

## CHEMICAL AND PHYSICAL PROPERTIES

**Boiling Point:** 101.7 °C

**Melting Point:** -77.8 °C

**Molecular Weight:** 86.13

**Dissociation Constants:**

**Log Octanol/Water Partition Coefficient:** 0.91 [9]

**Water Solubility:** 59,500 mg/L at 25 oC [29]

**Vapor Pressure:** 35.4 mm Hg at 25 °C [29]

**Henry's Law Constant:** $6.36 \times 10^{-5}$ atm-m³/mole at 25 °C [13]

## ENVIRONMENTAL FATE/EXPOSURE POTENTIAL

**Summary:** Methyl n-propyl ketone occurs naturally in plants and is released into the environment as a plant volatile as well as a product of combustion, photooxidation, and microbial degradation. It may also enter the environment as emissions or in wastewater from its production and use as a solvent. If released on land, methyl n-propyl ketone should volatilize from the soil surface and leach into the ground. If released in water, methyl n-propyl ketone will be lost by volatilization (half-life 11-17 hr in a model river). Although experimental data in environmental media are lacking, it is probable that biodegradation will occur both in water and in soil. Adsorption to sediment and bioconcentration in aquatic organisms should not be significant. In the atmosphere, methyl n-propyl ketone will slowly degrade by reaction with photochemically produced hydroxyl radicals (half-life 3.5 days) and be scavenged by rain. The general public will be exposed to methyl n-propyl ketone from ambient air, ingesting foods which contain it naturally or as a flavor ingredient, and possibly from drinking water. Occupational exposure via inhalation and dermal contact would also be expected.

**Natural Sources:** Methyl n-propyl ketone is released to the environment in forest fires [8]. It is also found in various foods [18]. Trees and other plants including bay-leafed willow, european fir, evergreen cypress, red bilberry shrub, bilberry shrub, and fern emit the ketone [14]. Methyl ketones are oxidation products of alkanes and methyl n-propyl ketone is formed during the photooxidation and biodegradation of n-pentane [8,22].

**Artificial Sources:** Methyl n-propyl ketone may be released into the environment in wastewater and as emissions during its production and use as a solvent and artificial flavor ingredient [10]. It also may be emitted from turbine and auto exhaust, wood pulping, and from energy-related processes such as coal gasification, shale oil production, and offshore oil and gas production [2,8,11,27,30].

**Terrestrial Fate:** If released on land, methyl n-propyl ketone should volatilize from the soil surface and leach into the ground. Although experimental data are lacking, it is probable that biodegradation will occur.

**Aquatic Fate:** If released in water, methyl n-propyl ketone will be lost by volatilization (half-life 11-17 hr in a model river) and possibly biodegrade. Adsorption to sediment should not be significant.

**Atmospheric Fate:** In the atmosphere, methyl n-propyl ketone will slowly degrade by reaction with photochemically produced hydroxyl radicals (half-life 3.5 days). Methyl n-propyl ketone is fairly soluble in water and would therefore be scavenged by rain.

**Biodegradation:** Ketones are generally more resistant to biodegradation than the corresponding aldehydes [7], but methyl n-propyl ketone is attacked by microorganisms isolated from soil and sludge [18]. In a screening test designed to simulate biodegradation in a polluted river, the 5-day BOD for methyl n-propyl ketone was 43% of theoretical [3]. Another short term screening study resulted in slow oxidation, 1.8% of theoretical, after 24 hr using activated sludge from wastewater treatment facilities [7]. Methyl ketones are formed during the biodegradation of alkanes [22] and therefore microorganisms adapted to grow on or oxidize pentane may be simultaneously adapted to degrading methyl n-propyl ketone.

**Abiotic Degradation:** In the gas phase, methyl n-propyl ketone reacts with photochemically produced hydroxyl radicals with a half-life of 3.5 days [1]. It has low to moderate reactivity compared with other solvents in terms of ozone forming potential [5,19].

**Bioconcentration:** No experimental data on the bioconcentration of methyl n-propyl ketone in aquatic organisms could be found in the literature. Using the octanol/water partition coefficient, one can estimate that the BCF is 3 using a recommended regression equation [23], indicating that bioconcentration in aquatic organisms is not important.

**Soil Adsorption/Mobility:** No experimental data on the adsorption of methyl n-propyl ketone to soil could be found in the literature. Using the octanol/water partition coefficient, one can estimate that the Koc is 74 using a recommended regression equation [23], indicating that adsorption to soil and sediment is not significant.

**Volatilization from Water/Soil:** Using the Henry's Law constant, one can estimate that the volatilization half-life of methyl n-propyl ketone from a model river 1 m deep with a 1 m/sec current and a 3 m/sec wind is 15.5 hr [23]. The volatilization half-life in a wind-wave tank with a 6 m/sec wind speed was 14.5 hr [24]. Due to its relatively high vapor pressure and low adsorption to soil, methyl n-propyl ketone would be expected to volatilize readily from soil and surfaces.

**Water Concentrations:** DRINKING WATER: In a survey of drinking waters of 10 cities, methyl n-propyl ketone was found only in water from Ottumwa, IA, concentration 0.1 ppb [16]. In a survey of 14 treated drinking water supplies of varied sources in England, methyl n-propyl ketone was detected in 4 supplies which came from both ground and surface sources [6]. SURFACE WATER: Detected, but not quantified, in the Glatt River in Switzerland [33].

**Effluent Concentrations:** In a comprehensive survey of wastewater from 4000 industrial and publicly owned treatment works (POTWs) sponsored by the Effluent Guidelines Division of the USEPA, methyl n-propyl ketone was identified in discharges of the following industrial category (frequency of occurrence, median concn in ppb): timber products - 1, 422.5; petroleum refining - 1, 8.3; organics and plastics - 5, 40.4; inorganic chemicals - 2, 108.1; plastics and synthetics - 1, 109.4; pulp and paper - 5, 10.7; oil and gas extraction - 2, 137.6; publicly owned treatment works - 3, 190.0 [31]. The highest effluent concn was 517.3 ppb in a publicly owned treatment works [31]. In another study involving 63 industrial effluents, methyl n-propyl ketone was identified in 1 discharge at a level <10 ppb [28].

**Sediment/Soil Concentrations:**

**Atmospheric Concentrations:** RURAL/REMOTE: Methyl n-propyl ketone has been identified in air from the Southern Black Forest in Germany [15]. URBAN/SUBURBAN: Methyl n-propyl ketone was detected, but not quantified, in the air of Pretoria, Johannesburg, and Durban in South Africa [20]. SOURCE RELATED: The concn of methyl n-propyl ketone near a pilot-scale shale oil wastewater facility at Logan Wash, CO was 0.14 ppb [12].

**Food Survey Values:** Dairy products that have quantified amounts of methyl n-propyl ketone include milk (7-26 ppb), cream (25-45 ppb), and swiss cheese (980 ppb) [18]. Methyl n-propyl ketone has been additionally detected, but not quantified, in the volatiles of Beaufort cheese [4], a Gruyere-type cheese produced in the French Alps, and in roasted filberts [17], other unspecified cheeses, evaporated milk, bananas, white bread, soybeans, potato chips, and toasted oats [18]. The concn of methyl n-propyl ketone in samples of dry beans was 45-245 ppb, while in split peas and lentils it was 59 and 29 ppb, respectively [21].

**Plant Concentrations:**

# Methyl n-Propyl Ketone

**Fish/Seafood Concentrations:**

**Animal Concentrations:**

**Milk Concentrations:**

**Other Environmental Concentrations:** Methyl n-propyl ketone was found in used machine cutting-fluid emulsion (0.217 ppm) but not in the fresh emulsion [32].

**Probable Routes of Human Exposure:** Humans are exposed to methyl n-propyl ketone in ambient air, ingesting foods which contain it naturally or as a flavor ingredient, and possibly from drinking water. Occupational exposure via inhalation and dermal contact would also be expected to occur.

**Average Daily Intake:**

**Occupational Exposure:** NIOSH (NOES Survey 1981-1983) has statistically estimated that 5900 workers are exposed to methyl n-propyl ketone in the United States [25]. NIOSH (NOHS Survey 1972-74) has statistically estimated that 791 workers are exposed to methyl n-propyl ketone in the United States [25]. The concn of methyl n-propyl ketone inside a pilot-scale shale oil wastewater facility at Logan Wash, CO was 0.85 ppb [12].

**Body Burdens:** Methyl n-propyl ketone was detected, but not quantified, in 2 of 12 samples of human milk from 4 urban/industrial areas in the US [26].

# REFERENCES

1. Atkinson R et al; Int J Chem Kinet 14: 839-47 (1982)
2. Brooks JJ; Edgewood Arsenal Spec Publ (US Dept Army) 6th Proc Annu Symp (Trace Anal Detect Environ) EO-SP-6001 pp. 89-104 (1976)
3. Dore M et al; Trib Cebedeau 28: 3-11 (1975)
4. Dumont JP, Adda J; J Agric Food Chem 26: 364-7 (1978)
5. Farley FF; pp. 713-27 in Inter Conf Photochemical Oxidant Pollut Control Dimitriades, B ed USEPA-600/3-77-001b (1977)
6. Fielding M et al; Organic Micropollutants In Drinking Water TR-159 Medmenham, Eng Water Res Cent pp. 49 (1981)
7. Gerhold RM, Malaney GW; J Water Pollut Contr Fed 38: 562-79 (1966)

8. Graedel TE; Chemical Compounds in the Atmosphere p. 183 Academic Press NY (1978)
9. Hansch C, Leo AJ; Medchem Project Issue No.26 Pomona College, Claremont CA (1985)
10. Hawley GG; p. 689 in Condensed Chem Dictionary 10th ed Von Nostrand Reinhold NY (1981)
11. Hawthorne SB et al; Environ Sci Tech 19: 922-7 (1985)
12. Hawthorne SB, Sievers RE; Environ Sci Technol 18: 483-90 (1984)
13. Hine J, Mookerjee PK; J Org Chem 40: 292-8 (1975)
14. Isidorov VA et al; Atmos Environ 19: 1-8 (1985)
15. Juttner F; Chemosphere 15: 985-92 (1986)
16. Keith LH et al; pp. 329-73 in Ident Anal Organic Pollut Water Keith, LH ed Ann Arbor, MI: Ann Arbor Press (1976)
17. Kinlin TE et al; J Agric Food Chem 20: 1021 (1972)
18. Lande SS et al; Investigation Of Selected Potential Environmental Contaminants: Ketonic Solvents USEPA-560/2-76-003 (1976)
19. Levy A; Solvent Theory And Practices American Chemical Society, Washington, DC: Adv Chem Ser 124: 70-94 (1973)
20. Louw CW et al; Atmos Environ 11: 703-17 (1977)
21. Lovegren NV et al; J Agric Food Chem 27: 851-3 (1979)
22. Lukins HB, Foster JW· J Bacteriol 85: 1074-87 (1963)
23. Lyman WJ et al; Handbook of Chemical Property Estimation Methods NY: McGraw-Hill (1982)
24. Mackay D, Yeun ATK; Environ Sci Technol 17: 211-7 (1983)
25. NIOSH; National Occupational Health Survey (1975): NIOSH; National Occupation Exposure Survey (1983)
26. Pellizzari ED et al; Bull Environ Contam Toxicol 28: 322-8 (1982)
27. Pellizzari ED et al; ASTM Spec Tech Publ; STP 686: 256-74 (1979)
28. Perry DL et al; Identification Of Organic Compounds In Industrial Effluent Discharges USEPA-600/4-79-016 (1979)
29. Riddick JA et al; Organic Solvents: Physical Properties and Methods of Purification. Techniques of Chemistry. 4th Ed. New York: Wiley-Interscience pp 1325 (1986)
30. Sauer TC Jr; Environ Sci Technol 15: 917-23 (1981)
31. Shackelford WM et al; Analyt Chim Acta 146: 15-27 (1983)
32. Yasuhara A et al; Agric Biol Chem 50: 1765-70 (1986)
33. Zuercher F, Giger W; Vom Wasser 47: 37-55 (1976)

# Morpholine

**Synonyms:**

**Structure:**

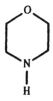

**CAS Registry Number:** 110-91-8

**Molecular Formula:** $C_4H_9NO$

**Wiswesser Line Notation:** T6M DOTJ

## CHEMICAL AND PHYSICAL PROPERTIES

**Boiling Point:** 128.9 °C at 760 mm Hg

**Melting Point:** -4.9 °C

**Molecular Weight:** 87.12

**Dissociation Constants:** pKa = 8.492 [16]

**Log Octanol/Water Partition Coefficient:** -0.86 [7]

**Water Solubility:** Miscible [17]

**Vapor Pressure:** 10.08 mm Hg at 25 °C [17]

**Henry's Law Constant:** 1.41 x $10^{-7}$ atm-m³/mole at 25 °C (calculated by the bond method) [9]

## ENVIRONMENTAL FATE/EXPOSURE POTENTIAL

**Summary:** Morpholine is used as a rubber accelerator, corrosion inhibitor in steam boiler systems, optical brightener, solvent, organic

355

intermediate (catalysts, antioxidants, pharmaceuticals, bactericides), in textile chemicals, photographic developers, hair conditioners, waxes, and polishes, and in the preservation of book paper. Total demand for morpholine was 11,000 metric tons (1975) and the U.S. estimated morpholine emissions to the atmosphere (as of 1978) were about 10 million pounds/year. If released to soil surfaces, morpholine will probably rapidly volatilize. Morpholine in soil will move with soil moisture and is expected to leach extensively. Based on screening test results, biodegradation may be significant; hydrolysis should not be important. Morpholine released to natural waters will tend not to bioconcentrate, volatilize, or sorb to organic particulate matter. Biodegradation may be the most important removal process while hydrolysis should not be significant. No information was found on photolysis. Morpholine in the atmosphere is not expected to be persistent (estimated half-life 4 hr). Morpholine has been found in baked ham, fish, cigarettes, and cigarette smoke condensate.

**Natural Sources:**

**Artificial Sources:** The breakdown in 1975 by end use of morpholine was estimated in metric tons: rubber chemicals, 3600; corrosion inhibitors, 2700; optical brighteners, 900; waxes and polishes, 900; alkyl morpholines, 900; miscellaneous, 900; and export, 900 [14]. Morpholine is used extensively as an intermediate in the production of delayed-action rubber accelerators and as a corrosion inhibitor in steam boiler systems [14]. Other uses include use as a solvent, in the preservation of book paper, as an organic intermediate (catalysts, antioxidants, pharmaceuticals and bactericides) [8], in textile chemicals, photographic developers, and hair conditioners [2]. Miscellaneous uses of morpholine may include use in shellac preparations, quick-setting printing inks and varnish, and propellants [14]. Estimated morpholine emissions (as of 1978) to the atmosphere were 10,028,000 pounds/year [1].

**Terrestrial Fate:** If released to moist soil surfaces, morpholine should only slowly volatilize. Morpholine that is in soil will tend not to sorb to organic matter and will thus move with soil moisture and leach to ground water. Biodegradation may be the most important removal mechanism for morpholine in soil. Hydrolysis should not be significant.

**Aquatic Fate:** Morpholine released to natural waters will tend not to bioconcentrate, volatilize, or sorb to organic matter. Biodegradation may be the most significant removal process for morpholine in natural waters. Hydrolysis should not be significant. No information was found on photolysis.

**Atmospheric Fate:** A computer estimated half-life for the reaction of morpholine with hydroxyl radicals is 4 hr. Washout of morpholine from the atmosphere may also be a significant removal process.

**Biodegradation:** When inoculated with a sewage seed, morpholine (250-900 ppm) did not exert any BOD after a 10 day incubation at 20 °C [13]. In another BOD test, morpholine exerted 0.9% and 5.1% of its theoretical BOD after 5 and 20 days (respectively) incubation in a medium inoculated with settled sewage [5]. In batch tests, no biodegradation of morpholine (10, 50, and 100 mg/L) was observed after 14 days incubation at 20 °C in media inoculated with river mud, treatment plant sludge bacteria, or adapted bacteria (bacteria exposed to morpholine for 15 days as the sole organic carbon source) [4]. In another study, however, two organisms were isolated that could grow on morpholine as the sole source of carbon, nitrogen, and energy [11]. The organisms were isolated from sludge taken from an industrial biological treatment plant and incubation was at 30 °C in agitated flasks. Initial morpholine concentration was 10 mmol/L (871 mg/L) in a mineral salts medium. In addition, >98% of influent morpholine (1 mmol/L) was removed in a laboratory scale activated sludge unit operated for about 6 months. Incubation temperature was 20-30 °C, feed pH was 7.5, and retention time of the liquor in the activated sludge unit was from 160 to 20 hr [11].

**Abiotic Degradation:** No information was found on photolysis or hydrolysis of morpholine. However, it should be resistant to hydrolysis since in one study, morpholine was autoclaved (121 °C for 15 min) without a significant change in concentration [11]. A computer estimated half-life for morpholine in the vapor phase with photochemically generated hydroxyl radicals is 4 hr assuming an average concentration of $5 \times 10^5$ radicals/cm$^3$ [3].

**Bioconcentration:** Because morpholine is miscible with water and has a low octanol/water partition coefficient, its tendency to bioconcentrate will be extremely low.

**Soil Adsorption/Mobility:** Using the octanol/water partition coefficient and a recommended regression equation [12], the estimated soil sorption coefficient for morpholine is 8. This indicates that morpholine will not strongly adsorb to soil [10] and is expected to leach extensively.

**Volatilization from Water/Soil:** From the Henry's Law constant, one can estimate a half-life for volatilization of 242 days from a model river 1 m deep with a 1 m/sec current and a 3 m/sec wind speed [12]. Therefore, morpholine should not volatilize from water or moist soils.

**Water Concentrations:**

**Effluent Concentrations:**

**Sediment/Soil Concentrations:**

**Atmospheric Concentrations:**

**Food Survey Values:** Average morpholine concentration in baked ham was 0.20 ppm [6].

**Plant Concentrations:**

**Fish/Seafood Concentrations:** Morpholine concentrations (ppm) in fish were: $\leq$ 0.6 in canned tuna, 9 in frozen ocean perch, 6 in spotted trout, $\leq$ 0.7 in small mouth bass, 1.0 in salmon, and trace (<0.3) in frozen cod [19].

**Animal Concentrations:**

**Milk Concentrations:**

**Other Environmental Concentrations:** Morpholine concentration in cigarette smoke condensate was <5 ppm (0.08 ug/cigarette) [18]. Morpholine concentration in an experimental cigarette was 0.3 ppm [18].

**Probable Routes of Human Exposure:** Since morpholine has been found in ham [6], fish [19], and cigarette smoke condensate [18], humans may be exposed to it by eating these foods or by breathing cigarette smoke.

**Average Daily Intake:**

# Morpholine

**Occupational Exposure:** NIOSH (NOES Survey 1981-1983) has statistically estimated that 153,545 workers are exposed to morpholine in the United States [15].

**Body Burdens:**

## REFERENCES

1. Anderson GE; Human Exposure to Atmospheric Concentrations of Selected Chemicals 1: 230 NTIS PB84-102540 (1983)
2. Anonymous; Chemical and Engineering News 64: 28 (1986)
3. Atkinson R; Internat J Chem Kinetics 19: 799-828 (1987)
4. Callamari D et al; Chemosphere 9: 753-62 (1980)
5. Ettinger MB; Ind Eng Chem 48: 256-9 (1956)
6. Hamano T et al; Agric Biol Chem 45: 2237-43 (1981)
7. Hansch C, Leo AJ; Medchem Project Issue No.26 Pomona College, Claremont CA (1985)
8. Hawley GG; Condensed Chemical Dictionary 10th ed Von Nostrand Reinhold NY (1981)
9. Hine J, Mookerjee PK; J Org Chem 40: 292-8 (1975)
10. Kenaga EE; Ecotox Env Safety 4: 26-38 (1980)
11. Knapp JS et al; J Appl Bacteriology 52: 5-13 (1982)
12. Lyman WJ et al; Handbook of Chemical Property Estimation Methods. Environmental Behavior of Organic Compounds. McGraw-Hill NY (1982)
13. Mills EJ Jr, Stack VT Jr; Proc 8th Industrial Waste Conf Eng Bull Purdue Univ Eng Ext Ser pp 492-517 (1954)
14. Mjos K; Kirk-Othmer Encyclopedia of Chemical Technology John Wiley and Sons NY 2: 295-308 (1978)
15. NIOSH; National Occupational Exposure Survey (1983)
16. Perrin DD; Dissociation Constants of Organic Bases in Aqueous Solution. IUPAC Chemical Data Series. 1972 Supplement Butterworth: London (1972)
17. Riddick JA et al; Organic Solvents: Physical Properties and Methods of Purification. Techniques of Chemistry. 4th Ed. New York: Wiley-Interscience pp 1325 (1986)
18. Singer GM, Lijinsky W; J Agric Food Chem 24: 553-5 (1976)
19. Singer GM, Lijinsky W; J Agric Food Chem 24: 550-3 (1976)

# 2-Nitropropane

**Synonyms:**

**Structure:**

$$CH_3$$
$$|$$
$$H - C - NO_2$$
$$|$$
$$CH_3$$

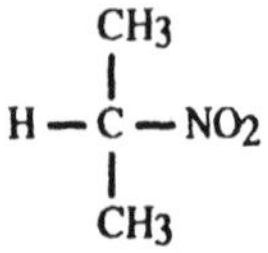

**CAS Registry Number:** 79-46-9

**Molecular Formula:** $C_3H_7NO_2$

**Wiswesser Line Notation:** WNY1&1

## CHEMICAL AND PHYSICAL PROPERTIES

**Boiling Point:** 120.3 °C at 760 mm Hg

**Melting Point:** -93 °C

**Molecular Weight:** 89.09

**Dissociation Constants:** pKa = 7.68 [12]

**Log Octanol/Water Partition Coefficient:** 0.554 [15]

**Water Solubility:** 170,000 mg/L at 25 °C [1]

**Vapor Pressure:** 18 mm Hg at 25 °C [1]

**Henry's Law Constant:** $1.23 \times 10^{-4}$ atm-m$^3$/mole (calculated from vapor pressure and water solubility)

## ENVIRONMENTAL FATE/EXPOSURE POTENTIAL

**Summary:** 2-Nitropropane enters the atmosphere connected with its manufacture and use as a solvent. These sources are primarily of

360

occupational concerns. Other uses such as octane or cetane boosters in fuels could lead to more widespread exposure, although no ambient monitoring data are available to document such exposure. 2-Nitropropane will photodegrade in the atmosphere although its rate of decay is speculative (hours to days). When spilled in water or on land, 2-nitropropane will be lost by evaporation, although photodegradation in surface water is possible. Since it is not adsorbed appreciably by soil, it may leach into ground water where its fate is unknown. 2-Nitropropane would not be expected to bioconcentrate in fish. Humans are exposed to 2-nitropropane in occupational settings.

**Natural Sources:**

**Artificial Sources:** Environmental emissions and spills are possible from its manufacture and use as a chemical intermediate, solvent in coatings, inks, and cellulose esters, stripping solvent for shellac and lacquer, in explosives, and rocket propellants, and in additives to fuel for racing cars and diesel fuels [1,5].

**Terrestrial Fate:** If spilled on land, 2-nitropropane would be expected to volatilize rapidly and also leach into soil where the rate of its degradation is unknown. Unless it is degraded in soil, there is a good possibility it will get into ground water.

**Aquatic Fate:** When spilled in water, 2-nitropropane will volatilize with a half-life of 9.5 hr from a model river. It would not be expected to adsorb appreciably to sediment. It could photolyze in surface water, but no estimates of this or other degradative processes are available.

**Atmospheric Fate:** 2-Nitropropane can degrade in the atmosphere by photolysis and reaction with photochemically produced hydroxyl radicals. Although the half-life for the latter has been estimated to be 3.3 hr, it is reported to have "low reactivity" under photochemical smog conditions.

**Biodegradation:** The biodegradability of 2-nitropropane is unknown. At a concentration of 0.5% it was toxic to 65 strains of bacteria, yeast, and fungi [4].

**Abiotic Degradation:** 2-Nitropropane absorbs UV light >290 nm [2,14] and by analogy with nitromethane and nitroethane undergoes a primary dissociative process forming free radicals [2]. In photochemical smog chamber experiments, 2-nitropropane was one of the least reactive of 45 solvents and solvent mixtures tested [6]. However, the rate of solvent degradation was not reported [6]. The atmospheric half-life due to reaction with hydroxyl radicals has been estimated to be 3.3 hr [3].

**Bioconcentration:** 2-Nitropropane is quite soluble in water and such compounds would not be expected to bioconcentrate appreciably in fish.

**Soil Adsorption/Mobility:** Using the reported water solubility, an estimated Koc of 5.8 was calculated [7]. Based on this estimated Koc value, 2-nitropropane will not be expected to adsorb to soil.

**Volatilization from Water/Soil:** From its Henry's Law constant, one can estimate 2-nitropropane's half-life for evaporation from a river 1 m deep flowing at 1 m/sec and with a wind velocity of 3 m/sec as 9.5 hr [7]. It has a high vapor pressure and would be expected to volatilize rapidly from spills.

**Water Concentrations:**

**Effluent Concentrations:**

**Sediment/Soil Concentrations:**

**Atmospheric Concentrations:**

**Food Survey Values:**

**Plant Concentrations:**

**Fish/Seafood Concentrations:**

**Animal Concentrations:**

**Milk Concentrations:**

**Other Environmental Concentrations:**

**Probable Routes of Human Exposure:** Occupational exposure to 2-nitropropane may occur in many industries including industrial construction and maintenance, printing, highway maintenance (traffic markings), shipbuilding and maintenance (marine coatings), and furniture and plastic products [11].

**Average Daily Intake:**

**Occupational Exposure:** NIOSH estimates that 185,000 workers are exposed to 2-nitropropane during its production and use [9]. NIOSH (NOES Survey 1981-1983) has statistically estimated that 8808 workers are exposed to 2-nitropropane in the United States [10]. Limited occupational monitoring data is available. In a 2-nitropropane production plant in Sterlington, LA, 141 of 144 samples were in the range 0.2-10 ppm whereas 3 samples ranged between 10-100 ppm (2 in spill areas) [8]. Earlier monitoring studies indicated that 580-1640 ppm were obtained during drum filling operations [8]. Lacquer spraying resulted in concn of 10-30 ppm [13].

**Body Burdens:**

## REFERENCES

1.  Baker PJ Jr, Bollmeier AF Jr; in Kirk-Othmer Encycl Chem Technol 3rd ed 15: 969-87 (1981)
2.  Calvert JG, Pitts JN Jr; Photochemistry; John Wiley & Sons New York NY pp.454-5 to 477-9 (1966)
3.  Cupitt LT; Fate of Toxic and Hazardous Materials in the Air Environment; USEPA-600/3-80-084 (1980)
4.  Kido T et al; Arch Microbiol 106: 165-9 (1975)
5.  Kirk-Othmer Encyclopedia of Chemical Technology; 3rd ed 16: 764 (1981)
6.  Levy A; Adv Chem Ser 124: 70-94 (1973)
7.  Lyman WJ et al; Handbook of Chemical Property Estimation Methods. Environmental Behavior of Organic Compounds. McGraw-Hill NY (1982)
8.  Miller ME, Temple GW; 2-NP Mortality Epidemiology Study of Sterlington, LA Employees 1-1-46 thru 6-30-77, International Minerals and Chemical Corp Mendellin, IL (1979)
9.  NIOSH; Amer Ind Hyg Assoc 41: A18 to A24 (1980)
10. NIOSH; National Occupational Exposure Survey (1983)
11. Parmeggiani L; Encyclopedia of Occupational Health and Safety; 3rd ed vol 2 p.1461 (1983)
12. Serjeant EP, Dempsey B; Ionisation Constants of Organic Acids in Aqueous Solution. IUPAC Chemical Data Series No. 23 New York, NY: Pergamon Press pp 989 (1979)
13. Skinner JB; Indust Medicine 16: 441-3 (1947)
14. Stadler Index 1 UV
15. U.S.EPA; Calculated using CLOGP-PCGEMS (1988)

# Pentachloroethane

**Synonyms:**

**Structure:**

```
        Cl    Cl
        |     |
Cl — C — C — H
        |     |
        Cl    Cl
```

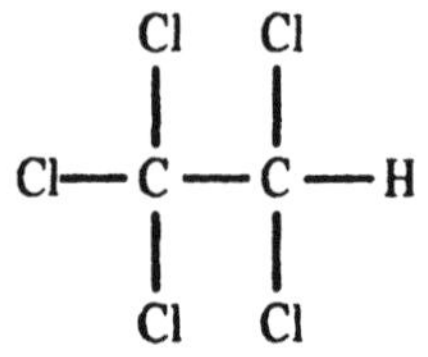

**CAS Registry Number:** 76-01-7

**Molecular Formula:** $C_2HCl_5$

**Wiswesser Line Notation:** GYGXGGG

## CHEMICAL AND PHYSICAL PROPERTIES

**Boiling Point:** 161-162 °C

**Melting Point:** -29 °C

**Molecular Weight:** 202.29

**Dissociation Constants:**

**Log Octanol/Water Partition Coefficient:** 3.05 [13]

**Water Solubility:** 480 mg/L at 25 °C [10]

**Vapor Pressure:** 3.5 mm Hg at 25 °C [5]

**Henry's Law Constant:** $1.94 \times 10^{-3}$ atm-m³/mole at 25 °C (calculated from vapor pressure and water solubility)

## ENVIRONMENTAL FATE/EXPOSURE POTENTIAL

**Summary:** Pentachloroethane is not currently produced commercially or imported in the United States. However, this compound may be

364

released to the environment as a combustion product of polyvinyl chloride (PVC). If released to moist soil, pentachloroethane is expected to have moderate to high mobility and it may undergo slow chemical hydrolysis. Pentachloroethane may volatilize slowly from dry soil surfaces. If released to water, volatilization appears to be an important, if not the dominant, removal mechanism (half-life 5 hr from a model river). This compound also has the potential to oxidize in the presence of light and form trichloroacetyl chloride. Moderate to slight adsorption of pentachloroethane to suspended solids and sediments may occur. Chemical hydrolysis is not expected to be environmentally important. If released to the atmosphere, pentachloroethane is expected to exist almost entirely in the vapor phase. It appears that reaction with photochemically generated hydroxyl radicals (half-life 1.2 years) would be the dominant fate process in the atmosphere. Potential products of this reaction include trichloroacetyl chloride and phosgene. Due to its persistence in the atmosphere, long-range transport of pentachloroethane is expected to occur.

**Natural Sources:** There is no evidence available which indicates that pentachloroethane is a natural product.

**Artificial Sources:** Pentachloroethane is not currently produced commercially or imported into the United States [23]. Formation during combustion or incineration of polyvinyl chloride (PVC) waste products is a potential source of pentachloroethane release to the environment [8].

**Terrestrial Fate:** If released to moist soil, pentachloroethane is expected to have moderate to high mobility. This compound may undergo slow chemical hydrolysis. Pentachloroethane may volatilize slowly from dry soil surfaces.

**Aquatic Fate:** If released to water, volatilization appears to be an important, if not the dominant, removal mechanism. The half-life for this compound volatilizing from a model river has been estimated to be 5 hr. This compound also has the potential to oxidize in the presence of light and form trichloroacetyl chloride. Moderate to slight adsorption of pentachloroethane to suspended solids and sediments may occur. Chemical hydrolysis probably occurs too slowly to be environmentally important.

# Pentachloroethane

**Atmospheric Fate:** Based on the vapor pressure, pentachloroethane is expected to exist almost entirely in the vapor phase in the atmosphere [11]. It appears that reaction with photochemically generated hydroxyl radicals in the atmosphere would be the dominant fate process. The half-life for this reaction has been estimated to be 1.8 years. Potential photooxidation products include trichloroacetyl chloride and phosgene. Due to its persistence in the atmosphere, long-range transport of pentachloroethane is expected to occur.

**Biodegradation:**

**Abiotic Degradation:** Pentachloroethane undergoes slow hydrolysis in water at normal temperatures [1]. This compound oxidizes in the presence of light to yield trichloroacetyl chloride [1]. The half-life for pentachloroethane vapor reacting with photochemically generated hydroxyl radicals in the atmosphere has been estimated to be 1.8 years based on an estimated reaction rate constant of $2.38 \times 10^{-14}$ cm$^3$/molecule-sec at 25 °C and an average ambient hydroxyl radical concentration of $5 \times 10^{+5}$ molecules/cm$^3$ [2,3]. Potential photooxidation products include trichloroacetyl chloride and phosgene [21].

**Bioconcentration:** A bioconcentration factor of 67 was measured for pentachloroethane in bluegill sunfish (14 day exposure), and this compound was found to have a half-life of <1 day in tissue [4]. These data suggest that pentachloroethane will not bioaccumulate significantly in aquatic organisms.

**Soil Adsorption/Mobility:** A soil adsorption coefficient (Koc) of 117 for pentachloroethane was estimated using a molecular topology and quantitative structure activity relationship [19], and a Koc of 244 was estimated using a linear regression equation based on the water solubility [15]. These Koc values suggest that pentachloroethane would be moderately to highly mobile in soil and that adsorption to suspended solids and sediments in water would be moderate to slight [22].

**Volatilization from Water/Soil:** Using the Henry's Law constant, the half-life of pentachloroethane volatilizing from a model river 1 m deep flowing 1 m/sec with a wind speed of 3 m/sec has been estimated to be 5 hr [15]. The volatilization half-life of a dilute solution of pentachloroethane in a beaker 6.5 cm deep, stirred 200 rpm in still air has been experimentally determined to be 46.5 minutes [9]. The relatively high vapor pressure of this compound suggests that it would volatilize slowly from dry soil surfaces.

**Water Concentrations:** SURFACE WATER: Detected in water taken from Fields Brook in Ashtabula, OH during 1976 at a concentration of 2 ug/L [14]. Identified in 2 out of 204 water samples collected from 14 heavily industrialized river basins located throughout the US, detection limit 1 ug/L [12]. DRINKING WATER: Detected in finished drinking water collected in the New Orleans/Baton Rouge area, mean concn <0.03 ug/L, sampling date not reported [17].

**Effluent Concentrations:** Identified in the spent chlorination liquor from the bleaching of sulphite pulp at a concentration corresponding to 0.1 g/ton pulp process [7]. During Aug 1972, pentachloroethane was detected in chlorinated effluent from a sewage treatment plant [20].

**Sediment/Soil Concentrations:**

**Atmospheric Concentrations:** Monitoring data obtained from samples collected in the Atlantic Ocean between 1982 and 1985 indicate that the average baseline level of pentachloroethane in the northern hemisphere is 0.1 ppt/volume. In tradewind systems and in the Southern hemisphere, the baseline level is below the detection limit (0.02 to 0.04 ppt/volume [8]). During 1976-77, pentachloroethane was detected in 0 out of 2 air samples collected in rural/remote areas, 0 out of 74 samples collected in urban/suburban areas, and 2 out of 54 samples in source dominated areas [6]. During Feb 1977, trace levels were found in ambient air of Iberville Parish, LA and 3,984 ng/m$^3$ was detected in ambient air collected from Dow Chemical Property in Freeport, TX during Aug 1986 [18].

**Food Survey Values:**

**Plant Concentrations:**

**Fish/Seafood Concentrations:**

**Animal Concentrations:**

**Milk Concentrations:**

**Other Environmental Concentrations:**

# Pentachloroethane

**Probable Routes of Human Exposure:**

**Average Daily Intake:**

**Occupational Exposure:** NIOSH has estimated that 231 workers are potentially exposed to pentachloroethane based on a survey conducted during 1981-83 in the United States [16]. Since there is currently no commercial production or importation of this compound in the United States [23], occupational exposure to pentachloroethane is probably only due to contamination in other products.

**Body Burdens:**

## REFERENCES

1.  Archer WL; Kirk-Othmer Encycl Chem Tech 3rd ed NY: Wiley 5: 737 (1979)
2.  Atkinson R; Internat J Chem Kinetics 19: 799-828 (1987)
3.  Atkinson R; Chem Rev 85: 69-201 (1985) .
4.  Barrows ME et al; Dyn Exposure Hazard Assess Toxic Chem Ann Arbor Science: Ann Arbor, MI pp. 379-92 (1980)
5.  Boublik T et al; The Vapor Pressure of Pure Substances. Vol 17 Elsevier Sci Pub Amsterdam, Netherlands p. 89 (1984)
6.  Brodzinsky R, Singh HB, Volatile Organic Chemicals in the Atmosphere: An Assessment of Available Data Menlo Park, CA: Atmospheric Science Center pp. 191-21 (1982)
7.  Carlberg GE et al; Sci Total Environ 48: 157-67 (1986)
8.  Class T, Ballschmiter K; Chemosphere 15: 413-27 (1986)
9.  Dilling WL; Environ Sci Tech 11: 405-9 (1977)
10.  Dilling WL et al; Environ Sci Technol 9: 833-8 (1975)
11.  Eisenreich SJ et al; Environ Sci Tech 15: 30-8 (1981)
12.  Ewing BB et al; Monitoring to Detect Previously Unrecognized Pollutants in Surface Waters p. 71 USEPA 560/6-77-015 (1977)
13.  Hansch C, Leo AJ; Medchem Project Issue No.26 Pomona College, Claremont CA (1985)
14.  Konasewich D et al; Status Report on Organics and Heavy Metal Contaminants in the Lakes Erie, Michigan, Huron and Superior Basins Great Lakes Water Quality Board p. 26 (1978)
15.  Lyman WJ et al: Handbook of Chemical Property Estimation Methods Environ Behavior of Organic Compounds McGraw-Hill NY (1982)
16.  NIOSH; National Occupational Exposure Survey (NOES) (1983)
17.  Pellizzari ED et al; Formulation of Preliminary Assessment of Halogenated Aromatic Compound in Man and Environmental Media p. 77 USEPA 560/13-79-006 (1979)
18.  Pellizzari ED; Quantification of Chlorinated Hydrocarbons in Previously Collected Air Samples pp. 41-43 USEPA 450/3-78-112 (1978)
19.  Sabljic A; J Agric Food Chem 32: 243-6 (1984)

20. Shackelford WM, Keith LH; Frequency of Organic Compounds Identified in Water p. 119 USEPA 600/4- 76-062 (1976)
21. Spence JW, Hanst PL; J Air Pollut Control Fed 18: 250-3 (1978)
22. Swann RL et al; Res Rev 85: 17-28 (1983)
23. USEPA; The Verification of the Production of 56 Chemicals p. 7 Office of Toxic Substances Contract No. 68-02-4209 (1985)

# Piperidine

## SUBSTANCE IDENTIFICATION

**Synonyms:** Hexahydropyridine

**Structure:**

**CAS Registry Number:** 110-89-4

**Molecular Formula:** $C_5H_{11}N$

**Wiswesser Line Notation:** T6MTJ

## CHEMICAL AND PHYSICAL PROPERTIES

**Boiling Point:** 106 °C at 760 mm Hg

**Melting Point:** -7 °C

**Molecular Weight:** 85.15

**Dissociation Constants:** pKa = 11.2 [18]

**Log Octanol/Water Partition Coefficient:** 0.84 [8]

**Water Solubility:** Miscible [20]

**Vapor Pressure:** 30.18 mm Hg at 25 °C [2]

**Henry's Law Constant:** 4.45 x $10^{-6}$ atm-m³/mole [10]

## ENVIRONMENTAL FATE/EXPOSURE POTENTIAL

**Summary:** Piperidine is used commercially as a solvent, chemical intermediate, curing agent, catalyst, and complexing agent and is also found naturally in food. It appears to be easily biodegraded and transported. In soil and water, it is likely to biodegrade and in soil it is likely to leach into ground water. In water at neutral conditions, piperidine will be dissociated and therefore evaporation will be pH dependent. On dry soil, its high vapor pressure would suggest significant evaporation. Because of its low octanol/water partition coefficient, bioconcentration of piperidine in fish is unlikely. Estimations of its fate in the atmosphere suggest a half-life of 3.4 days. Primary human exposure is due to food consumption.

**Natural Sources:** Found in small amounts in black pepper [9]. Commonly detected in food products [15,24]. Found in brain, skin, and urine of animals and in the brain, cerebrospinal fluid, and urine of humans [14].

**Artificial Sources:** Piperidine is used as a solvent and chemical intermediate, as a curing agent for rubber and epoxy chemical resins, a catalyst for condensation reactions, an ingredient in oils and fuels, and a complexing agent [9].

**Terrestrial Fate:** Piperidine released to soil is likely to biodegrade, evaporate, and leach into ground water. Relative rates of these processes are unknown. Hydrolysis and absorption to soil are not important processes.

**Aquatic Fate:** Piperidine released to aqueous systems will biodegrade but the relative rates are unknown. Adsorption to sediments or suspended solids and evaporation should not be important because piperidine is highly dissociated at neutral pH and has a low Henry's Law constant. It is unknown whether photodegradation will be important.

**Atmospheric Fate:** No experimental data relative to the environmental fate of piperidine in the atmosphere is available. Its estimated half-life in the atmosphere based upon reaction with hydroxyl radicals is 3.4 days.

**Biodegradation:** Piperidine is readily biodegradable in screening tests using mixed soil and sewage [21] or sewage [19] inocula. Greater than 30% BOD theoretical in 2 weeks was observed with the mixed inoculum BOD test [21]. There is some evidence from pure culture studies that suggests under certain situations microorganisms may form N-nitrosopiperidine [12]; however, other pure culture studies suggest that piperidine is not nitrosated [1].

**Abiotic Degradation:** Photolysis of piperidine with 254 nm light in anaerobic aqueous solutions of $NaNO_2$ produced N-nitrosamines and N-nitramines [3]. Whether similar reactions will occur in natural waters with sunlight is unknown. Estimates of the half-life in the atmosphere based upon reaction with hydroxyl radical is 3.4 days ($8 \times 10^{+5}$ hydroxyl radicals/$cm^3$) [4].

**Bioconcentration:** The low octanol/water partition coefficient and the high amount of dissociation at neutral pH suggests that piperidine will not bioconcentrate in aquatic organisms.

**Soil Adsorption/Mobility:** The low octanol/water partition coefficient and high amount of dissociation at neutral pH suggests that piperidine will not adsorb strongly to soils or sediments.

**Volatilization from Water/Soil:** The Henry's Law constant would suggest that piperidine will not evaporate rapidly from water. Its high vapor pressure and low adsorption to soil would suggest that piperidine would evaporate rapidly from dry soil.

**Water Concentrations:** DRINKING WATER: Detected, not quantitated in District of Columbia drinking water [22]. Identified, not quantified, in drinking water in the US from unspecified locations [13]. SURFACE WATER: Detected in rivers of Germany - 9 samples, 6 pos. 0.5-9 ppb [15].

**Effluent Concentrations:** Municipal wastewater, unchlorinated, primary effluent, typical concn 13 ppb [23].

**Sediment/Soil Concentrations:** Detected, not quantitated, in soils not under cultivation [5].

**Atmospheric Concentrations:**

**Food Survey Values:** Detected in preserved vegetables (10 samples, 1 pos., 0.1 ppm), pickles (12 samples, 6 pos., 0.1-5.6 ppm), stimulants (10 samples, 7 pos., 0.1-9 ppm) [15]. Baked ham - 0.2 ppm [24].

## Piperidine

Coffee - 1 ppm [24]. Fish sausage - 9 ppb [7]. Baked ham - 0.81 ppm [7]. Detected as a volatile component of boiled beef at 0.9 ug/kg of beef [6].

**Plant Concentrations:**

**Fish/Seafood Concentrations:** Detected in fish (6 samples, 4 pos., 0.2-0.7 ppm) [15]. Cod roe - 0.210 ppm [5].

**Animal Concentrations:**

**Milk Concentrations:** Canned milk - 0.3 ppm; whole milk - 0.11 ppm [24].

**Other Environmental Concentrations:** Detected (0.8-16 ppm) in cigarette smoke condensate; not detected in unburned tobacco [24]. Identified, not quantified, in distillate of ground Latakia tobacco leaves; leaves were steam distilled from a strongly alkaline solution into hydrochloric acid [11].

**Probable Routes of Human Exposure:** Primary human exposure appears to be from consumption of food containing natural amounts of piperidine.

**Average Daily Intake:** AIR INTAKE: insufficient data; WATER INTAKE: assume 1-10 ppb, 2-20 ug; FOOD INTAKE: assume 0.01-0.1 ppm, 16-160 ug.

**Occupational Exposures:** NIOSH (NOES Survey 1981-83) has statistically estimated that 5,800 workers are exposed to piperidine in the United States [16]. NIOSH (NOHS Survey 1972-74) has statistically estimated that 720 workers are exposed to piperidine in the United States [17].

**Body Burdens:**

### REFERENCES

1. Archer MC et al; IARC Sci Publ 19: 239-246 (1978)
2. Boublik T et al; The Vapor Pressures of Pure Substances Vol 17 Amsterdam, Netherlands: Elsevier Science Publ (1984)
3. Challis BC, Li BFL; IARC Sci Publ 41: 31-40 (1982)

# Piperidine

4.  Fates of Atmospheric Pollutants estimate on the Graphic Exposure Modeling System, Office of Toxic Substances EPA (1986)
5.  Golovnya RV et al; IARC Sci Publ 41: 327-335 (1982)
6.  Golovnya RV et al; Chem Senses Flavour 4: 97-105 (1979)
7.  Hamano T et al; Agric Biol Chem 45: 2237-2243 (1981)
8.  Hansch C, Leo AJ; Medchem Project Issue No 26. Claremont CA: Pomona College (1985)
9.  Hawley GG; The Condensed Chemical Dictionary; 10th ed. Van Nostrand Reinhold Company New York (1981)
10. Hine J, Mookerjee PK; J Org Chem 40: 292-8 (1975)
11. Irvine WJ, Saxby MJ; Phytochem 8: 473-6 (1969)
12. Klein D et al; Ann Nutr Aliment 32: 425-435 Merck Index; An encyclopedia of chemicals and drugs; 9th ed. 2) (1976)
13. Kopfler FC et al; Adv Environ Sci Technol 8: 419-33 (1977)
14. Merck Index; An encyclopedia of chemicals and drugs. 9th (1976)
15. Neurath GB et al; Food Cosmet Toxicol 15: 275-282 (1977)
16. NIOSH; The National Occupational Exposure Survey (NOES) (1983)
17. NIOSH; The National Occupational Hazard Survey (NOHS) (1974)
18. Perrin DD; Aust J Chem 17:484-8 (1964)
19. Pitter P, Simanova J; Sb Vys Sk Chem Technol Praze,Oddil, F: Technol Vody Pros Tredi F22: 93-113 (1978)
20. Riddick JA et al; Organic Solvents: Physical Properties and Methods of Purification, 4th Edit. New York: J Wiley & Sons (1986)
21. Sasaki S; p.283-298 in Aquatic Pollutants; Transformation and Biological Effects. Pergamon Press (1978)
22. Scheiman MA et al; Organic Contaminants in the District of Columbia Water Supply, Biomed Mass System 4: 209-211 (1974)
23. Scully FEJr et al; Am Chem Soc Div Environ Chem 193rd Natl Mtg 27: 156-8 (1987)
24. Singer GM, Lijinsky W; J Agric Food Chem 24: 550-3 (1976)

# 1,2-Propanediol

## SUBSTANCE IDENTIFICATION

**Synonyms:** Propylene glycol

**Structure:**

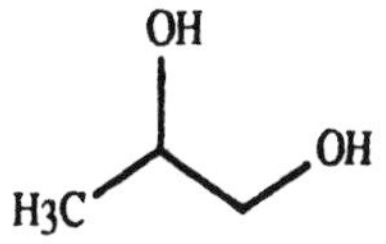

**CAS Registry Number:** 57-55-6

**Molecular Formula:** $C_3H_8O_2$

**Wiswesser Line Notation:** QY10

## CHEMICAL AND PHYSICAL PROPERTIES

**Boiling Point:** 189 °C

**Melting Point:**

**Molecular Weight:** 76.11

**Dissociation Constants:**

**Log Octanol/Water Partition Coefficient:** -0.92 [10]

**Water Solubility:** Miscible [16]

**Vapor Pressure:** 0.08 mm Hg at 20 °C [26]

**Henry's Law Constant:** $1.2 \times 10^{-8}$ atm-m³/mole [22]

## ENVIRONMENTAL FATE/EXPOSURE POTENTIAL

**Summary:** 1,2-Propanediol is released to the environment in wastewater effluents from its commercial production and use and by evaporation from paints, inks, and coatings in which it is used as a

375

solvent. Small amounts used for antifreeze and deicing applications may also be released. If released to the atmosphere, it is degraded rapidly by reaction with photochemically produced hydroxyl radicals (typical half-life of 32 hr). Physical removal from air by rainfall is possible. If released to water, 1,2-propanediol is expected to degrade relatively rapidly via biodegradation. If released to soil, relatively rapid biodegradation is also expected to occur. Significant leaching in soil can be predicted; however, concurrent biodegradation may proceed rapidly enough to diminish the importance of leaching. Exposure to the general population occurs through consumption of foods and drugs containing intentional additions of the compound. Dermal exposure occurs through use of cosmetics containing 1,2-propanediol. Occupational exposure occurs through inhalation and dermal contact.

**Natural Sources:** 1,2-Propanediol may form in the environment as a metabolite of propylene glycol dinitrate, which is a military propellant that may be found in wastewater streams from munitions plants and loading operations [12].

**Artificial Sources:** 1,2-Propanediol is present in various wastewater effluents generated at commercial sites of propylene glycols manufacture [14]. Use of 1,2-propanediol as a solvent in inks, paints, and coatings results in its direct evaporation to air. Small amounts of 1,2-propanediol may be used in antifreeze and deicing fluids [9] and result in environmental release.

**Terrestrial Fate:** 1,2-Propanediol has been shown to biodegrade readily by a number of biological screening studies, and is expected to biodegrade in soil. Based on its complete water solubility and log Kow, 1,2-propanediol can be expected to be susceptible to significant leaching. However, concurrent biodegradation may proceed rapidly enough to diminish the importance of leaching. Evaporation from dry surfaces is likely to be occur; however, volatilization from moist soils may not be significant.

**Aquatic Fate:** 1,2-Propanediol has been shown to biodegrade readily by a number of biological screening studies. Aquatic hydrolysis, oxidation, volatilization, bioconcentration, and adsorption to sediment are not expected to be significant fate processes. Therefore, when 1,2-propanediol is released to the aquatic environment, it is expected to be removed via biological processes.

# 1,2-Propanediol

**Atmospheric Fate:** 1,2-Propanediol is expected to exist almost entirely in the vapor phase in the ambient atmosphere based on the vapor pressure [7]. It is degraded rapidly in the vapor phase by reaction with photochemically produced hydroxyl radicals (estimated half-life of 32 hr in an average ambient atmosphere). The complete water solubility of 1,2-propanediol suggests that physical removal from the atmosphere via rainfall is possible.

**Biodegradation:** Standard dilution BOD water, 5-day 64% BODT, sewage inocula [3]. Warburg respirometer, 40-day 78% BODT, sewage inocula [11]. Nutrient broth, 100% degradation in 4 days (aerobic conditions), 100% degradation in 4-9 days (anaerobic conditions), activated sludge or digester sludge inocula, no significant degradation in sterile controls [12]. Standard dilution BOD water, 5-day 2.2% BODT, 10-day 56.7% BODT, 50-day 80% BODT, sewage inocula [13]. Standard dilution BOD water, 5-day 62% BODT, 20-day 79% BODT, sewage inocula [20]; synthetic seawater dilution, 5-day 55% BODT, 20-day 83% BODT, raw wastewater inocula [20]. Sewage die-away, 74.5% BODT in 5 days [25]. 1,2-Propanediol has been found to be degradable via anaerobic biotechnology [4,23]. Wastewater treatment, 95% removal in 6 hr, activated sludge inocula [8]. Standard dilution BOD water, 5-day 26.6% BODT; seawater dilution, 5-day 59.5% BODT [24].

**Abiotic Degradation:** The experimentally determined rate constant for the vapor-phase reaction of 1,2-propanediol with photochemically produced hydroxyl radicals has been reported to be $12 \times 10^{-12}$ cm$^3$/molecule-sec at 22 °C [2]; the atmospheric half-life for this reaction can be estimated to be 32 hours, assuming an average atmospheric hydroxyl radical concn of $5 \times 10^{+5}$ molecules/cm$^3$ [2]. The rate constant for the reaction of 1,2-propanediol with hydroxyl radicals in aqueous solution is approximately $0.94\text{-}1.68 \times 10^{+9}$ 1/mol-sec [6]; if the hydroxyl radical concn of sunlit natural water is assumed to be $1 \times 10^{-17}$ moles/L [17], the half-life would be approximately 1.3-2.3 years. Glycols are generally resistant to aqueous hydrolysis in the environment [15].

**Bioconcentration:** Based on the Kow, the BCF for 1,2-propanediol can be estimated to be <1 from a recommended regression-derived equation [15].

**Soil Adsorption/Mobility:** The complete miscibility in water and low Kow of 1,2-propanediol is indicative of very high mobility in soil.

**Volatilization from Water/Soil:** The value of the Henry's Law constant indicates that 1,2-propanediol is essentially not volatile from water [15].

**Water Concentrations:**

**Effluent Concentrations:** 1,2-Propanediol was qualitatively detected in a wastewater effluent from a chemical plant in Memphis, TN in Aug 1974 [21].

**Sediment/Soil Concentrations:**

**Atmospheric Concentrations:**

**Food Survey Values:**

**Plant Concentrations:**

**Fish/Seafood Concentrations:**

**Animal Concentrations:**

**Milk Concentrations:**

**Other Environmental Concentrations:**

**Probable Routes of Human Exposure:** 1,2-Propanediol is used as a solvent in foods, pharmaceuticals, and cosmetics [5]; exposure to the general population occurs through consumption of foods and drugs containing the compound [5]. Dermal exposure occurs through use of cosmetics containing 1,2-propanediol. Occupational exposure results from direct contact, or from inhalation of vapors and mists where the material is heated or violently agitated [5].

**Average Daily Intake:**

**Occupational Exposures:** 2,549,747 workers are potentially exposed to 1,2-propanediol based on statistical estimates derived from the NIOSH Survey conducted between 1972-74 in the United States [19]. 1,748,454 workers are potentially exposed to 1,2-propanediol based on preliminary statistical estimates derived from the NIOSH Survey conducted between 1981-83 in the United States [18]. Small amounts of 1,2-propanediol are used in deicing applications [9] and may result in occupational exposure.

## Body Burdens:

# REFERENCES

1. Anbar M, Neta P; Int J Appl Radiation and Isotopes 18: 493-523 (1967)
2. Atkinson RA; Chem Rev 85: 60-201 (1985)
3. Bridie AL et al; Water Res 13: 627-30 (1979)
4. Chou WL et al; Biotechnol Bioeng Symp 8: 391-414 (1979)
5. Clayton GD, Clayton FE; Patty's Industrial Hygiene and Toxicology, 3rd Rev Ed, NY: Wiley-Interscience 2C: 3852 (1982)
6. Dorfman LM, Adams GE; Reactivity of the Hydroxyl Radical in Aqueous Solution. NSRD-NBS-46, National Bureau of Standards (1973)
7. Eisenreich SJ et al; Environ Sci Technol 15: 30-8 (1981)
8. Grumwald A et al; Vodni Hospod 34: 247-52 (1984)
9. Fay RH; Kirk-Othmer Encycl Chem Tech 3rd Ed NY: Wiley 3: 79-94 (1978)
10. Hansch C, Leo AJ; Medchem Project Issue No 26. Claremont CA: Pomona College (1985)
11. Helfgott TB et al; An Index of Refractory Organics. USEPA-66/2-77-174 (1977)
12. Kaplan DL et al; Environ Sci Technol 16: 723-25 (1982)
13. Lamb CB, Jenkins GF; p.326-9 in Proc 8th Industrial Waste Conf, Purdue Univ (1952)
14. Liepins R et al; Industrial Process Profiles for Environmental Use. Chpt 6. USEPA-600/2-77-023f p.6-685 (1977)
15. Lyman WJ et al; Handbook of Chemical Property Estimation Methods NY:McGraw-Hill (1982)
16. Merck Index; An Encyclopedia of Chemicals, Drugs and Biologicals 10th ed p.1130-1 (1983)
17. Mill T et al; Science 207: 886-7 (1980)
18. NIOSH; National Occupational Exposure Survey (NOES) (1983)
19. NIOSH; National Occupational Hazard Survey (NOHS) (1974)
20. Price KS et al; J Water Pollut Control Fed 46: 63-77 (1974)
21. Shackelford WM, Keith JL; Frequency of Organic Compounds Identified in Water. USEPA-600/4-76-062 p.205 (1976)
22. Simmons P et al; p.212-7 in Book Pap, Int Tech Conf. Research Triangle Park, NC: Amer Assoc Text (1976)
23. Speece RE; Environ Sci Technol 17: 416A-427A (1983)
24. Takemoto S et al; Suishitsu Odaku Kenkyu 4: 80-90 (1981)
25. Wagner R; Vom Wasser 47: 241-65 (1976)
26. Weber RC et al; Vapor Pressure Distribution of Selected Organic Chemicals. USEPA-600/2-81-021 p.28 (1981)

# n-Propanol

SUBSTANCE IDENTIFICATION

**Synonyms:**

**Structure:**

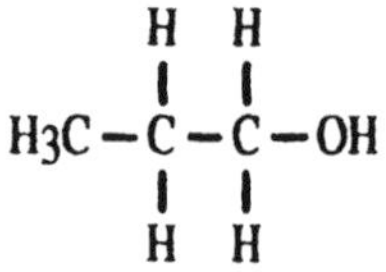

$$H_3C - \overset{\displaystyle \overset{H}{|}}{\underset{\displaystyle \underset{H}{|}}{C}} - \overset{\displaystyle \overset{H}{|}}{\underset{\displaystyle \underset{H}{|}}{C}} - OH$$

**CAS Registry Number:** 71-23-8

**Molecular Formula:** $C_3H_8O$

**Wiswesser Line Notation:** Q3

## CHEMICAL AND PHYSICAL PROPERTIES

**Boiling Point:** 97.2 °C

**Melting Point:** -126.5 °C

**Molecular Weight:** 60.09

**Dissociation Constants:**

**Log Octanol/Water Partition Coefficient:** 0.25 [6]

**Water Solubility:** Miscible [17]

**Vapor Pressure:** 20.99 mm Hg at 25 °C [17]

**Henry's Law Constant:** $6.85 \times 10^{-6}$ atm-m³/mole at 25 °C [9]

## ENVIRONMENTAL FATE/EXPOSURE POTENTIAL

**Summary:** n-Propanol will enter the environment as emissions from its manufacture and use as a solvent in industry and in many consumer products, and in emissions from fermentation and sewage treatment. It

380

naturally occurs as a plant volatile, and is released during microbial degradation of animal wastes and in the fermentation and spoilage of vegetable products. If spilled on land, it is apt to volatilize, leach into ground water, and biodegrade. Its fate in ground water in unknown. If released into water, it is apt to volatilize and biodegrade. It is not expected to adsorb to sediment or bioconcentrate in fish. In the atmosphere, it will photodegrade in days primarily by reaction with hydroxyl radicals. Rainout may be a significant process. Human exposure will be in occupational atmospheres, through use of products containing n-propanol, and proximity to sites where it is released, including industries and natural sources. Exposure will also be from ingestion of contaminated drinking water and foods containing n-propanol.

**Natural Sources:** Emissions occur from animal wastes and microbes, plant volatiles, forest fires, volcanos, and insects [5]. Emissions from fermentation and spoilage of many vegetable products, garbage, and landfills are also other sources. Natural sources are expected to release significant amounts of n-propanol [21].

**Artificial Sources:** Emissions occur from diesel exhaust, plastics combustion, printing, sewage treatment, whiskey production, and wood pulping [5]; emissions and wastewater from its manufacture of inks, nail polishes, polymers, degreasers, vegetable oils, and brake fluids as well as emissions from consumer products including lacquers, cosmetics, cleaners, polishes, and pharmaceuticals are likely sources of n-propyl alcohol [21].

**Terrestrial Fate:** If spilled on soil, n-propanol will volatilize and leach into the ground. Laboratory tests suggest that biodegradation in soil will be important. If degradation is not rapid, it is apt to leach into ground water. It may biodegrade in ground water.

**Aquatic Fate:** If released into water, n-propanol will volatilize (estimated half-life 6.5 days) and may biodegrade (based on laboratory tests). It is not expected to adsorb to sediment or bioconcentrate in fish.

**Atmospheric Fate:** If released into the atmosphere, n-propanol will photodegrade by reaction with hydroxyl radicals (estimated half-life is 6.7 days). Rainout may be a significant process based on water solubility.

**Biodegradation:** n-Propanol is readily degraded in laboratory tests using activated sludges, sewage seed, and wastewater inocula [4,15,16,22]. A 1-day theoretical BOD of 37% for 500 mg/L n-propanol using activated sludge at 20 °C was obtained [4]. Biodegradation of 3, 7, and 10 mg/L with filtered sewage seed in fresh water resulted in 64% BODT in 5 days and 75% in 20 days; in salt water, 43% BODT in 5 days and 73% in 20 days was observed [16]. n-Propanol was 99% removed using acclimated activated sludge at 20 °C (71 mg COD/g-hr rate) [15]. Although data are rather scant, available information concerning soil biodegradation [7] and anaerobic degradation [2,20,22] suggest that n-propanol will be degraded in these systems also.

**Abiotic Degradation:** The estimated half-life for n-propanol in the atmosphere is 2.6 days based on a rate constant of $3.8 \times 10^{-12}$ cm$^3$/mol sec [5] for the reaction with hydroxyl radicals and an estimated hydroxyl radical concentration of $8 \times 10^{+5}$ molecules/cm$^3$ [8]. Reaction with hydroxyl radicals in aquatic media will not likely be a significant process [1,3]. Alcohols are known to be resistant to hydrolysis [10].

**Bioconcentration:** No information on the bioconcentration factor for n-propanol could be found in the literature. However, its low octanol/water partition coefficient indicates that it will not bioconcentrate in fish.

**Soil Adsorption/Mobility:** No information on the soil or sediment adsorption of n-propanol could be found in the literature. Its low octanol/water partition coefficient indicates that its adsorption to soil will be low.

**Volatilization from Water/Soil:** The estimated half-life for evaporation of n-propanol from water 1 m deep with a 1 m/sec current and 3 m/sec wind is 4.3 days [10] based on the Henry's Law constant. The gas exchange rate plays a more dominant role than the liquid exchange rate [10]. n-Propanol's vapor pressure would suggest some evaporation from dry soil and other surfaces.

**Water Concentrations:** DRINKING WATER: Washington, DC - 0.001 ppm in water supply [19].

# n-Propanol

**Effluent Concentrations:** n-Propanol was detected in landfill leachates in Minnesota in the range of 76-37,000 ug/L (4 of 6 samples positive) [18]. It was also found in ground water suspected of leachate contamination (based on levels of inorganics) - 55 ug/L (2 of 13 samples pos), but not detected in ground water where inorganic levels indicated good or unknown water quality (all 7 samples negative) [18].

**Sediment/Soil Concentrations:**

**Atmospheric Concentrations:**

**Food Survey Values:** Identified in fermentation brine of pimentos used for stuffing olives [13]. n-Propanol found in 31 commercial brines - 180 to 380 mg n-propanol/L ethyl alcohol; Cahors wine - 33 mg n-propanol/L wine; Corbiers wine - 12 mg n-propanol/L wine; Jamaican Rum contained 31,300 mg n-propanol/L ethyl alcohol [13]. n-Propanol has also been found, not quantified, in beer and cistern room whiskey [21]. n-Propanol has been detected, not quantified, in milk products, including kefir and yogurt, and traces have been detected in fermented eggs [21].

**Plant Concentrations:** Plant volatile [5].

**Fish/Seafood Concentrations:** Detected in Japanese fish lances [12].

**Animal Concentrations:**

**Milk Concentrations:**

**Other Environmental Concentrations:**

**Probable Routes of Human Exposure:** Humans are most likely to be exposed to n-propanol in occupational settings associated with its use as a solvent or from fermentation process and sewage treatment plants. They also may be exposed when consuming products of fermentation or decomposition, such as hard liquor or beer [5].

**Average Daily Intake:**

**Occupational Exposure:** NIOSH (NOES Survey 1981-83) has statistically estimated that 112,488 workers are exposed to n-propanol in the United States [11].

# n-Propanol

**Body Burdens:** Detected in human milk for 4 urban areas (1 of 12 samples positive) [14].

# REFERENCES

1. Anbar M, Neta P; Int J Appl Rad Isotop 18: 493-523 (1967)
2. Chou WL et al; Biotech Bioeng Symp 8: 391-414 (1979)
3. Dorfman LM, Adams GE; Reactivity of the hydroxyl radical in aqueous solution p.51 NSRD-NBS-46 (1973)
4. Gerhold RM, Malaney GW; J Water Pollut Control Fed 38: 562-79 (1966)
5. Graedel TE; Chemical Compounds in the Atmosphere Academic Press NY p.249 (1978)
6. Hansch C, Leo AJ; Medchem Project Issue No.26 Pomona College, Claremont CA (1985)
7. Harada T, Nagashima Y; J Ferment Technol 53: 218-22 (1975)
8. Hewitt CN, Harrison RM; Atmos Environ 19: 545-54 (1985)
9. Hine J, Mookerjee PK; J Org Chem 40: 292-8 (1975)
10. Lyman WJ et al; Handbook of chemical property estimation methods. Environmental behavior of organic chemicals McGraw Hill NY (1982)
11. NIOSH; National Occupational Exposure Survey (1984)
12. Nonaha J et al; J Tokyo Univ Fish 62: 1-10 (1975)
13. Nosti VM, Fernandez GQ; Grasa Aceites 26: 369-77 (1975)
14. Pellizarri ED et al; Bull Environ Contam Toxicol 28: 322-8 (1982)
15. Pitter P; Water Res 10: 231-5 (1976)
16. Price KS et al; J Water Pollut Control Fed 46: 63-77 (1974)
17. Riddick JA et al; Organic Solvents: Physical Properties and Methods of Purification. Techniques of Chemistry. 4th Ed. New York: Wiley-Interscience pp 1325 (1986)
18. Sabel GV, Clark TP; Waste Manag Res 2: 119-30 (1984)
19. Scheiman MA et al; Biomed Mass Spect 4: 209-11 (1974)
20. Speece RE; Environ Sci Technol 17: 416A-27A (1983)
21. USEPA; CHIP (Draft) n-Propyl Alcohol (1983)
22. Wagner R; Vom Wasser 42: 271-305 (1974)

# Propionitrile

## SUBSTANCE IDENTIFICATION

**Synonyms:**

**Structure:**

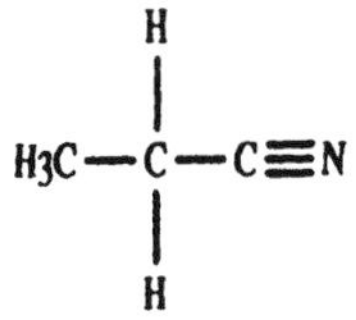

$$H_3C - \underset{\underset{H}{|}}{\overset{\overset{H}{|}}{C}} - C \equiv N$$

**CAS Registry Number:** 107-12-0

**Molecular Formula:** $C_3H_5N$

**Wiswesser Line Notation:**

## CHEMICAL AND PHYSICAL PROPERTIES

**Boiling Point:** 97.2 °C at 760 mm Hg

**Melting Point:** -91.8 °C

**Molecular Weight:** 55.08

**Dissociation Constants:**

**Log Octanol/Water Partition Coefficient:** 0.16 [5]

**Water Solubility:** 103,000 mg/L at 25 °C [12]

**Vapor Pressure:** 44.63 mm Hg at 25 °C [12]

**Henry's Law Constant:** 5.87 x $10^{-5}$ at 25 °C [8]

## ENVIRONMENTAL FATE/EXPOSURE POTENTIAL

**Summary:** Propionitrile may be released to the environment as fugitive emissions or in wastewater during its by-product formation during the electro-reduction of acrylonitrile to form adiponitrile. It will be slowly degraded in the atmosphere by reaction with hydroxyl radicals (half-life 83 days) and removal from the air by washout is possible. If released into water, it will volatilize and may biodegrade. Adsorption to sediment and bioconcentration in aquatic organisms will not be significant. If spilled on land, propionitrile will readily evaporate and may leach into ground water where its fate is unknown. Human exposure will most likely be via inhalation in occupational settings.

**Natural Sources:**

**Artificial Sources:** Propionitrile may be released in fugitive or wastewater releases during its by product formation during the electro-reduction of acrylonitrile to form adiponitrile [9]. It may also be emitted during shale oil processing [7,13] and from turbine combustion exhaust [4].

**Terrestrial Fate:** If released on land, propionitrile would be expected to leach readily into ground water and/or evaporate. It is reported to biodegrade rapidly in a screening study [14], and, therefore biodegradation in soil may be an important degradation process.

**Aquatic Fate:** If released into water, propionitrile would volatilize (half-life 13.3 hr in a model river). Biodegradation is possible but there is no supporting evidence from studies in environmental waters. Adsorption to the sediment or bioconcentration would not be significant.

**Atmospheric Fate:** If released into the atmosphere, propionitrile will degrade very slowly due to reaction with hydroxyl radicals (half-life 83 days [6]). Due to its high water solubility, it should be scavenged from the air by rain.

**Biodegradation:** In the only screening study available on its biodegradability, propionitrile was rapidly metabolized by activated sludge [14]. The indicated mechanism was by hydrolysis to the amide [14].

**Abiotic Degradation:** Propionitrile reacts slowly with photochemically produced hydroxyl radicals (half-life 83 days from experimental rate constant and assuming 12 hr of sunlight) [6]. Nitriles are unreactive with ozone and the expected half-life resulting from measured reaction with ozone is well over 100 days [1,6]. Nitriles are resistant to hydrolysis at neutral pH [11]. Alkyl nitriles do not absorb radiation >290 nm so direct photolysis should not be significant [2].

**Bioconcentration:** From its octanol/water partition coefficient, one can estimate a bioconcentration factor of 0.78 [10]. Propionitrile would therefore not be expected to bioconcentrate in aquatic organisms.

**Soil Adsorption/Mobility:** From its log octanol/water partition coefficient, one can estimate a Koc of 1.2 [10]. Propionitrile would therefore not be expected to adsorb significantly to soil.

**Volatilization from Water/Soil:** From the Henry's Law constant, one can estimate that the half-life of propionitrile in a model river (1 m deep with a 1 cm/sec current and 3 cm/sec wind) is 13.3 hr [10]. Volatilization from soil would be rapid due to its relatively high vapor pressure and low adsorption to soil.

**Water Concentrations:** DRINKING WATER: Detected, not quantified, in one sample of ground water-derived drinking water in a survey of 14 treated drinking waters from different sources [3]. Propionitrile was not found in a sample from the same supply after distribution [3].

**Effluent Concentrations:** Offgases from oil shale retort - Rio Blanco, Colorado 16-26 ppm (G/G) [13].

**Sediment/Soil Concentrations:**

**Atmospheric Concentrations:** Not detected in typical urban air or rural air in the oil shale region of western Colorado/Utah [7]. SOURCE AREA: detected, not quantified, in the air near a shale oil wastewater facility [7]. GC/MS detection limit was approx 0.5-1.0 ng/mL [7].

**Food Survey Values:**

**Plant Concentrations:**

**Fish/Seafood Concentrations:**

**Animal Concentrations:**

**Milk Concentrations:**

**Other Environmental Concentrations:**

**Probable Routes of Human Exposure:** Human exposure will most likely be by inhalation of trace contaminants of propionitrile in vapors of adiponitrile.

**Average Daily Intake:**

**Occupational Exposure:** Detected, not quantified, in an operating pilot-scale shale oil wastewater treatment plant [7]. GC/MS detection limit was approx 0.5-1.0 ng/mL [7].

**Body Burdens:**

## REFERENCES

1. Atkinson R, Carter WPL; Chem Rev 84: 437-70 (1984)
2. Calvert JG, Pitts JN Jr; Photochemistry Wiley p 455, 485 (1966)
3. Fielding M et al; Eng Water Res Cent p 49 TR-159 (1981)
4. Graedel TE; Chem Compounds in the Atmosphere Academic Press NY (1978)
5. Hansch C, Leo AJ; Medchem Project Issue No.26 Pomona College, Claremont CA (1985)
6. Harris GW et al; Chem Phys Lett 80: 479-83 (1981)
7. Hawthorne SB, Sievers RE; Environ Sci Technol 18: 483-90 (1984)
8. Hine J, Mookerjee PK; J Org Chem 40: 292-8 (1975)
9. Jarvis WF et al; Health and Effects Profile on Propionitrile. USEPA Contract 68-03-3228 Task II (1987)
10. Lyman WJ et al; Handbook of Chem Property Estimation Methods Environ Behavior of Org Compounds McGraw-Hill NY (1982)
11. Mill T; Structure Reactivity Correlation for Environ Reactions p 18 USEPA-500/11-79-012 (1979)
12. Riddick JA et al; Organic Solvents: Physical Properties and Methods of Purification. Techniques of Chemistry. 4th Ed. New York: Wiley-Interscience pp 1325 (1986)
13. Sklarew DS, Hayes DJ; Environ Sci Technol 18: 600-3 (1984)
14. Symons JM et al; J Air Water Pollut 4: 115 (1961)

# n-Propylamine

Synonyms:

Structure:

$$H_2N \diagdown\diagup CH_3$$

CAS Registry Number: 107-10-8

Molecular Formula: $C_3H_9N$

Wiswesser Line Notation: Z3

## CHEMICAL AND PHYSICAL PROPERTIES

Boiling Point: 48-49 °C

Melting Point: -83 °C

Molecular Weight: 59.11

Dissociation Constants: pKa = 10.708 at 20 °C [17]

Log Octanol/Water Partition Coefficient: 0.48 [8]

Water Solubility: Miscible [18]

Vapor Pressure: 307.9 mm Hg at 25 °C [18]

Henry's Law Constant: $1.2 \times 10^{-5}$ atm-m³/mole at 25 °C [9]

## ENVIRONMENTAL FATE/EXPOSURE POTENTIAL

Summary: n-Propylamine occurs naturally in various plants, such as tobacco and algae, as a bio-organic degradant, and as a volatile emission product from animal waste. It is released to the environment

389

through effluents from its industrial production and use. If released to the atmosphere, n-propylamine is rapidly degraded (estimated half-life of 12 hr) by reaction with photochemically produced hydroxyl radicals. If released to water, n-propylamine is physically removed by volatilization. Volatilization half-lives of 2.44 and 26.5 days have been estimated for a shallow (1 m deep) model river and environmental pond, respectively. Aquatic bioconcentration and adsorption to sediment are not expected to be important. If released to soil, n-propylamine is expected to be very mobile and easily leached based upon estimated Koc values of less than 50. Evaporation from dry soil is likely to occur. Several screening studies have demonstrated that n-propylamine is readily biodegraded by activated and non-activated sludges. The general population is primarily exposed through consumption of food products in which n-propylamine apparently occurs as a natural product. Inhalation exposure is possible in localized areas containing large quantities of animal waste, such as cattle yards. Occupational exposure is possible through inhalation and dermal contact at sites of commercial production and use.

**Natural Sources:** n-Propylamine has been reported to occur naturally in various species of marine algae [20] and in Latakia tobacco leaf [10]. n-Propylamine can occur naturally as a bio-organic degradant [1] and as a volatile emission product from animal waste [7].

**Artificial Sources:** An estimated 600 tons of n-propylamine are annually released in industrial effluents from industrial manufacturers and users of the compound [1].

**Terrestrial Fate:** When released to soil, n-propylamine is expected to be very mobile and easily leached based upon estimated Koc values of less than 50. Its relatively high vapor pressure suggests that rapid evaporation from dry surfaces may occur.

**Aquatic Fate:** n-Propylamine is physically removed from water by volatilization. Volatilization half-lives of 2.44 and 26.5 days have been estimated for a shallow (1 m deep) model river and environmental pond, respectively. Several screening studies have demonstrated that n-propylamine is readily biodegraded by activated and non-activated sludges; therefore, biodegradation in natural water may be possible (although not experimentally demonstrated). Aquatic bioconcentration and adsorption to sediment are not expected to be important.

# n-Propylamine

**Atmospheric Fate:** Based on the vapor pressure, n-propylamine is expected to exist almost entirely in the vapor phase in the ambient atmosphere [6]. Vapor-phase n-propylamine is degraded rapidly in the atmosphere by reaction with photochemically produced hydroxyl radicals (estimated half-life of 12 hr in an average atmosphere).

**Biodegradation:** n-Propylamine was readily bio-oxidized in Warburg respirometer studies using aniline-acclimated activated sludge [12]. A 13-day theoretical BOD of 102% was measured using a non-activated sludge inoculum [4]. n-Propylamine was readily acclimated to and metabolized by an activated sludge [22].

**Abiotic Degradation:** The rate constant for the vapor-phase reaction of n-propylamine with photochemically produced hydroxyl has been estimated to be $32.1 \times 10^{-12}$ cm$^3$/molecule-sec at 25 °C [2]; assuming an average atmospheric hydroxyl radical concn of $5 \times 10^{+5}$ molecules/cm$^3$, the half-life for this reaction is estimated to be 12 hr [2].

**Bioconcentration:** Based on the Kow, the log BCF for n-propylamine can be estimated to be 0.13 from a recommended regression-derived equation [11]. This low BCF value suggests little potential for bioconcentration.

**Soil Adsorption/Mobility:** Based on its miscibility and the log Kow, the Koc of n-propylamine can be estimated to be less than 50 from regression-derived equations [11]. These estimated Koc values are indicative of very high soil mobility [21].

**Volatilization from Water/Soil:** The Henry's Law constant of n-propylamine is indicative of potentially significant, but not rapid, volatilization from environmental waters [11]. The volatilization half-life from a model river (1 meter deep flowing 1 m/sec with a wind speed of 3 m/sec) has been estimated to be 2.44 days [11]. The volatilization half-life from an environmental pond is estimated to be 26.5 days [23].

**Water Concentrations:** DRINKING WATER: n-Propylamine was qualitatively detected in District of Columbia drinking water [19]. n-Propylamine was detected at a concn of 2.9 ppb in samples from the Elbe River in W Germany [14].

## n-Propylamine

**Effluent Concentrations:** An n-propylamine concn of 400 ppm was detected in a 1980 waste sample which was discharged into the ocean 74 km north of Arecibo, Puerto Rico from a pharmaceutical manufacturer [3].

**Sediment/Soil Concentrations:**

**Atmospheric Concentrations:** n-Propylamine has been qualitatively detected in air samples collected at a cattle feedyard [13]; the source of the amines in cattle feedyard air is likely the animal waste and/or decomposing manure [13].

**Food Survey Values:** The following naturally occurring concns (in mg/kg) were detected in various food products from W Germany: fresh rutabaga, 5.0; paprika, 2.3; paprika brine, 10.6; cucumber, 0.1-7.5; pepperoni, 1.4; pickled onions, 1.8; celery, 2.7; cheese, 2-8.7; brown bread, 1.6; freeze-dried coffee and coffee extract, trace-0.5 [14]. n-Propylamine has been qualitatively detected in wine, beer, rice, barley, malt, and corn [5,16].

**Plant Concentrations:** n-Propylamine has been identified in rice, corn, and barley [16]. n-Propylamine occurs naturally in various species of marine algae [20]. Latakia tobacco leaf has been found to contain n-propylamine [10].

**Fish/Seafood Concentrations:**

**Animal Concentrations:**

**Milk Concentrations:**

**Other Environmental Concentrations:**

**Probable Routes of Human Exposure:** The general population is primarily exposed to n-propylamine through consumption of food products. n-Propylamine apparently occurs as a natural product in many food items. Inhalation exposure is possible in localized areas containing large quantities of animal waste, such as cattle yards. Occupational exposure is possible through inhalation and dermal contact at sites of commercial production and use.

**Average Daily Intake:**

# n-Propylamine

**Occupational Exposure:** NIOSH (NOES Survey 1981-83) has statistically estimated that 28 workers are exposed to n-propylamine in the United States [15]. NIOSH (NOHS Survey 1972-74) has statistically estimated that 1806 workers are potentially exposed to n-propylamine in the United States [15].

**Body Burdens:**

# REFERENCES

1. Abrams EF et al; Identifications of Organic Compounds in Effluents from Industrial Sources. USEPA-560/3-75-002 p. 133 (1975)
2. Atkinson R; Inter J Chem Kinet 19: 799-828 (1987)
3. Brooks JM et al; pp. 171-98 in Wastes Ocean, Vol 1. Duedall IW ed. NY: Wiley (1983)
4. Chudoba J et al; Chem Prum 19: 76-80 (1969)
5. Drawert F; Vitis 5: 127-30 (1965)
6. Eisenreich SJ et al; Environ Sci Technol 15: 30-8 (1981)
7. Graedel TE et al; Atmospheric Chemical Compounds. Orlando, FL: Academic Press p. 394 (1986)
8. Hansch C, Leo AJ; Medchem Project Issue No.26 Pomona College, Claremont CA (1985)
9. Hine J, Mookerjee PK; J Org Chem 40: 292-8 (1975)
10. Irvine WJ, Saxby MJ; Phytochemistry 8: 473-6 (1969)
11. Lyman WJ et al; Handbook of Chemical Property Estimation Methods NY:McGraw-Hill (1982)
12. Malaney GW; J Water Pollut Control Fed 32: 1300-11 (1960)
13. Mosier AR et al; Environ Sci Technol 7: 642-5 (1973)
14. Neurath GB et al; Food Cosmet Toxicol 15: 275-82 (1977)
15. NIOSH; National Occupational Hazard Survey (1974): NIOSH; National Occupational Exposure Survey (1983)
16. Palamand SR et al; p. 54-8 in Amer Soc Brew Chem, Proc (1969)
17. Perrin DD; Dissociation Constants of Organic Bases in Aqueous Solution. IUPAC Chemical Data Series. Buttersworth: London (1965)
18. Riddick JA et al; Organic Solvents: Physical Properties and Methods of Purification. Techniques of Chemistry. 4th Ed. New York: Wiley-Interscience pp 1325 (1986)
19. Scheiman MA et al; Biomed Mass Spectrom 4: 209-11 (1974)
20. Steiner M, Hartmann T; Planta 79: 113-21 (1968)
21. Swann RL et al; Res Rev 85: 16-28 (1983)
22. Symons JM et al; Int J Air Water Pollut 4: 115-138 (1961)
23. USEPA; EXAMS II Computer Simulation (1987)

# Pyridine

**Synonyms:**

**Structure:**

**CAS Registry Number:** 110-86-1

**Molecular Formula:** C₅H₅N

**Wiswesser Line Notation:** T6NJ

## CHEMICAL AND PHYSICAL PROPERTIES

**Boiling Point:** 115-116 °C

**Melting Point:** -42 °C

**Molecular Weight:** 79.10

**Dissociation Constants:** pKa = 5.19 [6]

**Log Octanol/Water Partition Coefficient:** 0.65 [16]

**Water Solubility:** Miscible [35]

**Vapor Pressure:** 20 mm Hg at 25 °C [35]

**Henry's Law Constant:** $7.0 \times 10^{-3}$ atm-m³/mole [21]

## ENVIRONMENTAL FATE/EXPOSURE POTENTIAL

**Summary:** Pyridine is released to the environment in wastewater and as fugitive emissions during its production and use as a chemical intermediate and solvent. Energy-related processes such as coal and shale oil gasification is another important source of release. Several food items have been found to contain pyridine which is either in the food naturally or formed during cooking. Pyridine is contained in tobacco smoke and may contribute to its presence in indoor air. If released on land, pyridine will leach into the ground and biodegrade within approximately 8 days. If released into water, pyridine may be lost through biodegradation, photooxidation, and volatilization (half-life 90 hr for a model river). No biodegradation and photooxidation rates in natural waters are available. Bioconcentration in aquatic organisms should not be significant because of its high water solubility. In the atmosphere, pyridine will react slowly with photochemically produced hydroxyl radicals (half-life 32 and 16 days in clean and moderately polluted atmospheres, respectively) and be scavenged by rain. In polluted areas containing appreciable nitric acid vapor, reaction with nitric acid may be the major removal process. People are primarily exposed to pyridine in occupational settings, although the general public will be exposed from tobacco smoke and some food items.

**Natural Sources:**

**Artificial Sources:** Pyridine may be released to the environment in wastewater or in fugitive emissions during its production and use in the synthesis of agricultural chemicals, piperidine, pharmaceuticals, textile water repellents, and other chemicals, and as a solvent [10,17,38]. Pyridine is produced during shale oil gasification [34] and in coke ovens [13,38] where it is recovered from the resulting coal tar [38]. It is also found in tobacco smoke [13].

**Terrestrial Fate:** If released on land, pyridine will leach into the ground and biodegrade. It is adsorbed to acid clay to a moderate extent. Complete degradation in one soil occurred in less than 8 days.

**Aquatic Fate:** If released into water, pyridine should biodegrade after an acclimation period and be slowly lost through volatilization (half-life 90 hr for a model river). No estimates of biodegradation rates in natural waters are available. It may also be lost by photooxidation but no data containing actual rates were available. Adsorption to sediment or particulate matter in the water column should not be important because of the high water solubility.

**Atmospheric Fate:** If released into the atmosphere, pyridine will react slowly with photochemically produced hydroxyl radicals (half-life 32 and 16 days in clean and moderately polluted atmospheres, respectively) and be scavenged by rain. In situations where the atmosphere contains appreciable nitric acid vapor, reaction with nitric acid may be the major removal process.

**Biodegradation:** Results of biodegradability screening tests for pyridine using sewage or activated sludge inocula give mixed results ranging from rapid to no degradation [7,8,20,36,39,48]. Although biodegradability is improved in longer tests employing a more vigorous inocula, sometimes the same test gives disparate results [7,8,20,36,39]. Variations in the particular sewage or activated sludge inoculum would affect test results. One investigator obtained results ranging from 97% degradation in 6 days to no degradation in 30 days in 6 different standard tests [8]. After a preincubation step was added to one test in which no degradation was observed in 19 days, degradability increased to 91% [7]. However the closed bottle test which employs only a drop of sewage effluent as an inoculum, registered no degradation even with a preincubation step [7,8]. Tests that were designed to simulate biological treatment plants resulted in complete and rapid removal of pyridine [8,15]. In one test that employed a soil suspension as an inoculum, 100% degradation was obtained in 66-170 days [29]. When this test was repeated under anaerobic conditions, degradation was more rapid [29]. Complete degradation was obtained in 32-66 days [29]. When 2 micromole/g of pyridine was incubated with a silt loam soil, 11.7% remained after 4 days and none remained after 8 days [42].

**Abiotic Degradation:** Pyridine does not absorb light >290 nm [37]. It photolyzes by an indirect process in natural waters in the presence and absence of oxygen to form unidentified polar products [28]. No rates were reported for this process. Photolysis did not occur, however, in distilled water [28]. Pyridine was shown to react with the alkoxy radical to form pyridine N-oxide and the hydroxyl radical to form hydroxypyridines and more polar products [28]. Both of these radicals occur in natural waters. In the atmosphere, pyridine reacts with photochemically produced hydroxyl radicals with half-lives of 32 and 16 days in clean and moderately polluted atmospheres, respectively [1]. Reaction with ozone is negligible, but the reaction with gas phase nitric acid is significant (half-life <2 days with 1 ppb nitric acid) and could well be the major removal process in some situations [1].

**Bioconcentration:** Because of the high water solubility of pyridine, bioconcentration in water organisms is not expected.

Pyridine

**Soil Adsorption/Mobility:** The adsorption of pyridine to a basic subsoil (pH 8.15, 0.58% OC) is negligible, while in an acidic subsoil (pH 4.85, 0.24% OC), the Freundlich adsorption constant was measured to be 5.78 and the slope 0.679 [6]. This suggests a cationic adsorption mechanism as pyridine (pKa = 5.25 [6]) is predominantly in its protonated form. Pyridine adsorbs to colloidal particles of sodium montmorillonite and kaolinite, a process which is attributed to cation exchange and is a function of pH [6]. Adsorption is at a minimum at pH 1 and 11 and reaches a maximum at pH 4 for the montmorillonite and 5.5 for the kaolinite where the adsorption constants are 60 and 10, respectively [2]. The adsorption coefficient for the adsorption of pyridine to a sandstone from Bartlesville, OK is 0.035 at 38 °C [4].

**Volatilization from Water/Soil:** Using the Henry's Law constant, one can estimate a half-life for volatilization from a model river 1 m deep with a 1 m/sec current and a 3 m/sec wind of 90 hr [27].

**Water Concentrations:** DRINKING WATER: Pyridine was reported in drinking water in Cincinnati, OH [26]. SURFACE WATER: Pyridine was detected, not quantified, in the Cuyahoga River in the Lake Erie Basin [14]. Traces were found in the River Lee in England which receives effluents from many sewage treatment plants [46]. GROUND WATER: Two aquifers under the Hoe Creek coal gasification site contained 0.82-53 ppb of pyridine 15 months after gasification was completed [43]. Not detected in wells in Hanna and Gillete, WY prior to coal gasification [33].

**Effluent Concentrations:** In a survey of industrial effluents, pyridine was identified in discharges of the following industries (frequency of occurrence; median concn in ppb): timber products (1; 1032), paint and ink (5; 2), ore mining (3; 4), inorganic chemicals (3; 137), pharmaceuticals (2; 156), organic chemicals (11; 160), publicly owned treatment works (9; 77) [41]. Pyridine is contained in shale oil wastewater (7 ppm) and would be released to the atmosphere if the wastewater is heated as it would be when used to cool hot, retorted oil shale [3,18]. It was also found in the effluents from an advanced water treatment facility in Orange County, CA [26]. Wastewater from coal gasification contained an estimated 4.62 ppm of pyridine [9].

**Sediment/Soil Concentrations:** Pyridine was detected but not quantified in non-agricultural, loamy soil from the Moscow region [11].

Less than 0.22 ppm of the chemical was found in Eagle Harbor sediment, an area of Puget Sound that is contaminated with creosote [24].

**Atmospheric Concentrations:** Indoor and outdoor air in and near the shale oil wastewater treatment facility of Occidental Oil Shale Inc. at the Logan Wash site, CO contained 41 and 13 ug/m$^3$ of pyridine, respectively [19]. Rural air in an undeveloped area of the oil shale region as well as urban air (Boulder, CO) contained none of the chemical. Pyridine was detected in the air of residential homes in the Chicago area [23].

**Food Survey Values:** Pyridine has been identified as a volatile flavor compound in fried bacon [22] and boiled beef [12]. It is also a volatile component of Beaufort cheese, a Gruyere-type cheese manufactured in a limited area of the French Alps [5]. Volatile flavor components of fried chicken [44], roasted barely [47], and roasted coffee [45] contain pyridine.

**Plant Concentrations:**

**Fish/Seafood Concentrations:**

**Animal Concentrations:**

**Milk Concentrations:**

**Other Environmental Concentrations:** Pyridine has been identified as a component of tobacco [40], cigar [32], and marijuana [31] smoke.

**Probable Routes of Human Exposure:** People are primarily exposed to pyridine in occupational settings via dermal contact or inhalation with the vapor or aerosols. The general public will be exposed from tobacco smoke and some food items.

**Average Daily Intake:**

**Occupational Exposures:** According to the 1981-83 National Occupational Exposure Survey (NOES), 41,363 workers are potentially exposed to pyridine [30]. The 1972-74 National Occupational Hazard Survey (NOHS) reported that 29,000 workers are potentially exposed to pyridine [38]. This survey did not consider exposure from coke-ovens; the occupational categories listed for 88.7% of the estimated workers were laboratories (agricultural, biological, and chemical [38]). On the basis of a questionnaire sent out to major producers by the Pyridine Task Force in 1978, 813 persons were occupationally exposed to pyridine during its manufacture and first tier use (presumably as a synthetic intermediate [38]). Of the exposed workers, 557 were exposed for no longer than 8 hr/wk and that 8 hr TWA pyridine concentrations on the workplaces monitored were 0.008-1.0 ppm [38]. Pyridine is formed in the thermal decomposition of amine-cured epoxy powder paint and this could lead to occupational exposures if the epoxy resin is deposited on a surface hot enough to degrade the polymer (350 °C) [34].

**Body Burdens:** In a heterogenous, nonsmoking population of 62 subjects, pyridine was a common constituent in the expired air of the 14 prediabetic subjects [25].

## REFERENCES

1. Atkinson R et al; Environ Sci Technol 21: 64-72 (1987)
2. Baker RA, Lu MD; Water Res 5: 839-48 (1971)
3. Dobson KR et al; Water Res J 19: 849-56 (1985)
4. Donaldson EC et al; Adsorption Of Organic Compounds Cottage Grove Sandstone. BERC/RI - 75/4, Bartlesville, OK p. 16 (1975)
5. Dumont JP, Adda J; J Agric Food Chem 26: 364-7 (1978)
6. Felice LJ, et al; Quinoline Partitioning In Substance Materials Adsorption, Desorption, and Solute Competition PNL-SA-11728 Battelle Pacific NW Labs pp. 19 (1984)
7. Gerike P, Fisher WK; Ecotox Environ Safety 5: 45-55 (1981)
8. Gerike P, Fisher WK; Ecotox Environ Safety 3: 159-73 (1979)
9. Giabbai MF et al; Int J Environ Anal Chem 20: 113-29 (1983)
10. Goe GL, Kirk-Othmer Encyclopedia Chemical Technol 3rd ed Wiley-Interscience NY 19: 454-83 (1978)
11. Golovnya RV et al; Amines in Soil As Possible Precursors of N-Nitroso Compounds USSR Acad Med Sci pp. 327-35 (1982)
12. Golovnya RV et al; Chem Senses Flavour 4: 97-105 (1979)
13. Graedel TE; Chemical Compounds in the Atmosphere Academic Press, NY pp. 299 (1978)
14. Great Lakes Water Quality Board. An Inventory of Chemical Substances Identified in the Great Lakes Ecosystem Ontario, Canada 1: 195 (1983)
15. Gubser H; Gas Wasser Abwasser 49: 175-81 (1969)
16. Hansch C, Leo AJ; Medchem Project Issue No 26. Claremont CA: Pomona College (1985)
17. Hawley GG; Condensed Chem Dictionary 10th ed Van Nostrand Reinhold NY pp. 872 (1981)
18. Hawthorne SB et al; Environ Sci Tech 19: 922-7 (1985)
19. Hawthorne SB, Sievers RE; Environ Sci Technol 18: 483-90 (1984)

# Pyridine

20. Heukelekian H, Rand MC; J Water Pollut Contr Assoc 29: 1040-53 (1955)
21. Hine J, Mookerjee PK; J Org Chem 40: 292-8 (1975)
22. Ho CT et al; J Agric Food Chem 31: 336-42 (1983)
23. Jarke FH et al; ASHRAE Trans 87: 153-66 (1981)
24. Krone CA et al; Environ Sci Technol 20: 1144 (1986)
25. Krotoszynski BK, ONeill HJ; J Environ Sci 17: 855-83 (1982)
26. Lucas SV; GC/MS Analysis of Organics in Drinking Water Concentrations and Advanced Waste Treatment Concentrates EPA-600/1-84-020A Columbus, OH 1: 321 (1984)
27. Lyman WJ et al; Handbook of Chem Property Estimation Methods. Environ Behavior of Organic Compounds. McGraw-Hill NY pp 15.1-15.34 (1982)
28. Mill T et al; Science 207: 886-7 (1980)
29. Naik MN et al; Soil Biol. Biochem 4: 313-23 (1972)
30. NIOSH; National Occupational Exposure Survey (1985)
31. Novotny M et al; Chromatographia 15: 564-8 (1982)
32. Osman D, Barson J; Phytochemistry 3: 587-90 (1964)
33. Pellizzari ED et al; Astm Spec Tech Publ STP 686: 256-74 (1979)
34. Peltonen K; J Anal Appl Pyrolysis 10: 51-7 (1986)
35. Riddick JA et al; Organic Solvents: Physical Properties and Methods of Purification, 4th Edit. New York: J Wiley & Sons (1986)
36. Ruffo C et al; Ecotox Environ Safety 8: 275-9 (1984)
37. Sadlter; Sadlter Standard Spectra, Sadlter Res Lab, Philadelphia PA
38. Santodonato J et al; Monograph on Human Exposure to Chemicals in the Workplace SRC-TRC-84-1119 Syracuse Res Corp Syracuse NY (1984)
39. Sasaki S; pp. 283-98 in Aquatic Pollut, Hutzinger O, Von Letyoeld LH, Zoeteman BCJ eds Pergamon Press (1978)
40. Schmeltz I et al; J Agric Food Chem 27: 602-8 (1979)
41. Shackelford WM et al; Analyt Chem Acta 146: 15-27 (1983)
42. Sims GK, Sommers LE; J Environ Qual 14: 580-4 (1985)
43. Stuermer DH et al; Environ Sci Technol 16: 582-7 (1982)
44. Tang J et al; Agric Food Chem 31: 1287-1892 (1983)
45. Tijmensen WG; PT-Procestech 33: 575-80 (1978)
46. Waggot A; Chem Water Reuse 2: 55-9 (1981)
47. Wang PS et al; Agric Food Chem 33: 1775-81 (1969)
48. Young RHF et al; J Water Pollut Contr Fed 40: 354-68 (1968)

# 1,1,2,2-Tetrachloro-1,2-Difluoroethane

## SUBSTANCE IDENTIFICATION

**Synonyms:** Freon 112

**Structure:**

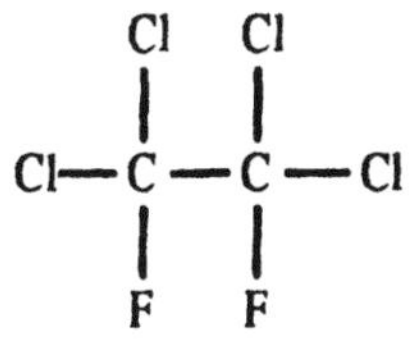

$$Cl—\underset{\underset{F}{|}}{\overset{\overset{Cl}{|}}{C}}—\underset{\underset{F}{|}}{\overset{\overset{Cl}{|}}{C}}—Cl$$

**CAS Registry Number:** 76-12-0

**Molecular Formula:** $C_2Cl_4F_2$

**Wiswesser Line Notation:**

## CHEMICAL AND PHYSICAL PROPERTIES

**Boiling Point:** 93 °C at 760 mm Hg

**Melting Point:** 25 °C

**Molecular Weight:** 203.83

**Dissociation Constants:**

**Log Octanol/Water Partition Coefficient:** 3.73 [11]

**Water Solubility:** 120 mg/L at 25 °C [12]

**Vapor Pressure:** 65.8 mm Hg at 20 °C [12]

**Henry's Law Constant:** 9.74 x $10^{-2}$ atm-m³/mole at 25 °C [7]

## ENVIRONMENTAL FATE/EXPOSURE POTENTIAL

**Summary:** Freon 112 may be released to the environment as emissions or in wastewater during its production, storage, transport, or use in dry cleaning and as a solvent. If released to soil, Freon 112 should rapidly volatilize from soil surfaces or leach through soil possibly into ground water. If released to water, essentially all Freon 112 is expected to be lost by volatilization (half-life 4 hr from a model river). If released to the atmosphere, Freon 112 will not degrade in the troposphere. This compound will gradually diffuse into the stratosphere (half-life 20 years). In the stratosphere Freon 112 will slowly photolyze or it will slowly react with singlet oxygen. As a result of its persistence in the atmosphere, this compound is transported long distances from its sources of emissions. The most probable routes of human exposure to Freon 112 are inhalation of contaminated air and dermal contact with this compound in occupational settings.

**Natural Sources:**

**Artificial Sources:** Freon 112 may be released to the environment as emissions during its production, storage, transport, or use in dry cleaning and as a solvent [6].

**Terrestrial Fate:** If released to soil, Freon 112 may rapidly volatilize from soil surfaces or leach through soil possibly into ground water.

**Aquatic Fate:** If released to water, essentially all Freon 112 is expected to be lost by volatilization (half-life 4 hr from a model river). Chemical hydrolysis, bioaccumulation, and adsorption to sediments are not significant fate processes in water.

**Atmospheric Fate:** If released to the atmosphere, essentially all Freon 112 is expected to exist in the vapor phase due to its high vapor pressure. The moderate water solubility of Freon 112 suggests that some loss by wet deposition occurs, but any loss by this mechanism is probably returned to the atmosphere by volatilization. Freon 112 will not degrade in the troposphere, thus diffusion into the stratosphere would be the ultimate removal process. The half-life for tropospheric to stratospheric diffusion of compounds is on the order of 20 years [4]. In the stratosphere, Freon 112 may undergo direct photolysis, producing chlorine atoms which in turn would participate in the catalytic removal of stratospheric ozone, or it may react with singlet oxygen. The stratospheric lifetime is predicted to be on the order of several decades. As a result of its persistence in the atmosphere, this compound is transported long distances from its sources of emissions.

**Biodegradation:**

# 1,1,2,2-Tetrachloro-1,2-Difluoroethane

**Abiotic Degradation:** Chemical hydrolysis of Freon 112 is not an environmentally significant fate process [5]. Freon 112 is essentially inert to reaction with photochemically generated radicals and ozone molecules [1,2,4]. This compound will not undergo direct photolysis in the troposphere [9]. Upon diffusion into the stratosphere, this compound will slowly photolyze to release chlorine atoms which in turn will participate in the catalytic removal of stratospheric ozone or it will slowly react with singlet oxygen [3,9]. By analogy to other Freon compounds, Freon 112 is predicted to have a stratospheric lifetime on the order of several decades [3].

**Bioconcentration:** Based on the water solubility, a bioconcentration factor (BCF) of 42 was estimated for Freon 112 [8]. This BCF value suggests that Freon 112 would not bioaccumulate significantly in aquatic organisms.

**Soil Adsorption/Mobility:** A soil adsorption coefficient (Koc) of 314 was estimated using a linear regression equation with the water solubility [8]. This Koc value suggests that Freon 112 would be moderately mobile in soil and that adsorption to suspended solids and sediments in water would be moderate [13].

**Volatilization from Water/Soil:** The value of Henry's Law constant suggests that Freon 112 would volatilize rapidly from all bodies of water and from soil surfaces [8]. Based on this value, the volatilization half-life of Freon 112 from a model river 1 m deep flowing 1 m/sec with a wind velocity of 3 m/sec is estimated to be 4 hr [8].

**Water Concentrations:**

**Effluent Concentrations:**

**Sediment/Soil Concentrations:**

**Atmospheric Concentrations:**

**Food Survey Values:**

**Plant Concentrations:**

**Fish/Seafood Concentrations:**

# 1,1,2,2-Tetrachloro-1,2-Difluoroethane

**Animal Concentrations:**

**Milk Concentrations:**

**Other Environmental Concentrations:**

**Probable Routes of Human Exposure:** The most probable route of exposure by the general population to Freon 112 is inhalation. Workers may be exposed by inhalation of contaminated air and dermal contact with this compound in occupational settings.

**Average Daily Intake:**

**Occupational Exposure:** NIOSH (NOES Survey 1981-83) has statistically estimated that 1433 workers are exposed to Freon 112 in the United States [10]. NIOSH has estimated that 25,166 workers are potentially exposed to Freon 112 based upon estimates derived from the NIOSH survey conducted 1972-74 in the United States [10].

**Body Burdens:**

## REFERENCES

1. Atkinson R; Chem Rev 85: 69-201 (1985)
2. Atkinson R; Internat J Chem Kinetics 19: 799-828 (1987)
3. Chou CC et al; J Phys Chem 82: 1-7 (1978)
4. Dilling WL; pp 154-97 in Environmental Risk Analysis for Chemicals; Conway RA ed NY: Van Nostrand Reinhold Co (1982)
5. Du Pont de Nemours Co; Freon Products Information B-2; A98825 12/80 (1980)
6. Graedel TE; Chemical Compounds in the Atmosphere NY: Academic Press p.327 (1978)
7. Hine J, Mookerjee PK; J Org Chem 40: 292-8 (1975)
8. Lyman WJ et al; Handbook of Chemical Property Estimation Methods. NY: McGraw-Hill (1982)
9. Makide T et al; Chem Lett 4: 355-8 (1979)
10. NIOSH; National Occupational Hazard Survey (1974): NIOSH; National Occupational Exposure Survey (1983)
11. PCGEMS; Graphical Exposure Modelling System, CLOGP3 Program, Office of Toxic Substances, U.S. EPA (1989)
12. Riddick JA et al; Organic Solvents: Physical Properties and Methods of Purification. Techniques of Chemistry. 4th Ed. New York: Wiley-Interscience pp 1325 (1986)
13. Swann RL et al; Res Rev 85: 17-28 (1983)

# 1,1,1,2-Tetrachloroethane

## SUBSTANCE IDENTIFICATION

**Synonyms:**

**Structure:**

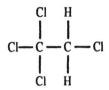

**CAS Registry Number:** 630-20-6

**Molecular Formula:** $C_2H_2Cl_4$

**Wiswesser Line Notation:**

## CHEMICAL AND PHYSICAL PROPERTIES

**Boiling Point:** 130.5 °C at 760 mm Hg

**Melting Point:** -70.2 °C

**Molecular Weight:** 167.85

**Dissociation Constants:**

**Log Octanol/Water Partition Coefficient:** 3.03 [8]

**Water Solubility:** 1100 mg/L at 25 °C [10]

**Vapor Pressure:** 12.03 at 25 °C [4]

**Henry's Law Constant:** 2.42 x $10^{-3}$ atm-m³/mole at 25 °C (calculated from vapor pressure and water solubility)

## ENVIRONMENTAL FATE/EXPOSURE POTENTIAL

**Summary:** It does not appear that 1,1,1,2-tetrachloroethane is presently produced in the United States or used commercially. It may, however,

405

be formed incidentally during the manufacture of other chlorinated ethanes and released into the environment as air emissions or in wastewater. If released on land, 1,1,1,2-tetrachloroethane would be expected to leach through soil and volatilize from the soil surface. If released into water, it would be primarily lost by volatilization (estimated half-life 4.2 hr from a model river). Adsorption to sediment is expected to be relatively low and bioconcentration in aquatic organisms would not be significant. 1,1,1,2-Tetrachloroethane is extremely stable in the atmosphere, reacting with photochemically produced hydroxyl radicals with an estimated half-life of 550 days. However, it will be susceptible to washout by rain. Due to its persistence, it will disperse over long distances and slowly diffuse into the stratosphere where it would be rapidly degraded.

**Natural Sources:**

**Artificial Sources:** 1,1,1,2-Tetrachloroethane has no commercial applications [9]. However, it is formed as an incidental product in the manufacture of other chloroethanes and, therefore, may be released to the environment as air emissions or in wastewater during the manufacture of these chloroethanes [1].

**Terrestrial Fate:** If released on land, 1,1,1,2-tetrachloroethane would be expected to leach into the ground and volatilize from the soil surface.

**Aquatic Fate:** If released into water, 1,1,1,2-tetrachloroethane would be primarily lost by volatilization (estimated half-life 4.2 hr from a model river). Adsorption to sediment is expected to be relatively low.

**Atmospheric Fate:** 1,1,1,2-Tetrachloroethane is extremely stable in the atmosphere (estimate half-life 550 days). It will be subject to washout by rain. Due to its persistence, it will disperse over long distances and slowly diffuse into the stratosphere where it would be rapidly degraded.

**Biodegradation:**

**Abiotic Degradation:** Hydrolysis of 1,1,1,2-tetrachloroethane is not significant at environmental temperatures and pHs; the half-life for this process exceeds 50 yr [13]. The estimated rate of reaction of 1,1,1,2-tetrachloroethane with photochemically produced hydroxyl radicals is $2.9 \times 10^{-14}$ cm$^3$/molecule-sec, resulting in a half-life of 550 days [2].

# 1,1,1,2-Tetrachloroethane

Degradation is rapid when irradiated in the presence of chlorine with 20% removal in 3.0 min [16]. Chlorine radicals are formed during the irradiation and are the reactive species. These radicals are also formed by the photolysis of chlorinated aliphatic compounds in the stratosphere.

**Bioconcentration:** No experimental data is available on the bioconcentration of 1,1,1,2-tetrachloroethane in aquatic organisms. Based on the water solubility, one can estimate a BCF of 12 using a recommended regression equation [12]. Confidence in this estimate is increased since similar compounds such as 1,1,1-trichloroethane and 1,1,2,2-tetrachloroethane have BCFs <10 and tissue half-lives of <1 day [3].

**Soil Adsorption/Mobility:** The experimentally determined Koc for 1,1,1,2-tetrachloroethane is reported to be 399 [14]. Based on the water solubility, one can estimate a Koc of 93 using a recommended regression equation [12]. Therefore, 1,1,1,2-tetrachloroethane should be moderately mobile in soil and may be expected to leach into ground water.

**Volatilization from Water/Soil:** The half-life for evaporation of the solute from a stirred dilute solution 6.5 cm deep into still air was 42.3 min [7] which translates into a half-life of 10.8 hr from a body of water 1 m deep. From the Henry's Law constant, one would estimate a volatilization half-life of 4.2 hr from a river 1 m deep with a 1 m/sec current with a wind speed of 3 m/sec with diffusion through the liquid phase controlling the rate of evaporation [12]. Since 1,1,1,2-tetrachloroethane has a moderate vapor pressure, and does not adsorb strongly to soil, it would be expected to volatilize from dry soil.

**Water Concentrations:** DRINKING WATER: US Ground Water Supply Survey (945 supplies derived from ground water chosen both randomly and on the basis that they may contain VOCs) - 1,1,1,2-tetrachloroethane was not detected in any supplies at a detection limit of 0.2 ppb [17]. It was confirmed to be in extracts from the Carrollton water plant in New Orleans which derives its water from the Mississippi River [11]. The concentration of 1,1,1,2-tetrachloroethane in a sample collected over 7 days was 0.11 ppb [11].

**Effluent Concentrations:** In a comprehensive survey of wastewater from 4000 industrial and publicly owned treatment works (POTWs) sponsored by the Effluent Guidelines Division of the US EPA, 1,1,1,2-tetrachloroethane was identified in discharges of the following industrial category (frequency of occurrence, median concn in ppb): organics and plastics (1, 27.4), inorganic chemicals (4, 14.8), and electronics (1, 272.6) [15]. The highest effluent concn was 272.6 ppb in the electronics industry [15].

**Sediment/Soil Concentrations:**

**Atmospheric Concentrations:** RURAL/REMOTE: Not detected in 2 sites in the US [5]. The mean concentrations of 1,1,1,2-tetrachloroethane in marine air at 8 sampling sites over the northern and southern Atlantic Ocean ranged from ND to 1.4 ppt, median 0.3 ppt [6]. URBAN/SUBURBAN: 602 sites/samples in US - 2.2 ppt median, 63 ppt maximum [5]. SOURCE AREAS: 43 sites/samples in US - ND in over 75% of samples, 0.071 ppt mean, 3.1 ppt maximum [5].

**Food Survey Values:**

**Plant Concentrations:**

**Fish/Seafood Concentrations:**

**Animal Concentrations:**

**Milk Concentrations:**

**Other Environmental Concentrations:**

**Probable Routes of Human Exposure:**

**Average Daily Intake:**

**Occupational Exposure:**

**Body Burdens:**

# 1,1,1,2-Tetrachloroethane

## REFERENCES

1. Archer WL; Kirk-Othmer Encyclopedia of Chemical Technology 3rd ed 5: 723-42 (1979)
2. Atkinson R; Int J Chem Kinet 19: 799-828 (1987)
3. Barrows ME et al; Dyn Exp Hazard Assess Toxic Chem Ann Arbor MI: Ann Arbor Sci pp. 379-92 (1980)
4. Boublik T et al; The Vapor Pressures of Pure Substances. Vol 17 Amsterdam, Netherlands: Elsevier (1984)
5. Brodzinsky R, Singh HB; Volatile Organic Chemicals in the Atmosphere. Menlo Park, CA: Atmospheric Science Center SRI International Contract 68-02-3452 (1982)
6. Class T, Ballschmiter K; Chemosphere 15: 413-27 (1986)
7. Dilling WL; Environ Sci Technol 11: 405-9 (1977)
8. Hansch C, Leo AJ; Medchem Project Issue No.26 Pomona College, Claremont CA (1985)
9. Hardie DWF; Kirk-Othmer Encyclopedia of Chemical Technology 2nd ed 5: 148-70 (1964)
10. Horvath AL; Halogenated Hydrocarbons: Solubility-Miscibility with Water. New York: Marcel Dekker pp 889 (1982)
11. Keith LH et al; pp. 329-73 in Ident Anal Organic Pollut Water Keith LH Ed Ann Arbor, MI: Ann Arbor Press (1976)
12. Lyman WJ et al; Handbook of Chem Property Estimation Methods. NY: McGraw-Hill (1982)
13. Mabey WR et al; in Symp Amer Chem Soc. Div Environ Chem 186th Natl Mtg Washington, DC 23: 359-61 (1983)
14. Rao PSC et al; J Environ Qual 14: 376-83 (1985)
15. Shackelford WM et al; Analyt Chim Acta 146: 15-27 (1983)
16. Spence JW, Hanst PL; J Air Pollut Control Fed 28: 250-3 (1978)
17. Westrick JJ et al; J Amer Water Works Assoc 76: 52-9 (1984)

# 1,1,2,2-Tetrachloroethane

## SUBSTANCE IDENTIFICATION

**Synonyms:**

**Structure:**

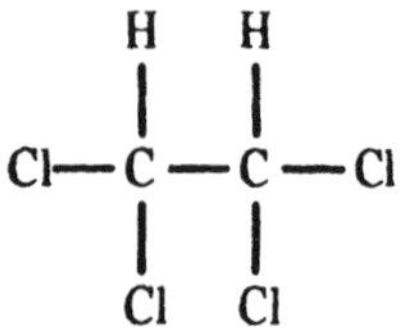

$$
\begin{array}{ccc}
& H & H \\
& | & | \\
Cl- & C - C & -Cl \\
& | & | \\
& Cl & Cl
\end{array}
$$

**CAS Registry Number:** 79-34-5

**Molecular Formula:** $C_2H_2Cl_4$

**Wiswesser Line Notation:**

## CHEMICAL AND PHYSICAL PROPERTIES

**Boiling Point:** 146.5 °C at 760 mm Hg

**Melting Point:** -36 °C

**Molecular Weight:** 167.86

**Dissociation Constants:**

**Log Octanol/Water Partition Coefficient:** 2.39 [16]

**Water Solubility:** 2962 mg/L at 25 °C [21]

**Vapor Pressure:** 6.1 mm Hg at 25 °C [11]

**Henry's Law Constant:** 4.55 x $10^{-4}$ atm-m³/mole at 25 °C (calculated from the water solubility and vapor pressure)

## ENVIRONMENTAL FATE/EXPOSURE POTENTIAL

**Summary:** Most of the released 1,1,2,2-tetrachloroethane enters the atmosphere where it is extremely stable (half-life >2 years). Some of the chemical will eventually diffuse into the stratosphere where it will rapidly photodegrade. 1,1,2,2-Tetrachloroethane which is released into water will primarily be lost by volatilization in a matter of days to weeks. The volatilization half-life in a model river and pond (the latter considers the effect of adsorption) has been estimated to be 6.3 hr and 3.5 days, respectively. 1,1,2,2-Tetrachloroethane is not expected to partition from the water column to organic matter contained in sediments and suspended solids. A measured Koc of 79 in a silt loam indicates 1,1,2,2-tetrachloroethane will be highly mobile in soil. When released to soil, part of the 1,1,2,2-tetrachloroethane may leach into ground water. There is evidence that 1,1,2,2-tetrachloroethane slowly biodegrades. A product of biodegradation under anaerobic conditions is 1,1,2-trichloroethane. Under alkaline conditions, 1,1,2,2-tetrachloroethane may be expected to hydrolyze. A measured aqueous hydrolysis rate constant of $KB = 2.3 \times 10^7 M^{-1}\text{-}yr^{-1}$ at pH of 9 and 25 °C corresponds to half-lives of 1.1 and 111 days at pH of 9 and 7. 1,1,2,2-Tetrachloroethane will not be expected to bioconcentrate in aquatic organisms. The major source of human exposure is from ambient air near industrial sources.

**Natural Sources:** 1,1,2,2-Tetrachloroethane is not known to occur as a natural product [22].

**Artificial Sources:** 1,1,2,2-Tetrachloroethane may be released into the atmosphere in connection with its manufacture of trichloroethylene from acetylene or from its use as a metal degreasing agent, paint, varnish, and rust remover, and as an extractant, solvent, and chemical intermediate [29,55]. 1,1,2,2-Tetrachloroethane can be emitted from hazardous waste landfills [18].

**Terrestrial Fate:** 1,1,2,2-Tetrachloroethane may undergo hydrolysis in alkaline soil. A measured aqueous hydrolysis rate constant of $KB = 2.3 \times 10^7 M^{-1}\text{-}yr^{-1}$ at pH of 9 and 25 °C corresponds to half-lives of 1.1 and 111 days of pH of 9 and 7 [25]. A measured Koc of 79 in a silt loam [6] suggests 1,1,2,2-tetrachloroethane will be highly mobile in soil [50] and therefore can leach into ground water. The calculated Henry's Law constant suggests volatilization of 1,1,2,2,-tetrachloroethane from moist soils should be important.

**Aquatic Fate:** Under alkaline conditions, 1,1,2,2-tetrachloroethane may be expected to hydrolyze. A measured aqueous hydrolysis rate constant of $KB = 2.3 \times 10^7 M^{-1}\text{-}yr^{-1}$ at pH of 9 and 25 °C corresponds to half-lives of 1.1 and 111 days at pH of 9 and 7 [25]. The primary loss of 1,1,2,2-tetrachloroethane from the aquatic compartment will be by evaporation which should have a half-life of days to weeks depending on the body of water in question. Based upon the calculated Henry's Law constant, the volatilization half-life from a model river can be estimated to be 6.3 hr [32]. Adsorption to sediment would not be a significant loss mechanism. The volatilization half-life from a model pond, which considers the effect of adsorption, has been estimated to be about 3.5 days [53]. Some biodegradation may occur in situations where evaporation is extremely slow and the body of water is rich in microorganisms, such as in an eutrophic lake. Biodegradation in ground water is possible under anaerobic conditions to form 1,1,2-trichloroethane.

**Atmospheric Fate:** Based upon the vapor pressure, 1,1,2,2-tetrachloroethane is expected to exist entirely in the vapor phase in ambient air [10]. 1,1,2,2-Tetrachloroethane is practically inert in the troposphere with a half-life exceeding 800 days. As such it will be transported long distances with some of it returning to earth in rain. It can be expected to diffuse slowly into the stratosphere where it will degrade rapidly by photodissociation. With continual release, one might expect to find increasing atmospheric concentrations.

**Biodegradation:** One investigator who incubated the tetrachloroethane with sewage seed for 7 days and followed that with three successive 7-day subcultures found no significant degradation under these conditions [51]. These results are in conflict with those of another investigator who obtained 41% degradation in 24 days in a modified shake flask biodegradability test using an unacclimated inoculum and 19% degradation in a river die-away test while 5 other chlorinated ethanes and ethenes tested were undegraded [34]. A continuous flow biofilm column operating under anaerobic conditions with a sewage inoculum achieved 97% steady state removal during 4 months of operation [2]. A product of the biodegradation was 1,1,2-trichloroethane [2]. The most commonly found products of microbial degradation of these compounds evidently come from reductive dehalogenation, while nonmicrobial degradations tend to involve hydrolysis and/or oxidation [46].

**Abiotic Degradation:** 1,1,2,2-Tetrachloroethane is virtually inert in the troposphere. The half-life for the reaction with photochemically produced hydroxyl radicals is >800 days or <0.1% loss per 12 hr sunlit day [44]. In the stratosphere it may photolyze [5] or degrade rapidly by reaction with chlorine radicals [48]. A measured aqueous hydrolysis rate constant of KB = 2.3 x $10^7$ $M^{-1}$-$yr^{-1}$ at pH of 9 and 25 °C corresponds to half-lives of 1.1 and 111 days at pH of 9 and 7 [25].

**Bioconcentration:** 1,1,2,2-Tetrachloroethane would not be expected to bioconcentrate in fish. The log of the bioconcentration factor in fish is reported to be 0.9-1 [1,24]. After 14 days exposure to an average water concn of 9.62 ug/L, the log bioconcentration factor of 1,1,2,2-tetrachloroethane in the tissue of bluegill sunfish (Lepomis macrochirus) was 0.9 [54].

**Soil Adsorption/Mobility:** A measured Koc of 79 in a silt loam [6], suggests 1,1,2,2-tetrachloroethane will be highly mobile in soil [50].

**Volatilization from Water/Soil:** Laboratory measurements of the rate of evaporation of 1,1,2,2-tetrachloroethane from water gave a half-life of 32-56 minutes [8,35]. In natural waters, one would expect a half-life for volatilization in the order of days to weeks depending on mixing conditions. Based upon the Henry's Law constant, volatilization from environmental waters should be important [32]. The volatilization half-life from a model river (1 meter deep flowing 1 m/sec with a wind speed of 3 m/sec) has been estimated to be 6.3 hr [32]. The volatilization half-life from an model pond, which considers the effect of adsorption, has been estimated to be 3.5 days [53]. Due to its moderate vapor pressure, volatilization from dry soil will be fairly rapid.

**Water Concentrations:** DRINKING WATER: Detected in 2 of 3 investigations of US drinking water [4,5,22]. In treated water from 30 Canadian treatment facilities - 1 site positive (1 ppb) in Aug/Sept, not detected in Nov/Dec [39]. Also found in 1 of 13 drinking water wells in Tacoma, WA [42]. 1,1,2,2-Tetrachloroethane was listed as a contaminant found in drinking water [28] for a survey of US cities including Pomona, Escondido, Lake Tahoe, and Orange Co, CA and Dallas, Washington, DC, Cincinnati, Philadelphia, Miami, New Orleans, Ottumwa, IA, and Seattle [31]. SURFACE WATER: Detected not quantified - River Glatt, Switzerland [56]. Trace to <1 ppb measured samples from the Ohio River [9,38]; 1 ppb detected in the Detroit River [26]. Trace to 1.9 ppb in the Schuylkill River at Philadelphia, PA [9]; 67 of 608 samples pos in representative New Jersey surface waters, max of 3 ppb measured [40]. Not detected in raw water for 30

## 1,1,2,2-Tetrachloroethane

Canadian potable drinking facilities in Aug/Sept and only one facility had detectable amounts in Nov/Dec - 12 ppb [39]. Only 12 of 204 sites near heavily industrialized areas across US were positive, positive sites ranged from 1 to 9 ppb [12]. 1,1,2,2-Tetrachloroethane is listed as a contaminant of Lakes Erie and Ontario, and the St Lawrence River [14]. GROUND WATER: New Jersey - 64 of 1072 representative ground water sources positive, 2.7 max [40]. Detected, not quantified, in 10 most polluted wells from a 408 well survey in New Jersey, with the wells being located under urban land use areas [15]. Ground water samples from near the Hooker Chemical and Plastics Corp disposal site at Love Canal, NY contained 1,1,2,2-tetrachloroethane [20]. Six of 7 ground water samples from nearby the "Valley of Drums", KY contained 1,1,2,2-tetrachloroethane at a concn of 6.4, 18, 12, 5.7, 0.2, and 6.2 ug/L [49].

**Effluent Concentrations:** Only the metal finishing industry has mean water effluents exceeding 20 ppb, the mean effluent level of 1,1,2,2-tetrachloroethane for this industry is 290 ppb and the maximum observed level is 570 ppb [52]. Detected in samples of effluents from 3 US chemical plants and a US sewage treatment plant [22]. Unidentified isomer from industrial effluent, Saint Clair River, Sarnia, Ontario detected at 5 sites [26]. The biotreatment and final effluents of a Class A oil refinery contained 1,1,2,2-tetrachloroethane at a concn of greater than 50 and less than 10 ug/L, respectively [47]. Wastewater from the gaseous diffusion plant operated by Union Carbide at Oak Ridge, TN contained 1,1,2,2-tetrachloroethane in the volatile fraction [33]. Leachate from Hooker Chemical and Plastics Corp disposal site at Love Canal, NY contained 1,1,2,2-tetrachloroethane [20]. An unspecified isomer of tetrachloroethane was identified as a product of coal combustion [23].

**Sediment/Soil Concentrations:** Sediment samples from near the Hooker Chemical and Plastics Corp disposal site at Love Canal, NY contained 1,1,2,2-tetrachloroethane [20].

**Atmospheric Concentrations:** RURAL: Not detected in 2 US samples [3]. URBAN/SUBURBAN: 853 US sites: 5.4 ppt median, 25% of samples exceed 8.9 ppt, max 4800 ppt, not detected in >25% of samples [3]. 0.01-9.4 ppb in urban atmospheres in Japan [22]. Trace to 57 ppb avg measured in studies covering major US cities [17,19,30,44,45]. INDUSTRIAL: 0 to 0.25 ppb in 5 industrial sites in US 3 of 5 pos, 2.70 ppb, 3 max with 1 site detected, not quantified

# 1,1,2,2-Tetrachloroethane

[41]. US - source dominated areas (60 samples) 0 ppt median, 25% of samples exceed 27 ppt, max 700 ppt [3]. 1,1,2,2-Tetrachlorethane was detected in the ambient air at Love Canal, Niagara Falls, NY [20].

**Food Survey Values:** Unspecified isomers of tetrachloroethane had been detected, concn not reported, in volatile flavor constituents of broiled beef [22]. The Food and Drug Administration's "market basket" collections were demarcated as fatty and non-fatty fractions at the 20% lipid point [7]. The high, low, and average 1,1,2,2-tetrachloroethane concn of the fatty and non-fatty food groups according to the extracted procedure were 118, 50, and 70 ng/g, and 122, 49, and 84 ng/g, respectively [7]. The high, low, and average 1,1,2,2-tetrachloroethane concn for the fatty and non-fatty food groups according to the cleaned up procedure were 85, 24, and 58 ng/g, and 89, 8, and 57 ng/g, respectively [7].

**Plant Concentrations:**

**Fish/Seafood Concentrations:**

**Animal Concentrations:**

**Milk Concentrations:**

**Other Environmental Concentrations:**

**Probable Routes of Human Exposure:** Humans are primarily exposed to 1,1,2,2-tetrachloroethane from ambient air or from contaminated drinking water.

**Average Daily Intake:** AIR INTAKE: (assume 5.4 ppt) 0.74 ug; WATER INTAKE: (assume 0-1 ppb) 0-2 ug.

**Occupational Exposure:** 1974 National Occupational Hazard Survey concluded that workers most likely to be exposed are those in industrial controls, toiletry preparations, and electric service industries, the latter exposure stemming from use of commercial solvent cleaners [22]. NIOSH has estimated that approximately 11,000 persons have occupational contact with 1,1,2,2-tetrachloroethane [27]. NIOSH (NOHS Survey 1972-74) has statistically estimated that 7201 workers are exposed to 1,1,2,2-tetrachloroethane in the United States [37]. NIOSH (NOES Survey 1981-83) has statistically estimated that 4,143 workers

## 1,1,2,2-Tetrachloroethane

are potentially exposed to 1,1,2,2-tetrachloroethane in the United States [36]. The average indoor air concn of 1,1,2,2-tetrachloroethane equaled the outdoor concn at 0.03 for an office building and 0.02 for a school [43].

**Body Burdens:** 1,1,2,2-Tetrachloroethane was detected in the adipose tissue, liver, and lungs of humans [13].

## REFERENCES

1. Barrows ME et al; Dyn Exposure Hazard Assess Toxic Chem Ann Arbor, MI: Ann Arbor Press p 379-92 (1980)
2. Bouwer EJ, McCarty PL; Appl Environ Microbiol 45: 1286-94 (1983)
3. Brodzinsky R, Singh HB; Volatile Organic Chemicals in the Atmosphere: An Assessment of Available Data. SRI Contract 68-02-3452 198 p (1982)
4. Callahan MA et al; Proc Natl Conf Munic Sludge Manage 8th p 55-61 (1979)
5. Callahan MA et al; Water-Related Environmental Fate of 129 Priority Pollutants. EPA-440/4-79-029b p 47-1 (1979)
6. Chiou CT et al; Science 206: 831-2 (1979)
7. Daft JL; J Agric Food Chem 37: 560-4 (1989)
8. Dilling WL; Environ Sci Technol 11: 405-9 (1977)
9. Dreisch R et al; Survey of the Huntington and Philadelphia River Water Supplies for Purgeable Organic Contaminants EPA-903/9-81-003 p. 14 (1981)
10. Eisenreich SJ et al; Environ Sci Technol 15: 30-8 (1981)
11. Engineering Sciences Data Unit; Vapor Pressures and Critical Points of Liquids. VII Halogenated Ethanes and Ethylenes Eng Sci Data Item 76004 p. 43 (1976)
12. Ewing BB et al; Monitoring to detect previously unrecognized pollutant in surface waters. EPA-560/6-77-015 p. 75 (1977)
13. Geyer HJ et al; Regul Toxicol Pharmacol 6: 313-47 (1986)
14. Great Lakes Water Quality Board; Report on Great Lakes Water Quality p. 195 (1983)
15. Greenberg M et al; Environ Sci Technol 16: 14-9 (1982)
16. Hansch C, Leo AJ; Medchem Project Issue No.26 Pomona College, Claremont CA (1985)
17. Harkov R et al; J Air Pollut Control Assoc 33: 1177-83 (1983)
18. Harkov R; J Environ Sci Health Part A Environ Sci Eng 20: 491-502 (1985)
19. Harkov R et al; Toxic and Carcinogenic Air Pollutants in New Jersey - Volatile Organic Substances. Unpublished work Trenton, NJ: Off Cancer Toxic Sub (1981)
20. Hauser TR, Bromberg SM; Environ Monit Assess 2: 249-72 (1982)
21. Horvath AL; Halogenated Hydrocarbons: Solubility-Miscibility with Water Marcel Dekker Inc NY NY p. 889 (1982)
22. IARC; Monograph. Some Halogenated Hydrocarbons 20: 477-89 (1979)
23. Junk GA et al; Organic Compounds from Coal Combustion. In: ACS Symp Ser 319 Fossil Fuels Util: 109-23 (1986)
24. Kawasaki M; Ecotox Environ Safety 4: 444-54 (1980)
25. Kolling HP et al; Hydrolysis Rate Constants, Partition Coefficients and Water Solubilities for 129 Chemical. A Summary of Fate Constants Provided for the

# 1,1,2,2-Tetrachloroethane

Concentration-Based Listing Program. U.S. EPA Environmental Research Lab, Athens GA pp 36 (1987)

26. Konasewich D et al; Great Lakes Water Quality Status Report on Organic and Heavy Metal Contaminants in the Lakes Erie, Michigan, Huron and Superior Basins. Windsor, Ontario, Great Lakes Quality Board 373 p. (1978)
27. Konietzko H; Hazard Asses Chem Dev 3: 401-48 (1984)
28. Kool HJ et al; Crit Rev Env Control 12: 307-57 (1982)
29. Kusz P et al; J Chromat 286: 287-291 (1984)
30. Lioy PJ et al; J Water Pollut Control Fed 33:649-57 (1983)
31. Lucas SV; GC/MS Anal of Org in Drinking Water Concentrations and Advanced Treatment ConcentratessVol 1 EPA-600/1-84-020A NTIS PB85-1282329 p. 397 (1984)
32. Lyman WJ et al; Handbook of Chemical Property Estimation Methods NY: McGraw-Hill pp. 15-15 to 15-29 (1982)
33. McMahon LW; Organic Priority Pollutants in Wastewater. NTIS DE83010817 Gatinburg, TN p. 220-49 (1983)
34. Mudder TI; Amer Chem Soc Div Environ Chem Presentation Kansas City, MO Sept p52-3 (1982)
35. Neely WB; Control Hazard Mater Proc Natl Conf 3rd p 197-200 (1976)
36. NIOSH; National Occupational Exposure Survey (NOES) (1989)
37. NIOSH; National Occupational Hazard Survey (NOHS) (1974)
38. Ohio River Valley Water Sanit Comm; Assessment of Water Quality Conditions. Ohio River Mainstream 1978-9 Cincinnati, OH p. 34 (1980)
39. Otson R et al; J Assoc Off Anal Chem 65: 1370-4 (1982)
40. Page GW; Environ Sci Technol 15: 1475-81 (1981)
41. Pellizzari ED; Environ Sci Technol 16:781-5 (1982)
42. Schilling RD; Pollut Engr 17: 25-27 (1985)
43. Sheldon LS et al; Indoor Air in Public Buildings Vol 1 p. 163 EPA/600 6-88 009a PB89-102503 (1988)
44. Singh HB et al; Atmos Environ 15: 601-12 (1981)
45. Singh HB et al; Environ Sci Technol 16: 872-80 (1982)
46. Smith LR, Dragun J; Environ Int 10 (4): 291-8 (1985)
47. Snider EH, Manning FS; Environ Int 7: 237-58 (1982)
48. Spence JW, Hanst PL; J Air Pollut Fed 28: 250-3 (1978)
49. Stonebreaker RD, Smith AJ; Containment and Treatment of a Mixed Chemical Discharge "Valley of Drums" Louisville KY, Contr Haz Mate Spills, Proc Natl Conf p. 1-10 (1980)
50. Swann RL et al; Res Rev 85: 16-28 (1983)
51. Tabak HH et al; J Water Pollut Control Fed 53: 1503-18 (1981)
52. USEPA; Treatability Manual. EPA-600/2-82-001a page I.12.10-1 to I.12.10-4 (1981)
53. USEPA; EXAMS II Computer Simulation (1987)
54. Veith GD et al; An Evaluation of Using Partition Coefficients and Water Solubility to Estimate Bioconcentration Factors for Organic Chemicals in Fish ASTM STP 707 Aquatic Toxicology Easton JG et al: (ed) Amer Soc Test Mater p. 116-29 (1980)
55. Verschueren K; Handbook of Environmental Data on Organic Chemicals 2nd ed New York, NY: Van Nostrand Reinhold Co. p 1075-6 (1983)
56. Zuercher I, Giger W; Vom Wasser 47: 37-55 (1976)

# Tetrachloroethylene

## SUBSTANCE IDENTIFICATION

**Synonyms:**

**Structure:**

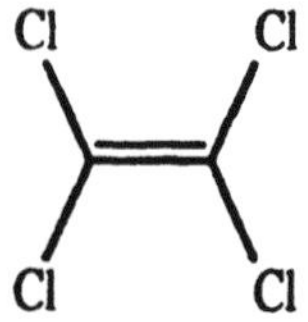

**CAS Registry Number:** 127-18-4

**Molecular Formula:** $C_2Cl_4$

**Wiswesser Line Notation:** GYGUYGG

## CHEMICAL AND PHYSICAL PROPERTIES

**Boiling Point:** 121 °C at 760 mm Hg

**Melting Point:** -19 °C

**Molecular Weight:** 165.82

**Dissociation Constants:**

**Log Octanol/Water Partition Coefficient:** 3.40 [37]

**Water Solubility:** 150.3 mg/L at 25 °C [41]

**Vapor Pressure:** 18.49 mm Hg at 25 °C [20]

**Henry's Law Constant:** 1.49 x $10^{-2}$ atm-m³/mole [75]

## ENVIRONMENTAL FATE/EXPOSURE POTENTIAL

**Summary:** Tetrachloroethylene (PCE) is likely to enter the environment by fugitive air emissions from dry cleaning and metal degreasing industries and by spills or accidental releases to air, soil, or water. If

418

# Tetrachloroethylene

PCE is released to soil, it will be subject to evaporation into the atmosphere and to leaching to the ground water. Biodegradation may be an important process in anaerobic soils based on laboratory tests with methanogenic columns. Slow biodegradation may occur in ground water where acclimated populations of microorganisms exist. If PCE is released to water, it will be subject to rapid volatilization with estimated half-lives ranging from <1 day to several weeks. It will not be expected to significantly biodegrade, bioconcentrate in aquatic organisms, or adsorb to sediment. PCE will not be expected to significantly hydrolyze in soil or water under normal environmental conditions. If PCE is released to the atmosphere, it will exist mainly in the gas-phase and it will be subject to photooxidation with estimates of degradation time scales ranging from an approximate half-life of 2 months to complete degradation in an hour. Some of the PCE in the atmosphere may be subject to washout in rain based on the solubility of PCE in water; PCE has been detected in rain. Major human exposure is from inhalation of contaminated urban air, especially near point sources such as dry cleaners, drinking contaminated water from contaminated aquifers and drinking water distributed in pipelines with vinyl liners, and inhalation of contaminated occupational atmospheres in metal degreasing and dry cleaning industries.

**Natural Sources:** PCE is not known to occur in nature.

**Artificial Sources:** Vaporization losses from dry cleaning and industrial metal cleaning [13]. Wastewater, particularly from metal finishing, laundries, aluminum forming, organic chemical/plastics manufacturing, and municipal treatment plants. It is also estimated that emissions account for approximately 90% of the PCE produced in the United States [82].

**Terrestrial Fate:** If PCE is released to soil, it will evaporate fairly rapidly into the atmosphere due to its high vapor pressure and low adsorption to soil. It can leach rapidly through sandy soil and therefore may reach ground water [79,91,101]. Biodegradation may be an important process in anaerobic soils based on laboratory tests with methanogenic columns. Slow biodegradation may occur in ground water where acclimated populations of microorganisms exist. There is some evidence of slow degradation in subsurface soils from a ground water recharge project. PCE should not hydrolyze under normal environmental conditions.

# Tetrachloroethylene

**Aquatic Fate:** If PCE is released in water, the primary loss will be by evaporation. The half-life for evaporation from water will depend on wind and mixing conditions and is estimated to range from 3 hours to 14 days in rivers, lakes, and ponds. Chemical and biological degradation are expected to be very slow. PCE will not be expected to significantly bioconcentrate in aquatic organisms or to adsorb to sediment. A mesocosm experiment was conducted to simulate Narragansett Bay during different seasons. Volatilization was the major removal process during all seasons and seasonal differences can be explained by hydrodynamics. The measured half-lives were 25 days in spring, 11 days in winter, and 14 days in summer [96]. In one study in which half-lives were calculated from concentration reduction between sampling points on the Rhine River and a lake in the Rhine basin, half-lives were 10 days and 32 days, respectively [102]. In a seawater aquarium, an 8-day half-life was demonstrated to be predominately the result of evaporation [46]. In a natural pond, PCE disappeared in 5 and 36 days at low (25 ppm) and high (250 ppm) dose levels, respectively [51].

**Atmospheric Fate:** If PCE is released to the atmosphere, it will be expected to exist in the vapor phase [27] based on a reported vapor pressure of 18.47 mm Hg at 25 °C [73]. Vapor phase PCE will be expected to degrade by reaction with photochemically produced hydroxyl radicals or chlorine atoms produced by photooxidation of PCE. Estimated photooxidation time scales range from an approximate half-life of 2 months [43,81] to complete degradation in an hour [24]. Some of the PCE in the atmosphere may be subject to washout in rain based on the solubility of PCE in water; PCE has been detected in rain.

**Biodegradation:** No degradation occurred in 21 days in 3 biodegradability tests with acclimated or unacclimated inocula or in a river die-away test [59]. Microbial degradation did not contribute to the removal of PCE in a mesocosm experiment which simulated Narragansett Bay, RI [96]. Under aerobic conditions there is no degradation in 25 weeks in a batch experiment with a sewage inoculum [6] or when low concentrations of PCE (16 ug/L) were circulated through an acclimated aerobic biofilm column over a period of 1 year [10]. While only 3.75% of the PCE treated by conventional, extended and 2-stage activated-sludge pilot plants appeared in the effluent, most of the PCE was discharged to the air from the extended aeration [98]. There is evidence that slow biodegradation of PCE occurs under anaerobic conditions when the microorganisms have been acclimated, yielding trichloroethylene (TCE) as a product [8,99]. An experiment in a continuous-flow laboratory methanogenic column using well-acclimated mixed culture and a 2-day detention time had an average PCE removal rate of 76% [9]. In a continuous-flow mixed-film methanogenic column with a liquid detention time of 4 days, mineralization of 24% of the PCE present occurred; TCE was the major intermediate formed (72%), but traces of dichloroethylene isomers and vinyl chloride (VC) were also found [95]. In other column studies under a different set of methanogenic conditions, nearly quantitative conversion of PCE to VC was found in 10 days [95]. Removal of 86% PCE occurred in a methanogenic biofilm column (8 weeks of activation followed by 9-12 weeks of acclimation [11]). A large reduction of PCE which had been recirculated through a soil column for 14 days was attributed to adsorption and volatilization [7]. In a microcosm containing muck from an aquifer recharge basin, 72.8% loss was observed in 21 days against 12-17% in controls, and the metabolites TCE, cis- and trans-1,2-dichloroethylene, dichloromethane, and chloroethene were identified [67]. However, when subsurface samples were aseptically removed from above and below the water table and incubated in the laboratory, no degradation occurred in 16 weeks [100]. In one field ground water recharge project, degradation was observed in the 50-day recharge period [6].

**Abiotic Degradation:** PCE reacts with hydroxyl radicals which are produced by sunlight in the troposphere with an estimated half-life of about 2 months or a loss of 1.5% per sunlit day [43,81]. Photooxidation in pure air with simulated tropospheric light is much faster than that predicted from the reaction with hydroxyl radicals with complete degradation occurring in 7 days in 1 report [84] and from 0.5% to 100% loss per hour in another [24]. The rate of loss is very sensitive to radiation in the 280-330 nm region and increases with increasing PCE concentration. The presence of nitrogen oxides has little effect on the rate of loss [24], and the main reaction product is phosgene (70-85%) with smaller amounts of carbon tetrachloride (8%), dichloroacetyl chloride, and trichloroacetyl chloride [84]. The proposed mechanism involved the molecular reaction with chlorine radicals produced by photooxidation of PCE [24]. Photodegradation in the stratosphere is rapid [60]. Some photodegradation occurs when PCE in air-saturated water is exposed to sunlight. In one year, 75% degradation occurred whereas 59-65% degradation was noted for dark controls [23]. When PCE adsorbed to silica gel is irradiated through a pyrex filter,

50-90% is lost in 6 days [33]. It is not clear whether PCE adsorbed on particulate matter will photodegrade as readily. Hydrolysis is not a significant degradative process (half-life 9 months at 25 °C in purified, de-ionized water) [23].

**Bioconcentration:** BCF: flathead minnow (Pimephales promelas), 38.9 [62]; bluegill sunfish (Lepomis macrochirus) 49 [4]. Based on the reported log Kow, a BCF of 226 was estimated [54]. Based on the reported and estimated BCF's, PCE will not be expected to significantly bioconcentrate in aquatic organisms.

**Soil Adsorption/Mobility:** Koc: 209 [77]; 210 [15]. In a laboratory system simulating a rapid-infiltration site, PCE appeared in the effluent but at significantly reduced concentration levels [44,45], although in a bank-infiltration system in Switzerland and the Netherlands, PCE was rapidly transported to ground water [34,71]. It is estimated that in a bay such as Narragansett Bay, RI, only about 0.01% of PCE is adsorbed to particulate matter [96]. A Koc of 238 was calculated [54] based on a reported Kom of 137.7 in a peaty soil [32]. Based on the reported log Kow, a Koc of 1,685 was estimated [54]. Based on the reported and estimated Koc's, PCE will be expected to exhibit low to medium mobility in soil [90] and therefore may leach slowly to ground water.

**Volatilization from Water/Soil:** PCE will evaporate rapidly from water based on estimates of half-life for the evaporation from water which range from fractions of an hour to several hours in laboratory experiments [16,22,54,85]. Two values of the ratio of the volatilization rate constant relative to the reaeration rate of oxygen are 0.52 [54] and 0.61 [74]. Using representative oxygen reaeration rates for various bodies of water, the half-lives for evaporation are as follows: pond 5-12 days; river 3 hr-7 days; lake 3.6-14 days [54]. Measured volatilization half-lives in a mesocosm simulating Narragansett Bay, RI were 11 days in winter, 25 days in spring, and 14 days in summer [96]. Due to its high vapor pressure and low adsorption to soil, volatilization of PCE from dry soil should be rapid.

**Water Concentrations:** DRINKING WATER: 180 US cities with finished surface water - 0.3 ppb median, 21 ppb max; 36 US cities with finished ground water - 3.0 ppb median; roughly 25% of the samples were positive [18]. Contaminated wells had much higher concentrations (a maximum of 1.5 ppm) [12,35]. 30 Canadian potable water treatment facilities (treated water) 1 ppb avg, 2 ppb max [66]; 230 ground water public drinking water sources in the Netherlands: 64 are >10 ppb, 12 are >100 ppb, 4 are >1 ppm, and 2 are >100 ppm [92]. Federal survey of finished waters in US: PCE occurred in 26.1% of ground water supplies, max concentration in ground water and surface water supplies 1500 and 21 ppb, respectively [25]. DRINKING WATER: Maximum concentration in tapwater from bank filtered Rhine water in the Netherlands 50 ppt [72]. Old Love Canal, Niagara Falls, NY (9 homes) 350-2900 ppt, 470 ppt median [3]. US surveys: State data, 1569 samples, 14% pos, trace to 3000 ppb, National Organics Monitoring Survey (NOMS, initiated in 1975), 113 samples, 42.4% pos, 0.2-3.1 ppb, National Screening Program (NSP, 1977-81), 142 samples, 16.9% pos, trace to 3.2 ppb, Community Water Supply Survey (CWSS, 1978), 452 samples, 4.9% pos, 0.5-30 ppb, Ground Water Supply Survey (GWS), 1982, finished drinking water, 466 samples selected at random from 1000 in survey, 7.3% pos, 0.5 ppb median, 23 ppb max [19]. GROUND WATER: 27 US cities, 0.6 ppb median (range 0.1-2 ppb) [18] San Fernando Valley, CA (1981-83) - 17 of 106 wells exceeded 4 ppb, max 130 ppb [14]. 10 British ground waters: equal or <2 ppb in 8 waters and higher levels at 2 sites where the aquifer was grossly polluted [31]. Ground water underlying 2 rapid infiltration sites 0.07 and 0.63 ppb [44]. Japan, national ground water survey, 1982, 1,083 shallow wells (most for domestic uses other than drinking water in private homes), 27% pos, 0.2-23,000 ppb, 277 deep wells (public, industrial, and commercial supplies), 30% pos, 0.2-150 ppb [55]. SURFACE WATER: 154 US cities - 2.0 ppb median, 13.6% positive [18]. Ohio R (1980-81, 11 stations, 4972 samples) - 49% positive, 340 basins in US (204 sites) - 77 sites above 1 ppb, 1 site above 11 ppb [30]. Lake Ontario (95 stations) 9 ppt, mean standard deviation 65 ppt [48]. Rhine R, km 865 (1976-82) 0.12-0.62 ppb with lower concentrations after 1978 [56]. Surface of Lake Zurich - 25-140 ppt, greater concentrations below the surface [36,78]. STORET data base, 9,323 data points, 38.0% pos, 0.100 ppb median [87]. SEAWATER: 0.1 to 0.8 ppt [61,68]. May be several orders of magnitude higher (10 ppb) near source, but concentration diminishes rapidly away from source [39]. Gulf of Mexico (open and coastal) 0-40 ppt where there is anthropogenic influence and <1 ppt in unpolluted areas [76]. Surface seawater Eastern Pacific Ocean 1981 (0-10 m depth), 30 samples, 90% pos, range of pos, 0.1-2.8 ppt, avg of all data, 0.7 ppt [83]. RAIN/SNOW: West Los Angeles (3/26/82) - 21 ppt [49]. Industrial city in England - 150 ppt [68]. La Jolla,

California - 5.7 ppt [89]. Central and Southern California - 1.4 and 2.3 ppt resp [89].

**Effluent Concentrations:** Industrial 1-20 ppb; Municipal treatment plants 1-10 ppb [88]; Baltimore Municipal Treatment Plant 8-129 ppb (higher levels in winter) [39]. Industries in which mean or maximum levels in raw wastewater exceeded 1 ppm are (number of samples, percent pos, mean pos, max ppm): raw wastewater: auto and other laundries (28 samples, 71.4% pos, <8.4 ppm mean, 93 ppm max), aluminum forming (4, 100%, <2.6, <4.0), metal finishing (96, 42.7%, 4.5, 110), organic chemical/plastics manufacturing (number of samples not reported, 19 pos, 5.1 mean, max concn not reported), and paint and ink formulation (36, 55.6%, 0.95, 4.9); treated wastewater: auto and other laundries (5 samples, 80% pos, 0.58 ppm mean, 1.0 ppm max), aluminum forming (16, 87.5%, <0.24, 3.0), metal finishing (not reported), organic chemical/plastics manufacturing (number of samples not reported, 14 pos, 0.047 mean, max concn not reported), and paint and ink formulation (24, 33.3%, 0.19, 0.70) [93]. Industrial effluent, STORET data base, 1,390 data points, 10.1% pos, 5.0 ppb median [87].

**Sediment/Soil Concentrations:** SEDIMENT: Liverpool Bay/172 stations - 4.8 ppt avg [68]. STORET data base, 359 data points, 7% pos, <0.050 ppb median [87].

**Atmospheric Concentrations:** US 577 sites, 1 ug/m$^3$ median [26]. BACKGROUND: Northern hemisphere background - 40 ppt [82]. RURAL/REMOTE: Point Barrow, Alaska - 128 ppt max (Feb), 56 ppt min (Sept) [50]. US - remote sites - typical levels 20-130 ppt [50,53,82]. Northern and Southern Atlantic, 7 sites, 85.7% pos, 84 samples, range of means of pos, 0.05-0.27 ppt [17]. Norwegian arctic air, 9 samples, July 1982, 0.0184 ppb avg, Spring 1983, 0.0382 ppb avg [42]. URBAN/SUBURBAN: Seven U.S. cities, 1980-81, range of means, 0.290-0.590 ppb, 7.60 ppb max, background, 0.050 ppb; 3 cities in New Jersey, 1981, 6 weeks in summer, range of means, 0.240-0.450 [1]. Avg worldwide distribution in 1978 (ppt): Northern Hemisphere, 56.0, Southern Hemisphere, 14.0, Global, 35.0 [40]. URBAN/INDUSTRIAL: US - Urban/Industrial Areas - typical levels 0.3-1.5 ppb but reaches 10 ppb and even higher [52,53,69,80,89]. INDUSTRIAL/SOURCE DOMINATED: Old Love Canal, Niagara, NY - Ambient air outside and inside 9 homes, 109 and 71 ppt median [3]. Classroom and playground in school situated near facility, 1.9 and 0.15 ppb, respectively [58]. Nursing home situated near former chemical waste dump, 1.2 and 0.2 ppb on first and second floors, respectively [58].

**Food Survey Values:** Chinese style sauce, 2 ppb; Quince jelly, 2.2 ppb; Crab apple jelly, 2.5 ppb; Grape jelly, 1.6 ppb; Chocolate sauce, 3.6 ppb. Not detected in seven market basket composites of meats (detection limit = 4.6 ppb), oils and fats (detection limit = 13 ppb), beverages (detection limit = 0.5 ppb), or dairy products (detection limit = 2.3 ppb) [28]. Various categories of food in England - 0.01-13 ppb, highest values in fats and oils [57]. US, wheat, 10 samples, 20% pos, 1.8-2.1 ppb, corn, 2 samples, 100% pos, 0.45-0.54 ppb; not detected in one sample each of oats and corn grits, 2 samples of corn meal [38].

**Plant Concentrations:** 13-23 ppb in marine algae [68].

**Fish/Seafood Concentrations:** 0.3-43 ppb in marine fish, 0.5-176 ppb in marine invertebrates in England [68], 250 ppb in American eel (Delaware River), 1050 ppb in American eel (Newark Bay), 77 ppb in carp (Delaware River), 108 ppb in striped bass (Raritan River), 88 ppb in spot fish (Houston Ship Channel) [21]. Rhine River from Strassburg to Lake Constance - a small number of fish 25-100 ppb, a few exceeded 100 ppb [5].

**Animal Concentrations:** 0.6-19 ppb in grey seal blubber (NE Coast of England); 1.4-39 ppb in marine and freshwater birds (coast of England) [68].

**Milk Concentrations:** Not detected (detection limit = 2.3 ppb) in seven market basket composites of dairy products [29].

**Other Environmental Concentrations:**

**Probable Routes of Human Exposure:** Human exposure to PCE will occur through inhalation of contaminated ambient air and ingestion of contaminated drinking water (especially from polluted ground water sources). Occupationally, exposure will occur from inhalation of contaminated air (especially in urban/industrial areas, in and around metal degreasing and dry cleaning industries). Food does not appear to be a major source, but the data are poor.

# Tetrachloroethylene

**Average Daily Intake:** AIR INTAKE: (assume 0.3-1.5 ppb [52,53,69,80,89] 41-207 ug; WATER INTAKE: (assume 0.3-3 ppb [18]) 0.6-6 ug; FOOD INTAKE: insufficient data.

**Occupational Exposures:** Time-weighted average (8-hr) exposures to PCE in the dry cleaning industry are reported as high as 178 ppm in air [65]. NIOSH (NOES Survey 1981-83) has statistically estimated that 536,688 workers are exposed to PCE in the US [63]. NIOSH (NOHS Survey 1972-74) has statistically estimated that 1,597,072 workers are exposed to PCE in the US [64].

**Body Burdens:** Has been detected in 7 of 8 samples in human milk from 4 urban areas in the US [70]. One hour after a visit to a dry cleaning plant, one sample of human milk contained 10 ppm PCE. This decreased to 3 ppm after 24 hr [47]. Old Love Canal, NY - 9 individuals: Human breath 600-4500 ng/m$^3$; Blood 0.35-260 ng/mL; Urine 120-690 ng/mL [3]. Human body fat (8 subjects) 0.4-29.2 ppb; Various human organs less than 6 ng/g [57]. Alveolar air geometric mean in 136 residents living near 12 dry-cleaning stores were: Living equal to or <5 floors above the stores 5 mg/m$^3$, adjacent houses 1 mg/m$^3$, one house away 0.2 mg/m$^3$, across street <.1 mg/m$^3$, whereas the mean concentration in 18 workers was 73 mg/m$^3$ [94]. Whole blood, US survey of 250 (121 males, 129 females), 0.7-23 ppb, 2.4 ppb avg [2]. Breath samples (ug/m$^3$, weighted statistics), Elizabeth and Bayonne, NJ, 1981, 295-339 samples, 93% pos, 280 max, 13.0 avg, 6.8 median [97]. Alveolar air in children and teachers in school situated near factory were 24 ug/m$^3$ avg for children and 11 and 47 ug/m$^3$ for the teachers [58]. The mean concentration of PCE in the classroom was 13 ug/m$^3$ [58]. Alveolar air of residents of a nursing home situated near a former chemical waste dump averaged 7.8 ug/m$^3$ first floor and 1.8 ug/m$^3$ on the second floor, where ambient concentrations averaged 8.2 and 1.6 ug/m$^3$, respectively [58]. US FY82 National Human Adipose Tissue Survey specimens, 46 composites, 61% pos (>3 ppb, wet tissue concn), 94 ppb max [86].

## REFERENCES

1. Andelman JB; Environ Health Persp 62: 313-8 (1985)
2. Antoine SR et al; Bull Environ Contam Toxicol 36: 364-71 (1986)
3. Barkley J et al; Biomed Mass Spectrom 7: 139-47 (1980)
4. Barrows ME et al; Dyn Exposure Hazard Assess Toxic Chem. Ann Arbor MI: Ann Arbor Sci. p. 379-92 (1980)

5. Binnemann PH et al; A Lebensm - Unters Forsch 176: 253-61 (1983)
6. Bouwer EJ et al; Environ Sci Technol 15: 596-9 (1981)
7. Bouwer EJ et al; Water Res 15: 151-59 (1981)
8. Bouwer EJ, McCarty PL; Appl Environ Microbiol 45: 1286-94 (1983)
9. Bouwer EJ, McCarty PL; Ground Water 22: 433-40 (1984)
10. Bouwer EJ, McCarty PL; Environ Sci Technol 16: 836-43 (1982)
11. Bouwer EJ, Wright JP; Am Chem Soc Div Environ Chem. 191st Natl Meet 26: 42-5 (1986)
12. Burmaster DE; Environ 24: 6-13, 33-6 (1982)
13. Chemical Marketing Reporter. Chemical Profile March 14, 1983 (1983)
14. Chemical Engineering 90: 35 (1983)
15. Chiou CT et al; Science 206: 831-2 (1979)
16. Chiou CT et al; Environ Inter 3: 231-6 (1980)
17. Class T, Ballschmiter K; Chemosphere 15: 413-27 (1986)
18. Coniglio WA et al; Occurrence of Volatile Organics in Drinking Water. p. 7 Unpublished EPA report (1980)
19. Cotruvo JA et al; pp. 511-30 in: Organic Carcinogens in Drinking Water (1986)
20. Daubert TE, Danner RP; Data Compilation Tables of Properties of Pure Compounds. American Institute of Chemical Engineers. pp 450 (1985)
21. Dickson AG, Riley JP; Mar Pollut Bull 7: 167-9 (1976)
22. Dilling WL; Environ Sci Technol 11: 405-9 (1977)
23. Dilling WL et al; Environ Sci Technol 9: 833-8 (1975)
24. Dimitriades B et al; J Air Pollut Control Assoc 33: 575-87 (1983)
25. Dyksen JE, Hess AF III; J Amer Water Works Assoc 74: 394-403 (1982)
26. Eichler DL, Mackey JH; Proc APCA 79th Ann Meeting pp. 17 (1986)
27. Eisenreich SJ et al; Environ Sci Technol 15: 30-8 (1981)
28. Entz RC, Hollifield HC; J Agric Food Chem 30: 84-8 (1982)
29. Entz RC et al; J Agric Food Chem 30: 846-9 (1982)
30. Ewing BB et al; Monitoring to Detect Previously Unrecognized Pollutants in Surface Water. EPA-560/6-77-015 & EPA-560/6-77-015A (1977)
31. Fielding M et al; Environ Technol Lett 2: 545-50 (1981)
32. Friesel P et al; Fresenius Z Anal Chem 319: 160-4 (1984)
33. Gaeb S et al; Nature 270: 331-3 (1977)
34. Giger W et al; Ges, Wasser, Abwasser 63: 517-31 (1983)
35. Giger W, Molnar-Kubica E; Bull Environ Contam Toxicol 19: 475-80 (1978)
36. Grob K, Grob G; J Chrom 90: 303-13 (1974)
37. Hansch C, Leo AJ; Medchem Project Issue No 26. Claremont CA: Pomona College (1985)
38. Heikes DL, Hopper ML; J Assoc Off Anal Chem 69: 990-8 (1986)
39. Helz GR, Hsu RY; Limnol Oceanogr 23: 858-69 (1978)
40. Herbert P et al; Chem Ind 24: 861-9 (1986)
41. Horvath AL; Halogenated Hydrocarbons: Solubility-Miscibility with Water. New York,NY: Marcel Dekker, Inc. pp 889 (1982)
42. Hov O et al; Geophys Res Lett 11: 425-8 (1984)
43. Howard CJ; J Chem Phys 65: 4771-7 (1976)
44. Hutchins SR et al; Environ Toxicol Chem 2: 195-216 (1983)
45. Hutchins SR, Ward CH; J Hydrol (Amsterdam) 67: 223-33 (1984)
46. Jensen S, Rosenberg R; Water Res 9: 659-61 (1975)
47. Jensen AA; Res Rev 89: 1-128 (1983)
48. Kaiser KLE et al; J Great Lakes Res 9: 212-23 (1983)
49. Kawamura K, Kaplan IR; Environ Sci Technol 17: 497-501 (1983)

50. Khalil MAK, Rasmussen RA; Environ Sci Technol 17: 157-64 (1983)
51. Lay JP et al; Arch Environ Contam Toxicol 13: 135-42 (1984)
52. Leoy PJ et al; Atmos Environ 17: 2321-30 (1983)
53. Lillian D et al; Amer Chem Soc Symp Ser 17: 152-8 (1975)
54. Lyman WL et al; Handbook of Chemical Property Estimation Methods NY: McGraw-Hill (1981)
55. Magara Y, Furuichi T; pp. 231-43 in: New Concepts and Development in Toxicol. Chambers PL et al eds. Elsevier Sci Publ (1986)
56. Malle KG; Z Wasser Abwasser Forsch 17: 75-81 (1984)
57. McConnell G et al; Endeavour 34: 13-18 (1975)
58. Monster AC, Smolders JFJ; Int Arch Occup Environ Health 53: 331-6 (1984)
59. Mudder TI; Amer Chem Soc Div Env Chem Conf p. 52-3 (1982)
60. Mueller JPH Korte F; Chemosphere 3: 195-8 (1977)
61. Murray AJ, Riley JP; Nature 242: 37-8 (1973)
62. Neely WB et al; Environ Sci Technol 8: 1113-15 (1974)
63. NIOSH; The National Occupational Exposure Survey (NOES) (1983)
64. NIOSH; The National Occupational Hazard Survey (NOHS) (1974)
65. NIOSH; Criteria for Recommended Standard. Occupational Exposure to Tetrachloroethylene. NIOSH Pub No 76-185 (1976)
66. Otson R et al; J Assoc Off Anal Chem 65: 1370-4 (1982)
67. Parsons F et al; J Amer Wat Works Assoc 76: 56-9 (1984)
68. Pearson CR, McConnell G; Proc Roy Soc London Ser B 189: 305-32 (1975)
69. Pellizzari ED; Quantation of Chlorinated Hydrocarbons in Previously Collected Air Samples. EPA-450/3-78-112 (1978)
70. Pellizzari ED et al; Bull Environ Contam Toxicol 28: 322-8 (1982)
71. Piet GJ et al; Studies Env Sci 17: 557-64 (1981)
72. Piet GJ, Morra CF; pp. 31-42 in Artificial Groundwater recharge; Huismon L, Olsthorn TN eds; Pitman Pub (1983)
73. Riddick JA et al; Organic Solvents: Physical Properties and Methods of Purification. 4th. Wiley-Interscience pp. 1325 (1986)
74. Roberts PV, Dandliker PG; Environ Sci Technol 17: 484-9 (1983)
75. Roberts PV et al; J Water Pollut Control Fed 56:157-63 (1984)
76. Sauer TC Jr; Org Geochem 3: 91-101 (1981)
77. Schwarzenbach RP, Westall J; Environ Sci Technol 15: 1360-67 (1981)
78. Schwarzenbach RP et al; Environ Sci Technol 13: 1367-73 (1979)
79. Schwarzenbach RP et al; Environ Sci Technol 17: 472-9 (1983)
80. Singh HB et al; Environ Sci Technol 16: 872-80 (1982)
81. Singh HB et al; Atmos Environ 15: 601-12 (1981)
82. Singh HB et al; Atmospheric distributions, sources and sinks of selected halocarbons, hydrocarbons, $SF_6$ and $N_2O$; p.34 USEPA-600/3-79-107 (1979)
83. Singh HB et al; J Geophys Res 88: 3675-83 (1983)
84. Singh HB et al; Environ Lett 10: 253-6 (1975)
85. Smith JH et al; Environ Sci Technol 14: 1332-7 (1980)
86. Stanley JS; Broad Scan Analysis of the FY82 National Human Adipose Tissue Survey Specimens Vol. I Executive Summary p. 5 USEPA-560/5-86-035 (1986)
87. Staples CA et al; Environ Toxicol Chem 4: 131-42 (1985)
88. STORET Data Base (1987)
89. Su C, Goldberg ED; Mar. Pollut Transfer pp. 353-74 (1976)
90. Swann RL et al; Res Rev 85: 17-28 (1983)
91. Tomson MB et al; Water Res 15: 1109-16 (1981)
92. Trouwborst T; Sci Total Environ 21: 41-6 (1981)

93. US EPA; Treatability Manual. p.I.12.26-1 to I.12.26-5 USEPA-600/2-82-001A (1981)
94. Verberk MM, Scheffers TML; Environ Res 21: 432-7 (1980)
95. Vogel TM, McCarty PL; Appl Environ Microbiol 49: 1080-3 (1985)
96. Wakeham SG; Environ Sci Technol 17: 611-7 (1983)
97. Wallace L et al; J Occup Med 28: 603-7 (1986)
98. Watanabe H; Gesuido Kyokaiski 20: 29-37 (1983)
99. Wilson JT et al; Devel Indust Microbiol 24: 225-33 (1983)
100. Wilson JT et al; Ground Water 21: 134-42 (1983)
101. Wilson JT et al; J Environ Qual 10: 501-6 (1981)
102. Zoeteman BCJ et al; Chemosphere 9: 231-49 (1980)

# Tetrahydrofuran

SUBSTANCE IDENTIFICATION

**Synonyms:**

**Structure:**

**CAS Registry Number:** 109-99-9

**Molecular Formula:** $C_4H_8O$

**Wiswesser Line Notation:** T5OTJ

## CHEMICAL AND PHYSICAL PROPERTIES

**Boiling Point:** 66 °C at 760 mm Hg

**Melting Point:** -108.5 °C

**Molecular Weight:** 72.12

**Dissociation Constants:**

**Log Octanol/Water Partition Coefficient:** 0.46 [6]

**Water Solubility:** Miscible [18]

**Vapor Pressure:** 162.3 mm Hg at 25 °C [2]

**Henry's Law Constant:** $9.63 \times 10^{-3}$ atm-m$^3$/mole [9]

## ENVIRONMENTAL FATE/EXPOSURE POTENTIAL

**Summary:** Tetrahydrofuran (THF) is used in large ($6.81 \times 10^{+10}$ grams/yr) quantities as a chemical intermediate and solvent. Because of its high vapor pressure and water solubility, significant amounts of

430

the THF used as solvents will be released to the environment and workers will be exposed to it. Once released to the environment its behavior is not well understood and very little monitoring data are available. In the atmosphere, THF should degrade rapidly (half-life - hours to days), especially under smog situations and should be removed by rain. THF in water may biodegrade (only screening studies with sewage inoculum available) but acclimation is probably important. Evaporation should be very important (half-life 1.57 hr) but it will not be removed by photodegradation or adsorption to sediment. Spills on soil are expected to evaporate rapidly and leach into ground water. THF is not expected to bioconcentrate in fish or other aquatic organisms.

**Natural Sources:** None.

**Artificial Sources:** THF is likely to be released to the environment by volatilization losses from its use as a solvent for resins, such as polyvinyl chloride, adhesives, printer's ink, lacquers, and other coatings [7,10], and wastewater from its production and use as a chemical intermediate [10].

**Terrestrial Fate:** THF would be expected to volatilize from soil as well as leach rapidly into the ground. Its biodegradation in soil is unknown.

**Aquatic Fate:** Once released into water, the fate of THF is uncertain. Based on screening study data using sewage inoculum, THF may biodegrade, but only where acclimated microorganisms are present. It is stable toward photodegradation and would not be expected to adsorb to sediment. Its rate of evaporation is estimated to be 1.57 hr.

**Atmospheric Fate:** THF released into the atmosphere will degrade by photochemical reactions with hydroxyl radicals. Data suggest that the half-life in the atmosphere will range from hours to a few days. A soluble chemical such as THF will be expected to washout in rain.

**Biodegradation:** THF is significantly biodegraded in a standard biodegradability test [20]. Biodegradation tests using activated sludge inoculum, data from 22 European laboratories, median delay time of 17 days, percent theoretical oxygen demand: delay time + 10 days, 0-72%, 34% median, 33% avg; after 14 days, 0-63%, 2% median, 11% avg; after 28 days, 0-74%, 38% median, 11% avg [15].

# Tetrahydrofuran

**Abiotic Degradation:** THF does not adsorb radiation above 220 nm and therefore cannot directly photolyze in the environment. It is, however, moderately reactive towards hydroxyl radicals with a half-life of 1.6 days in the atmosphere [23]. It is also moderately reactive in photochemical smog situations where nitrogen oxides are present - reactions occur in time scales of hours [4,12,17]. Acrolein and formaldehyde have been reported as reaction products [17]. THF is stable to photooxidation in water [1,8]. Hydrolysis will not be a significant degradation process under normal environmental conditions.

**Bioconcentration:** THF would not be expected to bioconcentrate due to its very low octanol/water partition coefficient.

**Soil Adsorption/Mobility:** Although experimental data are lacking, one would not expect THF to adsorb to soil because of its very high solubility in water.

**Volatilization from Water/Soil:** Using the Henry's Law constant, the volatilization half-life of THF from a shallow model river 1 m deep flowing 1 m/sec with a wind speed of 3 m/sec can be estimated to be approximately 1.57 hr.

**Water Concentrations:** DRINKING WATER: Identified, not quantified, in drinking water from unspecified sources [11]. Concn (ppm) in drinking water at laboratory in Florida, 6 mo after PVC pipe installation (residence time in PVC pipe, hr): 0 ppm (0 hr), 1.0 (4 hr), 1.7 (8 hr), 5.8 (16 hr), 12 (48 hr), 13 (64 hr), 13 (96 hr), 8 mo after installation: 0 ppm (0 hr), 0.7 (4 hr), 2.4 (16 hr), 6.8 (48 hr), 7.5 (72 hr), source of contamination was PVC pipe cement [22]. GROUND WATER: Max concn detected in contaminated ground water in The Netherlands, 3 ppb [24]. SURFACE WATER: 14 heavily industrialized river basins in US (201 sites), 29 sites (14%) >1 ppb, range 1-318 ppb [3]. Identified, not quantified, in the Cuyahoga River (empties into Lake Erie) [5].

**Effluent Concentrations:** Municipal solid waste landfills in Minnesota, leachates, 6 sites, 100% pos, 18-430 ppb, contaminated ground water where elevated levels of inorganics suggest leachate contamination, 13 sites, 46.2% pos, 24-3000 ppb, ground water from landfills or dumps with unknown or apparently good water quality based on inorganic parameters, 7 sites, 28.6% pos, 5.4-3900 ppb [19]. Leachate from

# Tetrahydrofuran

landfill sites in 5 Connecticut towns, 60% pos, 20-330 ppb, 137 ppb avg [21].

**Sediment/Soil Concentrations:**

**Atmospheric Concentrations:**

**Food Survey Values:**

**Plant Concentrations:**

**Fish/Seafood Concentrations:**

**Animal Concentrations:**

**Milk Concentrations:**

**Other Environmental Concentrations:**

**Probable Routes of Human Exposure:** People will be exposed to THF primarily from occupational exposure relating to its use as a solvent for resins, adhesives, printer's ink, and coatings. Some exposure is possible from contact with water near industrial outfalls or from drinking water; however, the Henry's Law constant would suggest that THF probably volatilizes during water treatment.

**Average Daily Intake:**

**Occupational Exposures:** NIOSH (NOES Survey 1981-83) has statistically estimated that 303,049 workers are exposed to THF in the United States [13]. NIOSH (NOHS Survey 1972-74) has statistically estimated that 95,027 workers are exposed to THF in the United States [14].

**Body Burdens:** Human milk - 4 urban areas - 1 of 8 samples positive [16].

## REFERENCES

1. Anbar M, Neta P; Int J Appl Radiation and Isotopes 18:493-523 (1967)
2. Boublik T et al; The Vapor Pressures of Pure Substances Vol 17 Amsterdam, Netherlands: Elsevier Science Publ (1984)
3. Ewing BB et al; Monitoring to Detect Previously Unrecognized Pollutants in

Surface Water p 74 EPA-560/6-77-015 (1977)
4. Farley FF; International Conference on Photochemical Oxidant Pollution and its Control p 713-27 EPA-560/6/77-015A (1977)
5. Great Lakes Water Quality Board; An Inventory of Chemical Substances Identified in the Great Lakes Ecosystem Vol 1 - Summary. Report to the Great Lakes Quality Review Board, Windsor Ontario, Canada pp. 195 (1983)
6. Hansch C, Leo AJ; Medchem Project Issue No 26. Claremont CA: Pomona College (1985)
7. Hawley GG; Condensed Chemical Dictionary p 1007 (1981)
8. Hendry DG et al; J Phys Chem Ref Data 3:944-78 (1974)
9. Hine J, Mookerjee PK; J Org Chem 40: 292-8 (1975)
10. Kirk-Othmer's Encyclopedia of Chemical Technology 2nd ed 10:249-250 (1966)
11. Kool HJ et al; Crit Rev Environ Control 12: 307-57 (1982)
12. Levy A; Adv Chem Ser 124:70-94 (1973)
13. NIOSH; National Occupation Exposure Survey (1983)
14. NIOSH; National Occupational Hazard Survey (1974)
15. Painter HA, King EF; Ring Test Program 1983-84. Assessment of Biodegradability of Chemicals in Water by Manometric Respirometry. Comm Eur Communities, EUR 9962, pp. 105 (1985)
16. Pellizzari ED et al; Bull Environ Toxicol 28:322-8 (1982)
17. Popav VA; Gig Sanit 36:7-10 (1971)
18. Riddick JA et al; Organic Solvents: Physical Properties and Methods of Purification, 4th Edit. New York: J Wiley & Sons (1986)
19. Sabel GV, Clark TP; Waste Manag Res 2: 119-30 (1984)
20. Sasaki S; Aquatic Pollutants: Transformation and Biological Effects p 283-98 (1978)
21. Sawhney BL, Kozloski RP; J Environ Qual 13: 349-52 (1984)
22. Wang TC, Bricker JL; Bull Environ Contam Toxicol 23: 620-3 (1979)
23. Winer AM et al; Chem Phys Lett 51:221-6 (1977)
24. Zoeteman BCJ et al; Sci Total Environ 21: 187-202 (1981)

# Toluene

## SUBSTANCE IDENTIFICATION

**Synonyms:** Methylbenzene

**Structure:**

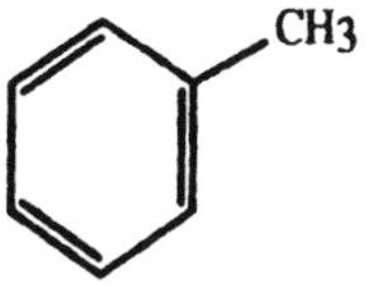

**CAS Registry Number:** 108-88-3

**Molecular Formula:** $C_7H_8$

**Wiswesser Line Notation:** 1R

## CHEMICAL AND PHYSICAL PROPERTIES

**Boiling Point:** 110.6 °C at 760 mm Hg

**Melting Point:** -95 °C

**Molecular Weight:** 92.13

**Dissociation Constants:**

**Log Octanol/Water Partition Coefficient:** 2.73 [32]

**Water Solubility:** 534.8 mg/L at 25 °C [90]

**Vapor Pressure:** 28.4 mm Hg at 25 °C [17]

**Henry's Law Constant:** 5.94 x $10^{-3}$ atm-m³/mole [82]

## ENVIRONMENTAL FATE/EXPOSURE POTENTIAL

**Summary:** Toluene is released into the atmosphere principally from the volatilization of petroleum fuels and toluene-based solvents and thinners and from motor vehicle exhaust. Considerable amounts are discharged into waterways or spilled on land during the storage, transport, and disposal of fuels and oils. If toluene is released to soil, it will be lost by evaporation from near-surface soil and by leaching to the ground water. Biodegradation occurs both in soil and ground water but it is apt to be slow, especially at high concentrations, which may be toxic to microorganisms. The presence of acclimated microbial populations may allow rapid biodegradation. It will not significantly hydrolyze in soil or water under normal environmental conditions. If toluene is released into water, its concn will decrease due to evaporation and biodegradation. This removal can be rapid or take several weeks, depending on temperature, mixing conditions, and acclimation of microorganisms. It will not significantly adsorb to sediment or bioconcentrate in aquatic organisms. If toluene is released to the atmosphere, it will degrade by reaction with photochemically produced hydroxyl radicals (half-life 3 hr to slightly over 1 day) or be washed out in rain. It will not be subject to direct photolysis. The primary source of human exposure is from inhalation of contaminated ambient air, especially in traffic or near filling stations, or in occupational atmospheres where toluene-based solvents are used.

**Natural Sources:** Volcanos, forest fires, and crude oil [29].

**Artificial Sources:** Motor vehicle exhaust. Emissions from gasoline storage tanks, filling stations, carburetors, etc. Petroleum spills and discharges on land and in waterways. Emissions and wastewater from its use as a solvent and thinner for paints, lacquers, etc. Emissions from its production from petroleum, coal, and as a by-product from styrene production. Emissions from its use as a chemical intermediate [29,106]. Tobacco smoke [29].

**Terrestrial Fate:** If toluene is released to soil, it will be lost by evaporation from near-surface soil and microbial degradation. In one study, 94% of the chemical added to a clay loam was lost by these processes [65]. Since it is relatively mobile in soil, it may leach into the ground water where slow biodegradation may occur. It will not significantly hydrolyze under normal environmental conditions.

**Aquatic Fate:** If toluene is released into water, it will be lost by both volatilization to the atmosphere and biodegradation. The predominant process will depend on water temperature, mixing conditions, and the existence of acclimated microorganisms at the site. The half-life will range from days to several weeks. It will not significantly hydrolyze, directly photolyze, adsorb to sediment, or bioconcentrate in aquatic organisms.

**Atmospheric Fate:** If toluene is released to the atmosphere, it will exist predominantly in the vapor phase [22]. It degrades moderately rapidly by reaction with photochemically produced hydroxyl radicals. Its half-life ranges from 3 hr to somewhat over a day [85,91]. It is very effectively washed out by rain [40,94]. It will not be subject to direct photolysis in sunlight, although a complex of toluene with molecular oxygen has been shown to absorb light at wavelengths >290 nm [65].

**Biodegradation:** Toluene is readily degradable in a variety of standard biodegradability tests using sewage seed or sludge inocula [8,18,42,52,53,71,93]. Degradation has been observed in several die-away tests using seawater or estuarine water [11,46,73,105]. The degradation rate is much faster in systems which have been contaminated by oil [11,46]. Complete degradation has been observed in 4 days and 22 days in a marine mesocosm with summer and spring conditions, respectively [105] and 10 days in a 1% gas oil mixture in a North Sea coast water inoculum [103]. A 90 day half-life in uncontaminated estuarine water was reduced in 30 days in oil-polluted water [46]. The half-life in water collected near Port Valdez, Alaska was 12 days [11]. 1.5 mM and 3 mM Ring-labeled toluene added to a methanogenic inoculum originally enriched from sewage sludge and incubated at 35 °C for 60 days resulted in 3.6% and 4.5% $^{14}$C final activity, respectively [30]. Toluene completely degraded in ground water in 8 days including a lag of 3-4 days while microbial populations became acclimated [39]. Other investigators found that only 1-2% of toluene degraded in the subsurface environment [109] and >90% degraded in 4 weeks in soil cores at various depths both above and below the water table [54,111]. Microbial attack proceeds via immediate hydroxylation of the benzene ring followed by ring-cleavage or oxidation of the side chain followed by hydroxylation and ring-cleavage [25].

**Abiotic Degradation:** Toluene does not adsorb radiation >290 nm [76] and therefore is not subject to direct photolysis. However, a complex is formed with molecular oxygen which does adsorb light in this region but no rate data were found [65]. No significant photodegradation of toluene was detected in deionized water, salt water, or pond water [85]. No mutagenic products are formed in the presence of nitrite ions [91].

It will not significantly hydrolyze under normal environmental conditions [50]. It is principally degraded in the atmosphere by reaction with photochemically produced hydroxyl radicals with a half-life of slightly over a day and a degradation rate of 41%/day [67,84]. In smog chamber experiments, degradation is somewhat faster with avg hourly degradation rates ranging from 5-17% [19,45,104,107,112]. Products formed include nitrophenols, nitrocresols, nitrotoluenes, cresols, benzaldehyde, and benzyl nitrate [1,5,38,60].

**Bioconcentration:** BCF: eels (Anguilla japonica), 13.2 [62]; Manila clam (Tapes semidecussata), 1.67 [61]; mussel (Mytilus edulis), 4.2 [27]; algae (Chlorella fusca), 380, golden ide fish (Leuciscus idus melanotus), 90 [26].

**Soil Adsorption/Mobility:** Koc: Wendover silty loam, 37; Grimsby silt loam, 160; Vaudreil sandy loam, 46 [57]; sandy soil, 178 [110]; soil, 100 and 151 [34]. Based on the reported Koc values, toluene will be expected to exhibit very high to moderate mobility in soil [92] and therefore may leach to the ground water. Field data from infiltration sites are conflicting; in one study toluene is eliminated during bank infiltration [79], while in other studies it penetrates infiltration sites [68,98]. These results may bear on site-related factors such as load, flow rate, soil characteristics, and other loss factors such as evaporation and biodegradation.

**Volatilization from Water/Soil:** Toluene evaporates rapidly from water having an experimentally determined half-life for evaporation from 1 m of water with moderate mixing conditions of 2.9-5.7 hr [50,51,72]. Using oxygen reaeration rates of typical bodies of water [55], and the reaeration rate relative to oxygen, 0.65 [72], one would expect the evaporation half-life of toluene from a river and lake to be 1 and 4 days, respectively. In a mesocosm experiment with simulated conditions for Narragansett Bay, RI, the loss was primarily by evaporation in winter with a half-life of 13 days [105]. Due to its high vapor pressure, toluene would be expected to volatilize fairly rapidly from dry surface soils.

**Water Concentrations:** DRINKING WATER: 30 Canadian treatment facilities, 2 ppb avg, 14 ppb max, 33% pos in summer; <1 ppb avg, 13 ppb max, 53% pos in winter [64]. 12 Great Lakes municipalities 0-2 ppb, 5 cities' supplies pos [108]. 17 US drinking waters <1 ppt, 14 pos [56]. In a federal survey of finished drinking water from ground water sources <5% occurrence [21]. 3 New Orleans area water supplies 0-10 ppb [43]. National Organics Reconnaissance Survey 60% occurrence [6]. Max concn in tapwater derived from bank-filtered Rhine R water 1 ppb [69]. Three contaminated drinking water wells in New Jersey and 55, 260, 6400 ppb whereas the highest concn in drinking water from surface water sources in 6.1 ppb [10]. Ground water supply in England 210 m from gasoline storage 0.15 ppb [97]. In 3 federal surveys of finished surface waters toluene was found in 19% of the samples [15]. In a 5-city survey in which the water supplies came from different types of sources with various sources of pollution, 2 contained toluene, one 0.1 ppb, and the other 0.7 ppb [13]. US Ground Water Supply Survey, 1982, 466 randomly selected drinking water supplies that used ground water as a source, 1.3% pos, 2.9 ppb max, 0.8 ppb median [16]; GROUND WATER: Ground water underlying 2 rapid infiltration sites 02 ppb [36]. Ground water under gasification site, 15 months after gasification performed 170-740 ppb [89]. Contaminated wells from gasoline storage tanks, etc. 0.55-6400 ppb [10,97]. In cluster well study under old industrial site mean levels in bedrock wells were 90 ppb while shallow and deep glacial wells were 10 ppb [74]. SURFACE WATER: 14 heavily industrialized river basins in US (204 sites 1-5 ppb, 15% pos [23]). Detected in various rivers [28,81,113]. USEPA STORET data base, 1,804 data points, 16.0% pos, 5.0 ppb median concn [87]. SEA WATER: Gulf of Mexico 3-376 ppt [77,78]. Vineland Sound, MA 10-54 ppt, 27 Mexico 3-376 ppt [77,78]. Vineland Sound, MA 10-54 ppt, 27 ppt avg, source, fuel from boats [31]. RAIN WATER: West Los Angeles 76 ppt [41]. 7 rain events, Portland, OR, Feb-Apr 1984, 71.4% pos, concn in rain (ppt) range of pos, 40-220, 88 avg; concn in gas phase (ng/m$^3$), 1800-8600, 3800 avg [47].

**Effluent Concentrations:** Industries in which the mean effluent levels exceed 1000 ppb are: auto and other laundries, iron and steel manufacturing, gum and wood chemicals, pharmaceuticals, organic chemicals/plastics manufacturing, and paint and ink formulation. The highest mean value is 52 ppm for pharmaceuticals and the highest maximum values are 230 and 260 ppm for pharmaceuticals and organic chemicals/plastics manufacturing [102]. Plume from General Motors Paint Plant, Janesville, WI 156 ppb [80]. Auto exhaust 196-718 mg/m$^3$ [100]. USEPA STORET data base, 1,498 data points, 19.7% pos, 5.0 ppb median concn [87]. Minnesota municipal solid waste landfills, leachates, 6 sites, 100% pos, 7.5-600, 1.5-8300 ppb, contaminated ground water (by inorganic indices), 13 sites, 46% pos, 7 sites, 14% pos, 3.8 ppb [75].

**Sediment/Soil Concentrations:** SEDIMENT: Detected, not quantified, in sediment in industrial rivers in US [28,88]. USEPA STORET data base, 397 data points, 17% pos, 5.0 ppb dry weight median concn [87]. Lake Pontchartrain at 3 Passes, 33.3% pos, 0.7 ppb wet weight [24].

**Atmospheric Concentrations:** RURAL/REMOTE: US, 115 samples 0.057-30 ppb, 0.66 ppb median [9]. Remote concn (ppb) 1978: Great Smoky Mountains National Park, TN, Sept, 2.5-18.2, 6.5 avg; Rio Blanco County, CO, July, 5.3-11.9 [4]. URBAN/SUBURBAN: US, 3195 samples 0-85 ppb, 11 ppb median [9]. Tulsa, OK, 1978, 14.23/16.6 ppb [4]. New Jersey, Summer 1981/Winter 1982, avg concn (ppb): Camden, 2.25/4.1, Newark, 11.39/8.23, Elizabeth, 4.82/5.51 [48]. 6 cities, New Jersey 1979, not detected - 85 ppb, 0.01 ppb avg of all samples [7]. SOURCE DOMINATED: US, 188 samples 0.037-5500 ppb, 4.6 ppb median [9]. Lipari and BF landfills in New Jersey - 0.40 and 310 ppb [7]. URBAN/SUBURBAN: Toronto, Aug 1971, 188 ppb max, 30 ppb avg, Los Angeles, 1966, 129 ppb max, 37 ppb avg [70]. Houston, industrial and urban sites, Sept 1973, 9 sites, 15.5-1110 ppb, Jan 1974, 7 sites, 204-452 ppb, Apr 1974, 5 sites, 18.5-98.6 ppb [49]. 3 cities 1979: Los Angeles, 1.14-53.3 ppb, 11.72 avg, Oakland, 0.15-16.9 ppb, 3.11 avg, Phoenix, AZ, 0.54-38.7 ppb, 8.63 avg [83]. Los Angeles, 1967, 120 samples, 30 ppb avg [2]. 4 US cities, May-Jul 1980, range of max concn, 6.45-65.65 ppb, range of avgs, 1.52-10.33 ppb, avg of avgs, 5.97 ppb [83]. SOURCE DOMINATED: 4-11 mi downwind from General Motors paint plant 14-22 ppb [99]. RURAL/REMOTE: Semi-rural Belgium, 6 sites, 100% pos, 0.044-3.74 ppb, daily variations in concentrations and ratios of toluene to benzene indicate that auto traffic is the most common source of atmospheric toluene [96]. Rural, UK, May-Aug 1983, 204 samples, 0.20-6.40 ppb, 1.27 ppb avg [12]. Norwegian Arctic air, July 1982, 9 samples from 2 sites, 10 ppt avg, Spring 1983, 10 samples from 1 site, 51.4 ppt [35]. URBAN/SUBURBAN: London, May-Aug 1983, 276 samples, 0.84-42.37 ppb, 15.54 ppb avg [12]. SOURCE DOMINATED: Rural motorway, UK, May-Aug 1983, 184 samples, 0.07-15.67 ppb, 2.69 ppb avg [12].

**Food Survey Values:** Identified, not quantified, in: baked potatoes [14]; Mountain cheese [20]; fried bacon [33]; and fried chicken [95].

## Toluene

**Plant Concentrations:**

**Fish/Seafood Concentrations:** Flesh of fish from petroleum contaminated harbor in Japan, 5 ppm [101]. Biota, Lake Pontchartrain, oysters, 1 sites, avg of 5 samples, 3.4 ppb (wet weight), clams, 2 sites, 18 and 11 ppb in composite samples, respectively [24].

**Animal Concentrations:**

**Milk Concentrations:**

**Other Environmental Concentrations:** Light oil from coal 12-20% toluene; crude oil 1.2-2.4% auto fuel 8-12% [106].

**Probable Routes of Human Exposure:** Humans are primarily exposed to toluene from ambient air, particularly in areas with heavy traffic and around filling stations. In addition, high concentrations may exist in enclosed areas where toluene is used in solvents or thinners. Other sources of exposure are from tobacco smoke and from glue sniffing.

**Average Daily Intake:** AIR INTAKE: (assume median concn 11 ppb [9]) 843 ug; WATER INTAKE: (assume 2 ppb [63]) 4 ug; FOOD INTAKE: insufficient data.

**Occupational Exposures:** Office air - 6 of 30 offices in Ottawa >0.2 ppb, max 8 ppm [64]. Air in spacecraft 29 ppm [44]. In 1969, 11 workshops operating chromic rotary processes for photogravure printing 4-240 ppm [37]. NIOSH (NOES Survey 1981-83) has statistically estimated that 1,625,598 workers are exposed to toluene in the United States [58]. NIOSH (NOHS Survey 1972-74) has statistically estimated that 3,972,080 workers are exposed to toluene in the United States [59].

**Body Burdens:** All 8 samples of human milk from 4 urban areas were positive [66]. Human blood, 250 specimens, 100% pos, 0.2-38 ppb, 1.5 ppb avg [3]. US FY82 National Human Adipose Tissue Survey specimens, 46 composites, 91% pos, ( >1 ppb, wet tissue concn), 250 ppb max [86].

# Toluene

## REFERENCES

1. Akimoto H et al; Photooxidation of toluene-$NO_2$-O2-N2 system in gas phase p 243-6 NBS SP-526 (1978)
2. Altshuller AP et al; Environ Sci Technol 5: 1009-16 (1971)
3. Antoine SR et al; Bull Environ Contam Toxicol 36: 364-71 (1986)
4. Arnts RR, Meeks SA; Biogenic hydrocarbon contribution to the ambient air of selected areas 128 USEPA 600/3-80-023 (1980)
5. Atkinson R et al; Int J Chem Kinet 12: 779-836 (1980)
6. Bedding ND et al; Sci Total Environ 25: 143-67 (1982)
7. Bozzelli JW et al; Analysis of selected toxic and carcinogenic substances in ambient air in New Jersey, NJ Dept Environ Protect (1980)
8. Bridie AL et al; Water Res 13: 627-30 (1979)
9. Brodzinsky R, Singh HB; Volatile organics in the atmosphere: an assessment of available data. p.126-7 SRI contract 68-02-3459 (1982)
10. Burmaster DE; Environ 24: 6-13, 33-6 (1982)
11. Button DK et al; Appl Environ Microbiol 42: 708-19 (1981)
12. Clark AI et al; Sci Total Environ 39: 265-79 (1984)
13. Coleman WE et al; p.305-27 in Analysis and identification of organic substances in water. Keith L ed, Ann Arbor, MI, Ann Arbor Sci (1976)
14. Coleman EC et al; J Agric Food Chem 29: 42-8 (1981)
15. Coniglio WA et al; The occurrence of volatile organics in drinking water. Exposure Assessment Project. Criteria and Standards Div. (1980)
16. Cotruvo JA; Sci Total Environ 47: 7-26 (1985)
17. Daubert TE, Danner RP; Data Compilation Tables of Properties of Pure Compounds. American Institute of Chemical Engineers pp 450 (1985)
18. Davis EM et al; Water Res 15: 1125-7 (1981)
19. Dilling WL et al; Environ Sci Technol 10: 351-6 (1976)
20. Dumont JP, Adda J; J Agric Food Chem 26: 364-7 (1978)
21. Dyksen JE, Hess AF III; J Am Water Works Assoc 74: 394-403 (1982)
22. Eisenreich SJ et al; Environ Sci Technol 15: 30-8 (1981)
23. Ewing BB et al; Monitoring to detect previously unrecognized pollutants in surface waters USEPA 560/6-77-015 (appendix USEPA 560/6-77-015a) (1977)
24. Ferrario JB et al; Bull Environ Contam Toxicol 34: 246-55 (1985)
25. Fewson CA; FEMS Symp 12 (Microb Deg Xenobiotics Recalcitrant Cmpds): 141-79 (1981)
26. Freitag D et al; Chemosphere 14: 1589-1616 (1985)
27. Geyer H et al; Chemosphere 11: 1121-34 (1982)
28. Goodley PC, Gordon M; KY Acad Sci 37: 11-5 (1976)
29. Graedel TE; p.108 in Chemical Compounds in the Atmosphere, New York, NY, Academic Press (1978)
30. Grbic-Galic D, Vogel TM; Appl Environ Microbiol 53: 254-60 (1987)
31. Gschwend PM et al; Environ Sci Technol 16: 31-8 (1982)
32. Hansch C, Leo AJ; Medchem Project Issue No 26. Claremont CA: Pomona College (1985)
33. Ho CT et al; J Agric Food Chem 31: 336-42 (1983)
34. Hodson J, Williams NA; Chemosphere 17: 67-77 (1988)
35. Hov O et al; Geophys Res Lett 11: 425-8 (1984)
36. Hutchins SR et al; Environ Toxicol Chem 2: 195-216 (1983)
37. Ikeda M, Ohtsuji H; Br J Ind Med 26: 244-6 (1969)

# Toluene

38. Ishikawa H et al; Bull Chem Soc Japan 51: 2173-4 (1978)
39. Kappeler T, Wuhrmann K; Water Res 12: 327-33 (1978)
40. Kato T et al; Yokohama Kokuritsu Daigaku Kankyo Kagaku Kenkyu Senta Kiyo 6: 11-20 (1980)
41. Kawamura K, Kaplan IR; Environ Sci Technol 17: 497-501 (1983)
42. Kawasaki M; Ecotox Environ Safety 4: 444-54 (1980)
43. Keith LH et al; p 329-73 in Identification and analysis of organic pollutants in Water, Keith LH ed, Ann Arbor Sci Publ (1976)
44. Korte F, Klein W; Ecotox Environ Saf 6: 311-27 (1982)
45. Laity JL; Environ Sci Technol 5: 1218-20 (1971)
46. Lee RF; 1977 Oil Spill Conf Am Petrol Inst p 611-6 (1977)
47. Ligocki MP et al; Atmos Environ 19: 1609-17 (1985)
48. Lioy PJ et al; J Water Pollut Control Fed 33: 649-57 (1983)
49. Lonneman WA et al; Hydrocarbons in Houston air 44 p.USEPA 600/3-79-018 (1979)
50. Lyman WJ et al; Handbook of chemical property estimation methods. Environmental behavior of organic chemicals, New York, NY, McGraw Hill (1982)
51. Mackay D, Yeun ATK; Environ Sci Technol 17: 611-4 (1983)
52. Malaney GW, McKinney RE; Water Sewage Works 113: 302-9 (1966)
53. Matsui S et al; Prog Water Technol 7: 645-59 (1975)
54. McNabb JF et al; 81st Ann Mtg Am Soc Microbiology p.213 (1981)
55. Mill T et al; Protocols for evaluating the fate of organic chemicals in air and water p.255 USEPA 600/3-82-022 (1982)
56. NAS; The Alkyl Benzene USEPA Contract No. 68-02-4655 (1980)
57. Nathwani JS, Phillips CR; Chemosphere 6: 157-62 (1977)
58. NIOSH; The National Occupational Exposure Survey (NOES) (1983)
59. NIOSH; The National Occupational Hazard Survey (NOHS) (1974)
60. Nojima K et al; Chemosphere 1: 25-30 (1976)
61. Nunes P, Benville PE Jr; Bull Environ Contam Toxicol 12: 719-24 (1979)
62. Ogata M, Miyake Y; Water Res 12: 1041-4 (1978)
63. Otson R et al; J Assoc Off Analyt Chem 65: 1370-4 (1982)
64. Otson R et al; Bull Environ Contam Toxicol 31: 222-9 (1983)
65. Overcash MR et al; Behavior of organic priority pollutants in the terrestrial system: di-n-butyl phthalate ester, toluene and 2,4-dinitrophenol Raleigh NC Water Res Inst Rept No 171 104 (1982)
66. Pellizzari ED et al; Bull Environ Contam Toxicol 28: 322-8 (1982)
67. Perry RA et al; J Phys Chem 81: 296-304 (1977)
68. Piet GJ et al; Int Symp Quality Groundwater Van Duijvenbooden W et al eds; Studies Environ Sci 17: 557-64 (1981)
69. Piet GJ, Morra CF; p.31-42 in Artificial Groundwater Recharge (Water Res Eng Ser) Huisman L, Olsthorn TN, eds, Pitman Publ (1983)
70. Pilar S, Graydon WF; Environ Sci Technol 7: 628-71 (1973)
71. Price KS et al; J Water Pollut Control Fed 46: 63-77 (1974)
72. Rathbun RE, Tai DY; Water Res 15: 243-50 (1981)
73. Reichardt PB et al; Environ Sci Technol 15: 75-9 (1981)
74. Rich CA; Stud Environ Sci 17(Qual Groundwater): 309-14 (1981)
75. Sabel GV, Clark TP; Waste Manag Res 2: 119-30 (1984)
76. Sadtler 155 UV (1988)
77. Sauer TC Jr et al; Mar Chem 7: 1-16 (1978)
78. Sauer TC Jr; Org Geochem 3: 91-101 (1981)

# Toluene

79. Schwarzenbach RP et al; Environ Sci Technol 17: 472-9 (1983)
80. Sexton K, Westberg H; Environ Sci Technol 14: 329-32 (1980)
81. Sheldon LS, Hites RA; Environ Sci Technol 12: 1188-94 (1978)
82. Shen TT; J Air Pollut Contr Assoc 32:79-82 (1982)
83. Singh HB et al; Atmospheric measurements of selected hazardous organic chemicals USEPA 600/53-81-032 (1981)
84. Singh HB et al; Atmos Environ 15: 601-12 (1981)
85. Smith JH, Harper JC; 12th Conf Environ Toxicol Airforce Aerospace Med Res Lab p.336-53 (1982)
86. Stanley JS; Broad Scan Analysis of the FY82 National Human Adipose Tissue Survey Specimens Vol. I Executive Summary p 5 USEPA-560/5-86-035 (1986)
87. Staples CA et al; Environ Toxicol Chem 4: 131-42 (1985)
88. Steinheimer TR et al; Anal Chim Acta 129: 57-67 (1981)
89. Stuermer DH et al; Environ Sci Technol 16: 582-7 (1982)
90. Sutton C, Calder JA; J Chem Eng Data 20:320-2 (1975)
91. Suzuki J et al; Bull Environ Contam Toxicol 31: 79-84 (1983)
92. Swann RL et al; Res Rev 85: 17-28 (1983)
93. Tabak HH et al; J Water Pollut Control Fed 53: 1503-8 (1981)
94. Tada T et al; Tokyo-Toritsu Eisei Kenkyusko Kenkyu Nempo 1979: 206-9 (1979)
95. Tang JT et al; J Agric Food Chem 31: 1287-92 (1983)
96. Termonia M; Comm Eur Comm EUR 7624 p 356-61 (1982)
97. Tester DJ, Harker RJ; Water Pollut Control 80: 614-31 (1981)
98. Tomson MB et al; Water Res 15: 1109-16 (1981)
99. Tsani-Bazaca E et al; Chemosphere 11: 11-23 (1982)
100. Tsani-Bazaca E et al; Environ Technol Lett 2: 303-16 (1981)
101. USEPA; Ambient Water Quality Criteria for Toluene p.C-1 to C-7 (1980)
102. USEPA; Treatability Manual p. I.9.10-1 to I.9.10-5 USEPA 600/2-82-001a (1981)
103. Van der Linden AC; Dev Biodegrad Hydrocarbons 1: 165-200 (1978)
104. Van Aalst RM et al; Comm Eur Comm EUR 6621 1: 136-49 (1980)
105. Wakeham SG et al; Environ Sci Technol 17: 611-7 (1983)
106. Walker P; Air Pollution Assessment of Toluene. p. 1-22, 46-92 MTR-7215 Mitre Corp, McLean VA (1976)
107. Washida N et al; Bull Chem Soc Japan 51: 2215-21 (1978)
108. Williams DT et al; Chemosphere 11: 263-76 (1982)
109. Wilson JT et al; Devel Indust Microbiol 24: 225-33 (1983)
110. Wilson JT et al; J Environ Qual 10: 501-6 (1981)
111. Wilson JT et al; Ground Water 21: 134-42 (1983)
112. Yanagihara S et al; 4th Int Clean Air Conf p.472-7 (1977)
113. Zuercher F, Giger W; Vom Wasser 47: 37-55 (1976)

# Tributylamine

SUBSTANCE IDENTIFICATION

**Synonyms:**

**Structure:**

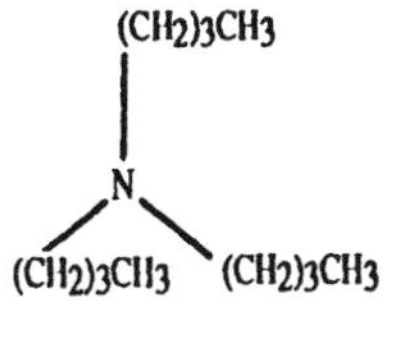

$(CH_2)_3CH_3$

N

$(CH_2)_3CH_3$  $(CH_2)_3CH_3$

**CAS Registry Number:** 102-82-9

**Molecular Formula:** $C_{12}H_{27}N$

**Wiswesser Line Notation:**

## CHEMICAL AND PHYSICAL PROPERTIES

**Boiling Point:** 216-217 °C

**Melting Point:** -70 °C

**Molecular Weight:** 185.34

**Dissociation Constants:** pKa = 10.89 [10]

**Log Octanol/Water Partition Coefficient:** 4.41 [8]

**Water Solubility:** 40 mg/L at 18 °C [10]

**Vapor Pressure:** 0.3 mm Hg at 25 °C [9]

**Henry's Law Constant:** $1.83 \times 10^{-3}$ atm-m³/mole at 25 °C [3]

## ENVIRONMENTAL FATE/EXPOSURE POTENTIAL

**Summary:** Tributylamine may be released to the environment from various waste streams generated at sites of its commercial production and use as a chemical intermediate. If released to the atmosphere, rapid

445

degradation will occur via reaction with photochemically produced hydroxyl radicals (estimated half-life of 3.9 hr). If released to water, volatilization is expected to be the dominant fate process. The volatilization half-life of tributylamine from a model river (1 m deep) has been estimated to be 4.7 hr. If released to soil, relatively low soil mobility is expected based on an estimated Koc of 570. Significant evaporation may occur from dry surfaces. Humans will be primarily exposed to tributylamine by inhalation or dermal contact in occupational settings.

**Natural Sources:**

**Artificial Sources:** Various waste effluent streams generated during the commercial production of the n-butylamines (mono-, di-, and tri-) contain the n-butylamines [4]. Wastewater and atmospheric vent effluents may also be generated at the industrial sites using tributylamine as a chemical intermediate.

**Terrestrial Fate:** No data are currently available to suggest that tributylamine is readily degraded in soil by chemical processes. A single screening study has indicated that tributylamine is not readily biodegraded. Tributylamine is expected to have relatively low soil mobility based on an estimated Koc value of 570. Based on the estimated vapor pressure, some evaporation from dry surfaces may occur.

**Aquatic Fate:** Volatilization appears to be the major environmental fate process for tributylamine in water. The estimated Henry's Law constant is indicative of relatively rapid volatilization. The volatilization half-life of tributylamine from a model river (1 m deep) has been estimated to be 4.7 hr. Aquatic hydrolysis, oxidation, direct photolysis, and bioconcentration are not important. A single screening study has indicated that tributylamine is not readily biodegraded.

**Atmospheric Fate:** Tributylamine is expected to exist almost entirely in the vapor phase in the ambient atmosphere based on its estimated high vapor pressure [2]. Tributylamine is very readily degraded in the vapor phase by reaction with photochemically produced hydroxyl radicals (estimated half-life of 3.9 hr in an average ambient atmosphere).

# Tributylamine

**Biodegradation:** Only slight bio-oxidation of tributylamine was observed (15-day theoretical BOD of approximately 5%) using the JIS (Japanese Industrial Standards) BOD dilution method with an activated sludge inoculum [12].

**Abiotic Degradation:** Tributylamine does not contain any functional groups that are hydrolyzable under environmental conditions; therefore, hydrolysis in water should not be an important process. Tributylamine is not expected to be degraded by direct photolysis or by aquatic oxidation in the environment. The rate constant for the vapor-phase reaction of tributylamine with photochemically produced hydroxyl radicals has been estimated to be $1 \times 10^{-10}$ cm$^3$/molecule-sec at 25 °C [1]; assuming an average hydroxyl radical concn of $5 \times 10^{+5}$ molecules/cm$^3$ is present in the ambient atmosphere, a reaction half-life of 3.9 hours can be calculated [1].

**Bioconcentration:** Based on the water solubility, the log BCF for tributylamine can be estimated to be 1.9 from a recommended regression-derived equation, and therefore, bioconcentration in aquatic organisms should not be an important fate process [5].

**Soil Adsorption/Mobility:** Based on the water solubility, the Koc value for tributylamine can be estimated to be 570 from a regression-derived equation [5]. This Koc value is indicative of low (but approaching medium) soil mobility [11].

**Volatilization from Water/Soil:** The value of the Henry's Law constant is indicative of relatively rapid volatilization from environmental waters [5]. Using the estimated value of Henry's Law constant, the volatilization half-life of tributylamine from a model river (1 m deep flowing 1 m/sec with a wind velocity of 3 m/sec) has been estimated to be 4.7 hr [5].

**Water Concentrations:**

**Effluent Concentrations:**

**Sediment/Soil Concentrations:**

**Atmospheric Concentrations:**

**Food Survey Values:**

447

# Tributylamine

**Plant Concentrations:**

**Fish/Seafood Concentrations:**

**Animal Concentrations:**

**Milk Concentrations:**

**Other Environmental Concentrations:**

**Probable Routes of Human Exposure:** Humans will be primarily exposed to tributylamine by inhalation or dermal contact in occupational settings.

**Average Daily Intake:**

**Occupational Exposure:** NIOSH has estimated that 28,265 workers are potentially exposed to tributylamine based on estimates derived from the survey conducted between 1972-74 in the United States [7]. NIOSH has estimated that 47,014 workers are potentially exposed to tributylamine based on estimates derived from the survey conducted between 1981-83 in the United States [6]

**Body Burdens:**

## REFERENCES

1. Atkinson RA; Inter J Chem Kinet 19: 799-828 (1987)
2. Eisenreich SJ et al; Environ Sci Technol 15: 30-8 (1981)
3. Hine J, Mookerjee PK; J Org Chem 40: 292-8 (1975)
4. Liepins R et al; Industrial Process Profiles for Environmental Use: Chapt 6. USEPA-600/2-77-023f p.6-667 (1977)
5. Lyman WJ et al; Handbook of Chemical Property Estimation Methods NY: McGraw-Hill (1982)
6. NIOSH; National Occupational Exposure Survey (NOES) (1983)
7. NIOSH; National Occupational Hazard Survey (NOHS) (1974)
8. PCGEMS; Graphical Exposure Modelling System, CLOGP3 Program, Office of Toxic Substances, U.S. EPA (1989)
9. PCGEMS; Graphical Exposure Modelling System, PCCHEM Program, Office of Toxic Substances, U.S. EPA (1989)

# Tributylamine

10. Riddick JA et al; Organic Solvents: Physical Properties and Methods of Purification. Techniques of Chemistry. 4th Ed. New York: Wiley-Interscience pp 1325 (1986)
11. Swann RL et al; Res Rev 85: 16-28 (1983)
12. Yoshimura K et al; J Amer Oil Chem Soc 57: 238-41 (1980)

# 1,1,1-Trichloroethane

## SUBSTANCE IDENTIFICATION

**Synonyms:** Methyl chloroform

**Structure:**

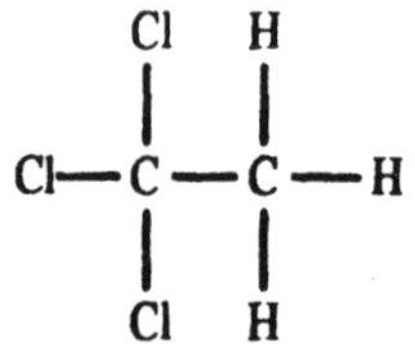

```
        Cl   H
        |    |
Cl — C — C — H
        |    |
        Cl   H
```

**CAS Registry Number:** 71-55-6

**Molecular Formula:** $C_2H_3Cl_3$

**Wiswesser Line Notation:** GXGG1

## CHEMICAL AND PHYSICAL PROPERTIES

**Boiling Point:** 74.1 °C at 760 mm Hg

**Melting Point:** -30.4 °C

**Molecular Weight:** 133.42

**Dissociation Constants:**

**Log Octanol/Water Partition Coefficient:** 2.49 [28]

**Water Solubility:** 1,495 mg/L at 25 °C [31]

**Vapor Pressure:** 123.7 mm Hg at 25 °C [50]

**Henry's Law Constant:** $8 \times 10^{-3}$ atm-m³/mole [36]

## ENVIRONMENTAL FATE/EXPOSURE POTENTIAL

**Summary:** 1,1,1-Trichloroethane is likely to enter the environment from air emissions, in wastewater from its production, or from use in vapor degreasing, metal cleaning, etc. Releases to surface water will decrease in concentration almost entirely due to evaporation. Spills on land will decrease in concentration almost entirely due to volatilization and percolation into ground water. Releases to air will be transported long distances and partially return to earth in rain. In the troposphere, 1,1,1-trichloroethane will degrade very slowly by photooxidation and also slowly diffuse to the stratosphere where photodegradation will be rapid. Major human exposure is from air and drinking water. Exposure can be high near sources of emission or where drinking water is contaminated.

**Natural Sources:** 1,1,1-Trichloroethane is not known to occur as a natural product [33].

**Artificial Sources:** Wastewater and stack and fugitive emissions from production. Volatilization losses from its use in the cold cleaning of metals, in vapor degreasing, and as a solvent and aerosol, etc. [63]. Mean emissions rate of 1,1,1-trichloroethane that would contribute to its presence in indoor air are (source - rate ng/min-sq m): cleaning agents and pesticides - 37,000; painted sheetrock - 31; glued wallpaper - 84; glued carpet - 260 [66].

**Terrestrial Fate:** Evaporates fairly rapidly into the atmosphere because of its high vapor pressure. Passes rapidly through soil into ground water [51].

**Aquatic Fate:** Primary loss will be by evaporation into the atmosphere. Half-life will range from hours to a few weeks depending on wind and mixing conditions. Half-lives in a mesocosm simulating the conditions in Narragansett Bay were 24, 12, and 11 days under spring, summer, and winter conditions, respectively [65]. Biodegradation and adsorption onto particulate matter will be insignificant relative to volatilization [65]. Turbulence in microcosm tanks are substantially less than in the bay or the open ocean so volatilization may be significantly (up to an order of magnitude) faster in the bay or open water than measured in the mesocosms.

**Atmospheric Fate:** 1,1,1-Trichloroethane is fairly stable in the atmosphere and is transported long distances, being found even at the South Pole [5,35,48]. It is transported to Point Barrow, Alaska from the mid-latitudes [35]. It is slowly degraded principally by reaction with hydroxyl radicals and has a half-life of 6 months to 25 years [5,12]. The rate of degradation is increased by the presence of chlorine radicals and nitrogen oxides. 15% of the 1,1,1-trichloroethane drifts into the stratosphere where it is rapidly degraded by photodissociation [5,12]. Due to the large input of 1,1,1-trichloroethane into the atmosphere and its slow degradation, the amount of 1,1,1-trichloroethane in the atmosphere is increasing by 12-17% a year [12,35]. Some of the 1,1,1-trichloroethane returns to earth in rain as is evidenced by its presence in rainwater and a 40% reduction in air concentrations on rainy days [42].

**Biodegradation:** No, or very slow, degradation in soils. No degradation has been observed in subsurface soils in 27 weeks; However in loamy sand, slow degradation has been observed under acclimated conditions [7,68]. Slow degradation may occur in water under anaerobic or aerated conditions. Degradation may take several weeks and acclimation is important [6,60]. In seawater, a half-life of 9 months has been determined and vinylidene chloride is the degradation product [45]. No degradation in river water has been found [39]. No utilization of 1,1,1-trichloroethane occurred in a continuously-fed aerobic biofilm reactor that utilized acetate as its primary substrate [37]. However, 98% removal was obtained in a similar anaerobic reactor with a 2-day retention time after 8 wk acclimation [37]. 1,1,1-Trichloroethane degraded to vinylidene chloride as a first step in its biotransformation in microcosms containing aquifer water and sediment collected from uncontaminated sites in the Everglades [44]. Considerable degradation occurred within two weeks [44]. Field evidence of biodegradation in aquifers was obtained by following the concentration of 1,1,1-trichloroethane in a confined aquifer after it was injected with reclaimed ground water [37]. The half-life of 1,1,1-trichloroethane was 231 days with biodegradation given as the probable cause of loss [37].

**Abiotic Degradation:** Hydrolysis is not a significant degradation process, having a half-life of approximately 6 months [12,19]. The product of hydrolysis is vinylidene chloride [25]. Direct photolysis is not important in the troposphere since 1,1,1-trichloroethane does not absorb light above 290 nm. In the stratosphere, photolysis is important and leads to the chemical's rapid degradation [19,32]. 1,1,1-Trichloroethane reacts slowly with hydroxyl radicals which are produced by sunlight in the atmosphere. The half-life for this reaction is 5 yr, assuming a diurnally averaged OH radical concentration of 5 x $10^{+5}$ radicals/cm$^3$ [2]. Estimates of half-life in the troposphere range from 0.5 to 2.2 years, much slower than unsaturated chloroalkanes, but much greater than completely chlorinated compounds such as carbon tetrachloride [12,27,55]. Products of photooxidation include phosgene, $Cl_2$, HCl, and $CO_2$ [45,58]. Degradation is reported to be greatly increased by exposure to ozone and chlorine but no actual data could be found in regard to 1,1,1-trichloroethane's reactivity with ozone [58]. On exposure to nitrogen oxide, less than 5% degradation occurs in 8 hr [20]. There is some evidence that photodegradation is catalyzed by surfaces which results in complete degradation within 2 weeks [10]. Indirect evidence of photooxidation comes from the fact that levels of 1,1,1-trichloroethane are lowest in the afternoon and 8% less on sunny days than cloudy ones [54]. Photodegradation is not observed in water [19].

**Bioconcentration:** The BCF in bluegill sunfish in a 28 day test was 8.9 [17]. This indicates that 1,1,1-trichloroethane has little tendency to bioconcentrate in fish. Although the amount of experimental data for 1,1,1-trichloroethane is limited, confidence in this result is increased because values of BCFs in related compounds are similar [4].

**Soil Adsorption/Mobility:** The adsorption of 1,1,1-trichloroethane to soil is proportional to the organic carbon content of the soil [24,49,61]. The mineral content of the soil is not a contributing factor [49]. The partition coefficient of 1,1,1-trichloroethane to 5 soils (organic carbon 0.1-4.9%) ranged from <0.05 to 0.5 while that adsorbed to sand and clay was too small to determine the isotherms [61]. The partition coefficient of 6 chlorinated alkanes including 1,1,1-trichloroethane between bentonite and spring water ranged from 27-76 and between Neckar River sediment and water, 2-108 [30]. 1,1,1-Trichloroethane is adsorbed strongly to peat moss, less strongly to clay, very slightly to dolomite limestone and not at all to sand [20]. It has a low adsorption to silt loam ($K_{oc}$ = 183) [13]. From the fact that it is not retained in the soil during bank infiltration, and that it is frequently found in ground water in high concentrations, one can safely conclude that it is not adsorbed strongly by soils, especially subsurface soils [51].

**Volatilization from Water/Soil:** 1,1,1-Trichloroethane has a high Henry's Law constant ($8 \times 10^{-3}$ atm-m$^3$/mole [36]) and will volatilize rapidly from water and soil with diffusion through the liquid phase controlling volatilization from water [36,53]. Half-life for evaporation from water obtained from laboratory systems range from a fraction of an hour to several hours [20]. Using the Henry's Law constant, one would calculate a half-life of 3.7 hr from a model river 1 m deep with a 1 m/sec current and a 3 m/sec wind [36]. Using the experimentally determined ratio of the volatilization rate constants of 1,1,1-trichloroethane relative to oxygen, 0.59 [43], and the oxygen reaeration coefficients for various bodies of water, one calculates that the volatilization half-lives range from 5.1-10.6 days for ponds, 3-29 hr for rivers, and 3.8-12 days for lakes [36]. Loss in a mesocosm is entirely due to evaporation and half-lives ranged from 24 days in spring to 11 days in winter [65].

**Water Concentrations:** DRINKING WATER: 133 US cities with finished surface water - 0.4 ppb median, 3.3 ppb max; 23 US cities with finished ground water - 2.1 ppb median, 3.0 max, 22% of the samples were positive [15]. Contaminated drinking water wells in New York, New Jersey, Connecticut, and Maine have values of 950-5440 ppb [11]. Results of the 1982 EPA Ground Water Supply Survey for 1,1,1-trichloroethane (466 samples) - 5.8% pos, 0.8 ppb median of positives, 18 ppb max [16]. As part of EPA's Total Exposure Assessment Methodology (TEAM) study, the concentration of various toxic substances in drinking water of sample populations was measured [67]. The mean (maximum) concentrations of 1,1,1-trichloroethane in Bayonne and Elizabeth, NJ, an industrial/chemical manufacturing area, were 0.6 (5.3), 0.2 (2.6), and 0.2 (1.6) ppb in the fall 1981, summer 1982, and winter 1983, respectively [67]. For comparison, the drinking water of a sample of residents of a manufacturing city without a chemical or petroleum refining industry, Greensboro, NC, and a small, rural, and agricultural town in North Dakota contained 0.03 (0.05) and 0.04 (0.07) ppb of 1,1,1-trichloroethane, respectively [67]. GROUND WATER: Raw ground water in 13 US cities - 1.1 ppb median, 13 ppb max, 23% were positive [15]. SURFACE WATER: Raw surface water in 105 US cities - 0.2 ppb median, 1.2 ppb max, 12% positive [15]. Large study of the Ohio River Basin in 1980-81 (4972 samples) reports 33.6% of samples above 0.1 ppb, 3.9% between 1.0 and 0.3% above 10 ppb [41]. In a study of 14 heavily industrialized river basins in 1975-76, 9% of the sites had values above 1 ppb, and 8 ppb was the maximum value measured [22]. At industrial sites, mean values are above 10 ppb with maximum values as high as 334 ppb [46]. Concentration 20-800 meters away from outfalls of four producing plants and 1 user was 0.1-169 ppm [5]. SEAWATER: Liverpool Bay seawater averaged <0.25 ppb, 3.3 ppb maximum [45]. RAIN/SNOW: West Los Angeles (Mar 26, 1982) - 69 ppt [34]; La Jolla, Ca - 8.1 ppt [59]; an industrial area of England - 0.9 ppt [45]. Southern California 6.2 ppt, central California - 0.6 ppt, Alaska 27 ppt [45].

# 1,1,1-Trichloroethane

**Effluent Concentrations:** Mean values in raw wastewater of 15 industries range from 3.6 to 38,000 ug/L with the maximum value range from 10 to 1,300,000 ug/L. The highest values were for the metal finishing industry [62]. Mean value of treated wastewater for 11 industries 0.6-89 ug/L with maximum values ranging from 0.6 to 7100 ug/L [62]. 18-344 ppb outfall from producing plants [5]. In a comprehensive survey of wastewater from 4000 industrial and publicly owned treatment works (POTWs) sponsored by the Effluent Guidelines Division of the USEPA, 1,1,1-trichloroethane was identified in discharges of the following industrial category (frequency of occurrence; median concn in ppb): timber products (2; 359.7), leather tanning (4; 2.7), iron and steel mfg (6; 34.4), petroleum refining (5; 13.4), nonferrous metals (12; 35.9), paint and ink (36; 9.7), printing and publishing (6; 28.3), ore mining (5; 2.3), coal mining (6; 5.7), organics and plastics (23; 8.5), inorganic chemicals (13; 5.2), textile mills (12; 6.0), plastics and synthetics (12; 1.6), pulp and paper (12; 7.0), rubber processing (10; 24.0), soaps and detergents (1; 26.3), auto and other laundries (10; 6.4), pesticides manufacture (4; 17.0), photographic industries (3; 3.9), pharmaceuticals (20; 3.9), explosives (7; 14.6), plastics mfg (1; 8.3), foundries (5; 54.0), electronics (36; 62.5), electroplating (2; 229.1), organic chemicals (15; 7.2), mechanical products (20; 98.0), transportation equipment (5; 706.3), amusements and athletic goods (4; 33.0), synfuels (8; 6.63), and publicly owned treatment works (302; 10.6) [52]. The highest effluent concns were 6397 and 6028 ppb in the mechanical products and electronics industry, respectively [52].

**Sediment/Soil Concentrations:** Liverpool Bay marine sediment <5.5 ppb [45]. Soil around production plants and user industry 0.06-0.94 ppb; sediment upstream and downstream of production plants and user industry 0.039-2.6 ppb; average background concentration in soil (St. Francis National Forest) 0.42 ppb; average background concentration in sediment (St. Francis National Forest) 0.45 ppb [5].

**Atmospheric Concentrations:** RURAL/REMOTE: Rural/remote sites in US (1977-80) - 60-156 ppt, 110 ppt avg [9,48,57]. Yearly rate of increase is 12-17%/year [48,57]. The baseline 1,1,1-trichloroethane level in the northern hemisphere (60 deg N to 40 deg N) is 200 ppt while in the northern hemisphere it is 140 ppt [14]. URBAN/SUBURBAN: Urban/suburban in US areas (1977-80) - 420 ppt avg, 700-8000 ppt maximum, <20% samples may be positive [8,9,55,56]. SOURCE AREAS: Source dominated areas in US

(1977-80) - 1200 ppt avg [9]. Although maximum values are usually under 10 ppb, one maximum value of 111 ppb has been reported in New Jersey [46]. INDOOR AIR: the concentration of 1,1,1-trichloroethane in a new office building before and after occupancy was 500 and 60 ug/m$^3$ (90 and 10.8 ppb), respectively [66]. OTHER: As part of EPA's Total Exposure Assessment Methodology (TEAM) study, the concentration of various toxic substances in the personal air (2 consecutive 12-hr periods) of sample populations was measured as well as the outdoor air near their residences [67]. The weighted median results for 1,1,1-trichloroethane in personal air in Bayonne and Elizabeth, NJ, an industrial/chemical manufacturing area, were 17, 9.3, and 22 ug/m$^3$ in the fall 1981, summer 1982, and winter 1983, respectively [67]. The corresponding results for outdoor air were 4.6, 5.1, and 1.4 ug/m$^3$ [67]. For comparison the personal air of a sample of residents of a manufacturing city without a chemical or petroleum refining industry, Greensboro, NC, and a small, rural, and agricultural town in North Dakota contained 32 and 25 ug/m$^3$ of 1,1,1-trichloroethane, respectively, and the outdoor air, 60 and 0.05 ug/m$^3$ [67].

**Food Survey Values:** 5-10 ng/g oils and fats; 1-4 ng/g fruits and vegetables; 2-7 ng/g meat, tea, and bread [38]. 1,1,1-Trichloroethane was not found in samples of wheat, corn, oats, corn meal, or corn grits [29]. Of the 9 samples of intermediate grain-based food analyzed, it was found in 3, namely, yellow corn meal (3.8 ppb), fudge brownie mix (3.0 ppb), and yellow cake mix (0.74 ppb) [38].

**Plant Concentrations:** <9.4-35 ppb (in analytical work CCl$_4$ was not separable from 1,1,1-trichloroethane) in marine algae [45].

**Fish/Seafood Concentrations:** Three species of fish, mollusks in Irish Sea - 2-16 ppb [18]. Flesh of nine samples of various fish from Liverpool Bay and Thames Estuary - 0-5 ppb, gut contained up to 26 ppb [45]. Marine invertebrates in bays and estuaries of Great Britain - 0-34 ppb [45].

**Animal Concentrations:** <16-30 ppb gray seal blubber, <2.3-7 ppb common shrew, <1.1-4.7 ppb in flesh or organs of fresh- and seawater birds, <4.2-43 ppb in eggs of fresh- and seawater birds (in analytical work CCl$_4$ was not separable from 1,1,1-trichloroethane) [45].

**Milk Concentrations:**

# 1,1,1-Trichloroethane

**Other Environmental Concentrations:** Of the 1026 brand samples of household products representing 67 product categories (cleaners, polishes, lubricants, and paint removers), 14.1% of samples and 47.8% of product categories contained 1,1,1-trichloroethane ranging from 3.3% to 100% [23].

**Probable Routes of Human Exposure:** Humans may be exposed to 1,1,1-trichloroethane dermally and by inhalation of air at occupational sites, from using household products containing the chemical, from ambient air, or ingestion of contaminated drinking water and food.

**Average Daily Intake:** AIR INTAKE: rural (assume 0.110 ppb) 12.2 ug; urban/suburban (assume 0.420 ppb) - 46.5 ug; residents in source dominated areas (assume 1.20 ppb) - 133.0 ug. WATER INTAKE: surface water source (assume 0.4 ppb) - 0.8 ug; ground water source (assume 2.1 ppb) - 4.2 ug. FOOD INTAKE: insufficient data.

**Occupational Exposures:** NIOSH (NOES Survey 1981-83) has statistically estimated that 392,805 workers are exposed to 1,1,1-trichloroethane in the United States [40]. Trichloroethane concn of 1.5-350 ppm in ambient air of various industries (degreasing, metals, electrical, etc.) [64]. 11% of 8 hr TWA concentration of 1,1,1-trichloroethane at 3 wastewater treatment plants serving greater Cincinnati were above the 6 ppb detection limit [21]. There were 2100 and 3400 ppb in the plant containing the highest loads of the chemical in its influent [21].

**Body Burdens:** Body fat of 8 subjects - 1.6-24 ng/g, various organs - <5.1 ng/g [38]. Sample residents of Old Love Canal, breath - 290 ng/m³ median, 2800 ng/m³ maximum; blood - 0.85 ng/mL median, 2.0 ng/mL maximum; urine - 80 ng/L median, 180 ng/L maximum [3]. Detected in all eight samples of human milk from four urban areas [47]. 59% of individuals (39 subjects, 23-54 years of age) from Dusseldorf, West Germany who were not occupationally exposed to 1,1,1-trichloroethane - whole blood levels of the chemical ranging from <0.1-3.4, ppb median 0.2 ppb [26]. Whole blood samples of those occupationally exposed contained 1,1,1-trichloroethane ranging from <0.1-0.2 ppb for motor vehicle mechanics, <0.1 for painters, 0.1-15.5 ppb for precision tool makers, 389.0-2497.9 ppb for dry cleaners using tetrachloroethylene as dry-cleaning agent, and 17.6-48.2 ppb for dry cleaners using trichlorofluoromethane as a dry-cleaning agent [26].

# 1,1,1-Trichloroethane

Blood samples were drawn during the work day after 4 to 7 hr exposure. In another study, 1,1,1-trichloroethane in the whole blood of 250 patients who suffered from a variety of symptoms that may have been related to exposure to environmental pollutants ranged from not detectable to 26 ppb, 1.0 ppb mean [1]. As part of EPA's Total Exposure Assessment Methodology (TEAM) study, the concentration of various toxic substances on the breath of sample populations was measured [67]. The weighted median results for 1,1,1-trichloroethane in Bayonne and Elizabeth, NJ, an industrial/chemical manufacturing area, were 6.6, 5.2, and 2.3 ug/m$^3$ in the fall 1981, summer 1982, and winter 1983, respectively [67]. For comparison, the breath of a sample of residents of a small, rural, agricultural town in North Dakota contained 9.3 ug/m$^3$ of 1,1,1-trichloroethane [67].

## REFERENCES

1.  Antoine SR et al; Bull Environ Contam Toxicol 36: 364-71 (1986)
2.  Atkinson R; Chem Rev 85: 69-201 (1985)
3.  Barkley J et al; Biomed Mass Spectrom 7: 139-47 (1980)
4.  Barrows ME et al; Dyn Exp Hazard Assess Toxic Chem Ann Arbor Mi: Ann Arbor Sci p. 379-92 (1980)
5.  Battelle Columbus Labs; Multimedia Levels Methylchloroform p 2.1-2.22 EPA-560/6-77-030 (1977)
6.  Bouwer EJ, McCarty PL; Appl Environ Microbiol 45:1286-94 (1983)
7.  Bouwer EJ et al; Water Res 15:151-9 (1981)
8.  Bozzelli JW, Kebbekus BB; Analysis of selected volatile organic substances in ambient air. Air Force Final Report, Apr-Nov 1978 Newark, NJ: NJ Inst Technol p. 80 (1979)
9.  Brodzinsky R, Singh HB; Volatile organic chemicals in the atmosphere: an assessment of available data. p. 12-5 SRI Inter contract 68-02-3452 (1982)
10. Buchardt O, Manscher OH; CEC Proceedings, 2nd Meeting (1978)
11. Burmaster DE; Environ 24: 6-13, 33-6 (1982)
12. Callahan MA et al; Water-Related Environmental Fate of 129 Priority Pollutants Vol II page 45-1 to 45-12 EPA-440/4-79-029B (1979)
13. Chiou CT et al; Science 206: 831-2 (1979)
14. Class T, Ballschmitter K; Chemosphere 15: 413-27 (1986)
15. Coniglio WA et al; EPA Briefing. Criteria and Standards Div. Sci Technol Branch. Exp Assess Proj p. 16 (1980)
16. Contruvo JA; Sci Tot Environ 47: 7-26 (1985)
17. Davies RP, Dobbs AJ; Atm Res 18: 1253-62 (1984)
18. Dickson AG, Riley JP; Marine Pollut Bull 7:167-70 (1976)
19. Dilling WL et al; Environ Technol 9:833-8 (1975)
20. Dilling WL et al; Environ Sci Technol 10;351-6 (1976)
21. Dunovant VS et al; J Water Pollut Control Fed 58: 886-95 (1986)

22. Ewing BB et al; Monitoring To Detect Previously Undetected Pollutants In Surface Waters. Appendix: Organic Analysis Data p 1-129 EPA-560/6-77-015A (1977)
23. Frankenberry M et al; Household products containing methylene chloride and other chlorinated solvents: A shelf survey. Rockville, MD: Westat Inc (1987)
24. Friesel P et al; Fresenius Z Anal Chim 319: 160-4 (1984)
25. Haag WR et al; Am Chem Soc Div Environ Chem Preprint 26: 248-53 (1986)
26. Hajimiragha H et al; Int Arch Occup Environ Health 58: 141-50 (1986)
27. Hampson RF; FAA-EE-80-17, US Dept of Transportation (1980)
28. Hansch C, Leo AJ; MEDCHEM Project Claremont CA: Pomona College (1985)
29. Heikes DL, Hopper ML; J Assoc Off Anal Chem 69: 990-8 (1986)
30. Hellmann H; Dtsch Gewaesserkd Mitt 29: 111-5 (1985)
31. Horvath AL; Halogenated Hydrocarbons: Solubility-Miscibility with Water New York: Marcel Dekker (1982)
32. Hubrich C, Stuhl F; J Photochem 12:93-107 (1980)
33. IARC; Monograph Some Halogenated Hydrocarbons 19: 515-31 (1979)
34. Kawamura K, Kaplan LR; Environ Sci Technol 17: 497-501 (1983)
35. Khalil MAK, Rasmussen RA; Environ Sci Technol 17:57-64 (1983)
36. Lyman WJ et al; Handbook of Chemical Property Estimation Methods New York: McGraw-Hill (1982)
37. McCarthy PL et al; Groundwater Pollut Microbiol pp. 89-115 (1984)
38. McConnell G et al; Endeavour 34: 13-8 (1975)
39. Mudder TI; Amer Chem Soc Div Environ Chem p 52-3, Kansas City, MO (1982)
40. NIOSH; National Occupational Exposure Survey (1983)
41. Ohio River Valley Water Sanit Comm; Assess of Water Qual Cond 1980-81 Cincinnati, OH (1982)
42. Ohta T et al; Atmos Environ 11:985-7 (1977)
43. Okouchi S; Wat Sci Tech 18: 137-8 (1986)
44. Parsons F, Lage GB; J Am Water Works Assoc 77: 52-9 (1985)
45. Pearson CR, McConnell G; Proc Roy Soc London B 189:305-32 (1975)
46. Pellizzari ED et al; Formulation of Preliminary Assessment of Halogenated Organic Compounds In Man and Environmental Media p. 38-94 EPA-560/13-79-006 (1979)
47. Pellizzari ED et al; Environ Sci Technol 16: 78-5 (1982)
48. Rasmussen RA et al; Science 211: 285-7 (1981)
49. Richter RO; Am Chem Soc Div Environ Chem Preprints 23: 193-4 (1983)
50. Riddick JA et al; Organic Solvents New York: Wiley Interscience (1986)
51. Schwarzenbach RP et al; Environ Sci Technol 7:472-9 (1983)
52. Shackelford WM et al; Analyt Chim Acta 146: 15-27 (1983)
53. Shen TT; J Air Pollut Control Assoc 32: 79-82 (1982)
54. Singh HB et al; Environ Sci Technol 16;872-80 (1982)
55. Singh HB et al; Atmos Environ 15:601-12 (1981)
56. Singh HB et al; Atmospheric measurement of selected hazardous organic chemicals EPA-600/S3-81-032 (1981)
57. Singh HB et al; Atmospheric distribution, sources and sinks of selected halocarbons, hydrocarbons, SF$_6$ and N$_2$O. p. 114-22 EPA-600/3-79-107 (1979)
58. Spence JW, Hanst PL; J Air Pollut Control Fed 28:250-3 (1978)
59. Su C, Goldberg ED; Marine Pollut Tfr HL Windom et al eds Lexington MA: DC Heath Co p. 353-74 (1976)
60. Tabak HH et al; J Water Pollut Control Fed 53:1503-18 (1981)

# 1,1,1-Trichloroethane

61. Urano K, Murata C; Chemosphere 14: 293-9 (1985)
62. USEPA; Treatability Manual page I.12.8-1 to I.12.8-4 EPA-600/2-82-001a (1982)
63. USEPA; Source Assessment: Chlorinated Hydrocarbon Manufacture USEPA-600/2-79-019G (1979)
64. USEPA; Ambient Water Quality Criteria for Chlorinated Ethane p. C5-C13 EPA-440/5-80-029 (1980)
65. Wakeham SG et al; Environ Sci Technol 17: 611-7 (1983)
66. Wallace LA et al; Atmos Environ 21: 385-93 (1987)
67. Wallace LA et al; Environ Res 43: 290-307 (1987)
68. Wilson JT et al; Devel Indust Microbiol 24:225-33 (1983)

# 1,1,2-Trichloroethane

## SUBSTANCE IDENTIFICATION

**Synonyms:**

**Structure:**

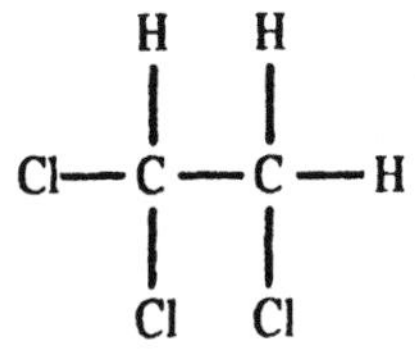

$$Cl—\overset{\displaystyle H}{\underset{\displaystyle Cl}{C}}—\overset{\displaystyle H}{\underset{\displaystyle Cl}{C}}—H$$

**CAS Registry Number:** 79-00-5

**Molecular Formula:** $C_2H_3Cl_3$

**Wiswesser Line Notation:** GYG1G

## CHEMICAL AND PHYSICAL PROPERTIES

**Boiling Point:** 113.8 °C at 760 mm Hg

**Melting Point:** -36.5 °C

**Molecular Weight:** 133.42

**Dissociation Constants:**

**Log Octanol/Water Partition Coefficient:** 2.07 (calculated) [12]

**Water Solubility:** 4420 mg/L at 20 °C [22]

**Vapor Pressure:** 30.3 mm Hg at 20 °C [22]

**Henry's Law Constant:** $1.2 \times 10^{-3}$ atm-m³/mole at 20 °C [22]

## ENVIRONMENTAL FATE/EXPOSURE POTENTIAL

**Summary:** 1,1,2-Trichloroethane will enter the atmosphere from its use in the manufacture of vinylidene chloride and its use as a solvent. It will also be discharged in wastewater associated with these uses.

461

# 1,1,2-Trichloroethane

Releases to water will primarily be lost through evaporation (half-life days to weeks). Once in the atmosphere, 1,1,2-trichloroethane will photodegrade by reaction with hydroxyl radicals (half-life 24 days in unpolluted atmospheres to a few days in polluted atmospheres). 1,1,2-Trichloroethane has a low soil partition coefficient and as such will not partition into sediment and will readily pass through soil into the ground water where biodegradation is unlikely to occur. Bioconcentration is not a significant process. Primary human exposure is from occupational exposure and from ambient air in the vicinity of industrial sources and contaminated drinking water.

**Natural Sources:** None [32]

**Artificial Sources:** Emissions from its use as an intermediate in vinylidene chloride synthesis and as a solvent for chlorinated rubber, fats, oils, resins, and adhesives [16,39].

**Terrestrial Fate:** When released to land 1,1,2-trichloroethane should partially volatilize and partially leach into the ground water. It is unlikely that biodegradation will be an important process.

**Aquatic Fate:** When released into water, 1,1,2-trichloroethane should primarily volatilize into the atmosphere with a half-life of days to weeks. Little of the chemical will be lost by adsorption to sediment or by biodegradation.

**Atmospheric Fate:** When released into the atmosphere, 1,1,2-trichloroethane should degrade by reaction with hydroxyl radicals with a half-life of 24 days. The half-life in polluted atmospheres is much shorter, being of the order of 16 hr. A relatively water-soluble chemical such as 1,1,2-trichloroethane will be expected to partially washout in rain.

**Biodegradation:** 1,1,2-Trichloroethane showed no biodegradation in both a 24-day modified shake flask test and a river die-away test [23]. Similar results were obtained in another screening biodegradability test [18]. When a solution containing 1,1,2-trichloroethane was applied to a column filled with sandy soil, no loss could be attributed to biodegradation [42]. One investigation reported very slow biodegradation with long acclimation [36]. Vinyl chloride was observed to be a biodegradation product of 1,1,2-trichloroethane when microbes from an anaerobic digester at a municipal wastewater treatment facility were used as inocula [11]. No significant degradation occurred over a 16-week incubation period in either sterilized or non-sterile subsurface (5 m deep) soil samples [41].

**Abiotic Degradation:** 1,1,2-Trichloroethane will react with photochemically produced hydroxyl radicals in the troposphere and degrade with a half-life of 24 days or a 2.8% loss per 12 hr sunlit day [31]. Its reaction under simulated smog conditions is much faster, the half-life being 16 hr [7]. Hydrolysis is expected to be slow, one experiment reporting no significant decrease in aqueous concentration in 8 days [3,17] and less than a 5% decrease in concentration in seawater in a closed system in darkness or daylight [17]. The half-life for the neutral aqueous hydrolysis of 1,1,2-trichloroethane at 25 °C has been estimated to be in excess of 50 years [21].

**Bioconcentration:** 1,1,2-Trichloroethane would not be expected to bioconcentrate in fish having a log bioconcentration factor of <1 [18].

**Soil Adsorption/Mobility:** There is little information on the measured adsorption of 1,1,2-trichloroethane to soil. The chemical moved readily through a column of sandy soil having a retardation factor (velocity of water through the soil divided by the velocity of pollutant) of <1.5, and the Koc is about 70 [42]. The log Kom (Koc = 1.724 Kom) of 1,1,2-trichloroethane has been estimated to range from 1.06-2.49 based on its log Kow value and five published prediction equations [30].

**Volatilization from Water/Soil:** Laboratory measurements of the rate of evaporation of 1,1,2-trichloroethane from water gave a half-life of 21 minutes [6]. In natural waters, half-lives would be expected to be of the order of days to weeks. Based upon field monitoring data, the half-life of 1,1,2-trichloroethane in a section of the Rhine river was 1.9 days which is probably the result of evaporative loss [43]. Due to its moderate vapor pressure, volatilization from soil will occur.

**Water Concentrations:** DRINKING WATER: Analysis of 945 finished water supplies nationwide (USEPA Ground Water Supply Survey) that use ground water sources did not detect any 1,1,2-trichloroethane at a quantification limit of 0.5 ppb [40]. 1,1,2-Trichloroethane has been detected in drinking water from samples of US cities [16,19,37] with 0.1-8.5 ppb being measured in the finished water from one metropolitan supply [37]. 20 ppb reported in a contaminated New York State well [2,4]; 30 Canadian water treatment facilities, 2 positive, 7 ppb max in

Aug/Sept; not detected in Nov/Dec [25]. SURFACE WATER: 1,1.2-Trichloroethane was detected in 2.0% of 1047 USEPA STORET water stations at a median concn below 5.0 ppb [35]. Not detected in raw water from 30 Canadian treatment facilities [25]. Detected, not qualified, River Glatt, Switzerland [44] 53 to 603 samples pos in representative New Jersey surface waters, max 18.7 ppb [27]; 2 tributaries on Ohio River, 3 of 7 samples pos, 0.6 ppb max, not found in 88 additional stations [24], Ohio River mainstream 0.4% of 246 samples pos, <1.0 ppb avg [8,24]. Not detected in Schuylkill River at Philadelphia, PA, DL = 0.01 ppb [8]. GROUND WATER: 1,1,2-Trichloroethane was detected in 2 of 13 ground water samples (associated with leaching from waste sites) from Minnesota at levels of 7.7 and 31 ppb [29]. Detected in 72 of 1069 samples in New Jersey, max 31.1 ppb, some of the most polluted being under urban land use areas [9,27]. SEA WATER: Shorewater concn of 1,1,2-trichloroethane off Point Reyes, CA 153 ppt [32].

**Effluent Concentrations:** Industries whose mean wastewater effluent exceeds 500 ppb are auto and other laundries and paint and ink formulations, max observed effluent concn is 3 ppm (auto and other laundries) [38] 5.4 ppm detected in industrial effluent discharge [16]. 1,1,2-Trichloroethane was detected in 2.8% of 1345 USEPA STORET effluent stations at a median concn below 2.0 ppb [35]. Positive identification was made for 1 of 13 effluent samples collected from a community septic tank serving 97 homes near Tacoma, WA [5].

**Sediment/Soil Concentrations:** 1,1,2-Trichloroethane has been qualitatively detected in the sediment/soil/water matrix of the Love Canal waste site near Niagara, NY [15].

**Atmospheric Concentrations:** URBAN/SUBURBAN: US (930 samples) 9.1 ppt median, 11,000 ppt max [1]. 1,1,2-Trichloroethane has been detected in New Jersey cities - 0.01-0.037 ppb avg but is not very prevalent (9 of 263 samples pos in one study) [13,14,20] 10 major US cities - 6-41 ppt avg [31,33]. Avg background concn at 2 California sites near San Francisco - inland site 14.6 ppt, 95% of 115 samples >6 ppt; nearshore site - 14.1 ppt, 48% of 225 samples >6 ppt [32]. SOURCE RELATED: US (97 samples) 45 ppt median, 25% of samples exceed 210 ppt, 2300 ppt max [1]; 2 of 5 industrial sites in US positive, trace to 1.06 ppb [28]; RURAL/REMOTE: US - not detected [1,10].

## 1,1,2-Trichloroethane

**Food Survey Values:**

**Plant Concentrations:**

**Fish/Seafood Concentrations:**

**Animal Concentrations:**

**Milk Concentrations:**

**Other Environmental Concentrations:**

**Probable Routes of Human Exposure:** Humans are exposed to 1,1,2-trichloroethane from ambient air, particularly near sources of emission and from contaminated drinking water supplies.

**Average Daily Intake:** AIR INTAKE: (assume 9.1 ppt) 1.0 ug; WATER INTAKE: (assume 0 ppb) 0 ug; FOOD INTAKE: insufficient data.

**Occupational Exposures:** Office air contained 5.7 ppm [26]. It is estimated that 112,000 workers are exposed to 1,1,2-trichloroethane [37].

**Body Burdens:** 1,1,2-Trichloroethane was not detected in any human adipose sample analyzed as part of the USEPA National Human Adipose Tissue Survey for fiscal year 1982 [34].

### REFERENCES

1.  Brodzinsky R, Singh HB; Volatile organic chemicals in the atmosphere: an assessment of available data. SRI Inter Contract 68-02-3452 Menlo Park, CA: Atmos Sci Ctr 198 p (1982)
2.  Burmaster DE; Environ 24: 6-13, 33-6 (1982)
3.  Callahan MA et al; Water-related environmental fate of 129 priority pollutants. EPA-440/4-79-029b p 46.1 to 46.9 (1979)
4.  Council on Environmental Quality; Contamination of ground water by toxic organic chemicals. Washington, DC p. 84 (1981)
5.  DeWalle TB et al; Determination of Toxic Chemicals in Effluent from Household Septic Tanks EPA-600/S2-85-050 (1985)
6.  Dilling WL; Environ Sci Technol 11: 405-9 (1977)
7.  Dilling WL et al; Environ Sci Technol 10: 351-6 (1976)
8.  Dreisch FA et al; Survey of the Huntington and Philadelphia river water supplies for purgeable organic contaminants. EPA-903/9-81-003 p. 14 (1980)
9.  Greenberg M et al; Environ Sci Technol 16: 14-9 (1982)

# 1,1,2-Trichloroethane

10. Grimsrud EP, Rasmussen RA; Atmos Environ 9: 1014-7 (1975)
11. Hallen RT et al; ACS Div Environ Chem 192nd Natl Mtg 26: 344-6 (1986)
12. Hansch C, Leo AJ; Medchem Project Issue No 26. Claremont CA: Pomona College (1985)
13. Harkov R et al; J Air Pollut Control Assoc 33: 1177-83 (1983)
14. Harkov R et al; Toxic and carcinogenic air pollutants in New Jersey - volatile organic substances. Unpublished work. Trenton, NJ: Office of Cancer and Toxic Substances Research (1981)
15. Hauser TR, Bromberg SM; Environ Monit Assess 2: 249-72 (1982)
16. IARC; Some halogenated hydrocarbons 20: 533-43 (1979)
17. Jensen S, Rosenberg R; Water Res 9: 659-61 (1975)
18. Kawasaki M; Ecotox Environ Safety 4: 444-54 (1980)
19. Keith LH et al; Identification and analyses of organic pollutants in water. Ann Arbor, MI: Ann Arbor Press p 329-73 (1976)
20. Lioy PJ et al; J Water Pollut Control Fed 33: 649-57 (1983)
21. Mabey WR et al Symp Amer Chem Soc, Div Environ Chem 186th Natl Mtg Washington DC 23: 359-61 (1983)
22. MacKay D, Shiu WY; J Phys Chem Ref Data 10: 1175-1199 (1981)
23. Mudder TI; Amer Chem Soc Div Environ Chem Presentation Kansas City, MO Sept 1982 p 52-3 (1982)
24. Ohio River Valley Water Sanit Comm; Assessment of water quality conditions. Ohio River Mainstream 1978-9 Cincinnati, OH (1980)
25. Otson R et al; J Assoc Off Anal Chem 65(6): 1370-4 (1982)
26. Otson R et al; Bull Environ Contam Toxicol 31: 222-9 (1983)
27. Page GW; Environ Sci Technol 15: 1475-81 (1981)
28. Pellizzari ED; Environ Sci Technol 16: 781-5 (1982)
29. Sabel GV, Clark TP; Waste Manage Res 2: 119-30 (1984)
30. Sabljic A; Environ Sci Technol 21: 358-66 (1987)
31. Singh HB et al; Atmos Environ 15: 601-12 (1981)
32. Singh HB et al; J Air Pollut Control Assoc 27: 332-6 (1977)
33. Singh HB et al; Environ Sci Technol 16: 872-80 (1982)
34. Stanley JS; Broad Scan Analysis of the FY82 National Human Adipose Tissue Survey Specimens. Volume I - Exec Sum. EPA-560/5-86-035 p. 5 (1986)
35. Staples CA et al; Environ Toxicol Chem 4: 131-42 (1985)
36. Tabak HH et al; J Water Pollut Control Fed 53: 1503-18 (1981)
37. USEPA; Ambient Water Quality Criteria for Chlorinated Ethanes. EPA-440/5-80-029 (1980)
38. USEPA; Treatability Manual. EPA-600/2-82-001a page I.12.9-1 to I.12.9-4 (1981)
39. Verschueren K; Handbook of environmental data on organic chemicals 2nd ed New York, NY: Van Nostrand Reinhold Co., Inc. p 1128-9 (1983)
40. Westrick JJ et al; J Amer Water Works Assoc 76: 52-9 (1984)
41. Wilson JT et al; Ground Water 21: 134-42 (1983)
42. Wilson JT et al; J Environ Qual 10: 501-6 (1981)
43. Zoeteman BCJ et al; Chemosphere 9: 231-49 (1980)
44. Zuercher F, Giger W; Vom Wasser 47: 37-55 (1976)

# Trichloroethylene

## SUBSTANCE IDENTIFICATION

**Synonyms:**

**Structure:**

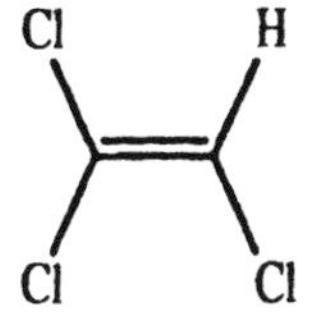

Cl          H

＼        ／

＼＝＝／

／        ＼

Cl          Cl

**CAS Registry Number:** 79-01-6

**Molecular Formula:** $C_2HCl_3$

**Wiswesser Line Notation:** GYGU1G

## CHEMICAL AND PHYSICAL PROPERTIES

**Boiling Point:** 87 °C

**Melting Point:** -73 °C

**Molecular Weight:** 131.40

**Dissociation Constants:**

**Log Octanol/Water Partition Coefficient:** 2.42 [27]

**Water Solubility:** 1100 mg/L at 25 °C [31]

**Vapor Pressure:** 69.0 mm Hg at 25 °C [4]

**Henry's Law Constant:** $1.03 \times 10^{-2}$ atm-m$^3$/mole [44]

## ENVIRONMENTAL FATE/EXPOSURE POTENTIAL

**Summary:** Over 155 million pounds of trichloroethylene are used for vapor degreasing of metals which should result in releases to the environment through evaporation, spills, and leaks in storage tanks.

467

# Trichloroethylene

Trichloroethylene released to the atmosphere will exist primarily in the vapor phase based on its relatively high vapor pressure. It will react fairly rapidly, especially under smog conditions. Atmospheric residence time of 5 days has been reported with formation of phosgene, dichloroacetyl chloride, and formyl chloride. It is not subject to direct photolysis. If trichloroethylene is released to water, the primary removal process will be evaporation with a half-life of minutes to hours, depending upon turbulence. Biodegradation, hydrolysis, and photooxidation are extremely slow by comparison. Adsorption to sediment and bioconcentration in aquatic organisms are not important processes. Releases to soil will be partially evaporated and partially leached into ground water, where it may remain for a long time. However, there is some monitoring data that suggests degradation to other chlorinated alkenes. High levels of exposure are expected for workers in degreasing plants due to inhalation of vapors or adsorption through the skin. Lower exposure by inhalation is expected in persons living near degreasing plants or at spill sites. Broad population exposure to low levels is expected from inhalation of contaminated ambient air and ingestion of contaminated drinking water.

**Natural Sources:**

**Artificial Sources:** Air emissions from metal degreasing plants [11] are likely. Wastewater from metal finishing, paint and ink formulation, electrical/electronic components, and rubber processing industries contain trichloroethylene [56].

**Terrestrial Fate:** Spills or releases of trichloroethylene to soil will evaporate rapidly due to its reasonably high vapor pressure. It will also leach into ground water rapidly. Trichloroethylene appears to be fairly stable in soil although one field study of ground water contamination from a leaking trichloroethylene tank has detected cis- and trans-1,2-dichloroethylene [39] which suggests that degradation in ground water can occur. Hydrolysis is not an important process.

**Aquatic Fate:** The primary removal process will be evaporation [18,57] with a half-life of minutes to hours, depending upon turbulence. Biodegradation, hydrolysis, and photooxidation are extremely slow by comparison. Adsorption to sediment and bioconcentration in aquatic organisms are not important processes [18].

# Trichloroethylene

**Atmospheric Fate:** Trichloroethylene released to the atmosphere will exist primarily in the vapor phase based on its relatively high vapor pressure [21]. It will react fairly rapidly, especially under smog conditions. Atmospheric residence time of 5 days has been reported with formation of phosgene, dichloroacetyl chloride, and formyl chloride. It is not subject to direct photolysis.

**Biodegradation:** Trichloroethylene biodegrades very slowly in water under most conditions. Only a few studies have noted significant aerobic biodegradation [46,55], but acclimation was slow [55]. Other studies found no biodegradation under aerobic conditions in screening studies [5] or in seawater [33,57]. Biodegradation under anaerobic conditions has been noted and ranged from very little after 12 weeks [5] to 40% after 8 weeks [6]. In laboratory studies, trichloroethylene does not appear to biodegrade in ground water [59,58]. However, in microcosms of authentic aquifer material known to support methanogenesis, the percent removal after 40 weeks ranged from 70% to >99% (avg 89%) compared to an avg 48% removal using autoclaved aquifer material; a long lag period was indicated by the 4% removal after 7 weeks [60]. Also, in field studies, cis- and trans-1,2-dichloroethylene have been detected near trichloroethylene contamination sites which suggests biodegradation [39].

**Abiotic Degradation:** Trichloroethylene is not hydrolyzed by water under normal conditions [9]. It does not adsorb light of less than 290 nm and therefore should not directly photodegrade [9]. However, slow (half-life - 10.7 months) photooxidation in water has been noted [19]. Trichloroethylene is relatively reactive under smog conditions [62] with 60% degradation in 140 min [24] and 50% degradation in 1 to 3.5 hr [18] reported. Atmospheric residence times based upon reaction with hydroxyl radicals is 5 days [10,16,50] with production of phosgene, dichloroacetyl chloride, and formyl chloride [16,24].

**Bioconcentration:** Marine monitoring data only suggest moderate bioconcentration (2-25 times the concentration in water) [17,40]. Bioconcentration factors of 17 to 39 have been reported in bluegill sunfish and rainbow trout [3,36].

**Soil Adsorption/Mobility:** Low adsorption coefficient (log Koc = 2.0) [59] to a number of soil types [19] indicates ready transport through soil and low potential adsorption to sediments. The mobility in soil is confirmed in soil column studies [59] and river bank infiltration studies

[48,52,63]. 4-6% of environmental concentrations of trichloroethylene adsorbed to two silty clay loams (Koc = 87 and 150) [45]. No adsorption to Ca-saturated montmorillonite and 17% adsorption to Al-saturated montmorillonite was observed [45].

**Volatilization from Water/Soil:** The high Henry's Law constant indicates rapid evaporation from water [36]. Half-lives of evaporation have been reported to be on the order of several minutes to hours, depending upon the turbulence [19,36]. Field studies also support rapid evaporation from water [57]. Relatively high vapor pressure indicates rapid evaporation from near-surface soil and other surfaces.

**Water Concentrations:** SURFACE WATER: 1-24 ppb industrial rivers in US, with Lake Erie - 188 ppb, 88 of 204 samples pos [22]; third most frequently detected compound in Ohio River - 2427 of 4972 samples pos, 86% 0.1-1.0 ppb [38]; Zurich, Switzerland lake surface - 38 ppb, 30 m depth - 65 ppb [26]; USEPA STORET data base, 9,295 data points, 28.0% pos, 0.10 ppb median [53]. DRINKING WATER: 28 of 113 US public water supplies pos, mean 2.1 ppb [8]; finished ground water mean 6.76 ppb, range 0.11-53.0 ppb in 36% of 25 US cities [12]. Love Canal, Niagara Falls, NY 7 of 9 samples pos, 10-250 ppt [2]; finished ground water, 466 random samples, 6.4% pos, 1 ppb median concn, 78 ppb max concn [13]. State data, 2894 samples, 28.0% pos, trace to 35,000 ppb; US National Screening Program, 142 samples, 25.4% pos, trace to 53 ppb; Community Water Supply Survey, 452 samples, 3.3% pos, 0.5-210 ppb [14]. GROUND WATER: Most frequently detected and in highest concentration, 28% of wells in 8 states sample pos max concn reported 35,000 ppb [20]; 38.5% of 13 US cities pos mean 29.72 ppb range 0.2-125 ppb [12]. New Jersey, 670 wells, 1.8% and 4.0% of wells had concn >100 ppb and > 10 ppb, respectively [61]. Ground water in the Netherlands 1976-78, 232 pumping stations, 67% pos (>0.01 ppb) [64]. MARINE: average 0.3 ppb, max 3.6 ppb [20]. RAIN WATER: La Jolla, CA 5 ppt, industrial area in England 150 ppt [54]. Portland, OR, Feb-Apr 1984, concn (ppt), 7 rain events, 100% pos, 0.78-16, 5.6 avg [34]. SNOW: Southern California 30 ppt, Central California <1.5 ppt, Alaska 39 ppt [54].

**Effluent Concentrations:** Detected, not quantified, in wastewater in vicinity of a specialty chemicals plant [30]. Industries with mean concentrations greater than 75 ppb, paint and ink formulation, electrical/electronic components, rubber processing, mean range, 7-530 ppb, max range 3-1600 ppb [56]. USEPA STORET data base, 1,480 data points, 19.6% pos, 5.0 ppb median [53]. Ground water at 178 CERCLA hazardous waste disposal sites, 51.3% pos [43]. Minnesota municipal solid waste landfills, leachates, 6 sites, 83.3% pos, 0.7-125 ppb, contaminated ground water (by inorganic indices), 13 sites, 69.2% pos, 0.2-144 ppb, other ground water (apparently not contaminated as indicated by inorganic indices), 7 sites, 28.6% pos, 0.2-6.8 ppb [47].

**Sediment/Soil Concentrations:** Not detected in sediment in vicinity of specialty chemicals plant [30]. Detected in marine sediments at a max of 9.9 ppb Liverpool Bay, England [40]. USEPA STORET data base, 338 data points, 6.0% pos, <5.0 ppb median concn [53]. Lake Pontchartrain at Passes, sediment from 3 sites, 66.7% pos, 0.1-0.2 ppb, wet weight [23].

**Atmospheric Concentrations:** Global avg 8 ppt, northern hemisphere 15-16 ppt, southern hemisphere <3 ppt [15,51]; major US cities mean 96-483 ppt, max 236-3097 ppt, min 5-36 ppt [49,50]. Portland, OR, Feb-Apr 1984, concn in air (ng/m$^3$) during 7 rain events, 100% pos, 240-3900, 1537 avg [34]. Industrial - 1.2 ppb mean; urban/suburban- 0.25 ppb mean, rural - trace-0.10 ppb [7,25,35]. England: industrial 40-60 ppb, suburban 1-20 ppb, rural 5 ppb [40]. Love Canal (Niagara Falls, NY): 2 of 3 samples pos (1.6 and 3.4 ppb), home basement level estimated at 0.83 ppb [2]. Waste disposal site (Edison, NJ) trace-61 ppb [42].

**Food Survey Values:** Intermediate grain-based foods (1984): 9 varieties, 44.4% pos, 0.77-2.7 ppb, 1.9 ppb max concn in yellow corn meal; wheat, corn, oats (1984), 10, 2, and 1 samples, respectively: not detected [28]. Table-ready foods: 19 varieties, 47% pos, 1.7-8.0 ppb, 1.5 ppb avg, max concn in plain granola; butter, 7 samples, 100% pos; 1.6-20 ppb, 9.7 ppb avg; margarine, 7 samples, 100% pos, 3.7-980 ppb, 135 ppb avg; cheese, 4 types, 8 samples, 87.5% pos, 1.2-9.5 ppb, 4.3 ppb avg of pos, max concn in mozzarella cheese [29]. Trace detected in extracted edible oils [32]. Also detected in meat, beverages, dairy products, fruits and vegetables, oil and fats, range 0.02-60 ug/kg [32].

**Plant Concentrations:**

**Fish/Seafood Concentrations:** marine fish, flesh - 0.04-1.1 ppm, liver - 0.66-20.0 ppb, mussels - 50 day exposure 1.37 ppm [40]. Lake Pontchartrain at Passes, oysters, 5 samples, 2.2 ppb avg; clams, composite samples from 2 sites, 5.7 and 0.8 ppb [23].

**Animal Concentrations:**

**Milk Concentrations:** Detected in dairy products [32].

**Other Environmental Concentrations:**

**Probable Routes of Human Exposure:** High levels of exposure are expected for workers in degreasing plants due to inhalation of vapors or adsorption through the skin. Lower inhalation exposure is expected in persons living near degreasing plants or at spill sites. Broad population exposure to low levels from inhalation of contaminated ambient air and ingestion of contaminated drinking water.

**Average Daily Intake:** AIR INTAKE: (assume typical concn of 100-500 ppt [49,50]) - 11-33 ug; WATER INTAKE: (assume 2-7 ppb [8,12]) 2-20 ug; FOOD INTAKE: insufficient data.

**Occupational Exposures:** The number of US workers exposed to trichloroethylene is estimated to be 283,000 [32]. Operating room levels range from 0.3-103 ppm, with an estimated 5000 medical, dental, and hospital personnel being routinely exposed [32]. Levels at a dial assembly workshop in Japan measured 25-100 ppm; degreasing room levels, 150-250 ppm [32]. NIOSH (NOES Survey 1981-83) has statistically estimated that 392,805 workers are exposed to trichloroethylene in the US [37].

**Body Burdens:** Human milk, 4 US urban areas, 8 of 8 samples pos [41]. Post-mortem wet tissue samples 1-32 ppb [32]. Love Canal, Niagara Falls, NY - breath - trace 4 of 9 samples pos, blood - 09-2.50 ppb, 6 of 9 samples pos, urine - 40-550 ppt, 9 of 9 samples pos [2]. Whole blood specimens from 250 subjects, not detected to 1.5 ppb, 0.4 ppb avg [1].

## REFERENCES

1. Antoine SR et al; Bull Environ Contam Toxicol 36: 364-71 (1986)
2. Barkley J et al; Biomed Mass Spectrom 7:139-47 (1980)
3. Barrows ME et al; Dynamics, Exposure, Hazard Assessment Toxic Chem p. 379-92 (1980)
4. Boublik T et al; The Vapor Pressures of Pure Substances Vol 17 Amsterdam, Netherlands: Elsevier Science Publ (1984)
5. Bouwer EJ et al; Environ Sci Technol 15:569-9 (1981)

# Trichloroethylene

6. Bouwer EJ, McCarty PL; Appl Environ Microbiol 45:1286-94 (1983)
7. Bozzelli JW, Kebbekus BB; J Environ Sci Health 17:693-13 (1982)
8. Brass HJ et al; Drinking Water Qual Enhancement Source Prot. p 393-416 (1977)
9. Callahan MA et al; Water-Related Environmental Fate of 129 Priority Pollutants - vol II EPA-440/4-79-029B (1979)
10. Chang JS, Kaufman F; J Chem Phys 66:4989-94 (1977)
11. Chemical Marketing Reporter; February 14, 1983, Chemical Profile (1983)
12. Council on Environmental Quality. Contamination of ground water by toxic organic chemicals. p 26-34 (1980)
13. Cotruvo JA; Sci Total Environ 47: 7-26 (1985)
14. Cotruvo JA et al; pp. 511-30 In: Organic Carcinogens in Drinking Water (1986)
15. Cox RA et al; Atmos Environ 10:305-8 (1976)
16. Cupitt LT; Fate of Toxic and Hazardous Materials in the Air Environment EPA-600/3-80-084 (1980)
17. Dickson AG, Riley JP; Marine Pollut Bull 7:167-9 (1976)
18. Dilling WL et al; Environ Sci Tech 10:351-6 (1976)
19. Dilling WL et al; Environ Sci Tech 9:833-8 (1975)
20. Dyksen JE, Hess AF III; J Amer Water Work Assoc 394-403 (1982)
21. Eisenreich SJ et al; Environ Sci Technol 15: 30-8 (1981)
22. Ewing BB et al; Monitoring to detect previously unrecognized pollutants in surface waters. EPA-560/6-77-015. p 74 (1977)
23. Ferrario JB et al; Bull Environ Contam Toxicol 34: 246-55 (1985)
24. Gay BW et al; Environ Sci Tech 10:58-67 (1976)
25. Grimsrud EP, Rasmussen RA; Atmos Environ 9:1014-7 (1975)
26. Grob K, Grob G; J Chromatogr 90:303-13 (1974)
27. Hansch C, Leo AJ; Medchem Project Issue No 26. Clarmont CA: Pomona College (1985)
28. Heikes DL, Hopper ML; J Assoc Off Anal Chem 69: 990-8 (1986)
29. Heikes DL; J Assoc Off anal Chem 70: 215-26 (1987)
30. Hites RA et al; ACS Symp Ser 94:63-90 (1979)
31. Horvath AL; Halogenated Hydrocarbons: Solubility-Miscibility with Water. New York,NY: Marcel Dekker, Inc. pp 889 (1982)
32. IARC Monographs on the Evaluation of Carcinogenic Risk of Chemicals to Man 11:263-76 (1976)
33. Jensen S, Rosenberg R; Water Res 9:659-61 (1975)
34. Ligocki MP et al; Atmos Environ 19: 1609-17 (1985)
35. Lillian D et al; Environ Sci Technol 9:1042-8 (1975)
36. Lyman WJ; Handbook of Chemical Property Estimation Methods Ann Arbor Sci, MI (1981)
37. NIOSH; The National Occupational Exposure Survey (NOES) (1983)
38. Ohio River Valley Water Sanit Comm 190-81. Assessment of Water Quality Conditions (1982)
39. Parsons F et al; J. Amer Water Works Assoc 76:56-9 (1984)
40. Pearson CR, McConnell G; Proc R Soc Lond B 189:305-32 (1975)
41. Pellizzari ED et al; Bull Environ Contam Toxicol 28:322-8 (1982)
42. Pellizzari ED; Environ Sci Technol 16:781-5 (1982)
43. Plumb RHJr; Ground Water Monit Rev 7: 94-100 (1987)
44. Roberts PV et al; J Water Pollut Control Fed 56: 157-63 (1984)
45. Rogers RD, McFarlane JC; Environ Monit Assess 1:155-62 (1981)
46. Rott B et al; Chemosphere 11:531-8 (1982)

# Trichloroethylene

47. Sabel GV, Clark TP; Waste Manag Res 2: 119-30 (1984)
48. Schwarzenbach RP et al; Environ Sci Technol 17:472-9 (1983)
49. Singh HB et al; Environ Sci Technol 16:872-80 (1982)
50. Singh HB et al; Atmos Environ 15:601-12 (1981)
51. Singh HB et al; Atmospheric Distributions, Sources and Sinks of Selected Hydrocarbons, Halocarbons, $SF_6$ and $NO_2$. EPA-600/3-79-107. p 4 (1979)
52. Sontheimer H; J Amer Water Works Assoc 72:386-90 (1980)
53. Staples CA et al; Environ Toxicol Chem 4: 131-42 (1985)
54. Su C, Goldberg ED; Mar Pollut Transfer 1976:353-74 (1976)
55. Tabak HH et al; J Water Pollut Contam Fed 53:1503-18 (1981)
56. USEPA; Treatability Manual page I.12.23-1 to I.12-23-5 EPA-600/2-82-001A (1981)
57. Wakeham SG et al; Environ Sci Technol 17:611-17 (1983)
58. Wilson JT et al; Devel Indust Microbiol 24:225-33 (1984)
59. Wilson JT et al; J Environ Qual 10:501-6 (1981)
60. Wilson BH et al; Environ Sci Technol 20: 997-1002 (1986)
61. Wilson JT, Wilson BH; Appl Environ Microbiol 49: 242-3 (1985)
62. Yanagihara S et al; Photochemical Reactivities of Hydrocarbons Proc Int Clean Air Congress, 4th page 472-7 (1977)
63. Zoeteman BCJ et al; Chemosphere 9:231-49 (1980)
64. Zoeteman BCJ et al; Sci Total Environ 21: 187-202 (1981)

# Trichlorofluoromethane

## SUBSTANCE IDENTIFICATION

**Synonyms:** Freon 11

**Structure:**

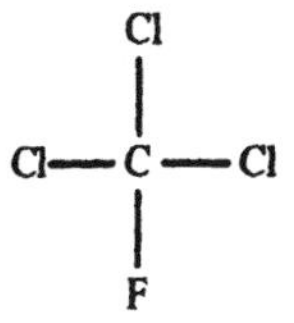

$$Cl—\underset{\underset{F}{|}}{\overset{\overset{Cl}{|}}{C}}—Cl$$

**CAS Registry Number:** 75-69-4

**Molecular Formula:** $CCl_3F$

**Wiswesser Line Notation:**

## CHEMICAL AND PHYSICAL PROPERTIES

**Boiling Point:** 23.7 °C at 760 mm Hg

**Melting Point:** -111 °C

**Molecular Weight:** 137.38

**Dissociation Constants:**

**Log Octanol/Water Partition Coefficient:** 2.53 [16]

**Water Solubility:** 1080 mg/L at 30 °C [17]

**Vapor Pressure:** 802.8 mm Hg at 25 °C [7]

**Henry's Law Constant:** 0.097 atm-m³/mole at 25 oC [50]

## ENVIRONMENTAL FATE/EXPOSURE POTENTIAL

**Summary:** Trichlorofluoromethane (Freon 11) was primarily released to the environment during its use as a propellant in aerosol sprays. However, this use was banned in the US on Dec 15, 1978. Other

475

sources of emissions include its use as a refrigerant, foaming agent for polyurethane foams, solvent and degreaser, and fire extinguishing agent. If released in water or on land, trichlorofluoromethane will be lost by volatilization (half-life 3.4 hr in a typical river). Concentration profiles in oceans show that trichlorofluoromethane is primarily in surface layers, suggesting that the oceans are not a sink for this chemical. Bioconcentration in fish is unlikely. If released on land, trichlorofluoromethane may also pass through the soil and into ground water where it is likely to persist for long periods of time. The troposphere is apparently also not a sink for trichlorofluoromethane since estimates of its half-life range from 52 to 207 years. In fact the only major sink for trichlorofluoromethane is from its slow diffusion into the stratosphere where photolysis occurs followed by subsequent reactions which destroy ozone. As a result of its stability, trichlorofluoromethane is transported long distances and its concentration is fairly uniform around the globe away from known sources. Since there are no major tropospheric sinks, the concentration of trichlorofluoromethane had been increasing by about 10% a year in the late 1970s, a trend that seems to be leveling off as a result of its ban in aerosols. Human exposure is mostly occupational and from ambient air, although drinking water from contaminated ground water sources may also be important.

**Natural Sources:** Volcanoes [14].

**Artificial Sources:** Trichlorofluoromethane may be released as emissions or in wastewater during its production, storage, transport, and use as a foaming agent for polyurethane foams [21,22], degreaser and solvent, especially in the aerospace and electronics industries, and fire extinguishing agent [14,24,48]. In the early 1970s the largest release of trichlorofluoromethane was from aerosols (75%) with refrigerants and foaming agent use coming next by contributing 14% and 21%, respectively [24]. Because its release into the atmosphere was believed to cause depletion of the ozone layer, production was curtailed, peaking in 1974 and declining after that [24]. Production of trichlorofluoromethane containing propellants was banned after Dec 15, 1978 in the US [24].

**Terrestrial Fate:** If released on soil, trichlorofluoromethane will rapidly evaporate into the atmosphere, because of its high vapor pressure, as well as readily pass through soil into ground water.

# Trichlorofluoromethane

**Aquatic Fate:** If released into water, trichlorofluoromethane will be lost almost exclusively by evaporation (half-life 4.3 hr in a typical river). Abiotic degradation, biodegradation, and adsorption to sediment will be insignificant.

**Atmospheric Fate:** Trichlorofluoromethane is very stable in the troposphere, having a half-life of 52-207 years. It is transported long distances, being found even at the South Pole. Some trichlorofluoromethane will be lost due to rainout, but this will revolatilize rapidly. The only degradation loss process is through diffusion to the stratosphere where photolysis will occur. As a result of this latter process, ozone is destroyed. The sharp decrease in trichlorofluoromethane concentrations above the tropopause supports this conclusion.

**Biodegradation:** No significant biodegradation was observed when incubated with a sewage seed for 7 days followed by three weekly subcultures [44]. Losses reported during biological treatment are probably the result of evaporative losses or adsorption to charcoal [34]. No information could be found concerning the biodegradation of trichlorofluoromethane under aerobic conditions.

**Abiotic Degradation:** Trichlorofluoromethane does not absorb UV radiation >290 nm [18] nor does it react appreciably with reactive atmospheric species such as hydroxyl radicals or singlet oxygen atoms [15] or degrade under photochemical smog conditions [19]. Estimates of tropospheric half-lives based on time series concentration measurements, range from 52-207 yr [1,29,36]. The only sink is diffusion to the stratosphere, where photolysis by short wavelength UV radiation occurs. Heterogeneous photolysis occurs when trichlorofluoromethane is adsorbed on silica gel with 23% degradation occurring in 170 hr when exposed to sunlight [13]; however, it is difficult to assess if this is a likely environmental process. Hydrolysis is not a significant loss process for trichlorofluoromethane; however, the rate is greatly affected by the presence of metals such as steel which act as catalysts [10].

**Bioconcentration:** The levels of trichlorofluoromethane in three species of mollusks and five species of fish are only slightly enriched (usually 2-25 times on a dry weight basis) over the seawater levels [9]. The usual order of enrichment was found to be brain > gill > liver > muscle [9].

**Soil Adsorption/Mobility:** Trichlorofluoromethane has negligible adsorption to soil and modeling studies which predict distributions between environmental compartments indicate that none of the chemical will reside in the soil or sediment [25,32].

**Volatilization from Water/Soil:** Trichlorofluoromethane has an high Henry's Law constant. From this value, it is predicted to have a volatilization half-life of 3.4 hr from a river 1 m deep with a 3 m/sec wind and 1 m/sec current [5,30]. The rate of evaporation will be limited by the rate of diffusion in the aqueous compartment [30]. From modeling studies, 99.97% of trichlorofluoromethane should partition into the air compartment [32]. Due to its extremely high vapor pressure and negligible soil adsorption, trichlorofluoromethane will evaporate rapidly from soil.

**Water Concentrations:** DRINKING WATER: Detected, not quantified [27]. GROUND WATER: Cluster wells at a manufacturing research facility in the northeastern US: glacial shallow wells 80 ppb mean, median not detectable; glacial deep wells 5 ppb mean, 4 ppb median; bedrock wells 650 ppb mean, 23 ppb median [38]. SURFACE WATER: Ohio River Basin (11 stations, 4972 samples) - 5.3% of samples contained >0.1 ppb including 46 samples between 1 and 10 ppb with 3 > 10 ppb [33]. 14 heavily industrialized river basins in US (204 sites) - 11 sites positive all in the Chicago area and Illinois River Basin 3-20 ppb [11]. Lake Erie central and eastern basins - 34 and 46 ppt average in 1977 and 1978, respectively, with concentration uniform throughout the basin [20]. Delaware River Basin (30 sites, depth integrating samples) - 3% of sites had values >1 ppb [8]. Lake Michigan (9 sites) - 5 sites positive, 1-20 ppb [26]. SEAWATER: Pacific Ocean 0.13 ppt average at surface. 0.06 ppt average at 300 m depth [41]. Point Reyes, CA nearshore 43 ppt [42]. Trichlorofluoromethane is most abundant in the surface layers of the sea as opposed to the depths as was demonstrated in concentration profiles of the Greenland and Norwegian Seas [4].

**Effluent Concentrations:** Industries whose raw or treated wastewater exceed an average of 10 ug/L include: auto and other laundries, electrical components, nonferrous metal manufacturing, coal mining, photographic equipment/supplies, and textile mills [46]. Maximum levels at or above 100 ppb occurred in textile mills (2100 ppb), auto and other laundries (120 ppb), and nonferrous metal manufacturing

(100 ppb) [46]. Industrial effluents (343 sites representing all STORET stations) 0.6% positive, median <5 ppb [43]. National Urban Runoff Program (19 cities including 11 of the 18 river basins in the contiguous US - 86 samples) - 6% frequency of detection, 0.6-27 ppb [6]. Los Angeles municipal wastewater <0.3 ppb [51].

**Sediment/Soil Concentrations:** Bottom sediment from the submarine outfall of Los Angeles sewage treatment plant <0.5 ppb [51].

**Atmospheric Concentrations:** RURAL/REMOTE: US (1977-80, 431 samples) 120 ppt median, 230 ppt maximum [2]. South Pole (1975-80) 90-166 ppt, average annual concentration increase 8-12% [37]. US Pacific NW (1975-80) 125-188 ppt, average annual concentration increase 8-12% [37]. Harwell, England (1/78-6/81) 207-272 ppt monthly average, annual rate of increase between 1975-81 - 10-11 ppt, rate of increase slowing down [1]. Southern Hemisphere Background: 182 ppt average in 1981 with an annual growth 11.5 ppt for 1979-81 as measured at Cape Point, South Africa [3] and 130 ppt in June 1977 with an increase of nearly 20% in the previous 12 months as measured at Cape Grim, Australia [49]. Point Barrow, Alaska 192.3-202.6 ppt with the highest concentration during the winter [23]. The source of the trichlorofluoromethane is anthropogenic in the mid-latitudes [23]. Concentration measurements as a function of latitude show that the concentration decreases with altitude in the troposphere followed by a sharper decrease through the tropopause and into the stratosphere [40]. Mean concentration in troposphere over southern France, June-Sept 1977, 130 ppt [12]. Three altitudes over Pacific Ocean, Hawaii to Alaska, Oct-Nov 1974, 6 locations (km altitude, range ppt) - 15.2, 65-106; 18.3, 29-96; 21.3, 5.5-86 [47]. Northern stratosphere, April 1974 - Nov 1976, 49-90 ppt, 65.3 avg; upper troposphere, May-Nov 1976, 115-126 ppt, 122 avg [28]. Tropospheric concn range of 127-149 ppt, May 1975 - April 1977, from 4 sample sites between California and Alaska [39]. URBAN/SUBURBAN: US (903 samples) 100-4900 ppt, 380 ppt median [2]. SOURCE AREA: US (1 site) 260 ppt [2]. INDOOR AIR: 0.306-52 ppb [45].

**Food Survey Values:**

**Plant Concentrations:**

# Trichlorofluoromethane

**Fish/Seafood Concentrations:** Isle of Man - Irish Sea: 3 species of mollusks 0.2-1.4 ppb dry weight; 5 species of fish 0.1-5.0 ppb dry weight [9].

**Animal Concentrations:**

**Milk Concentrations:**

**Other Environmental Concentrations:**

**Probable Routes of Human Exposure:** Humans are exposed to trichlorofluoromethane in the ambient air as well as in the workplace. Exposure from indoor air where fluorocarbon-pressured spray cans were used was high; however, the use of trichlorofluoromethane as a propellant for aerosol cans has been banned since Dec 1978. Exposure may also occur from drinking water, especially that which originated from contaminated ground water.

**Average Daily Intake:** AIR INTAKE: (assume concentration 380 ppt) 43 ug.

**Occupational Exposure:** Beauty shop where fluorocarbon-pressured cosmetic sprays are used 50.4 ppb [45]. NIOSH (NOES Survey 1981-1983) has statistically estimated that 232,769 workers are exposed to trichlorofluoromethane in the United States [31].

**Body Burdens:** Human milk from 4 urban sites in US - 7 of 8 samples positive [35].

## REFERENCES

1.  Brice KA et al; Atmos Environ 16: 2543-54 (1982)
2.  Brodzinsky R, Singh HB; Volatile organic chemicals in the atmosphere: an assessment of available data SRI International p.198 contract 68-02-3452 (1982)
3.  Brunke EG, Halliday EC; Atmos Environ 17: 823-6 (1983)
4.  Bullister JL, Weiss RF; Science 221: 265-8 (1983)
5.  Cadina F et al; J Water Pollut Control Fed 56: 460-3 (1984)
6.  Cole RH et al; J Water Pollut Control Fed 56: 898-908 (1984)
7.  Daubert TE, Danner RP; Data Compilation Tables of Properties of Pure Compounds. Amer Inst Chem Engn pp 450 (1985)
8.  Dewalle FB, Chian ESK; Proc Ind Waste Conf 32: 908-19 (1978)
9.  Dickson AG, Riley JP; Mar Pollut Bull 7: 167-9 (1976)
10. DuPont de Nemours Co; Freon products information B-2; A98825 12/80 (1980)

11. Ewing BB et al; Monitoring to detect previously unrecognized pollutants in surface waters USEPA-560/6-77-015 (1977)
12. Fabian P et al; J Geophys Res 84: 3149-54 (1979)
13. Gaeb S et al; Angew Chem 90: 398-9 (1978)
14. Graedel TE; Chemical compounds in the atmosphere Academic Press NY (1978)
15. Hampson RF; Chemical kinetic and photochemical data sheets for atmospheric reactions; FAA-EE-80-17 (1980)
16. Hansch C, Leo AJ; Medchem Project Issue No.26 Pomona College, Claremont CA (1985)
17. Horvath AL; Halogenated Hydrocarbons: Solubility-Miscibility with Water. New York: Marcel Dekker pp 889 (1982)
18. Hubrich C, Stuhl F; J Photochem 12: 93-107 (1980)
19. Japar S et al; Univ Calif Riverside CA Personal Comm (1974)
20. Kaiser KLE, Valdamis I; J Great Lakes Res 5: 160-9 (1979)
21. Khalil MAK, Rasmussen RA; J Air Pollut Contr Assoc 36: 159-63 (1986)
22. Khalil MAK, Rasmussen RA; Chemosphere 16: 759-75 (1987)
23. Khalil MAK, Rasmussen RA; Environ Sci Technol 17: 157-64 (1983)
24. Kirk-Othmer; Encyclopedia of Chemical Technology 3rd Ed 10:856-74 (1980)
25. Kloepffer W et al; Ecotox Environ Safety 6: 294-301 (1982)
26. Konasewich D et al; Status report on organic and heavy metal contaminants in the Lake Erie, Michigan, Huron and Superior Basins. Great Lakes Qual Rev Board (1978)
27. Kool HJ et al; Crit Rev Environ Control 12: 307-57 (1982)
28. Leifer R et al; J Geophys Res 85: 1096-72 (1980)
29. Lillian D et al; Environ Sci Technol 9: 1042-8 (1975)
30. Lyman WJ et al; Handbook of chemical property estimation methods. Environmental behavior of organic compounds McGraw Hill NY (1982)
31. NIOSH; National Occupational Exposure Survey (1983)
32. Neely WB; Environ Toxicol Contam 1: 259-66 (1982)
33. Ohio R. Valley Water Sanit Comm; Assessment of water quality conditions. Ohio River Mainstream 1980-81 Cincinnati OH (1982)
34. Patterson JW, Kodukala PS; Chem Eng Prog 77: 48-55 (1981)
35. Pellizzari ED et al; Bull Environ Contam Toxicol 28: 322-8 (1982)
36. Penkett SA et al; Nature 286: 793-5 (1980)
37. Rasmussen RA et al; Science 211: 285-7 (1981)
38. Rich CA; Stud Environ Sci 17, Qual Groundwater: 309-14 (1981)
39. Saunder WD et al; Water Air Pollut 10: 421-39 (1978)
40. Schmeltekopf PD et al; Geophys Res Lett 2: 393-6 (1975)
41. Singh HB et al; Atmospheric distributions, sources and sinks of selected halocarbons, hydrocarbons, $SF_6$ and $N_2O$ pp.134 USEPA-600/3-79-107 (1979)
42. Singh HB et al; J Air Pollut Control Assoc 27: 332-6 (1977)
43. Staples CA et al; Environ Toxicol Chem 4: 131-42 (1985)
44. Tabak HH et al; J Water Pollut Control Fed 53: 1503-18 (1981)
45. USEPA; Ambient Water Quality Criteria for Halomethanes p.C-18 PB81-117624 Oct (1980)
46. USEPA; Treatability Manual p.1.12.22-1 to 1.12.22-4 USEPA-600/2-82-001a (1981)
47. Vedder JF et al; Geophys Res Lett 5: 33-6 (1978)

48. Verschuren K; Handbook of environmental data on organic chemicals 2nd ed Van Nostrand Reinhold NY p.673-71 (1983)
49. Wainwright L; Clean Air 13: 5-10 (1979)
50. Warner MJ, Weiss RF; Deep-Sea Res Part A 32: 1485-97 (1985)
51. Young DR et al; pp.87-84 in Water chlorination: environ impact health effects book 2 vol 4 (1983)

# 1,1,2-Trichloro-1,2,2-Trifluoroethane

## SUBSTANCE IDENTIFICATION

**Synonyms:** Freon 113

**Structure:**

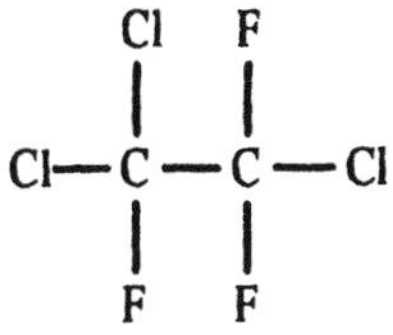

```
        Cl   F
        |    |
Cl— C — C — Cl
        |    |
        F    F
```

**CAS Registry Number:** 76-13-1

**Molecular Formula:** $C_2Cl_3F_3$

**Wiswesser Line Notation:**

## CHEMICAL AND PHYSICAL PROPERTIES

**Boiling Point:** 47.7 °C at 760 mm Hg

**Melting Point:** -36.4 °C

**Molecular Weight:** 187.38

**Dissociation Constants:**

**Log Octanol/Water Partition Coefficient:** 3.16 [15]

**Water Solubility:** 170 mg/L at 25 °C [18]

**Vapor Pressure:** 362.5 mm Hg at 25 °C [3]

**Henry's Law Constant:** 0.526 atm-m³/mole at 25 °C (calculated from vapor pressure and water solubility)

## ENVIRONMENTAL FATE/EXPOSURE POTENTIAL

**Summary:** Freon 113 may be released to the environment as emissions from production, storage, transport, turbine engines, use as a foaming agent, refrigerant, and solvent, or use in the manufacture of fluoropolymers, and it may be released to soil from the disposal of products containing this compound (e.g. commercial/industrial refrigeration units). The global release rate of Freon 113 is estimated to be $9.1 \times 10^{+4}$ tons per year, which corresponds with a 15% annual increase in the abundance of Freon 113 in the atmosphere. If released to soil, Freon 113 would rapidly volatilize from soil surfaces or leach through soil possibly into ground water. If released to water, essentially all Freon 113 is expected to be lost by volatilization (half-life 4 hr from a model river). If released to the atmosphere, Freon 113 will not degrade in the troposphere. This compound will gradually diffuse into the stratosphere (half-life 20 years). In the stratosphere, the dominant removal mechanism is photolysis with reaction with singlet oxygen of secondary importance (stratospheric lifetime 63-122 years). Due to its stability, detection long distances from its sources of emissions has occurred. General population exposure occurs by inhalation of Freon 113 found in ambient air. Occupational exposure may occur by inhalation of contaminated air or dermal contact.

**Natural Sources:** No data available which indicate that Freon 113 is a naturally occurring compound.

**Artificial Sources:** Freon 113 may be released to the environment as emissions from production, storage, transport, turbine engines, use as a foaming agent, refrigerant, and solvent, or use in the manufacture of fluoropolymers [10,11]. Freon 113 together with Freon 114, Freon 115, and Freon 13 contain about 3% of the organically bound chlorine present in the atmosphere [10]. The global release rate of Freon 113 is estimated to be $9.1 \times 10^{+4}$ tons per year, which corresponds with a 15% annual increase in the abundance of Freon 113 in the atmosphere [10]. This compound may be released to soil from the disposal of products containing this compound. These products include mobile air conditioners, retail food refrigeration units, and centrifugal and reciprocating chillers [6].

**Terrestrial Fate:** If released to soil, Freon 113 would rapidly volatilize from soil surfaces or leach through soil possibly into ground water.

**Aquatic Fate:** If released to water, essentially all Freon 113 is expected to be lost by volatilization (half-life 4 hr from a model river). Chemical hydrolysis, bioaccumulation, and adsorption to sediments would not be significant fate processes in water.

**Atmospheric Fate:** If released to the atmosphere, essentially all Freon 113 is expected to exist in the vapor phase due to its high vapor pressure. The moderate water solubility of Freon 113 suggests that some loss by wet deposition occurs, but any loss by this mechanism is probably returned to the atmosphere by volatilization. Freon 113 will not degrade in the troposphere, thus diffusion from the troposphere to the stratosphere would be the sole removal mechanism (half-life 20 years [7]). In the stratosphere, direct photolysis would be the dominant removal mechanism and reaction with singlet oxygen would be of secondary importance. The stratospheric lifetime of this compound ranges between 63 to 122 years. As a result of its persistence in the atmosphere, this compound is transported long distances and its concentration should be fairly uniform throughout the globe even away from known sources.

**Biodegradation:**

**Abiotic Degradation:** Chemical hydrolysis of Freon 113 is not an environmentally significant fate process [8]. Freon 113 is essentially inert to reaction with photochemically generated radicals and ozone molecules [1,2]. This compound will not undergo direct photolysis in the troposphere [14]. The stratospheric lifetime of Freon 113 has been estimated to range from 63 and 122 years with direct photolysis being the dominant removal mechanism and reaction with singlet oxygen being the secondary removal mechanism [5]. In the stratosphere, this compound will slowly photolyze to release chlorine atoms which in turn participates in the catalytic removal of stratospheric ozone [14].

**Bioconcentration:** Based on a water solubility and the Kow value, bioconcentration factors (BCF) of 34 and 11, respectively, were estimated for Freon 113 [13]. These BCF values suggest that Freon 113 would not bioaccumulate significantly in aquatic organisms.

**Soil Adsorption/Mobility:** Soil adsorption coefficient (Koc) values of 191 and 259 were estimated using linear regression equations based on the Kow and water solubility, respectively [13]. These Koc values suggest that Freon 113 would be moderately mobile in soil and that moderate adsorption to suspended solids and sediments in water would take place [21].

# 1,1,2-Trichloro-1,2,2-Trifluoroethane

**Volatilization from Water/Soil:** Based upon the Henry's Law constant, Freon 113 would volatilize rapidly from all bodies of water and from soil surfaces [13]. Using the Henry's Law constant, the volatilization half-life of Freon 113 from a model river 1 m deep flowing 1 m/sec with a wind velocity of 3 m/sec has been estimated to be 4 hr [13].

**Water Concentrations:** Detected in water samples taken from the Niagara River and Cayuhoga River [12]. Not detected in water samples taken from Lake Ontario [12].

**Effluent Concentrations:**

**Sediment/Soil Concentrations:**

**Atmospheric Concentrations:** During 1978, the average concentrations of Freon 113 in ambient air in the Northern and Southern hemispheres were 13 and 12 ppt, respectively [9]. Freon 113 was detected in air samples collected throughout the US between 1973 and 1980: rural/remote locations, 284 data points, median concn 31 ppt, mean concn 28 ppt; urban/suburban areas, 851 data points, median concn 170 ppt, mean concn 220 ppt [4]. Detected in "clean air" samples collected in central California in May 1975, avg concn 19.9 ppt [19]. Detected in air samples collected in the San Francisco area during winter 1975, 274 samples, 100% pos, avg concn 16.9 ppt [20].

**Food Survey Values:**

**Plant Concentrations:**

**Fish/Seafood Concentrations:**

**Animal Concentrations:**

**Milk Concentrations:**

**Other Environmental Concentrations:**

**Probable Routes of Human Exposure:** The general population is exposed to Freon 113 in ambient air. In occupational settings, it is expected that exposure occurs by inhalation of contaminated air and dermal contact with this compound.

# 1,1,2-Trichloro-1,2,2-Trifluoroethane

**Average Daily Intake:** AIR INTAKE: (assume 13-31 ppt [4,9]) 2.0-4.8 ug/day.

**Occupational Exposure:** NIOSH has estimated that 225,117 workers are potentially exposed to Freon 113 based upon estimates derived from a survey conducted during 1981-83 in the United States [16]. NIOSH earlier estimated that 975,189 workers are potentially exposed to Freon 113 based upon estimates derived from a survey conducted in 1972-74 in the United States [17].

**Body Burdens:**

## REFERENCES

1. Atkinson R; Chem Rev 85: 69-201 (1985)
2. Atkinson R; Internat J Chem Kinetics 19: 799-828 (1987)
3. Boublik T et al; The Vapor Pressure of Pure Substances Vol 17 Amsterdam, Netherlands: Elsevier (1984)
4. Brodzinsky R, Singh HB; pp. 23-184 in Volatile Organic Chemicals in the Atmosphere: An Assessment of Available Data Menlo Park, CA: SRI International (1982)
5. Chou CC et al; J Phys Chem 82: 1-7 (1978)
6. Clayton GD, Clayton FE eds; pp. 3102-3 in Patty's Industrial Hygiene and Toxicology Vol IIb 3rd ed NY: Wiley and Sons (1981)
7. Dilling WL; Environmental Risk Analysis for Chemicals; Conway RA ed NY: Van Nostrand Reinhold Co pp. 154-97 (1982)
8. Du Pont de Nemours Co; Freon Products Information B-2; A98825 12/80 (1980)
9. Fabian P et al; J Geophys Res 90: 13091-3 (1985)
10. Fabian P; pp. 23-51 in The Handbook of Environmental of Environmental Chemistry, Vol 4/Part A; Hutzinger O ed NY: Springer-Verlag (1986)
11. Graedel TE; Chemical Compounds in the Atmosphere NY: Academic Press p. 327 (1978)
12. Great Lakes Water Quality Board; in An Inventory of Chemical Substances Identified in the Great Lakes Ecosystem Vol.1- Summary Windsor, Ontario: Great Lakes Quality Board (1983)
13. Lyman WJ et al; Handbook of Chemical Property Estimation Methods. NY: McGraw-Hill (1982)
14. Makide T et al; Chem Lett 4: 355-8 (1979)
15. McDuffie B; Chemosphere 10: 73-82 (1981)
16. NIOSH; National Occupational Exposure Survey (NOES) (1974)
17. NIOSH; National Occupational Hazard Survey (NOHS) (1974)
18. Riddick JA et al; Organic Solvents: Physical Propertiis and Methods of Purification. Techniques of Chemistry. 4th Ed. New York: Wiley-Interscience pp 1325 (1986)
19. Singh HB et al; Atmos Environ 11: 819-28 (1977)
20. Singh HB et al; J Air Pollut Control Assoc 27: 332-6 (1977)
21. Swann RL et al; Res Rev 85: 17-28 (1983)

# Triethanolamine

## SUBSTANCE IDENTIFICATION

**Synonyms:**

**Structure:**

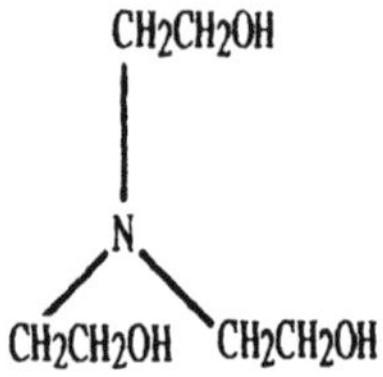

CH2CH2OH
N
CH2CH2OH
CH2CH2OH


**CAS Registry Number:** 102-71-6

**Molecular Formula:** $C_6H_{15}NO_3$

**Wiswesser Line Notation:** Q2N2Q2Q

## CHEMICAL AND PHYSICAL PROPERTIES

**Boiling Point:** 335.4 °C at 760 mm Hg

**Melting Point:** 20-21 °C

**Molecular Weight:** 149.19

**Dissociation Constants:** pKa = 7.92 [20]

**Log Octanol/Water Partition Coefficient:** -1.59 [19]

**Water Solubility:** Miscible [5]

**Vapor Pressure:** $3.59 \times 10^{-6}$ mm Hg at 25 °C [5]

**Henry's Law Constant:** $3.38 \times 10^{-19}$ atm-m$^3$/mole at 25 °C [10]

## ENVIRONMENTAL FATE/EXPOSURE POTENTIAL

**Summary:** Triethanolamine may be released to the environment in emissions or effluents from sites of its manufacture or industrial use, from disposal of consumer products which contain this compound, from application of agricultural chemicals in which this compound is used as a dispersing agent, and during use of copper triethanolamine complex as an aquatic herbicide. In soil and water, triethanolamine will biodegrade fairly rapidly following acclimation (half-life on the order of days to weeks). In soil, residual triethanolamine may leach into ground water. In the atmosphere, triethanolamine is expected to exist partly in the vapor phase and partly in particulate form. Triethanolamine vapor is expected to react with photochemically generated hydroxyl radicals in the atmosphere (half-life 4 hr). Wet and dry deposition may also be important removal processes. The most probable route of exposure to triethanolamine is dermal contact with personal care products (i.e. soaps, cosmetics, emollients), household detergents, and other surfactants which contain this compound.

**Natural Sources:**

**Artificial Sources:** Triethanolamine may be released to the environment in emissions or effluents from sites of its manufacture or industrial use, from disposal of consumer products which contain this compound, from application of agricultural chemicals in which this compound is used as a dispersing agent, and during use of copper triethanolamine complex as an aquatic herbicide [9,13,16].

**Terrestrial Fate:** If released to soil, triethanolamine is expected to biodegrade fairly rapidly following acclimation (half-life on the order of days to weeks). Residual triethanolamine may leach into ground water. Volatilization from soil surfaces is not expected to be an important fate process.

**Aquatic Fate:** If released to water, triethanolamine should biodegrade. The half-life of this compound is expected to range from a few days to a few weeks depending, in large part, on the degree of acclimation of the system. Bioconcentration in aquatic organisms, adsorption to suspended solids and sediments, and volatilization are not expected to be important fate processes in water.

**Atmospheric Fate:** Based on the vapor pressure, triethanolamine is expected to exist partly in the vapor phase and to be partly adsorbed to particulates in the atmosphere [6]. Triethanolamine vapor is expected to react with photochemically generated hydroxyl radicals in the atmosphere (half-life 4 hr). The complete solubility of triethanolamine in water suggests that this compound may also be removed from the atmosphere in precipitation. Dry deposition may be an important removal process for triethanolamine adsorbed on particles.

**Biodegradation:** 10 day die-away, initial concn 50 ppm - 70% BODT, acclimated Kanawha River water as seed, sewage as inoculum [15]. BOD water, initial concn 2.5 ppm, 5, 10, 15, and 20 days - 0, 0.8, 3.5, and 6.8% BODT, respectively, sewage inoculum [12]. BOD water, 20 day - 66% BODT, sewage inoculum [21]. Synthetic sea water, 20 day - 69% BODT, sewage inoculum [21]. BOD water, 5 day - 5% BODT (unadapted) and 28% BODT (adapted), inoculum was effluent from a biological sanitary waste treatment plant [3]. BOD water, initial concn 500 ppm, 15 day acclimation period, 10 day - 22% BODT, activated sludge inoculum [7]. Zahn-Wellens, initial concn equivalent to 1000 mg/L COD, 14 day - 89% degradation, non-adapted activated sludge inoculum [23]. Zahn-Wellens, initial concn equivalent to 400 ppm C, 8 day - 82% DOC removal, activated sludge inoculum [8]. Sturm - $CO_2$ evolution, initial concn equivalent to 10 ppm C, 14 day acclimation period, 28 days - 91% $CO_2$ evolution and 100% DOC removal, sewage inoculum [8]. OECD, initial concn equivalent to 3-20 ppm C, 19 day 96% - DOC removal, sewage inoculum [8]. Modified Closed Bottle, 2 ppm, 30 day - 9% BODT, enriched sewage inoculum [8]. French AFNOR, initial concn equivalent to 40 ppm C, 42 day - 97% DOC removal, sewage inoculum [8]. Japanese MITI, 100 ppm, 14 days - <30% BODT, activated sludge inoculum [11].

**Abiotic Degradation:** The half-life for triethanolamine vapor reacting with photochemically generated hydroxyl radicals in the atmosphere has been estimated to be 4 hr based on an estimated reaction rate constant of $10.4 \times 10^{-11}$ $cm^3$/molecules-sec at 25 °C and an average ambient hydroxyl concentration of $5 \times 10^{+5}$ molecules/$cm^3$ [2].

**Bioconcentration:** A bioconcentration factor (BCF) of <1 was estimated for triethanolamine based on the log Kow [14]. This BCF value and complete solubility of triethanolamine in water suggest that this compound does not bioconcentrate significantly in aquatic organisms.

**Soil Adsorption/Mobility:** A soil adsorption coefficient (Koc) of 3 was estimated for triethanolamine based on the log Kow [14]. This Koc value and the complete solubility of triethanolamine in water suggests that this compound would be extremely mobile in soil and would not adsorb appreciably to suspended solids and sediments in water [22].

**Volatilization from Water/Soil:** The Henry's Law constant value suggests that volatilization of triethanolamine from water and moist soil surfaces would be negligible [14].

**Water Concentrations:**

**Effluent Concentrations:**

**Sediment/Soil Concentrations:**

**Atmospheric Concentrations:**

**Food Survey Values:**

**Plant Concentrations:**

**Fish/Seafood Concentrations:**

**Animal Concentrations:**

**Milk Concentrations:**

**Other Environmental Concentrations:**

**Probable Routes of Human Exposure:** The most probable route of exposure to triethanolamine is dermal contact with personal care products (i.e., soaps, cosmetics, emollients), household detergents, and other surfactants which contain this compound [4,9]. The chief risk in industry would be from direct local contact of the skin or eyes with the undiluted, unneutralized fluid [1].

**Average Daily Intake:**

**Occupational Exposure:** NIOSH (NOHS Survey 1972-74) has statistically estimated that 1,658,822 workers are potentially exposed to triethanolamine in the United States [17]. NIOSH (NOES Survey 1981-83) has statistically estimated that 1,198,609 workers are potentially exposed to triethanolamine in the United States [18].

**Body Burdens:**

# REFERENCES

1. Am Conf Gov Ind Hyg; Appendix: Documentation of the Threshold Limit Values and Biological Exposure Indices 5th ed. p. 3169 Cincinnati, OH (1986)
2. Atkinson R; Inter J Chem Kinet 19: 799-828 (1987)
3. Bridie AL et al; Water Res 13: 627-30 (1979)
4. Chemical Marketing Reporter; Chemical Profile: Ethanolamine NY: Schnell Publishing Nov 10 (1986)
5. Dow Chemical Co; The Alkanolamines Handbook Midland, MI: Dow Chemical Co. (1980)
6. Eisenreich SJ et al; Environ Sci Tech 15: 30-8 (1981)
7. Gannon JE et al; Microbios 23: 7-18 (1978)
8. Gerike P, Fischer WK; Ecotox Environ Safety 3: 159-43 (1979)
9. Hawley GG; The Condensed Chemical Dictionary 10th ed NY: Van Nostrand Reinhold p. 1045 (1981)
10. Hine J, Mookerjee PK; J Org Chem 40: 292-8 (1975)
11. Kawasaki M, Ecotox Environ Safety 4: 444-54 (1980)
12. Lamb CB, Jenkins GF; pp. 326-39 in Proc 8th Ind Waste Conf Purdue Univ (1952)
13. Liepins R et al; Industrial Process Profiles for Environmental Use. USEPA-600/2-77-023f NTIS PB-281 478 pp. 6-386 to 6-387 (1977)
14. Lyman WJ et al; Handbook of Chemical Property Estimation Methods NY: McGraw-Hill (1982)
15. Mills EJ, Stack VT; Proc 9th Ind Waste Conf Eng Bull Purdue Univ: Ext Ser 9: 449-64 (1955)
16. Mullins RM; Kirk-Othmer Encycl Chem Tech 3rd ed NY: Wiley-Interscience 1: 959 (1978)
17. NIOSH; National Occupational Hazards Survey (NOHS) (1974)
18. NIOSH; National Occupational Exposure Survey (NOES) (1983)
19. PCGEMS; Graphical Exposure Modelling System, CLOGP3 Program, Office of Toxic Substances, U.S. EPA (1989)
20. Perrin DD; Dissociation Constants of Organic Bases in Aqueous Solution. 1972 Supplement. IUPAC Chemical Data Series. Buttersworth: London (1972)
21. Price KS et al; J Water Poll Contr Fed 46: 63-77 (1974)
22. Swann RL et al; Res Rev 85: 17-28 (1983)
23. Zahn R, Wellens H, Z Wasser Abwasser Forsch 13: 1-7 (1980)

# Triethylamine

## SUBSTANCE IDENTIFICATION

**Synonyms:**

**Structure:**

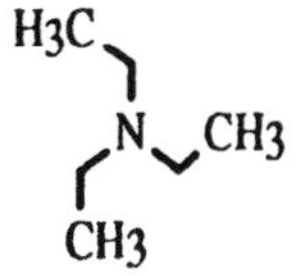

$$H_3C$$
$$N \quad CH_3$$
$$CH_3$$

**CAS Registry Number:** 121-44-8

**Molecular Formula:** $C_6H_{15}N$

**Wiswesser Line Notation:** 2N2&2

## CHEMICAL AND PHYSICAL PROPERTIES

**Boiling Point:** 89.3 °C

**Melting Point:** -115 °C

**Molecular Weight:** 101.19

**Dissociation Constants:** pKa = 10.778 at 25 °C [17]

**Log Octanol/Water Partition Coefficient:** 1.45 [9]

**Water Solubility:** 55,000 mg/L at 25 °C [17]

**Vapor Pressure:** 57.07 mm Hg at 25 °C [17]

**Henry's Law Constant:** $1.38 \times 10^{-4}$ at 25 °C (calculated from vapor pressure and water solubility)

## ENVIRONMENTAL FATE/EXPOSURE POTENTIAL

**Summary:** Triethylamine is produced in large quantities and large amounts of the chemical will be released, primarily as emissions during

493

its production and use. It also occurs naturally in some food and as a metabolic product. If released on land, triethylamine should volatilize moderately from moist soil and should leach into the soil. It has not been established whether it biodegrades in soil or water. If released to water, it will be removed slowly by volatilization (half-life of 9.3 hr in a model river). Adsorption to sediment and bioconcentration in aquatic organisms will not be appreciable. In the atmosphere, triethylamine will react with photochemically produced hydroxyl radicals with an estimated half-life of 4.5 hr. In polluted atmospheres, triethylamine completely degrades photochemically in 90 min. It will also be scavenged by rain. Triethylamine is a precursor of dimethylnitrosamine. The latter is formed in the atmosphere in the presence of nitrogen oxides. Human occupational exposure to triethylamine is via inhalation and dermal contact with the vapor. The general public is exposed primarily by ingesting food in which it occurs naturally.

**Natural Sources:**

**Artificial Sources:** Triethylamine is produced in large quantities, 17.3 million lbs in 1985 [22], and will be released in emissions and wastewater during its production and use as a catalytic solvent in chemical synthesis, accelerator activator for rubber, corrosion inhibitor, curing and hardening agent for polymers, propellant, in the manufacture of wetting, penetrating, and waterproofing agents of the quaternary ammonium type, and the desalination of seawater [10,18]. It is also emitted from sewage treatment plants [8].

**Terrestrial Fate:** If released on soil, triethylamine will probably volatilize moderately from moist soil and leach through the soil. There are insufficient data available to ascertain whether biodegradation will be an important loss process.

**Aquatic Fate:** If released into water, triethylamine will volatilize slowly (estimated half-life 9.3 hr in a model river). There are insufficient data to ascertain whether biodegradation will be an important removal process in natural waters. Direct photolysis and adsorption to sediment will not be significant.

**Atmospheric Fate:** Triethylamine, released into the atmosphere, should react rapidly with photochemically produced hydroxyl radicals with an estimated half-life of 4.5 hr. Under polluted atmospheric conditions when nitrogen oxide concentrations are high, diethylamine degrades within 90 min. Due to its high water solubility, washout by rain will also be an important removal process.

**Biodegradation:** Only one screening study could be located on the biodegradability of triethylamine. This study concluded that triethylamine was not degraded by activated sludge even when acclimatized (BOD 5.3% of theoretical after 13 days) [3]. It was, however, completely degraded by an <u>Aerobacter</u> sp. in 11 hr [21]. The concentration of amine used in the first study was not reported. From work on other aliphatic amines, degradation is rapid but inhibition is noted at concentrations as low as 50 mg/L [2,4]. It is possible that at the concentration employed in the screening study, inhibition was occurring.

**Abiotic Degradation:** Triethylamine is a strong base and undergoes the typical reactions of primary amines [18]. Many of these reactions may occur in the environment, but there is little documentation as to what reactions take place in the environment and at what rates. Reaction of aliphatic amines with photochemically produced hydroxyl radicals are estimated to be 4.5 hr at a concentration of $5 \times 10^{+5}$ radicals/cm$^3$ [1]. Experiments show that triethylamine reacts with NO-$NO_2$-$H_2O$ mixtures to form diethylnitroamine both in the dark and on irradiation [16]. On irradiation, triethylamine is highly reactive forming ozone, PAN, acetaldehyde, diethylnitroamine, diethylformamide, ethylacetamide, and diethylacetamide and aerosols [16]. These experiments were performed in large outdoor chambers under natural conditions of temperature, humidity, and illumination. Initially the mixture was allowed to react for two hours in the dark and then exposed to sunlight. The triethylamine completely disappeared after 90 minutes of illumination [16]. Based on data obtained from smog chambers, triethylamine has been classified as having negligible reactivity based on its rate of photochemical reaction or ozone-forming potential [5,11]. The ability of amines to form complexes with metallic ions is well known [20]. Metallic ions in soils or natural waters may therefore combine with triethylamine but no information could be found on reactions with soil components. Humic acids that occur in natural waters contain carbonyl groups that could also potentially react with the amine groups to form adducts but again data in natural systems are lacking [18]. Triethylamine does not contain any chromophores which absorb radiation >290 nm so direct photolysis will not be significant.

# Triethylamine

**Bioconcentration:** Using the octanol/water partition coefficient, an estimated BCF of 7.45 was calculated for triethylamine using a recommended regression equation [12]. Triethylamine would therefore not be expected to bioconcentrate in aquatic organisms.

**Soil Adsorption/Mobility:** Using the octanol/water partition coefficient or the water solubility, an estimated Koc of 11-146 was obtained using a recommended regression equation [12]. Based on these Koc values, triethylamine will not absorb appreciably to soils and sediments.

**Volatilization from Water/Soil:** Using the Henry's Law constant, a half-life of 9.3 hr was estimated for a model river 1 m deep flowing at 1 m/sec with a wind velocity of 3 m/sec [12]. The rate of evaporation was 1.58 cm/hr in an experiment in which air is bubbled through a 100 ppm aqueous solution of triethylamine [13]. The experimental half-life derived from this rate for a 1 m deep body of water is 43.9 hr. Triethylamine has a moderate vapor pressure and does not adsorb strongly to soil; therefore it should volatilize slowly from dry soil.

**Water Concentrations:**

**Effluent Concentrations:** Triethylamine has been reported in an effluent sample from the plastics and synthetics industry at a concentration of 356.5 mg/L [19].

**Sediment/Soil Concentrations:** Identified in uncultivated loamy soil from the Moscow region [6]. Since this soil is uncultivated, it is possible that the amines are formed naturally rather than being a contaminant or a metabolite of a fertilizer or pesticide [6].

**Atmospheric Concentrations:**

**Food Survey Values:** Triethylamine has been identified as a volatile component of boiled beef [7]. The interest for the presence of amines in food arises in part because they are regarded as possible precursors of carcinogenic N-nitroso compounds [7].

**Plant Concentrations:**

**Fish/Seafood Concentrations:**

**Animal Concentrations:**

**Milk Concentrations:**

**Other Environmental Concentrations:**

**Probable Routes of Human Exposure:** Triethylamine has been identified in at least one common food item, boiled beef, therefore the general population may be exposed to triethylamine from ingesting food in which it occurs naturally. Occupational exposure will occur primarily via inhalation and dermal contact with the vapor.

**Average Daily Intake:**

**Occupational Exposure:** According to the 1981-83 National Occupational Exposure Survey (NOES), 51,070 workers are potentially exposed to triethylamine [15]. The 1972-74 National Occupational Hazard Survey (NOHS) reported that 11,157 workers are exposed to triethylamine [14].

**Body Burdens:**

## REFERENCES

1. Atkinson R; Internat J Chem Kinetics 19: 799-828 (1987)
2. Calamari D et al; Chemosphere 9: 753-62 (1980)
3. Chudoba J et al; Chem Prum 19: 76-80 (1969)
4. Dojlido JR; Investigations of Biodegradability and Toxicity of Organic Compounds, Final Report 1975-9 EPA-600/2-79-163 p. 118 (1979)
5. Farley FF; Photochemical Reactivity Classification of Hydrocarbons and Other Organic Compounds EPA-600/3-77-001B pp. 713-27 (1977)
6. Golovnya RV et al; USSR Acad Med Sci pp. 327-35 (1982)
7. Golovnya RV et al; Chem Senses Flavour 4: 97-105 (1979)
8. Graedel TE; Chemical Compounds in the Atmosphere. Academic Press NY pp. 289 (1978)
9. Hansch C, Leo AJ; Medchem Project Issue No.26 Pomona College, Claremont CA (1985)
10. Hawley GG; Condensed Chem Dictionary 10th ed Von Nostrand Reinhold NY (1981)
11. Levy A; The Photochemical Smog Reactivity of Organic Solvents, Solvent Theory and Practices, Am Chem Soc Adv Chem Ser 124: 70-94 (1973)
12. Lyman WJ et al; Handbook of Chem Property Estimation Methods. Environ Behavior of Organic Compounds. McGraw-Hill NY pp. 15-1 to 15-34 (1982)
13. Neely WB; Control Hazard Mater Spills Proc Nat Conf 3rd pp. 196-200 (1976)
14. NIOSH; National Occupational Health Survey (1975)

# Triethylamine

15. NIOSH; National Occupational Exposure Survey (1985)
16. Pitts JN Jr. et al; Environ Sci Technol 12: 946-53 (1978)
17. Riddick JA et al; Organic Solvents: Physical Properties and Methods of Purification. Techniques of Chemistry. 4th Ed. New York: Wiley-Interscience pp 1325 (1986)
18. Schweizer AE et al; Encycl Chem Technol 3nd ed. 2: 272-83 (1978)
19. Shackelford WM et al; Analyt Chem Acta 146: 15-27 (1983)
20. Stumm W, Morgan JJ; Aquatic Chemistry 2nd edition pp. 356-63 (1981)
21. USEPA; Treatability Manual Vol 1 EPA-600/8-80-042 (1980)
22. USITC; United States International Trade Commission, USTIC 1892 (1986)

# Trimethylamine

## SUBSTANCE IDENTIFICATION

**Synonyms:**

**Structure:**

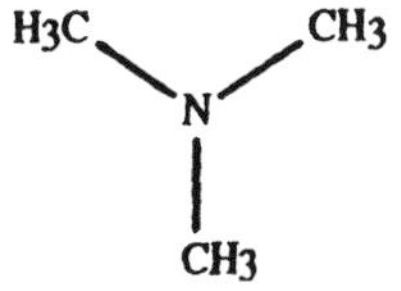

**CAS Registry Number:** 75-50-3

**Molecular Formula:** $C_3H_9N$

**Wiswesser Line Notation:** 1N1&1

## CHEMICAL AND PHYSICAL PROPERTIES

**Boiling Point:** 2.87 °C at 760 mm Hg

**Melting Point:** -117.2 °C

**Molecular Weight:** 59.11

**Dissociation Constants:** pKa = 9.801 [26]

**Log Octanol/Water Partition Coefficient:** 0.16 [14]

**Water Solubility:** 890,000 mg/L at 30 °C [27]

**Vapor Pressure:** 1607 mm Hg at 25 °C [7]

**Henry's Law Constant:** 1.04 x $10^{-4}$ atm-m³/mole at 25 °C [15]

## ENVIRONMENTAL FATE/EXPOSURE POTENTIAL

**Summary:** Trimethylamine is widely distributed throughout the environment as a decomposition product of plants, animals, fish, and animal waste. Major sources of anthropogenic release appear to be

499

effluents and emissions from manufacturing plants and facilities which use this compound as an intermediate, especially choline chloride manufacturing plants. If released to moist soil, trimethylamine would be susceptible to biodegradation. Potential biodegradation products include dimethylamine, formaldehyde, formate, and $CO_2$ under aerobic conditions and dimethylamine, $NH_3$, and $CH_4$ under anaerobic conditions. Chemical hydrolysis is not expected to be an important fate process. If released to dry soil, trimethylamine is expected to volatilize rapidly. It is not certain whether trimethylamine would adsorb strongly to soil. If released to water, trimethylamine is expected to undergo partial removal by volatilization (half-life about 11 hr from a model river) and partial removal by biodegradation. Bioaccumulation in aquatic organisms, chemical hydrolysis, and reaction with photochemically generated hydroxyl radicals are not expected to be important fate processes. Trimethylamine should exist almost entirely in the vapor phase in the atmosphere. Reaction with photochemically generated hydroxyl radicals is expected to be the dominant removal mechanisms (half-life approximately 4 hr). The most probable routes of human exposure to trimethylamine are inhalation of tobacco smoke and ingestion of foods in which this compound occurs. Worker exposure may occur by dermal contact and/or inhalation.

**Natural Sources:** Trimethylamine is widely distributed in the environment as a result of its formation during the decay of organic matter in plants, animals, fish, sewage, and animal waste [16,25,32]. Trimethylamine forms as a result of microbial breakdown of both choline and betaine, common constituents of plants and animals, and from bacterial reduction of trimethylamine N-oxide, a common metabolite and excretory product of aquatic organisms [3,16,25].

**Artificial Sources:** Trimethylamine may be released to the environment in the effluent or emissions from its manufacturing or use facilities.

**Terrestrial Fate:** If released to moist soil, trimethylamine would be susceptible to biodegradation under both aerobic and anaerobic conditions. Potential biodegradation products include dimethylamine, formaldehyde, formate, and $CO_2$ under aerobic conditions and dimethylamine, ammonia, and methane under anaerobic conditions. Chemical hydrolysis is not expected to be an important fate process. If released to dry soil, trimethylamine is expected to volatilize rapidly. Data on adsorption to soil were not available.

# Trimethylamine

**Aquatic Fate:** If released to water, trimethylamine is expected to undergo partial removal by volatilization (half-life 11 hr from a model river) and partial removal by biodegradation. Potential biodegradation products include dimethylamine, formaldehyde, formate, and $CO_2$ under aerobic conditions and dimethylamine, $NH_3$, and $CH_4$ under anaerobic conditions. Bioaccumulation in aquatic organisms, chemical hydrolysis, and reaction with photochemically generated hydroxyl radicals are not expected to be important fate processes.

**Atmospheric Fate:** Based on the vapor pressure, trimethylamine should exist almost entirely in the vapor phase in the atmosphere [8]. Reaction with photochemically generated hydroxyl radicals is expected to be the dominant removal mechanism (half-life 4 hr).

**Biodegradation:** Numerous strains of bacteria isolated from seawater, lake water, mud, garden soil, and activated sludge have been found capable of growth on trimethylamine [4,6,17,31]. 77.2% BODT uptake was observed when trimethylamine was incubated in activated sludge for 13 days [5]. A mixed culture of microorganisms in a mineral salt medium degraded trimethylamine to dimethylamine. Production of dimethylamine after 36 hours corresponded to 24% of the initial trimethylamine added [4]. Trimethylamine incubated in a marine sediment slurry underwent about 35% removal, as measured by production of $CO_2$ and $CH_4$, in 12 hours [18]. Fermentation of trimethylamine accounted for 35.1% to 61.1% of total $CH_4$ produced by in situ concentrations of trimethylamine is marine sediments [18] and 90% of total $CH_4$ produced in anoxic salt marsh sediment amended with Spartina foliosa [25]. Microbial production of dimethylamine from trimethylamine in soil was found to be greater under acidic conditions than at near neutral pH and greater under aerobic conditions than anaerobic conditions [30]. Degradation products formed under aerobic conditions include dimethylamine, formaldehyde, formate, and $CO_2$ [20], and products formed under anaerobic conditions include dimethylamine, $NH_3$, and $CH_4$ [16].

**Abiotic Degradation:** The half-life for the reaction of trimethylamine with photochemically generated hydroxyl radicals in water has been estimated to be 62 days based on a reaction rate constant of $1.3 \times 10^{+10}$ 1/mole-sec and an ambient hydroxyl concentration of $1 \times 10^{-17}$ mole/L [13,21]. Based on the lack of hydrolyzable functional groups, trimethylamine is not expected to be susceptible to chemical hydrolysis under environmental conditions. The principal chemical loss mechanisms for amines in the atmosphere is expected to be reaction with photochemically generated hydroxyl radicals [12]. The half-life for the reaction of trimethylamine with hydroxyl radicals has been estimated to be approximately 4 hours based on a reaction rate constant of 6.09 x $10^{-11}$ cm³/molecule-sec at 25.5 °C and assuming an average ambient hydroxyl radical concentration of 8.0 x $10^{+5}$ molecules/cm³ [2,9]. Trimethylamine has an estimated half-life of 1.4 days with ozone based on a measured reaction rate constant of 9.73 x $10^{-18}$ cm³/molecule-sec at 23 °C and an ambient ozone concentration of 6.0 x $10^{+11}$ molecules/cm³. It has an estimated half-life with singlet oxygen of 15 days based on an experimentally determined reaction rate constant of 22.0 x $10^{-12}$ cm³/molecule-sec at 24.4 °C and an ambient singlet oxygen concentration of 2.5 x $10^{+4}$ molecules/cm³. Therefore, reaction with ozone or singlet oxygen are not expected to be important fate processes in the atmosphere [1,3,9,12].

**Bioconcentration:** A bioconcentration factor (BCF) of <1 was estimated for trimethylamine using a linear regression equation based of the octanol/water partition coefficient [19]. This BCF value and the relatively high water solubility of trimethylamine suggest that bioaccumulation in aquatic organisms would not be significant.

**Soil Adsorption/Mobility:** Soil adsorption coefficients (Koc) of 4 and 29 were estimated [19] for trimethylamine based on the water solubility and the octanol/water coefficient [2]. These low Koc values suggest that this compound would not adsorb significantly to suspended solids and sediments in water and would be highly mobile in soil [29]. However, trimethylamine is a base and should exist primarily as a cation under environmental conditions (pH 5-9). As a result trimethylamine may have greater adsorption and less mobility than its estimated Koc values indicate.

**Volatilization from Water/Soil:** Based on the Henry's Law constant, the volatilization half-life of trimethylamine from a model river 1 m deep, flowing 1 m/sec has been estimated to be 11 hr [19]. The vapor pressure suggests that trimethylamine would volatilize rapidly from dry soil surfaces.

**Water Concentrations:**

**Effluent Concentrations:**

# Trimethylamine

**Sediment/Soil Concentrations:** Trimethylamine has been identified in uncultivated soil [10].

**Atmospheric Concentrations:**

**Food Survey Values:** Trimethylamine has been identified as a volatile component of boiled beef [11].

**Plant Concentrations:** Trimethylamine has been found to be a volatile constituent of marine algae [28].

**Fish/Seafood Concentrations:**

**Animal Concentrations:**

**Milk Concentrations:**

**Other Environmental Concentrations:** Trimethylamine is a constituent of tobacco smoke [12]. This compound has been identified as a volatile component of cattle feed lots, probably from the decomposition of manure [22].

**Probable Routes of Human Exposure:** The most probable routes of human exposure to trimethylamine are inhalation of tobacco smoke and ingestion of food in which this compound occurs. Worker exposure may occur by dermal contact and/or inhalation.

**Average Daily Intake:**

**Occupational Exposure:** NIOSH has estimated that 5261 workers are potentially exposed to trimethylamine based on estimates derived from the survey conducted in 1981-83 in the United States [23]. NIOSH has estimated that 4491 workers are potentially exposed to trimethylamine based on estimates derived from the survey conducted in 1972-74 in the United States [24].

**Body Burdens:**

# Trimethylamine

## REFERENCES

1. Atkinson R, Pitts JN; Chem Phys 68: 911-15 (1978)
2. Atkinson R; Chem Rev 85: 69-201 (1985)
3. Atkinson R, Carter UP; Chem Rev 84: 437-70 (1984)
4. Ayanabe A, Alexander M; Appl Microbiol 25: 862-69 (1973)
5. Chudoba J et al; Chem Prum 20: 2079-100 (1969)
6. Colby J, Zatman LJ; Biochem J 132: 101-12 (1973)
7. Daubert TE, Danner RP; Data Compilation Tables of Properties of Pure Compounds. Amer Inst Chem Engn pp 450 (1985)
8. Eisenreich SJ et al; Environ Sci Tech 15: 30-38 (1981)
9. GEMS; Graphical Exposure Modeling System FAP Fate of Atmos Pollut (1986)
10. Golovnya RV et al; USSR Acad Med Sci pp.327-35 (1982)
11. Golovnya RV et al; Chem Sense Flavor 4: 97-105 (1974)
12. Graedel TE; Chemical Compounds in the Atmosphere Academic Press New York, NY p.228 (1978)
13. Guesten H et al; Atmos Environ 15: 1763-5 (1981)
14. Hansch C, Leo AJ; Medchem Project Issue No.26 Pomona College, Claremont CA (1985)
15. Hine J, Mookerjee PK; J Org Chem 40: 292-8 (1975)
16. Hippe H et al; Proc Natl Acad Sci USA 76: 494-98 (1977)
17. Kimura T et al; Bull Fac Fish 4: 1-9 (1977)
18. King GM et al; Appl Environ Microbiol 45: 1848-53 (1983)
19. Lyman WJ et al; Handbook of Chem Property Estimation Methods. Environ Behavior of Organic Compounds McGraw-Hill NY (1982)
20. Meiberg JBM, Harber W; J Gen Microbiol 106: 265-76 (1978)
21. Mill T et al; Science 207: 886-7 (1980)
22. Mosier AR et al; Environ Sci Tech 7: 642-4 (1973)
23. NIOSH; National Occupational Exposure Survey (NOES) (1983)
24. NIOSH; National Occupational Hazard Survey (1974)
25. Oremland RS et al; Nature 296: 143-5 (1982)
26. Perrin DD; Dissociation Constants of Organic Bases in Aqueous Solution. 1972 Supplement IUPAC Chemical Data Series. Buttersworth: London (1972)
27. Schweizer AE et al; Kirk-Othmer Encycl Chem Technol 3rd Ed. 2: 272-83 (1978)
28. Steiner M, Hartman T; Planta 79: 113-21 (1968)
29. Swann RL et al; Res Rev 85: 17-18 (1984)
30. Tate RL, Alexander M; Appl Environ Microbiol 31: 399-403 (1976)
31. Troyan OS et al; Samooch Bioindik Zagraz Vod Moscow, USSR pp.196-99 (1977)
32. Windholz M; The Merck Index 10th Ed Merck and Co Rahway, NY pp.168 and 313 (1983)

# 1,2-Xylene

## SUBSTANCE IDENTIFICATION

**Synonyms:** o-Xylene

**Structure:**

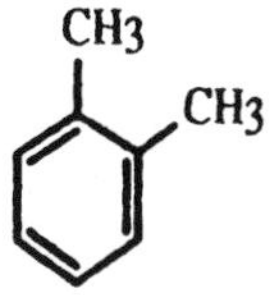

**CAS Registry Number:** 95-47-6

**Molecular Formula:** $C_8H_{10}$

**Wiswesser Line Notation:** IR B1

## CHEMICAL AND PHYSICAL PROPERTIES

**Boiling Point:** 144.4 °C

**Melting Point:** -25 °C

**Molecular Weight:** 106.16

**Dissociation Constants:**

**Log Octanol/Water Partition Coefficient:** 3.12 [25]

**Water Solubility:** 175 mg/L at 25 °C [60]

**Vapor Pressure:** 6.6 mm Hg at 25 °C [60]

**Henry's Law Constant:** 5.1 x $10^{-3}$ atm cu-m/mol [41]

## ENVIRONMENTAL FATE/EXPOSURE POTENTIAL

**Summary:** o-Xylene will enter the atmosphere primarily from fugitive emissions and exhaust connected with its use in gasoline. Industrial sources include emissions from petroleum refining, coal tar and coal gas distillation, and from its use as a solvent. Discharges and spills on land and in waterways result from its use in diesel fuel and gasoline. Most of the o-xylene is released into the atmosphere where it may photochemically degrade by reaction with hydroxyl radicals (half-life 1.5-15 hr). The dominant removal process in water is volatilization. o-Xylene is moderately mobile in soil and may leach into ground water where it has been known to be detectable for several years, although there is evidence that it biodegrades in both soil and ground water. This may be because o-xylene degrades under aerobic conditions and denitrifying conditions appear to be required when oxygen is lacking. Bioconcentration is not expected to be significant. The primary source of exposure is from the air, especially where emissions from petroleum products or motor vehicles are high or solvents containing p-xylene are used.

**Natural Sources:** Coal tar, petroleum [79], forest fire [20], and plant volatile [20].

**Artificial Sources:** Emissions from petroleum refining, coal tar and coal gas distillation, and emissions from its use as a chemical intermediate [48,79]. Emissions from its use as a solvent, and evaporative losses during the transport and storage of gasoline and from carburetors [20,48,79]. Auto emissions, tobacco smoke [20,48]. Emissions from residential wood-burning stoves and fireplaces [16]. Gasoline spills on land or in waterways during transport or from boats and other vehicles. Agricultural spraying [48]. o-Xylene emission rates from glued wallpaper and glued carpet have been measured to be 6.5 and 98 ng/min-sq m [81].

**Terrestrial Fate:** When spilled on land, o-xylene will volatilize and leach into the ground. o-Xylene degrades in soil under aerobic or anaerobic denitrifying conditions. Under aerobic conditions, 70% degradation after 10 days have occurred. Under anaerobic conditions, a lag period of six months or more may be required before degradation commences. The extent of the degradation will depend on its concentration, residence time in the soil, the nature of the soil, and whether resident microbial populations have been acclimated.

**Aquatic Fate:** In surface waters, volatilization appears to be the dominant removal process (half-life 1-5 days). Some adsorption to sediment will occur. Although o-xylene is biodegradable and has been observed to degrade in seawater, there are insufficient data to assess the rate of this process in natural aquatic systems.

**Atmospheric Fate:** When released into the air, o-xylene may degrade by reaction with photochemically produced hydroxyl radicals (half-life 1.5 hr in summer and 15 hr in winter). However, ambient levels are detected because of large emissions.

**Biodegradation:** o-Xylene is degraded in standard biodegradability tests using various inocula including sewage, activated sludge, and sea water [8,35,46,78]. It was completely degraded in 8 days in ground water in a gas-oil mixture; the acclimation period was 3-4 days [32]. In laboratory experiments designed to simulate saturated-flow conditions typical of a river water/ground water infiltration system, degradation was rapid with 70% removal in the first 1.5 cm of the column after 10 days of operation under aerobic conditions [37]. Degradation also occurred anaerobically, but required denitrifying conditions and occurred only after the other xylene isomers had been removed [37]. After six months lag required for the microorganism to establish the necessary enzymes for denitrification, biodegradation was rapid [37]. Another investigator found that o-xylene was readily biodegraded (33 mg/day loss) in shallow ground water in an unconfined sand aquifer when oxygen was present [4]. As the available oxygen was consumed, the rate of degradation decreased [4]. o-Xylene also degraded after a long lag period in a microcosm study that used anaerobic aquifer material from a site known to receive municipal landfill leachate and support methanogenesis [87]. A lag period of >20 wk was needed before there was any appreciable biodegradation and 78% and >99% degradation were reported at week 40 and 120, respectively [87].

**Abiotic Degradation:** o-Xylene reacts with hydroxyl radicals in the troposphere [11,26,59,62] with a half-life ranging from 1.5 hr in summer to 15 hr in winter [59] or a typical loss of 71%/day [67]. Many investigators have measured the degradation of o-xylene alone or in mixtures of hydrocarbons typical of auto exhaust and have obtained loss rates ranging from 6%-24% per hr [2,14,17,28,36,84,89], rates typical for reaction with hydroxyl radicals [14,31]. Photooxidation products that have been identified include glyoxal, methylglyoxal and biacetyl [76]. Xylenes are resistant to hydrolysis since they lack hydrolyzable groups.

**Bioconcentration:** Little bioconcentration has been observed; log BCF = 1.33 and 0.79 for eels [54] and clams [52], respectively. Based on the measured octanol/water partition coefficient, one would estimate that the log BCF in fish would be 2.12 [44].

**Soil Adsorption/Mobility:** o-Xylene has a low to moderate adsorption to soil and the Koc for these soils ranged from 48 to 68 [49]. The Koc for p-xylene in surface sediment (2-10 cm) collected from the central Tamar estuary in the UK was 25.4 [80]. Batch equilibrium measurements with soil from 3 aquifers gave a Koc of 129 [1]. Although the permeability of xylene is lower than for less hydrophobic solvents, the permeability in fire clay has been observed to sharply increase over a 4-day period after 24 days of normal behavior. This has been attributed to slow shrinkage of the clay which was responsible for the breakthrough [22]. This type of breakthrough in clay would be a problem unless such materials are co-disposed with materials with higher dielectric constants [22]. Concentration enhancement has been observed for o-xylene in a dune-infiltration project on the Rhine River [57]; however, no o-xylene reached ground water under a rapid infiltration site [75].

**Volatilization from Water/Soil:** The half-life for the evaporation of o-xylene from water with a wind speed of 3 m/sec, a current of 1 m/sec and a depth of 1 m is 3.2 hr [44]. The evaporation rate is much more sensitive to the mixing conditions in the water than in the air [45]. An experiment which measured the rate of evaporation of o-xylene from a mixed 1:1000 fuel-water mixture found that it averaged 0.55 times the oxygen reaeration rate [69]. Combining this ratio with oxygen reaeration rates for typical bodies of water [44] one estimates that its half-life for evaporation from a typical river or pond are 31 and 125 hr respectively.

**Water Concentrations:** DRINKING WATER: In treated water from 30 Canadian water treatment facilities in which o- or m-xylene combined were measured, 27% and 20% were positive in summer and winter respectively, with maximum values of 7 and 2 ppb, respectively, and mean values below 1 ppb [55]. In 12 Great Lakes municipalities tested on one or two days, 5 communities were free of m- and o-xylene combined with median community levels being 1.0 ppb and the highest sample of the combined isomers being 12 ppb [86]. o-Xylene was detected but not quantified in drinking water from Philadelphia, PA [71], Washington, DC [63], Tuscaloosa, AL [5], and

Houston, TX [5]. In 3 New Orleans treatment plants, the avg concn of water collected over 2 days was 3.4 ppb [33]. Bank-filtered drinking water in the Netherlands contained a max of 30 ppt o-xylene [56]. In 5 drinking water wells near and down gradient from a landfill contained 0.2-1.5 ppb, control well upstream 0.8 ppb [13]. In a survey of organics in drinking water derived from ground water sources, o- and p-xylene combined were found in 2.1% of 280 sample sites supplying <10,000 persons and 1.1% of 186 sites supplying >10,000 persons. The maximum combined concentrations were 0.59 and 0.91 ppb, respectively [85]. Detected in all 14 drinking water supplies studied, 10 from surface sources and 4 from ground sources in the lowland of Great Britain [18]. GROUND WATER: In the Chalk Aquifer in England 1550 ppb o-xylene was measured 10 m from a leaking petrol storage tank, 115 ppb was found 100-120 m from the tank and 0.02 ppb was found in a public water supply 210 m from the tank [74]. At the Hoe Creek Coal gasification site in Wyoming, 260-590 ppb was found 15 months after gasification was completed [70]. 6 ppb was found in landfill ground water [15]. Detected in recovery well from landfill 7 years after closure [12]. SURFACE WATER: In the raw water for 30 Canadian water treatment facilities, 7% and 17% of plants contained combined o- and m-xylene in summer and winter respectively with max levels being under 1 ppb [55]. Detected, not quantified, in the Black Warrior River in Tuscaloosa, AL [5], the Niagara River in the Lake Ontario basin [21], and the Glatt River in Switzerland [89]. The concn of o-xylene at 2 stations in the Tees River, an industrial estuary in the UK at 1.5 m depth, ranged from not detectable to 0.29 ppb and 0.23-1.1 ppb [27]. SEAWATER: In Vineyard Sound, MA samples taken from March through June ranged from 1.8 to 2.5 ppt and averaged 9.4 ppt [23]. In open ocean and coastal sections of the Gulf of Mexico contained 1-30 ppt [62]. A uniform pattern of alkylated benzenes are found in the Baltic Sea and central Atlantic which is believed to be the result of large scale transport in the atmosphere and water masses [12]. RAIN/SNOW: The concn of o-xylene in rain in Portland ranged from 12-110 ppt, 45 ppt, mean [38].

**Effluent Concentrations:** In a comprehensive survey of wastewater from 4000 industrial and publicly owned treatment works (POTWs) sponsored by the Effluent Guidelines Division of the US EPA, o-xylene was identified in discharges of the following industrial category (positive occurrences; median concn in ppb): timber products (8; 171.4), iron and steel mfg (4; 16.5), petroleum refining (11; 60.9), nonferrous metals (3; 0.9), paving and roofing (2; 33.3), paint and ink (39; 148.4), printing and publishing (2; 306.8), coal mining (11; 26.3), organics and plastics (13; 10.8), inorganic chemicals (7; 1054.9), textile mills (2; 118.2), plastics and synthetics (8; 115.7), pulp and paper (1; 56.3), rubber processing (4; 9.4), soaps and detergents (1; 132.5), auto and other laundries (8; 200.3), pesticides manufacture (2; 132.7), pharmaceuticals (6; 93.9), porcelain/enameling (2; 27.5), electronics (7; 60.1), oil and gas extraction (6; 22.6), organic chemicals (4; 290.5), mechanical products (7; 375.3), transportation equipment (2; 24.5), amusements and athletic goods (2; 1019.4), and publicly owned treatment works (45; 39.2) [65]. Maximum effluents above 5,000 ppb were observed in the industry (max effluent in ppb): timber products (6,157), petroleum refining (5,893), pain and ink (7,639), organics and plastics (6,668), and mechanical products (10,897) [65]. The average combined exhaust emissions of o-xylene/C9 paraffins from a 1975 to 1985 model year test fleet were 1.8% and 1.4% of total hydrocarbon emissions at normal urban and normal high speed driving [66]. Evaporative hydrocarbon emissions contained 1.3% o-xylene/C9 paraffins [66].

**Sediment/Soil Concentrations:** Detected, not quantified, in sediment in lake 8 km downstream from Napawin, Saskatchewan, a source of agricultural, mining, petrochemical, pulp and paper and municipal wastes [61]. In unspecified sediment 500 ppb [77].

**Atmospheric Concentrations:** RURAL/REMOTE: 114 areas in US 0.094 ppb median, 0.50 ppb avg, 37 ppb max [9], Northern Hemisphere background (35 samples) 0.014 ppb [53], Southern Hemisphere background (20 samples) 0.007 ppb [53]. URBAN/SUBURBAN: 1885 areas in US, 1.2 ppb median, 1.9 ppb avg, 89 ppb max [9]. Avg values in individual cities reported in the literature range from 0.02-6.5 ppb [6,7,9,39,41,42,67,68] and frequently maximum values reported are in the double digits [7,9,41,42], up to 266 ppb in Hamburg, Germany [24]. The vapor concn of o-xylene during 7 rain events in Portland, OR was 0.18-0.64 ppb, 0.30 ppb mean [38]. SOURCE AREAS: 183 areas in US, 0.81 ppb median, 122 ppb avg, 25% of samples exceed 1.8 ppb, max 8800 ppb [9]. 2 landfills in New Jersey 0-35 ppb, 6.3 ppb avg [7]. 6 and 16.5 km downwind from a large auto manufacturing plant, 1.6 and 1.1 ppb resp [64]. Concns of o-xylene in the Lincoln Tunnel in New York City in 1982 and 1970 were 82 and 321 ppb, respectively [40]. The decline is due to decreased organic carbon emission resulting from catalytic controls [40]. INDOOR AIR: The median concn of o-xylene in 340 New Jersey residences (6 pm to 6 am) was 1.1 ppb [82]. Concns of o-xylene in Japanese high-rise apartments ranged from 9-16 ppb [72]. OTHER: As part of EPA's Total Exposure Assessment Methodology (TEAM) study, the concentration of various toxic substances in the personal air (2 consecutive 12-hr periods) of sample populations was measured as well as the outdoor air near their residences [83]. The weighted median results for o-xylene in personal air in Bayonne and Elizabeth, NJ, an industrial/chemical manufacturing area, were 5.4, 5.2, and 8.0 ug/m³ in the fall 1981, summer 1982, and winter 1983, respectively [83]. The corresponding results for outdoor air were 3.0, 3.6, and 3.4 ug/m³ [83]. For comparison, the personal air of a sample of residents of a manufacturing city without a chemical or petroleum refining industry, Greensboro, NC, and a small, rural, agricultural town in North Dakota contained 3.6 and 2.7 ug/m³, respectively of o-xylene, and the outdoor air 0.6 ug/m³ and not detectable [83].

**Food Survey Values:** o-Xylene has been identified as a volatile component of baked potatoes [10], fried bacon [30], dried legumes [43], fried chicken [73], and roasted filberts [34].

**Plant Concentrations:**

**Fish/Seafood Concentrations:** o-Xylene was detected, but not quantitated, in rainbow trout from the Colorado River below Hoover Dam and carp taken from Las Vegas Wash, NV [29].

**Animal Concentrations:**

**Milk Concentrations:**

**Other Environmental Concentrations:** The percent of o-xylene in leaded, unleaded, and super unleaded gasoline sold in the New York-New Jersey area in 1982 were 1.1, 1.6 and 2.2, respectively, which is similar to the percentage of o-xylene (1.7%) of non methane organic hydrocarbons in tunnel air [40]. A premium unleaded gasoline contained 5.2% by weight of o-xylene [19]. Composite gasoline sample from Los Angeles 2.86% (wt) [48]; cigarette smoke <6-48 ug/cigarette [48].

**Probable Routes of Human Exposure:** Humans are exposed to o-xylene primarily from the air, particularly in areas with heavy traffic, near filling stations, near industrial sources such as refineries, or where o-xylene is used in solvents. Exposure may also arise from contaminated drinking water such as well water near leaking underground gasoline storage tanks.

**Average Daily Intake:** AIR INTAKE: (assume typical concn 1.2 ppb) 106 ug; WATER INTAKE: (assume typical concn 0-1 ppb) 2 ug; FOOD INTAKE: insufficient data.

**Occupational Exposures:** No data could be found relating to human exposure to o-xylene, however, both benzene and xylenes are components of gasoline. The US population exposed to o-xylene from petroleum related sources can be assumed to be the same as for benzene, namely: people choosing self-service at gasoline service stations 37,000,000; people living in the vicinity of gasoline service stations 118,000,000; petroleum refineries 6,597,000; urban exposure (auto emissions) 113,690,000 [47]. NIOSH (NOHS Survey, 1972-74) has statistically estimated that 5,774 workers are exposed to o-xylene in the United States [51]. NIOSH (NOES Survey, 1981-83) has statistically estimated that 8904 workers are exposed to o-xylene in the United States [50]. Mean concn of o-xylene in 2 plants where spray varnishing of vehicles and special metal pieces in spray cabins is performed was 2.5 mg/m$^3$ (0.58 ppm) [3]. In the petroleum industry, outside operators, transport drivers, and service attendents are exposed to mean levels of o-xylene of 0.110, 0.200, and 0.091 mg/m$^3$, respectively [58].

**Body Burdens:** In the USEPA's Total Exposure Methodology (TEAM) Study, 83% of the 358 breath samples of residents of Elizabeth and Bayonne, NJ tested contained measurable quantities of o-xylene; medium and maximum concns 2.2 and 220 ug/m$^3$, respectively [82]. People recently exposed to potential sources of o-xylene such as from paint, furniture refinishing, printing, science laboratories, plastics manufacturing, and chemical plants had significantly higher levels of o-xylene in their breath [82]. Blood of 35 men (masks worn) who spray varnish vehicles and special metal pieces in spray cabins contained 26.6 ug/L average of o-xylene [3].

# 1,2-Xylene

## REFERENCES

1.  Abdul AS et al; Hazard Waste Hazard Mat 4: 211-21 (1987)
2.  Altshuller AP et al; Environ Sci Technol 4: 503-6 (1970)
3.  Angerer J, Wulf H; Int Arch Occup Environ Health 56: 307-21 (1985)
4.  Barker JF et al; Groundwater Monit Rev 7: 64-72 (1987)
5.  Bertsch W et al; J Chromatogr 112: 701-18 (1975)
6.  Bos R et al; Sci Total Environ 7: 269-81 (1977)
7.  Bozzelli JW et al; Analysis of selected toxic and carcinogenic substances in ambient air in New Jersey, NJ Dept Environ Protect (1980)
8.  Bridie AL et al; Water Res 13: 627-30 (1979)
9.  Brodzinsky R, Singh HB; Volatile organic chemicals in the atmosphere: an assessment of available data SRI contract 68-02-3452 (1982)
10. Coleman EC et al; J Agric Food Chem 29: 42-8 (1981)
11. Cox RA et al; Environ Sci Technol 14: 57-61 (1980)
12. Derenbach JB; Mar Chem 15: 295-303 (1985)
13. DeWalle FB, Chian ESK; J Amer Water Works Assoc 73: 206-11 (1981)
14. Doyle GJ et al; Environ Sci Technol 9: 237-41 (1975)
15. Dunlap WJ et al; Organic pollutants contributed to ground water by a landfill pp.96-110 USEPA 600/9-76-004 (1976)
16. Edgerton SA et al; Environ Sci Technol 20: 803-7 (1986)
17. Farley FF; Inter Conf Photochem Oxid Pollut Control Dimitriades B ed pp.713-27 USEPA 600/3-77-001b (1977)
18. Fielding M et al; in Organic Micropollutants in Drinking Water TR159 Medmenham Great Britain Water Research Centre (1981)
19. Fishbein L; Sci Tot Environ 43: 165-83 (1985)
20. Graedel TE; Chemical compounds in the atmosphere New York NY Academic Press (1978)
21. Great Lakes Water Quality Board; An inventory of chemical substances identified in the Great Lakes Ecosystem, Windsor Ontario: Great Lakes Water Quality Board (1983)
22. Green WJ et al; J Water Pollut Control Fed 53: 1347-54 (1981)
23. Gschwend PM et al; Environ Sci Technol 16: 31-8 (1982)
24. Halket JM, Angerer J; pp.211-7 in Anal Chem Symp Ser Vol 3 Recent Dev Chromat Electroph (1980)
25. Hansch C, Leo AJ; MEDCHEM Project Claremont CA: Pomona College (1985)
26. Hansen DA et al; J Phys Chem 79: 1763-6 (1975)
27. Harland BJ et al; J Environ Anal Chem 20: 295-311 (1985)
28. Heuss JM, Glasson WA; Environ Sci Technol 2: 1109-16 (1968)
29. Hiatt MH; Anal Chem 55: 506-16 (1988)
30. Ho CT et al; J Agric Food Chem 31: 336-42 (1983)
31. Hustert K, Parlar H; Chemosphere 10: 1045-50 (1981)
32. Kappeler T, Wuhrmann K; Water Res 12: 327-33 (1978)
33. Keith LH et al; Identification and analysis of organic pollutants in water Keith LD ed Ann Arbor MI Ann Arbor Science pp.329-73 (1976)
34. Kinlin TE et al; J Agric Food Chem 20: 1021 (1972)
35. Kitano M; Biodegradation and bioaccumulation test on chemical substances. OECD Tokyo Mtg TSU-No. 3 (1978)
36. Kopczynski SL et al; Environ Sci Technol 6: 342 (1972)
37. Kuhn EP et al; Environ Sci Technol 19: 961-8 (1985)

# 1,2-Xylene

38. Ligocki M et al; Atmos Environ 19: 1609-17 (1985)
39. Lonneman WA et al; Hydrocarbons in Houston Air USEPA 600/3-79-018 (1979)
40. Lonneman WA et al; Environ Sci Technol 20: 790-6 (1986)
41. Lonneman WA et al; Environ Sci Technol 2: 1017-20 (1968)
42. Louw CW et al; Atmos Environ 11: 703-17 (1977)
43. Lovegren NV et al; J Agric Food Chem; 27: 851-3 (1979)
44. Lyman WJ et al; Handbook of Chemical Property Estimation Methods New York: McGraw-Hill (1982)
45. Mackay D, Leinonen PJ; Environ Sci Technol 9:1178-80 (1975)
46. Malaney GW, McKinney RE; Water Sewage Works 113: 302-9 (1966)
47. Mara SJ, Lee SS; Human exposure to atmospheric benzene. Center for Resource and Environmental Studies Report No. 30 p.3 Menlo Park, CA: SRI (1977)
48. NAS; The Alkyl Benzenes page I-1 to I-99 (1980)
49. Nathwani JS, Phillip CR; Chemosphere 6: 157-62 (1977)
50. NIOSH; National Occupational Exposure Survey (1985)
51. NIOSH; National Occupational Health Survey (1975)
52. Nunes P, Benville PE JR; Bull Environ Contam Toxicol 21: 719-24 (1979)
53. Nutmagul W, Cronn DR; J Atmos Chem 2: 415-33 (1985)
54. Ogata M, Miyaka Y; Water Res 12: 1041-4 (1978)
55. Otson R et al; J Assoc Off Analyt Chem 65: 1370-4 (1982)
56. Piet GJ, Morra CF; Artificial groundwater recharge (Water Res Eng Ser) Huisman L, Olsthorn TN eds Pitman Pub pp.31-42 (1983)
57. Piet GJ et al; Quality of groundwater Van Dwjvenbooden W et al eds; Studies in Environ Sci 17: 557-64 (1981)
58. Rappaport SM et al; Appl Ind Hyg 2: 148-54 (1987)
59. Ravishankara AR et al; Int J Chem Kinetics 10: 783-804 (1978)
60. Riddick JA et al; Organic Solvents New York: Wiley Interscience (1986)
61. Samolloff MR; Envron Sci Technol 17: 329-34 (1983)
62. Sauer TC Jr; Org Geochem 3: 91-101 (1981)
63. Saunders RA et al; Water Res 9: 1143-5 (1975)
64. Sexton K, Westberg H; Environ Sci Technol 14: 329-32 (1980)
65. Shackelford WM et al; Analyt Chim Acta 146: 15-27 (supplemental data) (1983)
66. Sigsby JE Jr et al; Environ Sci Technol 21: 466-75 (1987)
67. Singh HB et al; Atmos Environ 15: 601-12 (1981)
68. Singh HB et al; Atmospheric measurements of selected hazardous organic chemicals USEPA 600/S3-81-032 (1981)
69. Smith JH, Harper JC; 12th Conf on Environ Toxicol Airforce Aerospace Med Res Lab OH pp.336-53 (1982)
70. Stuermer DH et al; Environ Sci Technol 16: 582-7 (1982)
71. Suffet IH et al; pp.375-97 in Identification and analysis of organic pollutants in water Keith LD ed Ann Arbor MI Ann Arbor Science (1976)
72. Tanaka T; Kogai 19: 121-8 (1984)
73. Tang J et al; J Agric Food Chem 31: 1287-92 (1983)
74. Tester DJ, Harker RJ; Water Pollut Control 80: 614-31 (1981)
75. Tomson MB et al; Water Res 15: 1109-16 (1981)
76. Tuazon EC et al; Environ Sci Technol 20: 383-7 (1986)
77. USEPA; STORET Data Base
78. Van der Linden AC; Dev Biodeg Hydrocarbons 1: 165-200 (1978)
79. Verschueren K; Handbook on environmental data on organic chemicals 2nd ed New York NY, Van Nostrand Reinhold Co pp.118-91 (1983)
80. Vowles PD, Mantoura RFC; Chemosphere 16: 109-16 (1987)

# 1,2-Xylene

81.  Wallace LA et al; Atmos Environ 21: 385-93 (1987)
82.  Wallace L et al; J Occup Med 28: 603-7 (1986)
83.  Wallace LA et al; Environ Res 43: 290-307 (1987)
84.  Washida N et al; Bull Chem Soc Japan 51: 2215-21 (1978)
85.  Westrick JJ; J Amer Water Works Assoc 76: 52-9 (1984)
86.  Williams DT et al; Chemosphere 11: 263-76 (1982)
87.  Wilson BH et al; Environ Sci Technol 20: 97-1002 (1986)
88.  Yanagihara S et al; 4th Clean Air Congress pp.472-7 (1977)
89.  Zuercher F, Giger W; Vom Wasser 47: 37-55 (1976)

# 1,3-Xylene

## SUBSTANCE IDENTIFICATION

**Synonyms:** m-Xylene

**Structure:**

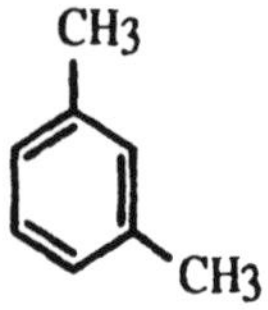

**CAS Registry Number:** 108-38-3

**Molecular Formula:** $C_8H_{10}$

**Wiswesser Line Notation:** 1R C1

## CHEMICAL AND PHYSICAL PROPERTIES

**Boiling Point:** 139.3 °C

**Melting Point:** -47.4 °C

**Molecular Weight:** 106.17

**Dissociation Constants:**

**Log Octanol/Water Partition Coefficient:** 3.20 [18]

**Water Solubility:** 146 mg/L at 25 °C [44]

**Vapor Pressure:** 8.3 mm Hg at 25 °C [44]

**Henry's Law Constant:** 7.68 x $10^{-3}$ atm-m$^3$/mole [34]

## ENVIRONMENTAL FATE/EXPOSURE POTENTIAL

**Summary:** m-Xylene will enter the atmosphere primarily from fuel emissions and exhausts linked with its use in gasoline. Industrial sources include emissions from petroleum refining and its use as a

516

solvent and chemical intermediate. The primary source of exposure is from air, especially in areas with high vehicular traffic. Discharges and spills on land and waterways result from its use in diesel fuel and gasoline and the storage and transport of petroleum products. Most of the m-xylene is released into the atmosphere where it may photochemically degrade by reaction with hydroxyl radicals (half-life 1-10 hr). The dominant removal process in water is volatilization. m-Xylene is moderately mobile in soil and may leach into ground water where it is known to persist for several years despite evidence that it biodegrades in both soil and ground water. This may be because m-xylene degrades under aerobic conditions and denitrifying conditions appear to be required when oxygen is lacking. Bioconcentration is not expected to be significant. The primary source of exposure is from air, especially where emissions from petroleum products or motor vehicles are high or solvents containing p-xylene are used.

**Natural Sources:** Petroleum [34].

**Artificial Sources:** Emissions from petroleum refining, gasoline, and diesel engines [34]. Emissions from its use as a solvent for alkyl resins, lacquers, enamels, rubber cement, pesticidal sprays, and in organic synthesis [21,34]. Many household products, especially aerosols of paints, varnishes, shellac, and rust preventatives contain xylenes [15]. Leaks and evaporation losses during the transport and storage of gasoline and other fuels from carburetor losses [34]. Emissions from residential wood-burning stoves and fireplaces [13].

**Terrestrial Fate:** When spilled on land, m-xylene will volatilize and leach into the ground. m-Xylene degrades in soil under aerobic or anaerobic, denitrifying conditions. Under aerobic conditions 70% degradation after 10 days have occurred. Under anaerobic conditions, a lag period of several months may be required before degradation commences. The extent of the degradation will depend on its concentration, residence time in the soil, the nature of the soil, and whether resident microbial populations have been acclimated. Biodegradation has generally only been observed under aerobic conditions, although recently it has also been observed under denitrifying conditions when oxygen is lacking. As a result of these factors, m-xylene may biodegrade fairly readily in the subsurface or it may persist for many years.

**Aquatic Fate:** In surface waters, volatilization appears to be the dominant removal process (half-life 1-5.5 days). Some adsorption to sediment will occur. Although m-xylene is biodegradable and has been observed to degrade in seawater, there are insufficient data to assess the rate of this process in surface waters.

**Atmospheric Fate:** When released into the atmosphere, m-xylene may degrade by reaction with photochemically produced hydroxyl radicals (half-life 1.0 hr in summer and 10 hr in winter). It will also be scavenged by rain.

**Biodegradation:** m-Xylene is degraded in standard biodegradability test using a variety of inocula including sewage, activated sludge, and seawater [6,26,32,57]. It was completely degraded within 8 days in ground water in a gas-oil mixture; the acclimation period being 3-4 days [24]. In laboratory experiments designed to simulate saturated-flow conditions typical of a river water/ground water infiltration system, degradation was rapid with 70% removal in the first 1.5 cm of the column after 10 days of operation under aerobic conditions [28]. Degradation also occurred anaerobically, but required denitrifying conditions [28]. After several months lag required for the microorganism to establish the necessary enzymes for denitrification, biodegradation was rapid [28]. In another laboratory experiment, m-xylene added to a 21.5 cm continuous-flow anaerobic aquifer column inoculated with denitrifying bacteria and acclimated to m-xylene was completely removed (80% mineralized) after 3 weeks of continuous operation [64]. The half-life was 1.5 hr [64]. The oxidation of the m-xylene was coupled with the reduction of nitrate to nitrite and ceased when nitrate was removed from the nutrient media [64] These results are in agreement with field experiments in which the concn of m-xylene was reduced by a factor of 10 in 1 day (distance of 2.5 m) in the River Glatt infiltration zone under mean flow condition [28]. Another investigator found that m-xylene was readily biodegraded (47 mg/day loss) in shallow ground water in an unconfined sand aquifer when oxygen was present [3]. As the available oxygen was consumed, the rate of degradation decreased [3].

**Abiotic Degradation:** m-Xylene reacts with hydroxyl radicals in the troposphere [9,19,43,50] with a half-life ranging from 1.0 in summer to 10 hr in winter [43] or a typical loss of 86%/day [50]. It has a moderately high photochemical reactivity under smog conditions, higher than the other xylene isomers, with loss rates varying from 9-42% per hr [12,27,58,63]. Photooxidation products that have been identified include glyoxal and methylglyoxal [56]. Xylenes are resistant to hydrolysis based upon the lack of hydrolyzable groups.

**Bioconcentration:** Little bioconcentration is expected; log BCF = 1.37 for eels [39] and 0.78 for clams [37]. Based on the octanol water partition coefficient, one can estimate the log BCF in fish to be 2.2 [31].

**Soil Adsorption/Mobility:** No measured values for Koc of m-xylene could be found in the literature. However, low to moderate adsorption would be expected [31] based on the Kow. Batch equilibrium measurements with soil from 3 aquifers gave a Koc of 166 [1]. It has been observed to pass through soil unchanged in concentration at a dune-infiltration site on the Rhine River [41].

**Volatilization from Water/Soil:** Using the Henry's law constant, 7.68 x $10^{-3}$ atm-m$^3$/mole [34], the half-life for evaporation of m-xylene from water with a wind speed of 3 m/sec, a current of 1 m/sec, and a depth of 1 m is calculated to be 3.1 hr [31]. An experiment which measured the rate of evaporation of m- and p-xylene from a 1:1000 jet fuel:water mixture found that it averaged 0.64 times the oxygen reaeration rate [51]. Combining this ratio with the oxygen reaeration rates of typical bodies of water [31], one estimates that the half-life for evaporation from a typical river and pond is 27 and 135 hr, respectively.

**Water Concentrations:** DRINKING WATER: In 12 Great Lakes municipalities tested on one or two days, 5 communities were free of m- and o-xylene combined, with median community levels of the combined isomers in the remaining 7 municipalities being 12 ppb [62]. In a survey of 30 Canadian water treatment facilities, the average value of m- and o-xylene combined in the treated water was <1 ppb, with maximums of 8 ppb being found in the summer and 2 ppb in the winter [40]. The frequency of occurrence was 27% and the maximum values in the raw water was <1 ppb [40]. m-Xylene has been identified but not quantified in the drinking water in Washington, DC [46] and Philadelphia, PA [53]. In a survey of occurrences of m-xylene in drinking water from ground water sources, 2.1% of 280 supplies serving fewer than 10,000 persons were positive with the median and maximum values of positive supplies being 0.32 and 1.5 ppb [61]. Of 186 supplies serving more than 10,000 persons, 1.1% were positive with median and maximum values of 0.46 and 0.61 ppb respectively

[61]. The maximum combined amount of m- and p-xylene in bank-filtered Rhine water in the Netherlands was 0.1 ppb [42]. In 4 drinking water wells near a landfill 0.1-0.8 ppb of m-xylene was determined [11]. Detected in all 14 drinking water supplies studied, 10 surface and 4 ground in the lowlands of Great Britain [14]. GROUND WATER: In ground water under a coal gasification site in Wyoming, 15 months after gasification was completed, 240-830 ppb [52] and in a recovery well under a landfill 7 years after landfill closed 0.4 ppb [11]. SURFACE WATER: In the raw water for 30 Canadian water treatment facilities; 7 and 17% of plants contained m- and o-xylene in summer and winter respectively with maximum levels being under 1 ppb [40]. Detected, not quantified, in the Black Warrior River in Tuscaloosa, AL [4], the Niagara River in the Lake Ontario basin [16] and the Glatt River in Switzerland [65]. The combined concn of m- and p-xylene at 2 stations in the Tees River, an industrial estuary in the UK at 1.5 m depth ranged from not detectable to 0.54 ppb and 0.37-1.5 ppb [20]. SEA WATER: In Vineland Sound, MA, samples taken over 15 months ranged from 4.5-66 ppt for the m- and p-xylene combined [17]. In open and coastal sections of the Gulf of Mexico 2.7-24.4 ppt for the m- and p-isomers combined [45]. A uniform pattern of alkylated benzenes are found in the Baltic Sea and central Atlantic which is believed to be the result of large scale transport in the atmosphere and water masses [10]. RAIN WATER: West Los Angeles 2 ppt [23]. The combined concn of m- and p-xylene in rain in Portland ranged from 34-260 ppt, 110 ppt, mean [29].

**Effluent Concentrations:** In a comprehensive survey of wastewater from 4000 industrial and publicly owned treatment works (POTWs) sponsored by the Effluent Guidelines Division of the US EPA, m-xylene was identified in discharges of the following industrial category (positive occurrences, median concn in ppb): timber products (13; 44.4), steam electric (2; 6.7), iron and steel mfg (1; 10.1), petroleum refining (11; 829.4), nonferrous metals (3; 39.5), paving and roofing (2; 23.6), paint and ink (50; 100.8), printing and publishing (2; 93.6), ore mining (4; 18.2), coal mining (12; 49.2), organics and plastics (16; 285.2), inorganic chemicals (9; 133.2), textile mills (4; 4.3), plastics and synthetics (11; 100.4), pulp and paper (1; 4.1), rubber processing (3; 16.8), auto and other laundries (9; 137.9), pesticides manufacture (4; 20.1), photographic industries (1; 307.6), gum and wood industries (2; 39.4), pharmaceuticals (5; 8.9), plastics mfg (4; 2250.6), electronics (13; 62.7), electroplating (1; 10.2), oil and gas extraction (17; 35.6), organic chemicals (9; 188.8), mechanical products (12; 651.7), transportation equipment (1; 11.0), amusements and athletic goods (3; 479.1), synfuels (2; 10.7), and publicly owned treatment works (107; 22.3) [48]. Industries whose maximum effluent exceeded 5,000 ppb were (industry (max concn in ppb)): timber products (6,002), petroleum refining (6,483), paint and ink (13,776), organics and plastics (15,417), and plastics mfg (26,829) [48]. The average combined exhaust emissions of m- and p-xylene from a 1975 to 1985 model year test fleet was 2.7 and 2.1% of total hydrocarbon emissions at normal urban and normal high speed driving [49]. Evaporative hydrocarbon emissions contained 2.5% m- and p-xylene [49].

**Sediment/Soil Concentrations:** Concns of m-and p-xylene combined in the Tees Estuary, a heavily industrial estuary in the UK, were 14-250 ppb (wet wt) at one station and 3-5 ppb at another station [20]. This was the major volatile aromatic hydrocarbon present in the sediment [20].

**Atmospheric Concentrations:** RURAL/REMOTE: 115 areas in U.S. (concn including p-xylene) 0.088 ppb median, 41 ppb max [7]. 6 semi-rural sites in Belgium (concn includes p-isomer) 0.018-2.46 ppb [55], Northern Hemisphere background (35 samples, includes p-isomer) 0.025 ppb [38], Southern Hemisphere background (21 samples, includes p-isomer) 0.013 ppb [38]. URBAN/SUBURBAN: 1911 areas in U.S. (concn including p-xylene) 2.8 ppb median, 4.1 ppb avg, 180 ppb max [7]. The concn of m- and p-xylene combined during 7 rain events in Portland, OR was 0.46-1.79 ppb, 0.78 ppb mean [29]. SOURCE DOMINATED AREAS: 186 areas in U.S. (concn including p-isomer) 1.7 ppb median, 155 ppb avg, 25% of samples exceed 4.4 ppb, max 10,000 ppb [7]. 2 landfills in New Jersey (concn including p-xylene) 0.16-99 ppb [5] 4-11 mi downwind from auto paint plant in Jonesville, WI (concn includes p-xylene) 3-5 ppb [47]. Concns of m-xylene in the Lincoln Tunnel in New York City in 1982 and 1970 were 126 and 784 ppb, respectively [30]. The decline is due to decreased organic carbon emission resulting from catalytic controls [30]. INDOOR AIR: the median combined concentration of m- and o-xylene in 340 New Jersey residences (6 pm to 6 am) was 3.2 ppb [59]. Concn of m-xylene in Japanese high-rise apartments ranged from 10-21 ppb [54]. OTHER: As part of EPA's Total Exposure Assessment Methodology (TEAM) study, the concentration of various toxic substances in the personal air (2 consecutive 12-hr periods) of sample populations was measured as well as the outdoor air near their residences [60]. The weighted median results for m- and p-xylene in personal air in Bayonne and Elizabeth,

NJ, an industrial/chemical manufacturing area, were 16, 13, and 22 ug/m$^3$ in the fall 1981, summer 1982, and winter 1983, respectively [60]. The corresponding results for outdoor air were 9.0, 10, and 10 ug/m$^3$ [60]. For comparison the personal air of a sample of residents of a manufacturing city without a chemical or petroleum refining industry, Greensboro, NC, and a small, rural, agricultural town in North Dakota contained 6.9 and 6.2 ug/m$^3$, respectively of m- and p-xylene, and the outdoor air 1.5 ug/m$^3$ and not detectable [60].

**Food Survey Values:** m-Xylene has been identified as a volatile component of baked potatoes [8] and roasted filberts [25].

**Plant Concentrations:**

**Fish/Seafood Concentrations:** The combined concentration of m- and p-xylene in rainbow trout from the Colorado River below Hoover Dam and carp taken from Las Vegas Wash, NV was 50 and 120 ppb, respectively [22].

**Animal Concentrations:**

**Milk Concentrations:**

**Other Environmental Concentrations:** Composite gasoline samples from Los Angeles m- and p-xylene combined are 6.73 wt% [34]. Cigarette smoke in various countries 1.6-48 ug/cigarette [34]. The percent of m-xylene in leaded, unleaded, and super unleaded gasoline sold in the New York-New Jersey area in 1982 was 1.9, 2.9, and 3.6, respectively, which is similar to the percentage of m-xylene (2.7%) of non methane organic hydrocarbons in tunnel air [30]. Another premium unleaded gasoline contained 12.2% by weight of m-xylene [15].

**Probable Routes of Human Exposure:** Humans are exposed to m-xylene primarily from air, particularly in areas with heavy traffic, near filling stations, near industrial sources such as refineries or where m-xylene is used as a solvent. Exposure may also arise from drinking contaminated well water such as might occur near leaking underground gasoline storage tanks or from spills of petroleum products.

**Average. Daily Intake:** AIR INTAKE: (assume typical concn of 2.0 ppb which is the median urban/suburban concn, 2.8 ppb for m- and p-xylene combined multiplied by the fraction of m-xylene in the m-plus p-xylene component of gasoline) 176 ug; WATER INTAKE: (assume typical concn 0-1 ppb) 0-2 ug; FOOD INTAKE: insufficient data.

**Occupational Exposures:** Benzene and xylenes are components of gasoline. The US population exposed to xylenes from petroleum related sources can be assumed to be the same as for benzene, namely: people choosing self-service at gasoline service stations 37,000,000; people living in the vicinity of gasoline service stations 118,000,000; petroleum refineries 6,597,000; urban exposure (auto emissions) 113,690,000 [33]. NIOSH (NOHS Survey, 1972-74) has statistically estimated that 4,615 workers are exposed to m-xylene in the United States [35]. NIOSH (NOES Survey, 1981-83) has statistically estimated that 19,868 workers are exposed to m-xylene in the United States [36]. Mean concn of m-xylene in 2 plants where spray varnishing of vehicles and special metal pieces in spray cabins is performed was 10.6 mg/m³ (2.44 ppm) [2].

**Body Burdens:** In the USEPA's Total Exposure Methodology (TEAM) Study, 95% of the 358 breath samples of residents of Elizabeth and Bayonne, NJ tested contained measurable quantities of m- and p-xylene; medium and maximum concns 6.4 and 350 ug/m³, respectively [59]. People recently exposed to potential sources of m- and p-xylene such as from paint, plastics manufacturing, and chemical plants had significantly higher levels of these chemicals in their breaths [59]. Blood of 35 men (masks worn) who spray varnish vehicles and special metal pieces in spray cabins contained 109.9 ug/L average of m-xylene [2].

## REFERENCES

1. Abdul AS et al; Hazard Waste Hazard Mat 4: 211-21 (1987)
2. Angerer J, Wulf H; Int Arch Occup Environ Health 56: 307-21 (1985)
3. Barker JF et al; Groundwater Monit Rev 7: 64-72 (1987)
4. Bertsch W et al; J Chromatogr 112: 701-8 (1975)
5. Bozzelli JW et al; Analysis of Selected Toxic and Carcinogenic Substances in Ambient Air in New Jersey; NJ Dept Environ Prot (1980)
6. Bridie AL et al; Water Res 13: 627-30 (1979)
7. Brodzinsky R, Singh HB; Volatile Organic Chemical in the Atmosphere: An assessment of available data; Menlo Park CA SRI Contract 68-02-3452 pp.128-9 (1982)
8. Coleman EC et al; J Agric Food Chem 29: 42-8 (1981)
9. Cox RA et al; Environ Sci Technol 14: 57-61 (1980)

# 1,3-Xylene

10. Derenbach JB; Mar Chem 15: 295-303 (1985)
11. DeWalle FB, Chian ESK; J Amer Water Works Assoc 73: 206-11 (1981)
12. Dilling WL et al; Environ Sci Technol 10: 351-6 (1976)
13. Edgerton SA et al; Environ Sci Technol 20: 803-7 (1986)
14. Fielding M et al; Organic Micropollutants in Drinking Water; TR 159 Medmenham U.K. Water Res Ctre (1981)
15. Fishbein L; Sci Tot Environ 43: 165-83 (1985)
16. Great Lakes Water Quality Board; An inventory of chemical substances identified in the Great Lakes Ecosystem, Windsor Ontario: Great Lakes Water Quality Board (1983)
17. Gschwend PM et al; Environ Sci Technol 16: 31-8 (1982)
18. Hansch C, Leo AJ; MEDCHEM Project Claremont CA: Pomona College (1985)
19. Hansen DA et al; J Phys Chem 78: 1763-6 (1975)
20. Harland BJ et al; J Environ Anal Chem 20: 295-311 (1985)
21. Hawley GG; The Condensed Chem Dictionary. Ninth ed. (1977)
22. Hiatt MH; Anal Chem 55: 506-16 (1988)
23. Kamamura K, Kaplan IR; Environ Sci Technol 17: 497-501 (1983)
24. Kappeler T, Wuhrmann K; Water Res 12: 327-33 (1978)
25. Kinlin TE et al; J Agric Food Chem 20: 1021 (1972)
26. Kitano M; Biodegradation and Bioaccumulation Tests on Chemical Substances; OECD Tokyo Mtg TSU-No.3 (1978)
27. Kopczynski SL et al; Environ Sci Technol 6: 342 (1972)
28. Kuhn EP et al; Environ Sci Technol 19: 961-8 (1985)
29. Ligocki M et al; Atmos Environ 19: 1609-17 (1985)
30. Lonneman WA et al; Environ Sci Technol 20: 790-6 (1986)
31. Lyman WJ et al; Handbook of Chemical Property Estimation Methods. McGraw Hill New York NY (1982)
32. Malaney GW, McKinney RE; Water Sewage Works 113: 302-9 (1966)
33. Mara SJ, Lee SS; Human Exposure to Atmos Benzene, Center for Resource Environ Studies Rep No.30 pp 3 Menlo Park CA: SRI (1977)
34. NAS; The Alkyl Benzenes; pp.II-1 to II-51 (1980)
35. NIOSH; National Occupational Health Survey (1975)
36. NIOSH; National Occupational Exposure Survey (1985)
37. Nunes P, Benville PE Jr; Bull Environ Contam Toxicol 21: 719-24 (1979)
38. Nutmagul W, Cronn DR; J Atmos Chem 2: 415-33 (1985)
39. Ogata M, Miyaka Y; Water Res 12: 1041-4 (1978)
40. Otson R et al; J Assoc Off Anal Chem 65: 1370-4 (1982)
41. Piet GJ et al; Quality of Groundwater Int Symp Proc Von Duijvanbouden W et al eds: Studies Environ Sci 17: 557-64 (1981)
42. Piet FJ, Morra CF; pp.31-42 in Artificial Groundwater Recharge Water Res Eng Ser; Huisman L, Olsthorn TN eds; Pitman Sci Publ (1983)
43. Ravishankara AR et al; Int J Chem Kinetics 10: 783-804 (1978)
44. Riddick JA et al; Organic Solvents New York: Wiley Interscience (1986)
45. Sauer TC Jr et al; Mar Chem 7: 1-16 (1978)
46. Saunders RA et al; Water Res 9: 1143-5 (1975)
47. Sexton K, Westberg HM; Environ Sci Technol 14: 329-32 (1980)
48. Shackelford WM et al; Analyt Chim Acta 146: 15-27 (supplemental data) (1983)
49. Sigsby JE Jr et al; Environ Sci Technol 21: 466-75 (1987)
50. Singh HB et al; Atmos Environ 15: 601-12 (1981)
51. Smith JH, Harper JC; 12th Conf Environ Toxicol: Behavior of hydrocarbon fuels in the aquatic environment; pp.336-53 (1980)

# 1,3-Xylene

52. Stuermer DH et al; Environ Sci Technol 16: 582-7 (1982)
53. Suffet IH et al; pp.375-97 in Identification and Analysis of Organic Pollutants in Water; Keith LD ed; Ann Arbor Science Ann Arbor MI (1976)
54. Tanaka T; Kogai 19: 121-8 (1984)
55. Termonia M; Comm Eur Comm Symp Phys Chem Behav Atmos Pollute; pp.356-61 EUR 7624 (1982)
56. Tuazon EC et al; Environ Sci Technol 20: 383-7 (1986)
57. Van der Linden AC; Dev Biodeg Hydrocarbons 1: 165-200 (1978)
58. Van Aalst RM et al; Comm Eur Comm Symp Phys Chem Behav Atmos Pollut; EUR6621 1: 136-49 (1980)
59. Wallace LA et al; J Occu Med 28: 603-7 (1986)
60. Wallace LA et al; Environ Res 43: 290-307 (1987)
61. Westrick JJ et al; J Amer Water Works Assoc 76: 52-9 (1984)
62. Williams DT et al; Chemosphere 11: 263-76 (1982)
63. Yanagihara S et al; 4th Int Clean Air Conf: Photochemical Reactivities of Hydrocarbons; pp.472-7 (1977)
64. Zeyer Y et al; Appl Environ Microbiol; 52: 944-7 (1986)
65. Zuercher F, Giger W; Vom Wasser 47: 37-55 (1976)

# 1,4-Xylene

## SUBSTANCE IDENTIFICATION

**Synonyms:** p-Xylene

**Structure:**

**CAS Registry Number:** 106-42-3

**Molecular Formula:** $C_8H_{10}$

**Wiswesser Line Notation:** 1R D1

## CHEMICAL AND PHYSICAL PROPERTIES

**Boiling Point:** 137-138 °C

**Melting Point:** 13-14 °C

**Molecular Weight:** 106.17

**Dissociation Constants:**

**Log Octanol/Water Partition Coefficient:** 3.15 [24]

**Water Solubility:** 156 mg/L at 25 °C [48]

**Vapor Pressure:** 8.7 mm Hg at 25 °C [48]

**Henry's Law Constant:** 7.68 x $10^{-3}$ atm-m³/mole [38]

## ENVIRONMENTAL FATE/EXPOSURE POTENTIAL

**Summary:** p-Xylene will enter into the atmosphere primarily from fugitive emissions and exhaust connected with its use in gasoline and as a solvent. Industrial sources include emissions from petroleum refining and its use as a solvent and chemical intermediate. Discharges and spills on land and waterways result from its use in diesel fuel and gasoline and the storage and transport of petroleum products. Most of the p-xylene is released into the atmosphere where it may photochemically degrade by reaction with hydroxyl radicals (half-life 1.7-18 hr). The dominant removal process in water is volatilization. p-Xylene is moderately mobile in soil and may leach into the ground where it is known to persist for several years despite evidence that it biodegrades in both soil and ground water. This may be because p-xylene degrades under aerobic conditions and denitrifying conditions appear to be required for biodegradation when oxygen is lacking. Bioconcentration is not expected to be significant. The primary source of exposure is from air, especially where emissions from petroleum products or motor vehicles are high or solvents containing p-xylene are used.

**Natural Sources:** Petroleum [38].

**Artificial Sources:** Emissions from petroleum refining, gasoline and diesel engines [38]; evaporative losses during the transport and storage of gasoline and carburetor losses [20,38]. Emissions from its use as a chemical intermediate in the production of dimethyl terephthalate and terephthalic acid for polyester production [9]. Evaporative losses from its use as a solvent [20,38]. Many household products, especially aerosols of paints, varnishes, shellac, and rust preventatives contain xylenes [19]. Emissions from residential wood-burning stoves and fireplaces [17].

**Terrestrial Fate:** When spilled on land, p-xylene will volatilize and leach into the ground. p-Xylene degrades in soil under aerobic or anaerobic denitrifying conditions. Under aerobic conditions 70% degradation after 10 days have occurred. Under anaerobic conditions, a lag period of several months may be required before degradation commences. The extent of p-xylene degradation will depend on its concentration, residence time in the soil, the nature of the soil, and whether resident microbial populations have been acclimated. Biodegradation has generally only been observed under aerobic conditions, although recently it has also been observed under denitrifying conditions when oxygen is lacking. As a result of these factors, p-xylene may biodegrade fairly readily in the subsurface or it may persist for many years.

# 1,4-Xylene

**Aquatic Fate:** In surface waters, volatilization appears to be the dominant removal process (half-life 1-5.5 days). Some adsorption to sediment will occur. Although p-xylene is biodegradable and has been observed to degrade in seawater, there are insufficient data to assess the rate of this process in surface waters.

**Atmospheric Fate:** When released into the atmosphere, p-xylene may degrade by reaction with photochemically produced hydroxyl radicals (half-life 1.7 hr in summer and 18 hr in winter). It will also be scavenged by rain.

**Biodegradation:** p-Xylene is degraded in standard biodegradability tests using a variety of inocula including sewage, activated sludge, and seawater [7,31,37,65]. It was completely degraded within 8 days in ground water in a gas-oil mixture; the acclimation period was 3-4 days [28]. In laboratory experiments designed to simulate saturated-flow conditions typical of a river water/ground water infiltration system, degradation was rapid with 70% removal in the first 1.5 cm of the column after 10 days of operation under aerobic conditions [33]. Degradation also occurred anaerobically, but required denitrifying conditions [33]. After several months lag required for the microorganisms to establish the necessary enzymes for denitrification, biodegradation was rapid [33]. Another investigator found that p-xylene was readily biodegraded (55 mg/day loss) in shallow ground water in an unconfined sand aquifer when oxygen was present [4]. As the available oxygen was consumed, the rate of degradation decreased [4].

**Abiotic Degradation:** p-Xylene reacts with hydroxyl radicals in the troposphere [25,47,55] with a half-life ranging from 1.7 hr in summer to 18 hr in winter [47] or a typical loss of 67%/day [55]. It is moderately reactive under photochemical smog conditions with reported loss rates varying from 4-25% per hr [2,14,15,32,70,72], rates typical of its reaction with hydroxyl radicals [15]. Photooxidation products that have been identified include glyoxal, and methylglyoxal [62]. Xylenes are resistant to hydrolysis based upon their lack of hydrolyzable groups.

**Bioconcentration:** Little bioconcentration is expected, the log BCF=1.37 for eels [42]. Based on the octanol/water partition coefficient, one can estimate the log BCF in fish to be 2.2 [36].

**Soil Adsorption/Mobility:** No measured values for Koc of p-xylene on surface soil could be found in the literature. However, low to moderate adsorption would be expected based on the log Kow value of 3.15 [24]. The Koc for p-xylene in surface sediment (2-10 cm) collected from the central Tamar estuary in the UK was 25.4 [66]. Batch equilibrium measurements with soil from 3 aquifers gave a Koc of 204 [1]. The measured permeability for fire clay is $1 \times 10^{-9}$ cm/sec and a factor of 4 and 50 times faster in ranger shale and kaoline respectively [22]. It has been detected in ground water under a rapid infiltration site [61] and passed through soil at a dune-infiltration site on the Rhine River unchanged in concentration [45].

**Volatilization from Water/Soil:** Using the Henry's Law constant, the half-life for evaporation of p-xylene from water with a wind speed of 3 m/sec, a current of 1 m/sec, and a depth of 1 m is calculated to be 3.1 hr [36]. An experiment which measured the rate of evaporation of m- and p-xylene from a 1:1000 jet fuel:water mixture found that it averaged 0.64 times the oxygen reaeration rate [57]. Combining this ratio with the oxygen reaeration rates of typical bodies of water [36], one estimates that the half-life for evaporation from a typical river or pond is 27 and 135 hr resp. Estimated half-lives in the 300 cm mean water column (mixed depth) of Narragansett Bay for p-xylene with respect to volatilization are 5.5 and 6.4 days under summer and winter conditions, respectively [67].

**Water Concentrations:** DRINKING WATER: In a survey of 30 Canadian water treatment facilities, the average value of p-xylene combined with ethylbenzene was <1 ppb with a maximum value of 10 ppb and 30% of the supplies positive [43]. The raw water for the supplies had a lower maximum concentration of <1 ppb [43]. p-Xylene has been identified but not quantitated in the municipal drinking water supplies of Cleveland, OH [49], Philadelphia, PA [59], Washington, DC [51], Tuscaloosa, AL and Houston, TX [5]. In a survey of organics in drinking water derived from ground water sources, p- and o-xylene combined were found in 2.1% of 280 sample sites supplying <10,000 persons and 1.1% of 186 sites supplying >10,000 persons. The maximum combined concentrations were 0.59 and 0.91 ppb respectively [71]. The maximum combined amount of m- and p-xylene in bank-filtered Rhine water in the Netherlands was 0.1 ppb [44]. In 6 drinking water wells near a landfill 0.3-2.1 ppb of p-xylene was found [13]. Detected in all 14 drinking water studies in the lowland of Great Britain, 10 from surface water sources and 4 from ground supplies [18]. GROUND WATER: In ground water under a landfill in Norman,

OK 0.9 ppb [16] and under a rapid infiltration site in Phoenix, 0.10-49 ppb [61] under a coal gasification site in Wyoming 15 months after gasification complete 240-830 ppb [58]. In a recovery well from landfill 7 years after closing 2.9 ppb [13]. SURFACE WATER: In the raw water supplies for 30 Canadian treatment facilities, 23% contained a combination of p-xylene and ethyl benzene which averaged <1 ppb and whose max value was <1 ppm in summer and 2 ppb in winter [43]. Detected, not quantified, in the Black Warrior River in Tuscaloosa, AL [5], the Niagara River in the Lake Ontario basin [21] and the Glatt River in Switzerland [73]. The combined concn of m- and p-xylene at 2 stations in the Tees River, an industrial estuary in the UK at 1.5 m depth ranged from not detectable to 0.54 ppb and 0.37-1.5 ppb [26]. SEAWATER: In Vineland Sound, MA, samples taken over 15 months ranged from 4.5-66 ppt for the p- and m-xylene combined [23]. In open and coastal sections of the Gulf of Mexico, 2.7-24.4 ppt for the p- and m-isomers combined was detected [50]. A uniform pattern of alkylated benzenes are found in the Baltic Sea and central Atlantic which is believed to be the result of large scale transport in the atmosphere and water masses [12]. RAINWATER: West Los Angeles, 9 ppt [29]. The combined concn of m- and p-xylene in rain in Portland ranged from 34-260 ppt, 110 ppt, mean [34].

**Effluent Concentrations:** In a comprehensive survey of wastewater from 4000 industrial and publicly owned treatment works (POTWs) sponsored by the Effluent Guidelines Division of the US EPA, p-xylene was identified in discharges of the following industrial category (positive occurrences; median concn in ppb): timber products (18; 44.4), steam electric (1; 3.3), iron and steel mfg (1; 76.5), petroleum refining (17; 44.4), nonferrous metals (12; 33.2), paving and roofing (4; 15.9), paint and ink (71; 143.4), printing and publishing (5; 64.6), ore mining (6; 23.4), coal mining (17; 29.0), organics and plastics (25; 12.2), inorganic chemicals (7; 29.0), textile mills (8; 64.0), plastics and synthetics (12; 152.3), pulp and paper (5; 19.2), rubber processing (1; 12.5), auto and other laundries (11; 121.0), pesticides manufacture (1; 4.1), photographic industries (2; 430.2), gum and wood industries (4; 18.7), pharmaceuticals (3; 66.9), plastics mfg (4; 2184.0), porcelain/enameling (2; 18.0), electronics (17; 193.4), oil and gas extraction (12; 41.0), organic chemicals (10; 159.5), mechanical products (23; 73.1), transportation equipment (6; 17.5), amusements and athletic goods (2; 49.4), and publicly owned treatment works (101; 31.6) [53]. Maximum effluents above 5,000 ppb were observed in industry (maximum concn in ppb): timber products (6,880), paints and ink (18,525), plastics and synthetics (7,502), auto and other laundries (6,000), and plastics mfg (8,180) [53]. The average combined exhaust emissions of m- and p-xylene from a 1975 to 1985 model year test fleet was 2.7% and 2.1% of total hydrocarbon emissions at normal urban and normal high speed driving, respectively [54]. Evaporative hydrocarbon emissions contained 2.5% m- and p-xylene [54].

**Sediment/Soil Concentrations:** Concn of m- and p-xylene combined in the Tees Estuary, a heavily industrial estuary in the UK, were 14-250 ppb (wet wt) at one station and 3-5 ppb at another station [26]. This was the major volatile aromatic hydrocarbon present in the sediment [26]. In unspecified sediment 100 ppb [63].

**Atmospheric Concentrations:** RURAL/REMOTE: 115 areas in US (concn including m-xylene) 0.088 ppb median, 41 ppb max [8]; 6 semi-rural sites in Belgium (concn includes m-isomer) 0.018-2.46 ppb [60]. Atlantic Ocean air, 0.01-0.02 ppb, Pacific Ocean air, 0.02-0.04 ppb, Indian Ocean air, 0.003-0.005 ppb [56], Northern Hemisphere background (35 samples, includes m-isomer) 0.025 ppb [41], Southern Hemisphere background (21 samples, includes m-isomer) 0.013 ppb [41]. URBAN/SUBURBAN: 1911 areas in US (concn including m-xylene) 2.8 ppb median, 4.1 ppb avg, 180 ppb max [8]. 15 urban areas in US 1979-84, (concn including m-xylene) not detected - 50 ppb, range of site avg, 1.0-10.2 ppb [56]. The vapor concn of m- and p-xylene combined during 7 rain events in Portland, OR was 0.46-1.79 ppb, 0.78 ppb mean [34]. SOURCE DOMINATED AREAS: 186 areas in US (concn including m-isomer) 1.7 ppb median, 155 ppb avg, 25% of samples exceed 4.4 ppb max 10,000 ppb [8]. 2 landfills in New Jersey (concn including m-xylene) 0.16-99 ppb [6] 4-11 mi downwind from auto paint plant in Janesville, WI (concn includes m-xylene) 3-5 ppb [52]. Concns of p-xylene in the Lincoln Tunnel in New York City in 1982 and 1970 were 54 and 278 ppb, respectively [35]. The decline is due to decreased organic carbon emissions resulting from catalytic controls [35]. INDOOR AIR: The median combined concentration of m- and p-xylene in 340 New Jersey residences (6 pm to 6 am) was 3.2 ppb [68]. OTHER: As part of EPA's Total Exposure Assessment Methodology (TEAM) study, the concentration of various toxic substances in the personal air (2 consecutive 12-hr periods) of sample populations was measured as well as the outdoor air near their residences [69]. The weighted median results for m- and p-xylene in personal air in Bayonne and Elizabeth, NJ, an industrial/chemical manufacturing area, were 16, 13, and 22 ug/m$^3$ in the fall 1981, summer 1982, and winter 1983, respectively [69]. The corresponding results for outdoor air were 9.0, 10, and 10 ug/m$^3$ [69]. For comparison, the personal air of a sample of residents of a manufacturing city without a chemical or petroleum refining industry, Greensboro, NC, and a small, rural, and agricultural town in North Dakota contained 6.9 and 6.2 ug/m$^3$, respectively of m- and p-xylene and the outdoor air 1.5 ug/m$^3$ and not detectable [69].

**Food Survey Values:** p-Xylene has been identified as a volatile component of baked potatoes [11] and roasted filberts [30].

**Plant Concentrations:**

**Fish/Seafood Concentrations:** The combined concentration of m- and p-xylene in rainbow trout from the Colorado River below Hoover Dam and carp taken from Las Vegas Wash, NV was 50 and 120 ppb, respectively [27].

**Animal Concentrations:**

**Milk Concentrations:**

**Other Environmental Concentrations:** Composite gasoline samples from Los Angeles m- and p-xylene combined are 6.73 wt% [38]. A premium unleaded gasoline contained 4.2% by weight of p-xylene [19].

**Probable Routes of Human Exposure:** Humans are exposed to p-xylene primarily from air, particularly in areas with heavy traffic, near filling stations, near industrial sources such as refineries or where p-xylene is used as a solvent. Exposure may also arise from drinking contaminated well water such as might occur near leaking underground gasoline storage tanks or from spills of petroleum products.

**Average Daily Intake:** AIR INTAKE: (assume typical concn of 0.8 ppb which is the median urban/suburban concn for m- and p-xylene combined multiplied by the fraction of p-xylene in the m- plus p-xylene component of gasoline) 71 ug. WATER INTAKE: (assume typical concn of 0-1 ppb) 0-2 ug. FOOD INTAKE: insufficient data.

**Occupational Exposures:** NIOSH (NOHS Survey, 1972-75) has statistically estimated that 1,908 workers are exposed to p-xylene in the United States [40]. NIOSH (NOES Survey, 1981-83) has statistically estimated that 5723 workers are exposed to p-xylene in the United States [39]. Concns of p-xylene observed in various rubber manufacturing plants in Italy include: shoe-sole factory (vulcanization area) 3.5-9.2 ppb, tire retreading factory (vulcanization area) 1.6-27.6 ppb, tire retreading factory (extrusion area) 0.5-2.3 ppb, electrical cables insulation plant (extrusion area) 0-25.3 ppb [10]. Mean concn of p-xylene in 2 plants where spray varnishing of vehicles and special metal pieces in spray cabins is performed was 3.4 mg/m$^3$ (0.78 ppm) [3]. In the petroleum industry, outside operators, transport drivers, and service attendants are exposed to mean levels of p-xylene of 0.391, 0.641, and 0.332 mg/m$^3$, respectively [46].

**Body Burdens:** In the USEPA's Total Exposure Methodology (TEAM) Study, 95% of the 358 breath samples of residents of Elizabeth and Bayonne, NJ tested contained measurable quantities of m- and p-xylene; medium and maximum concns 6.4 and 350 ug/m$^3$, respectively [68]. People recently exposed to potential sources of m- and p-xylene such as from paint, plastics manufacturing, and chemical plants had significantly higher levels of these chemicals in their breaths [68]. Blood of 35 men (masks worn) who spray varnish vehicles and special metal pieces in spray cabins contained 39.7 ug/L average of p-xylene [3].

## REFERENCES

1.  Abdul AS et al; Hazard Waste Hazard Mat 4: 211-21 (1987)
2.  Altshuller AP et al; Environ Sci Technol 4: 503-6 (1970)
3.  Angerer J, Wulf H; Int Arch Occup Environ Health 56: 307-21 (1985)
4.  Barker JF et al; Groundwater Monit Rev 7: 64-72 (1987)
5.  Bertsch W et al; J Chromatogr 112: 701-8 (1975)
6.  Bozzelli JW et al; Analysis of selected toxic and carcinogenic substances in ambient air in New Jersey. New Jersey Dept Environ Prot (1980)
7.  Bridie AL et al; Water Res 13: 627-30 (1979)
8.  Brodzinsky R, Singh HB; Volatile organic chemicals in the atmosphere: an assessment of available data. Menlo Park, CA: SRI contract 68-02-3452 pp. 128-9 (1982)
9.  Chemical Marketing Reporter Aug 22 (1983)
10. Cocheo V et al; Amer Ind Hyg J 44: 521-7 (1983)
11. Coleman EC et al; J Agric Food Chem 29: 42-8 (1981)
12. Derenbach JB; Mar Chem 15: 295-303 (1985)
13. DeWalle FB, Chain ESK; J Amer Water Works Assoc 73: 206-11 (1981)
14. Dilling WL et al; Environ Sci Technol 10: 351-6 (1976)
15. Doyle GJ et al; Environ Sci Technol 9: 237-41 (1975)
16. Dunlap WJ et al; Organic pollutants contributed to groundwater by a landfill pp. 96-110 USEPA 600/9-76-004 (1976)
17. Edgerton SA et al; Environ Sci Technol 20: 803-7 (1986)

18. Fielding M et al; in Organic Micropollutants in Drinking Water TR 159 Medmenham Great Britain Water Research Centre (1981)
19. Fishbein L; Sci Tot Environ 43: 165-83 (1985)
20. Graedel TE; Chemical Compounds in the Atmosphere. New York Academic Press p. 109 (1978)
21. Great Lakes Water Quality Board; An inventory of chemical substances identified in the Great Lakes Ecosystem, Windsor Ontario: Great Lakes Water Quality Board (1983)
22. Green WJ et al; J Water Pollut Control Fed 53: 1347-54 (1981)
23. Gschwend PM et al; Mar Chem 7: 1-16 (1982)
24. Hansch C, Leo AJ; Medchem Project Claremont CA: Pomona College (1985)
25. Hansen DA et al; J Phys Chem 79: 1763-6 (1975)
26. Harland BJ et al; J Environ Anal Chem 20: 295-311 (1985)
27. Hiatt MH; Anal Chem 55: 506-16 (1988)
28. Kappeler T, Wuhrmann K; Water Res 12: 327-33 (1978)
29. Kawamura K, Kaplan IR; Environ Sci Technol 17: 497-501 (1983)
30. Kinlin TE et al; J Agric Food Chem 20: 1021 (1972)
31. Kitano M; Biodegradation and bioaccumulation test on chemical substances, OECD Tokyo Mtg TSU-No. 3 (1978)
32. Kopczynski SL et al; Environ Sci Technol 6: 342 (1972)
33. Kuhn EP et al; Environ Sci Technol 19: 961-8 (1985)
34. Ligocki M et al; Atmos Environ 19: 1609-17 (1985)
35. Lonneman WA et al; Environ Sci Technol 20: 790-6 (1986)
36. Lyman WJ et al; Handbook of Chemical Estimation Methods. New York, NY McGraw Hill (1982)
37. Malaney GW, McKinney RE; Water Sewage Works 113: 302-9 (1966)
38. NAS; The Alkyl Benzenes page I-1 to II-51 (1980)
39. NIOSH; National Occupational Exposure Survey (1985)
40. NIOSH; National Occupational Health Survey (1975)
41. Nutmagul W, Cronn DR; J Atmos Chem 2: 415-33 (1985)
42. Ogata M, Miyake Y; Water Res 12: 1041-4 (1978)
43. Otson R et al; J Assoc Off Anal Chem 65: 1370-4 (1982)
44. Piet GJ, Morra CF; in Artificial groundwater recharge (Water Res Eng Ser) Huisman L, Olsthorn, TN eds; Pitman Pub pp. 31-42 (1983)
45. Piet GJ et al; in Quality of groundwater. Int Symp Proc Van Duijvenbooden W et al; Studies Environ Sci 17: 557-64 (1981)
46. Rappaport SM et al; Appl Ind Hyg 2: 148-54 (1987)
47. Ravishankara AR et al; Int J Chem Kinetics 10: 783-804 (1978)
48. Riddick JA et al; Organic Solvents New York: Wiley Interscience (1986)
49. Sanjivamurthy VA; Water Res 12: 31-3 (1978)
50. Sauer TC, Jr et al; Mar Chem 7: 1-16 (1978)
51. Saunders RA et al; Water Res 1143-5 (1975)
52. Sexton K, Westberg H; Environ Sci Technol 14: 329-32 (1980)
53. Shackelford WM et al; Analyt Chim Acta 146: 15-27 (supplemental data) (1983)
54. Sigsby JE Jr et al; Environ Sci Technol 21: 466-75 (1987)
55. Singh HB et al; Atmos Environ 15: 601-12 (1981)
56. Singh HB et al; Atmos Environ 19: 1911-9 (1985)
57. Smith JH, Harper JC; 12th Conf on Environ Toxicol: Behavior of hydrocarbon fuels in the aquatic environment pp.336-53 (1980)
58. Stuermer DH et al; Environ Sci Technol 16: 582-7 (1982)

59. Suffit IH et al; Identification and analysis of organic pollutants in water. Keith H ed Ann Arbor MI Ann Arbor Sci Publ pp. 375-97 (1976)
60. Termonia, M; Comm Eur Comm Symp Phys Behav Atmos Pollut EUR7624 pp. 356-61 (1982)
61. Tomson MB et al; Water Res 15: 1109-16 (1981)
62. Tuazon EC et al; Environ Sci Technol 20: 383-7 (1986)
63. USEPA; Storet Data Base
64. VanAalst RM et al; Comm Eur Comm Symp Phys Chem Behav Atmos Poll EUR6621 1: 136-49 (1980)
65. VanderLinden AC; Dev Biodeg Hydrocarbons 1: 165-200 (1978)
66. Vowles PD, Mantoura RFC; Chemosphere 16: 109-16 (1987)
67. Wakeham SG et al; Can J Fish Aquat Sci Suppl 2 40: 304-21 (1983)
68. Wallace LA et al; J Occu Med 28: 603-7 (1986)
69. Wallace LA et al; Environ Res 43: 290-307 (1987)
70. Washida N et al; Bull Chem Soc Japan 51: 2215-21 (1978)
71. Westrick JJ; J Amer Water Works Assoc 76: 52-9 (1984)
72. Yanagihara S et al; 4th Int Clean Air Conf: Photochemical reactivities of hydrocarbons pp. 472-7 (1977)
73. Zuercher F, Giger W; Vom Wasser 47: 37-55 (1976)

# Cumulative Index of Synonyms

# Cumulative Index by CAS Registry Number

# Cumulative Index by Chemical Formula